Human Pathobiochemistry

Toshitaka Oohashi • Hirokazu Tsukahara
Francesco Ramirez • Chad L. Barber
Fumio Otsuka
Editors

Human Pathobiochemistry

From Clinical Studies to Molecular Mechanisms

Springer

Editors
Toshitaka Oohashi
Department of Molecular Biology and Biochemistry, Graduate School of Medicine, Dentistry and Pharmaceutical Sciences
Okayama University
Okayama, Japan

Francesco Ramirez
Department of Pharmacological Sciences
Icahn School of Medicine at Mount Sinai
New York, NY, USA

Fumio Otsuka
Department of General Medicine, Graduate School of Medicine, Dentistry and Pharmaceutical Sciences
Okayama University
Okayama, Japan

Hirokazu Tsukahara
Department of Pediatrics, Graduate School of Medicine, Dentistry and Pharmaceutical Sciences
Okayama University
Okayama, Japan

Chad L. Barber
Department of Biology
College of Arts and Sciences
California Lutheran University
Thousand Oaks, CA, USA

ISBN 978-981-13-2976-0 ISBN 978-981-13-2977-7 (eBook)
https://doi.org/10.1007/978-981-13-2977-7

Library of Congress Control Number: 2019933569

This Springer imprint is published by the registered company Springer Nature Singapore Pte Ltd.
The registered company address is: 152 Beach Road, #21-01/04 Gateway East, Singapore 189721, Singapore

Preface

In just the past few decades, advances in molecular biology, cell biology, and biochemistry have allowed biomedical researchers to identify the mechanisms of action that belie many processes in our physiology. So much of what has been learned about these biochemical processes or mechanisms has been learned from studying human disease. More than ever, we need researchers and clinicians to be trained to think past the symptoms and seek to understand the underlying molecular or biochemical pathology that presents in a patient. Much of the research done on these diseases in the clinic and laboratory have led to the discovery of novel, targeted therapies. The shift in clinical practice is already happening. The question being asked is no longer only "how do I treat these symptoms" but also "how can we target the cause of the disease?" This culture shift in medicine will surely beget a generation of clinicians who practice medicine in a more informed way and contribute to elucidating new therapies. The purpose of this text is to teach clinicians-in-training to understand the pathology not only on a topical level but a deep understanding of the molecular underpinnings of human disease.

This textbook uses a case study approach to present the core principles of biochemistry and molecular biology in the context of human disease to students who will be involved in patient care. The 29 clinical cases are carefully selected to cover key scientific concepts and some common, and other not so common, diseases. This textbook covers many topics on metabolic disease as a major part but contains other topics on connective tissue disorders, neurological disorders, auto-inflammatory disorders, infective diseases, cancer, etc. Each chapter provides a specific patient report that includes the natural history, pertinent clinical laboratory data, physical findings, subsequent diagnosis, and therapy. This is followed by a comprehensive discussion of the normal biochemical processes and reactions pertaining to the case, along with the pathophysiological mechanisms of the disease.

The book is organized into three parts, the first of which considers diseases of metabolism. The diseases range from those that might already be familiar to a student, such as familial hypercholesterolemia and phenylketonuria (PKU), to others that are more rare, such as Fabry or Wilson disease. The second part of the book focuses on genetic disorders. The manifestations of the genetic disorders included are myriad. In some cases, the genetic insult results in physical deformities, such as achondroplasia, Marfan syndrome, and cherubism. In others, the result can be malignant disease, such as acute myeloid leukemia. The last section encompasses a host of other diseases, infections, and conditions which can now also be studied to the molecular level. Some of these have known etiologies; others are still under investigation. Hopefully, clinicians-in-training will appreciate the opportunity to think about all conditions as having a molecular basis and thereby have a target to which a therapy can be developed. Again, this book would hope to emphasize the importance of the *process* of diagnosis, as it relates to the biochemistry of the patient, rather than memorizing a list of symptoms for each disease.

This textbook has several unique features which we hope will enable instructors to get the most out of using it in a classroom setting. The chapter authors have provided discussion questions meant to help students in identifying important information or consider why therapies will or will not work based on the understanding of the pertinent biochemistry. The text is also meant to be usable by many different clinical students or even those in basic research. It provides various clinical cases for preclinical students in medical school and schools for other health professionals, including nurses, nutritionists, physiotherapists, or radiographers, for example. In all of these settings, we hope the instructors will use the discussion questions in various modalities, including techniques such as active deep learning, small group discussion (SGD), and/or problem-based learning (PBL). These teaching methods have been experimentally shown to improve student learning. In its design, this text has an advantage over a standard textbook where content delivery is emphasized, in that this text includes studying the *approach* clinicians and researchers used to discover the mechanism of disease. Instructors and students can also study similarities and differences in techniques and methodology between cases, in addition to content targeted to each molecular pathology being investigated. Inherently, it emphasizes the role and importance in "bench-to-bedside" training of health-care professionals. We hope that it encourages independent inquiry outside of the textbook, in that ongoing research of disease mechanisms and of novel therapeutic approaches are constantly being updated, changed, and improved. We hope that a new text with this approach will have wide appeal in many disciplines related to pathobiochemistry.

Okayama University Medical School has a long history of experience in delivering a pathobiochemistry tutorial course. Since its initiation in the 1990s by Prof. Ninomiya, it has conducted the course in English for approximately 30% of the students each year. The students and instructors use English in all discussions, written assignments, and presentations. The faculty who have been involved in the course have unanimously agreed on the need of a new pathobiochemistry text dedicated to a *project*-based learning for the clinician in training. Recently our school has begun accepting undergraduate or graduate (Master's level) students from other Asian countries (e.g., China, Korea, Indonesia, and Myanmar) as part of exchange programs. Their inclusion in the English-speaking student groups provides a positive stimulus for Japanese students to communicate in English. Furthermore, the faculty from those countries have also joined the course as tutors. It is therefore likely that their experience may translate in using the same English textbook for their own teaching. This international collaboration has clearly benefitted all involved.

Finally, we would like to express thanks to all of the contributors to this book for their expertise and hard work on their respective chapters. In addition, we would like to acknowledge Dr. Yasutaka Okazaki, Ms. Chieko Watanabe, Ms. Nobuko Hirota, Dr. Sue Lee at Springer Japan and Ms. Madona Samuel and and Mrs. Kandrakota Maadhuri at SPi Global.

Okayama, Japan — Toshitaka Oohashi
Okayama, Japan — Hirokazu Tsukahara
New York, NY, USA — Francesco Ramirez
Thousand Oaks, CA, USA — Chad L. Barber
Okayama, Japan — Fumio Otsuka

Contents

Part I

Metabolic Disorders

1 Citrin Deficiency

Yuan-Zong Song, Masahide Yazaki,
and Takeyori Saheki

Keywords

CTLN2 (adult-onset type II citrullinemia) · NICCD (neonatal intrahepatic cholestasis caused by citrin deficiency) · FTTDCD (failure to thrive and dyslipidemia caused by citrin deficiency) · Aspartate-glutamate carrier · Malate-aspartate shuttle

Abbreviations

4-HPL	4-Hydroxyphenyllactate
4-HPPV	4-Hydroxyphenylpyruvate
ABC	ATP-binding cassette
ABCG5/8	ATP-binding cassette, subfamily G, member 5/8
AFP	α-Fetoprotein
Ag	Antigen
AGC	Aspartate-glutamate carrier
ALP	Alkaline phosphatase
ALT	Alanine aminotransferase
ASS	Argininosuccinate synthetase
AST	Aspartate aminotransferase
BSEP	Bile salt export pump
CMV	Cytomegalovirus
CTLN1	Argininosuccinate synthetase deficiency or classical citrullinemia
CTLN2	Adult-onset type II citrullinemia
FTTDCD	Failure to thrive and dyslipidemia caused by citrin deficiency
GC-MS	Gas chromatography-mass spectrometry
GGT	γ-Glutamyltransferase
MCFA	Medium-chain fatty acid
MCT	Medium-chain triglyceride
MRD3	Multidrug-resistant protein 3
MS-MS	Tandem mass spectrometry
NICCD	Neonatal intrahepatic cholestasis caused by citrin deficiency
ORF	Open reading frame
TBA	Total bile acids
Tbil/Dbil	Total bilirubin/direct bilirubin

Y.-Z. Song
Department of Pediatrics, The First Affiliated Hospital, Jinan University, Guangzhou, Guangdong, China

M. Yazaki
Institute for Biomedical Sciences, Shinshu University, Matsumoto, Japan

Department of Clinical Laboratory Medicines, Shinshu University School of Health Sciences, Matsumoto, Japan
e-mail: mayazaki@shinshu-u.ac.jp

T. Saheki (✉)
Department of Hygiene and Health Promotion Medicine, Kagoshima University Graduate School of Medical and Dental Sciences, Kagoshima, Japan

T. Oohashi et al. (eds.), *Human Pathobiochemistry*, https://doi.org/10.1007/978-981-13-2977-7_1

1.1 Case Report

1.1.1 Patient 1

A 6-month-old male Chinese infant was referred to our hospital with jaundiced skin and sclera for nearly 6 months. Jaundice was observed on the third day after birth and then became gradually aggravated. At age 1.5 months, the infant underwent a physical examination at a local hospital that revealed a slightly enlarged liver 3 cm below the right subcostal margin. Subsequently, the baby was referred to several different hospitals in the local city, but the etiology of his jaundice remained unclear. At age 4 months, an anti-cytomegalovirus (CMV) IgM test at another local hospital was positive. Thus, CMV infection was diagnosed and intravenous ganciclovir was given for 1 month, but the therapeutic effect was unsatisfactory. At age 5 months, he was referred to a provincial hospital, where laboratory tests revealed abnormal liver function indices, increased levels of ammonia and lactate (Table 1.1), a reduced blood glucose level of 1.80 mmol/L (reference range 3.89–6.11 mmol/L), and positive CMV-Ag, CMV-IgM, and CMV-IgG. Once again, the infant was diagnosed with a CMV infection, and ganciclovir was given intravenously for an additional 20 days. As a result, CMV-Ag became negative, but the jaundice and abnormal liver function indices persisted. Therefore, the baby was referred to our hospital for further examination.

As the second baby of a non-consanguineous couple, the patient was born at the gestational age of 38 weeks after an uneventful pregnancy. The birth weight was 3.1 kg and the body length 50 cm. The baby was fed with breast milk. His

Table 1.1 Laboratory findings over time in the patient with NICCD

	Age at tests									
Indices (reference range)	5M	6M[a]	6.5M	7M	8M	11M	1Y4M	3Y8M	4Y6M	11Y7M
ALT (5–40 U/L)	96	42	26	46	39	23	27	19	14	13
AST (8–40 U/L)	200	140	77	147	74	44	47	36	35	21
GGT (7–50 U/L)	330	279	377	283	164	70	47	21	17	20
ALP (10–500 U/L)	437	399	300	350	385	216	194	247	300	485
TP (60–80 g/L)	56	72	78	52	69	73	64	72	70	67
Alb (35–55 g/L)	35	40	45	32	48	42	42	49	50	48
Glb (20–35 g/L)	21	32	33	21	22	31	21	23	20	19
Tbil(5.1–23.0 μmol/L)	94.8	107.4	50.2	43.7	12.2	2.3	3.0	7.2	11.0	8.7
DBil(0.6–6.8 μmol/L)	67.5	57.0	27.0	25.3	4.2	0.7	1.2	1.8	6.2	1.7
IBil(1.7–17.0 μmol/L)	27.3	50.4	23.2	18.4	8.0	1.6	1.8	5.4	4.8	7.0
TBA (0-10 μmol/L)	152	196	86	104	15	10	11	8	11	3
AFP (0-20 ng/mL)	–	320,000	140,000	19,000	450	10	–	–	–	–
TG(0.56–1.70 mmol/L)	–	2.82	1.63	–	–	–	–	2.64	0.96	1.12
Tchol(3.10–5.70 mmol/L)	–	6.76	4.03	–	–	–	–	3.94	4.74	5.00
Lactate (1.1–2.2 mmol/L)	2.70	1.54	–	–	–	2.11	–	–	–	–
Ammonia (<75 μg/dL)	95	113	73	94	67	57	–	–	–	–

[a]When the infant was referred to our hospital

Abbreviations: *ALT* alanine aminotransferase, *AST* aspartate aminotransferase, *GGT* γ-glutamyltranspeptidase, *ALP* alkaline phosphatase, *TP* total protein, *Alb* albumin, *Glb* globulin, *Tbil* total bilirubin, *Dbil* direct bilirubin, *Ibil* indirect bilirubin, *TBA* total bile acids, *TG* triglycerides, *Tchol* total cholesterol, *AFP* alpha-fetoprotein, Y and M represent year(s) and month(s) of age, respectively; –, not tested

elder brother was healthy. There was no family history of any genetic diseases.

Physical examination at admission revealed a body weight of 6.2 kg (−2.45SD), body length of 63.0 cm (−2.25SD), and head circumference of 40 cm (−2.77SD). His skin and sclera were mildly jaundiced. A chubby face was noticed. The lungs were clear on auscultation, and no abnormal cardiac sounds or murmurs were heard. The abdomen was distended, and the liver and spleen were palpable 5 cm and 2 cm below the right and left subcostal margins, respectively. His limb muscular tone was adequate, and the bilateral Babinski's, Brudzinski's, and Kernig's signs were all negative.

Liver function tests revealed increased levels of alanine aminotransferase (ALT), aspartate aminotransferase (AST), gamma-glutamyltranspeptidase (GGT), alkaline phosphatase (ALP), total bile acids (TBA), and total/direct bilirubin (Tbil/Dbil), as well as decreased total protein and albumin. Blood ammonia, lactate, cholesterol, and triglyceride levels were also increased, and the serum alpha-fetoprotein (AFP) level reached 320,000 ng/mL, an extremely elevated level rarely seen in clinical practice (Table 1.1). A tandem mass spectrometry (MS-MS) analysis detected elevated citrulline, threonine, tyrosine, and methionine in dried blood samples. Large quantities of galactose, galactitol, galactonate, 4-hydroxyphenyllactate (4-HPL), and 4-hydroxyphenylpyruvate (4-HPPV) were detected in urine by gas chromatography-mass spectrometry (GC-MS). A *SLC25A13* gene analysis revealed biallelic mutations of c.851_854del4 and c.1638_1660dup (Fig. 1.1).

Neonatal intrahepatic cholestasis caused by citrin deficiency (NICCD) was definitively diagnosed. Fat-soluble vitamins (A, D, E, and K) were given and breastfeeding was stopped. A lactose-free and medium-chain triglyceride (MCT)-enriched formula was introduced. As a result, his jaundice subsided, while the laboratory alterations recovered gradually. The patient has been followed up at our clinic for over 11 years. He has developed well with satisfactory anthropometric indices and social performance.

1.1.2 Patient 2

The patient was a 51-year-old Japanese male who was admitted to a local hospital due to consciousness disturbances. He had been well until he noticed a hand tremor when he climbed a mountain at age 51. After descending the mountain, he visited a neurosurgical clinic. His neurological examination was normal, and no abnormal findings were shown on a brain CT. Approximately a month later, his speech became rudimentary and garbled. The next morning, he was found lying unconscious on the floor of the entrance to his home, and he was immediately transferred to a local hospital. A brain CT was normal, and a laboratory examination revealed a highly elevated level of plasma ammonia (355 μg/dL, normal <70 μg/dL). He was thought to have hepatic encephalopathy and was treated with an infusion of branched amino acids followed by the hospital's low-protein diet (total calories 1600 kcal/day, protein 40 g/day) to reduce nitrogen sources. In addition, oral administration of lactulose and antibiotics (kanamycin) was started. After this treatment, the consciousness disturbances occurred frequently. An EEG recorded diffuse slow waves with occasional triphasic waves (Fig. 1.2A). Because his plasma levels of citrulline and arginine were elevated (408.9 nmol/mL, normal <40 nmol/mL and 186.4 nmol/mL, normal<120 nmol/mL, respectively), a urea cycle disorder was suspected, and he was admitted to Shinshu University Hospital. He did not have a family history suggestive of adult-onset type II citrullinemia (CTLN2). He had a food fondness for peanuts, milk, meat, and fish and had disliked sweets from childhood. On physical examination, he was very irritable and confused, and flapping tremor was noted in his hands. Icterus was not found and no hepatosplenomegaly was observed.

Laboratory data revealed a mild elevation of serum transaminases (AST 52 IU/L, normal <37 IU/L; ALT 96 IU/L, normal <45 IU/L). While serum GGT was moderately raised (138 IU/L, normal <50 IU/L), the levels of Tbil, albumin, and total cholesterol were within nor-

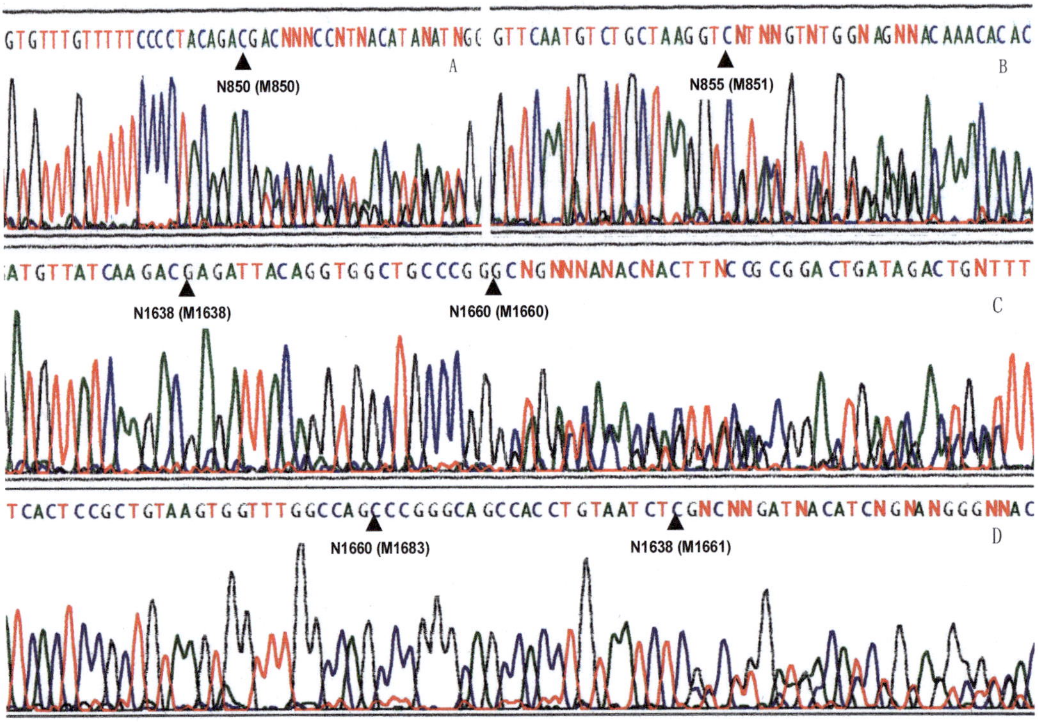

Fig. 1.1 Sanger sequencing results of exons 9 and 16 in the *SLC25A13* gene of the NICCD patient. Figures A and B represents the sequencing findings of exon 9, while figures C and D, of exon 16, in the coding and template strands of the *SLC25A13* gene, respectively. The digits in this figure denote the positions of the arrowhead-indicated bases within the normal (N) and mutated (M) *SLC25A13* alleles, respectively

mal values. There was no abnormality of coagulation function.

He was first thought to have hepatic encephalopathy. Laboratory data did not indicate the presence of hepatic failure. In addition, there was no serological evidence of hepatitis-related viral infection. The abdominal CT and MR images demonstrated neither liver cirrhosis nor extrahepatic portovenous shunt. Due to high plasma levels of citrulline and arginine, he was suspected to have CTLN2. DNA analysis of the *SLC25A13* gene, responsible for CTLN2, revealed that he was a compound heterozygote for the mutations of 851del 4 and IVS13+1 G > A, and a definite diagnosis of CTLN2 was made.

The patient began oral arginine (3 g/day) and a carbohydrate-restricted diet with a high fat content [total calories 1340 kcal/day, protein 50 g/day, carbohydrate 150 g/day, PFC (protein/fat/carbohydrate) ratio 15%:40%:45%]. His condition gradually ameliorated and his plasma ammonia level also decreased. After that, the daily dose of arginine was increased to 9 g/day, and total dietary calories were gradually increased to 1800 kcal in 3 months with the ratio of carbohydrate in the total dietary calories restricted to approximately 45% (Fukushima et al. 2010). After obtaining approval from the Institute Review Board of Shinshu University and written informed consent of the patient, oral intake of sodium pyruvate was added (6–9 g/day). Liver transplantation was not performed as there was no donor candidate for a live donor liver transplantation. At present, his condition has been stable for 8 years without receiving liver transplantation. An EEG recorded at age 59 years was normal, except for diffuse slow alpha waves (Fig. 1.2b).

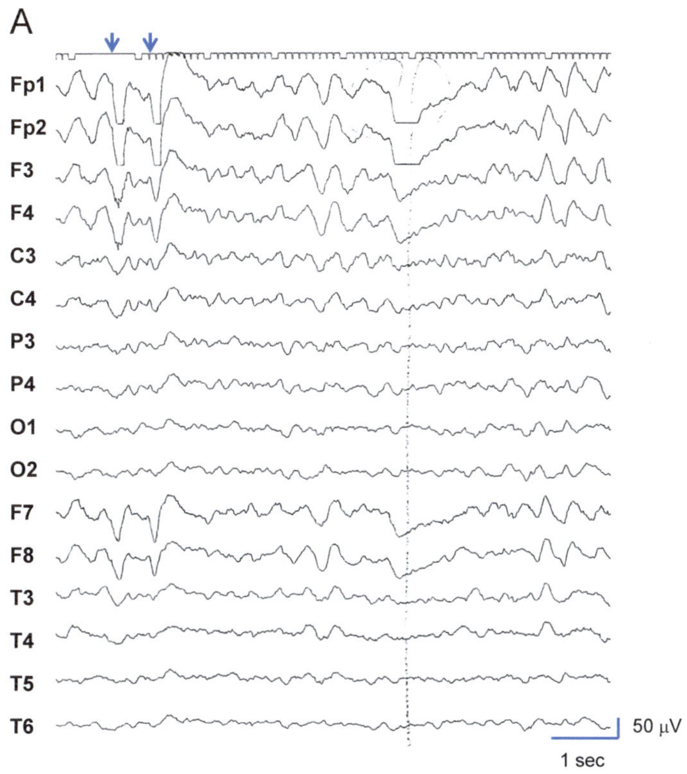

Fig. 1.2 EEG findings in the CTLN2 patient. (**a**) EEG recorded during treatment with a low-protein diet shows diffuse slow waves with occasional triphasic waves (arrows). (**b**) EEG at age 59 (7 years after starting sodium pyruvate therapy) appears almost normal, except for diffuse slow α-waves

1.2 Diagnosis

Patient 1 was a NICCD patient showing typical clinical and molecular manifestations. Prolonged jaundice and gradually aggravated hepatosplenomegaly in this patient suggested a chronic liver disease. Also, the growth retardation, as indicated by the body weight, body length, and head circumference at admission, suggested a long course of the disease. Unfortunately, none of these positive signs were pathognomonic. Moreover, although CMV infection did exist, this was not the major cause for this patient, as judged by the responses to ganciclovir treatment.

Cholestasis is a salient biochemical feature in NICCD. In this patient, the elevated GGT, TBA, and Tbil/Dbil all indicated a cholestatic liver disease. Reduced plasma levels of total protein and albumin, as well as increased blood ammonia, lactate, cholesterol, and triglyceride, as shown in Table 1.1, are common in NICCD subjects. The serum AFP level is usually dramatically increased. Once again, these biochemical alterations are neither specific for NICCD nor reliable for the definite diagnosis of such patients. There are no well-recognized clinical or biochemical diagnostic criteria for NICCD patients thus far.

Metabolome investigation often provides valuable clues for NICCD diagnosis. On GC-MS urinary analysis, coexistence of the markers for galactosemia (galactose, galactitol, and galactonate) and tyrosinemia (4-HPL and 4-HPP) is quite common in such cases. On MS-MS analysis of blood samples, typical NICCD patients have elevated citrulline, methionine, threonine, tyrosine, lysine, arginine, and ornithine, along with raised long-chain acylcarnitine levels. These metabolome alterations are usually transient, and not specific, either. Some NICCD patients have no such changes, especially in those who received dietary therapy, while some patients with other inborn errors of metabolism, such as galactosemia or mitochondrial DNA depletion syndrome, exhibit similar metabolome changes sometimes.

Thus far, *SLC25A13* gene and/or its expression product analysis has been taken as the reliable diagnostic tools for NICCD. Usually, the identification of biallelic *SLC25A13* mutations concludes a NICCD diagnosis (Lin et al. 2016).

In the diagnosis of CTLN2 (patient 2), detection of raised plasma ammonia level may be the first key to suspect CTLN2. Therefore, plasma ammonia should always be checked for patients with disturbed consciousness. However, as plasma ammonia level is not always high in CTLN2 patients, ammonia level should be measured several times, even if the first evaluation of ammonia shows no abnormality. An elevation of ammonia tends to be observed after a meal. Also, plasma amino acids assay including citrulline and arginine are mandatory. By abdominal US, CT, or MRI, the presence of other causes of hyperammonemia should be carefully checked (e.g., liver cirrhosis and portovenous shunt). The unique food preference is also a good clue to suspect citrin deficiency. The confirmation of diagnosis is made with a DNA analysis of *SLC25A13* (Kobayashi et al. 1999; Kikuchi et al. 2012; Lin et al. 2016).

1.3 Biochemical Perspectives

1.3.1 Historical Perspectives and Clinical Phenotypes

In 1981, Saheki et al. described two types of citrullinemia, qualitative and quantitative. The former is classical citrullinemia or CTLN1, caused by mutations in argininosuccinate synthetase (ASS) that catalyzes formation of argininosuccinate from citrulline and aspartate. The latter is named adult-onset type II citrullinemia or now CTLN2, in which ASS protein with normal enzymatic properties is liver specifically decreased. The gene (*SLC25A13*) causative for CTLN2 was described by Kobayashi et al. in 1999 and found to encode a liver-type mitochondrial aspartate-glutamate carrier, designated citrin. The human gene *SLC25A13* is localized at chromosome 7q21.3, which consists of 18 exons and encodes a 3.4 kb transcript with a predicted open reading frame (ORF) of 2025 bp. Citrin, as liver-type aspartate-glutamate carrier isoform 2 (AGC2), functions to export aspartate from the mitochondrial matrix in exchange for cytosolic glutamate and H^+(Palmieri et al. 2001), playing important roles in supplying aspartate to the cytosol for protein, nucleotide, and urea syntheses, and as a member of the malate-aspartate shuttle. Biallelic *SLC25A13* mutations result in citrin deficiency, a disease entity with three age-dependent clinical phenotypes, i.e., NICCD (OMIM#605814) in neonates or infants, CTLN2 (OMIM#603471) in adolescents/adults, and failure to thrive and dyslipidemia caused by citrin deficiency (FTTDCD) between NICCD and CTLN2 stages (Kobayashi et al. 2014). In addition, it is well-known that nonalcoholic fatty liver disease (NAFLD) including steatohepatitis (NASH) is frequently seen in citrin-deficient patients (Komatsu et al. 2008). Also, approximately 10–20% of citrin-deficient patients accompany pancreatitis or hepatoma in the pre-CTLN2 stage (Ikeda et al. 2001; Soeda et al. 2008). Hence, careful differential diagnosis may be also needed for patients with juvenile-onset, nonalcoholic pancreatitis or hepatic cancer unassociated with hepatitis virus infection. However, precise clinical pictures of citrin-deficient patients, especially in the pre-CTLN2 stage, remain obscure. While this disease is distributed worldwide, citrin deficiency is relatively more common among East Asian populations; however, some cases in western countries have been reported.

1.3.2 Structure and Function of Citrin

Citrin (AGC2) consists of 675 amino acid residues with a molecular weight of approximately 74 kD. The carboxy-terminal portion demonstrates similarity (20–30% amino acid identity) with proteins of the mitochondrial solute-carrier family (Kobayashi et al. 1999). The amino-terminal portion contains eight EF-hand domains that are conserved in calcium-binding proteins (Thangaratnarajah et al. 2014). The overall structure of citrin is most similar to that of the isoform, aralar (77.8% identity), which is encoded by the gene *SLC25A12*. AGC catalyzes unidirectional transport of glutamate and H^+ from cytosol to mitochondria and aspartate from mitochondria to cytosol under in vivo conditions. Properties, such as calcium activation and unidirectionality, indicate the metabolic significance of both AGCs.

1.3.3 Metabolic Functions of Citrin

Citrin plays at least three important roles in metabolism.

1. Citrin supplies aspartate from mitochondria to cytosol, where aspartate is needed for the synthesis of protein, nucleotides, and urea.
2. Citrin plays a role in transport of cytosolic NADH-reducing equivalent and energy metabolism as a member of the malate-aspartate shuttle.
3. Citrin is essential for gluconeogenesis from lactate and some other reduced substrates, such as glycerol and sorbitol. Citrin plays a role in maintaining stoichiometry of cytosolic NADH between production and consumption.

The transport of NADH-reducing equivalent from the cytosol to mitochondria, or the NADH shuttle, is mediated mainly by two systems, the malate-aspartate and glycerophosphate shuttles. The former is distributed in most tissues, while the latter is predominant in the skeletal muscles, brown adipose tissue, and brain. In the human liver, glycerophosphate shuttle activity is much less than that of the malate-aspartate shuttle, while the two shuttles are almost equally active in the rodent liver. This is why there were virtually no symptoms of human citrin deficiency in *Slc25a13* gene defect or citrin-KO mice.

1.3.4 Pathophysiology of Citrin Deficiency

The increased cytosolic NADH/NAD^+ ratio in hepatocytes, caused by the cytosolic metabolism of glucose, glycerol, and ethanol, all known precipitating factors of CTLN2, has been well-recognized as a key biochemical alteration. The high cytosolic NADH inhibits glycolysis at the step of glyceraldehyde 3-phosphate dehydrogenase, leading to a low supply of pyruvate into the mitochondria. This, together with no NADH-reducing equivalent from the cytosol to mitochondria, reduces the cytosolic and mitochondrial energy production from carbohydrates. Moreover, in citrin-deficient individuals, the mRNAs encoding enzymes/proteins involved in fatty acid oxidation and fatty acid transport are markedly suppressed, and the expression of peroxisome proliferator-activated receptorα (PPARα), a master regulator of hepatic lipid metabolism, is significantly suppressed (Komatsu et al. 2015). Serum concentrations of ketone bodies are also decreased in these patients (Imamura et al. 2003), suggesting reduced mitochondrial β-oxidation activity. All these findings suggest energy deficiency.

Fatty liver has been found in most citrin deficiency patients including NICCD and CTLN2 (Komatsu et al. 2008). It is most likely because, as stated above, the expression of PPARα is significantly suppressed, although the expression of hepatic genes associated with lipogenesis and triglyceride hydrolysis is not changed. It is also possible that the citrate-pyruvate shuttle, which is originally involved in fatty acid synthesis and induces a decrease in cytosolic NADH, may be activated, resulting in the stimulation of fatty acid synthesis.

1.3.5 Hyperammonemia in CTLN2 Is Different from Other Urea Cycle Enzyme Defects

It should be noted that the mechanism of hyperammonemia in citrin deficiency is different from other hyperammonemia caused by urea cycle

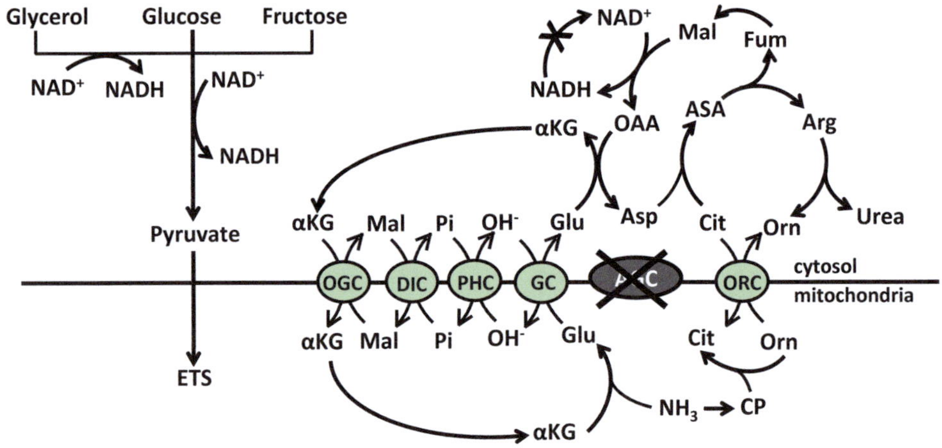

Fig. 1.3 Effect of carbohydrate metabolism on ureagenesis under citrin deficiency. *Arg* arginine, *Asp* aspartate, *ASA* argininosuccinate, *Cit* citrulline, *CP* carbomoylphosphate, *Fum* fumarate, *μKG* α-ketoglutarate ; *Mal* malate, *Pi* inorganic phosphate, *OAA* oxaloacetate, *Orn* ornithine, *AGC2* aspartate-glutamate carrier 2 or citrin, *DIC* dicarboxylate carrier, *ETS* electron transport system, *GC* glutamate carrier, *OGC* α-ketoglutarate malate carrier, *ORC* ornithine carrier, *PHC* phosphate carrier, *ASL* argininosuccinate lyase, *ASS* argininosuccinate synthetase, *AST* aspartate aminotransferase, *CPS1* carbamoylphosphate synthetase 1, *OCT* ornithine carbamoyltransferase, *MDH* malate dehydrogenase

enzyme deficiencies or conventional hyperammonemia. The intake of protein is harmful to patients with conventional hyperammonemia, while carbohydrate intake is harmful for citrin deficiency. The mechanism of hyperammonemia in citrin deficiency is as follows.

Urea can be synthesized from ammonia in the absence of citrin (Fig. 1.3). Glutamate formed from ammonia leaves the mitochondria via a glutamate carrier and is converted to aspartate in the cytosol, and urea is formed via argininosuccinate and arginine. In this metabolic pathway, oxaloacetate should be formed from fumarate via malate, the reaction in which NADH is formed. The NADH formed should be oxidized back to NAD^+. Under active metabolism from glucose, fructose, glycerol, and ethanol, NADH is formed. If the NADH is not oxidized back to NAD^+, oxaloacetate and aspartate cannot be formed, resulting in the inhibition of the ASS reaction, leading to citrullinemia and finally hyperammonemia.

Active carbohydrate metabolism inhibits urea synthesis, and an increase in NADH inhibits glycolysis at the step of glyceraldehyde-3-phosphate dehydrogenase, resulting in an energy deficit. It should be also noted that hyperammonemia in citrin deficiency is observed after eating or in the afternoon to evening, but not in the morning or after fasting.

Citrin-deficient subjects dislike carbohydrates, such as cooked rice and sweets, but enjoy protein- and fat-rich diets. This peculiar food preference could be related to the abovementioned pathophysiology and is important for diagnosis and treatment (Saheki et al. 2008).

At the end stage of CTLN2, hepatic ASS is decreased to less than 10% of the control on average, although the mechanism is not known. This decrease in hepatic ASS makes the intake of protein difficult as in CTLN1, rendering the treatment of CTLN2 more difficult.

1.3.6 Possible Mechanisms Related to Aberrant Metabolisms Under Citrin Deficiency

Cholestasis is defined as an impairment of bile secretion and flow followed by a lack of bile in the intestine and accumulation of potentially toxic cholephiles (substances that are soluble in or extracted by bile acid) in the liver and the systemic circulation. The canalicular bile secretion involves many energy-dependent molecular processes. As examples, the cross-canalicular excretion of bile acids, phospholipids, and cholesterol, the three major organic solutes in bile, is mediated by the ATP-binding cassette (ABC) transporters BSEP, MDR3, and ABCG5/8, respectively. Since citrin deficiency results in energy-lacking hepatocytes, it is not surprising for NICCD patients to have intrahepatic cholestasis that presents as late-onset/prolonged jaundice, hepatosplenomegaly, and/or abnormal biochemical changes including elevated aminotransferases, Tbil, Dbil, and TBA levels.

Dramatically elevated AFP with reduced albumin is quite common in NICCD patients. AFP and albumin have highly homologous and conserved primary structures, and their genes belong to the same family on human chromosome 4. AFP decreases dramatically and immediately after birth, resulting in the substitution of AFP by albumin. Since the expression pattern of the proteins mirrors development and maturation of the liver, the delayed switch of AFP to albumin may reflect the immature liver development in NICCD, although other factors, such as hepatocellular destruction and regeneration, are also likely to contribute to low albumin and high AFP.

Coexistence of the markers for galactosemia (galactose, galactitol, and galactonate) and tyrosinemia (4-HPL and 4-HPP) is quite common in urine samples of NICCD subjects, suggesting secondary defects in galactose and tyrosine metabolisms. UDP-galactose epimerase, one of the enzymes of galactose metabolism, has been reported to be inhibited by NADH as a competitor of NAD^+ bound to the enzyme. NAD^+ plays a role in the enzyme mechanism, although the enzyme is not oxidoreductase. This inhibition by NADH may be the cause of galactosemia in NICCD.

In blood samples of NICCD infants, in addition to elevated long-chain acylcarnitines, the levels of some amino acids (including citrulline, ornithine, arginine, threonine, methionine, and tyrosine) are also increased. Deficiency of citrin affects the liver cytosol concentration of aspartate for ASS, leading to an accumulation of citrulline, an upstream substrate in the same cycle. An elevated level of arginine, a downstream product of ASS in the urea cycle, can be explained by the fact that arginine is mainly synthesized in the kidney and small intestine (during neonatal period) from citrulline formed in the small intestine, and the accumulation of citrulline in NICCD accelerates this process. The increased threonine might reflect an inhibited degradation of this amino acid via threonine dehydrogenase, a biochemical reaction giving rise to NADH. The elevated long-chain acylcarnitine levels might result from the impaired mitochondrial β-oxidation of fatty acids.

1.3.7 Molecular Perspectives

Analysis of the *SLC25A13* gene and/or its expression product is a reliable diagnostic tool for NICCD. To date, 106 pathogenic *SLC25A13* mutations/variations have been reported (Lin et al. 2016), most of which are point mutations or short insertions/deletions. Among the *SLC25A13* mutations detected in Chinese NICCD patients, c.851_854del, c.1638_1660dup, c.615+5G>A, and IVS16ins3kb account for approximately 85% of all mutated alleles. In Japan, 11 mutations including c.851_854del, c.1019_1177del, c.1231_1311del, and c.675C > A account for approximately 95% (Kikuchi et al. 2012). *SLC25A13* cDNA cloning analysis with human peripheral blood lymphocytes (PBLs) may be a less invasive and more feasible diagnostic tool in cases where no positive results are available with conventional methods, such as polymerase chain reaction (PCR), long and accurate PCR (LA-PCR), PCR-restriction fragment length polymorphism (PCR-RFLP), and Sanger sequencing. In such cases, detection of citrin protein in biopsied liver specimens or cultured skin fibroblasts using Western blotting is an alternative diagnostic tool. In addition, both fresh and expanded PBLs have been tried as the protein source for Western blotting analysis of citrin molecules. Figure 1.4 illustrates a representative molecular diagnostic approach for NICCD using Western blotting to detect citrin protein in mitochondrial proteins extracted from cultured PBLs.

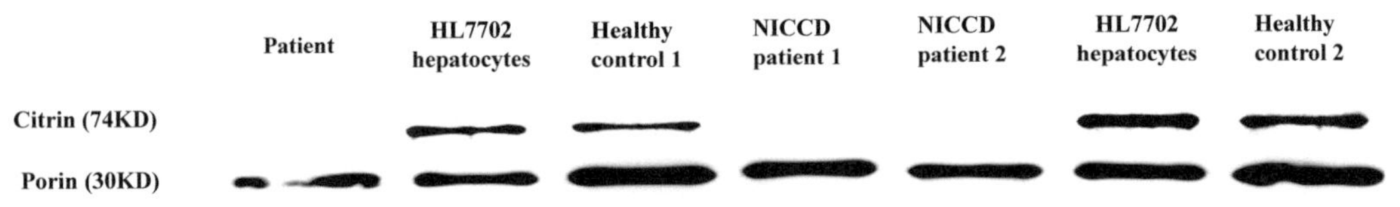

Fig. 1.4 Western blotting analysis of mitochondrial proteins. Citrin signal was detected in the cultured PBLs of two healthy controls and the HL7702 hepatocyte line, but absent in the cultured PBLs of this patient as well as two NICCD patients diagnosed previously

1.4 Treatment

1.4.1 Treatment of NICCD

To our knowledge, no specific medication has been developed to treat NICCD. Fat-soluble vitamins and zinc supplementation might be beneficial since these nutrients are usually deficient in NICCD infants. Notably, galactose-free and MCT-enriched formulas have been reported to be clinically and biochemically effective for NICCD patients in an increasing number of clinical cases.

Breast milk or common formula contains high carbohydrate in the form of lactose. In the gut, this carbohydrate is digested into galactose and glucose, which are then absorbed into the blood to serve as major substrates for energy production and fuel storage in human neonates or infants, and galactose is a well-known precursor of glucose. In the cytosol of hepatocytes of NICCD patients, the catabolic metabolism of glucose (glycolysis) produces NADH, and the resultant elevated NADH/NAD^+ ratio inhibits glycolysis, causing an energy shortage and leading to cholestasis and liver damage in NICCD patients. In addition, the energy shortage might distort the mitochondrial structures in citrin-deficient hepatocytes, thus impairing the mitochondrial β-oxidation, giving rise to lipid accumulation and resulting in fatty liver, and aggravating the energy shortage. Moreover, as previously stated, secondary galactosemia due to citrin deficiency might also be involved in NICCD pathogenesis, leading to accumulation of a large quantity of galactitol and galactonate. Galactitol has been suggested to be a substrate that causes jaundice, hepatosplenomegaly, hepatocellular insufficiency, and cataracts.

Energy shortage in the liver caused by an impairment of glycolysis due to an increased NADH/NAD^+ ratio has been proposed as an important pathophysiology of citrin deficiency. MCT is quickly hydrolyzed and absorbed mainly as medium-chain fatty acids (MCFA). The absorption of MCFA is not bile acid-dependent, which might reduce the burden of the liver to synthesize and excrete bile salt into the gut. Moreover, MCFA can be better absorbed and transported via the portal vein into the liver and is mainly metabolized to acetyl-CoA along with $FADH_2$ and NADH, which could supply more such substrates to hepatic cells as energy sources. Also, it is important to note that MCT administration is known to stimulate de novo hepatic lipogenesis, which consumes cytosolic NADH to produce NAD^+, thus decreasing the cytosolic NADH/NAD^+ ratio.

Taking all these factors together, it is not surprising that NICCD infants respond well to galactose-free and MCT-enriched therapeutic formulas.

1.4.2 Treatment of CTLN2

1.4.2.1 Liver Transplantation

To replace citrin and support (compensate for) citrin function are of utmost importance in the therapy for patients with CTLN2. Hence, liver transplantation is one of the most promising therapies for CTLN2 patients (Ikeda et al. 2001). After liver transplantation, abnormal laboratory findings including plasma ammonia and citrulline can be rapidly normalized. Also, neuropsychotic symptoms are completely absent after liver translation. However, the shortage of donors is a serious challenge, especially in Japan, and other therapeutic options should be developed.

1.4.2.2 Nonsurgical Treatment

In citrin deficiency, the cytosolic ratio of NADH/NAD^+ in the hepatocytes dramatically increases after glucose metabolism (Saheki and Kobayashi 2002). Most likely to evade this condition, citrin-deficient patients are assumed to avoid carbohydrates by their unique food preferences. The PFC ratios in the diet calories of patients with citrin deficiency are different from those in healthy individuals, and the energy ratio of carbohydrates in citrin-deficient patients (approximately 40%) is much lower than in healthy individuals (54–58%) (Saheki et al. 2008; Nakamura et al. 2011). Therefore, a carbohydrate-restricted diet with an energy ratio of approximately 40% carbohydrate is effective for CTLN2 patients (Imamura et al. 2003; Fukushima et al. 2010; Nakamura et al. 2011). In a hospital protein-restrictive diet for liver disorders (40 g/day), the carbohydrate ratio exceeds 70%; therefore, this diet should be avoided in CTLN2 patients (Fukushima et al. 2010). At Shinshu University Hospital, a carbohydrate-restriction diet (with a PFC ratio 15%:45%:40%, total 1720 kcal) has been recently served as a basic hospital diet for CTLN2 patients (Nakamura et al. 2011).

Several previous studies have reported the therapeutic efficacy of oral intake of MCT in CTLN2 patients (Hayasaka et al. 2014) to compensate for the shortage of energy in the hepatocytes caused by citrin deficiency.

Like patient 2, the efficacy of oral intake of sodium pyruvate has been described in other CTLN2 patients. Sodium pyruvate was first tried in patients with mitochondrial disorders (Tanaka et al. 2007). This chemical is presumed to play an important role in decreasing the cytosolic NADH/NAD^+ ratio (Moriyama et al. 2006). Long-term therapeutic efficacy of sodium pyruvate in CTLN2 patients is still being examined. Approximately 60% patients treated with sodium pyruvate have had a favorable response so far (Mutoh et al. 2008; Yazaki et al. 2013).

As for other therapeutic options, the efficacy of arginine granule has been described (Imamura et al. 2003). To reduce production of ammonia by intestinal bacteria, oral administration of lactulose or antibiotics may be also effective; however, this treatment remains controversial.

1.4.2.3 Unfavorable Treatments

High-carbohydrate consumption and drinking alcohol should be avoided. As demonstrated by patient 2, a hospital diet for liver disease (protein-restriction diet) may also be risky for CTLN2 patients because of the high-carbohydrate content (Fukushima et al. 2010).

We previously reported the risk of encephalopathy by glycerol infusion, which is a hyperosmotic diuretic solution consisting of 10% glycerol and 5% fructose used mainly to treat brain edema (Yazaki et al. 2005). Since then, its administration has been prohibited for CTLN2 patients in Japan. Instead, D-mannitol is recommended for brain edema in CTLN2 patients. Although a detailed pathophysiology is not available, glycerol metabolism seems to further facilitate cytosolic NADH accumulation (Yazaki et al. 2005). Most likely due to a similar mechanism, the infusion of hyperalimentation fluid containing high levels of glucose is also risky for CTLN2 patients.

End-of-Chapter Questions

1. Citrin deficiency has different clinical presentations among patients of different ages. What are the three age-dependent phenotypes?
2. NICCD patients usually have colorful but nonpathognomonic clinical and laboratory presentations. What are the major clinical and laboratory features for this phenotype?
3. Why do NICCD patients respond well to galactose-free and MCT-enriched formulas?
4. Why is a conventional hospital diet for liver disease or an intravenous infusion of glycerol solution risky for CTLN2 patients?
5. Why does carbohydrate intake result in hyperammonemia in citrin deficiency?

References

Fukushima K, Yazaki M, Nakamura M et al (2010) Conventional diet therapy for hyperammonemia is risky in the treatment of hepatic encephalopathy associated with citrin deficiency. Intern Med 49:1–20

Hayasaka K, Numakura C, Toyota K et al (2014) Medium-chain triglyceride supplementation under a low-carbohydrate formula is a promising therapy for adult-onset type II citrullinemia. Mol Genet Metab Rep 1:42–50

Ikeda S, Yazaki M, Takei Y et al (2001) Type II (adult onset) citrullinemia: clinical pictures and the therapeutic effect of liver transplantation. J Neurol Neurosug Psychiatr 71:663–670

Imamura Y, Kobayashi K, Shibatou T et al (2003) Effectiveness of carbohydrate-restrictive diet and arginine granules therapy for adult-onset type II citrullinemia: a case report of siblings showing homozygous *SLC25A13* mutation with and without the disease. Hepatol Res 26:68–72

Kikuchi A, Arai-Ichinoi N, Sakamoto O et al (2012) Simple and rapid genetic testing for citrin deficiency by. Screening 11 prevalent mutations in *SLC25A13*. Mol Genet Metab 105:553–558

Kobayashi K, Sinasac DS, Iijima M et al (1999) The gene mutated in adult-onset type II citrullinaemia encodes a putative mitochondrial carrier protein. Nat Genet 22:159–163

Kobayashi K, Saheki T, Song YZ (2014) Citrin deficiency. In: Pagon RA, Bird TD, Dolan CR, Stephens K, Adam MP (eds) Gene reviews [Internet]. University of Washington, Seattle; 1993–2005 Sept 16

Komatsu M, Yazaki M, Tanaka N et al (2008) Citrin deficiency as a cause of chronic liver disorder mimicking non-alcoholic fatty liver disease. J Hepatol 49:810–820

Komatsu M, Kimura T, Yazaki M et al (2015) Steatogenesis in adult-onset type II citrullinemia is associated with down-regulation of PPARα. Biochim Biophys Acta 1852:473–481

Lin WX, Zeng HS, Zhang ZH et al (2016) Molecular diagnosis of pediatric patients with citrin deficiency in China: *SLC25A13* mutation spectrum and the geographic distribution. Sci Rep 6:29732

Moriyama M, Li MX, Kobayashi K et al (2006) Pyruvate ameliorates the defect in ureogenesis from ammonia in citrin-deficient mice. J Hepatol 44:930–938

Mutoh K, Kurokawa K, Kobayashi K, Saheki T (2008) Treatment of a citrin-deficient patient at the early stage of adult-onset type II citrullinaemia with arginine and sodium pyruvate. J Inherit Metab Dis 31(Suppl 2):S343–S347

Nakamura M, Yazaki M, Kobayashi Y et al (2011) The characteristics of food intake in patients with type II citrullinemia. J Nutr Sci Vitaminol 57:239–245

Palmieri L, Pardo B, Lasorsa FM et al (2001) Citrin and aralar 1 are Ca^{2+}-stimulated aspartate/glutamate transporters in mitochondria. EMBO J 20:5060–5069

Saheki T, Kobayashi K (2002) Mitochondrial aspartate glutamate carrier (citrin) deficiency as the cause of adult-onset type II citrullinemia (CTLN2) and idiopathic neonatal hepatitis (NICCD). J Hum Genet 47:333–341

Saheki T, Kobayashi K, Terashi M et al (2008) Reduced carbohydrate intake in citrin-deficient subjects. J Inherit Metab Dis 31:386–394

Soeda J, Yazaki M, Nakata T et al (2008) Primary liver carcinoma exhibiting dual hepatocellular-biliary epithelial differentiations associated with citrin deficiency: a case report and review of the literature. J Clin Gastroenterol 42:855–860

Tanaka M, Nishigaki Y, Fuku N et al (2007) Therapeutic potential of pyruvate therapy for mitochondrial diseases. Mitochondrion 7:399–401

Thangaratnarajah C, Ruprecht JJ, Kunji ER (2014) Calcium-induced conformational changes of the regulatory domain of human mitochondrial aspartate/glutamate carriers. Nat Commun 5:5491

Yazaki M, Takei Y, Kobayashi K et al (2005) Risk of worsened encephalopathy after intravenous glycerol therapy in patients with adult-onset type II citrullinemia (CTLN2). Intern Med 44:188–195

Yazaki M, Kinoshita M, Ogawa S et al (2013) A 73-year-old patient with adult-onset type II citrullinemia successfully treated by sodium pyruvate and arginine. Clin Neurol Neurosurg 115:1542–1545

Aspartoacylase Deficiency (Canavan Disease, *N*-Acetylaspartic Aciduria)

2

Jörn Oliver Sass and Ina Knerr

Keywords
Aminoacylase · Neurometabolic disease · Organic aciduria · Enzyme activity assays · Magnetic resonance imaging (MRI)

2.1 Case Report

A baby girl presented at 3 months of age with the following symptoms: global developmental delay with regression and irritability along with feeding difficulties and poor sleep. On examination, her weight and length were in the lower normal range, but she was noted to have a relative macrocephaly. The baby had a normal eye examination at the time, but it was noticed that she was not fixing and following. Subsequently, over the next few months, the patient was further noted to have an abnormally increasing head circumference which was crossing percentiles, and at 6 months of age, her head circumference plotted above the 99.6th percentile.

A brain magnetic resonance imaging (MRI) scan was performed along with magnetic resonance spectroscopy (MRS) as part of her work-up. The MRI revealed diffuse white matter changes, e.g., of supratentorial white matter and also involvement of the cerebellum (Fig. 2.1a+b). MRS, too, was abnormal, with a strong increase in the *N*-acetyl aspartate (NAA) peak (Fig. 2.1c). In addition to these tests, a lumbar puncture had been performed. It showed a slightly elevated total protein concentration in the cerebrospinal fluid (CSF) with normal results for cells, glucose, lactate, and amino acids. Her plasma amino acids were normal, as were her urinary glycosaminoglycans, oligosaccharides, and plasma/white blood cell lysosomal enzymes.

Metabolic work-up, including urinary organic acid analysis, revealed large excretion of NAA in the patient's urine (*N*-acetylaspartic aciduria). This metabolite, NAA, is the biochemical hallmark for a rare neurometabolic disorder, Canavan disease (CD), which is caused by deficiency of the enzyme aspartoacylase. The biochemical laboratory contacted the primary physician immediately to request a repeat urine sample in this patient to confirm their finding. The repeat sample also revealed a grossly elevated excretion of NAA (Fig. 2.2). The diagnosis of CD was confirmed at the enzyme level, using cultured skin fibroblast cells and

J. O. Sass (✉)
Department of Natural Sciences, Inborn Errors of Metabolism, Bonn-Rhein-Sieg University of Applied Sciences, Rheinbach, Germany
e-mail: joern.oliver.sass@h-brs.de

I. Knerr
National Centre for Inherited Metabolic Disorders, Temple Street Children's University Hospital, Dublin, Ireland
e-mail: ina.knerr@cuh.ie

T. Oohashi et al. (eds.), *Human Pathobiochemistry*, https://doi.org/10.1007/978-981-13-2977-7_2

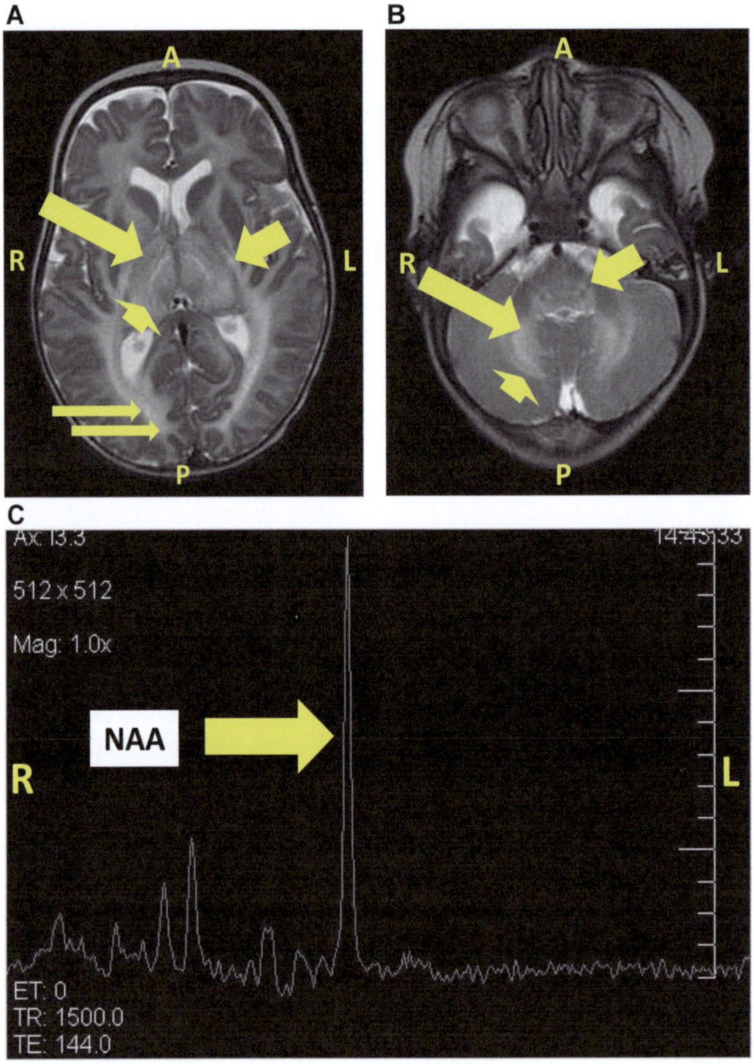

Fig. 2.1 (**a**) Magnetic resonance imaging (MRI) of the brain in a baby with Canavan disease (CD) (R, right; L, left; A, anterior; P, posterior). Axial T2 at the level of the basal ganglia shows abnormal signal in the thalamus (arrowhead) and globus pallidus (arrow) bilaterally as well as focally in the posterior limbs of the internal capsules (short arrow) and the posterior deep white matter (double arrow). Characteristic sparing of putamina, claustrum, caudate, corpus callosum, and most of the internal capsules
(**b**) MRI brain in a baby with CD. Axial T2 at the level of the pons shows abnormal signal in the cerebellar white matter (arrowhead), the dentate nuclei (arrow), and the pons (short arrow). There was extensive abnormality also in the midbrain and posterior medulla (*not shown*)
(**c**) Magnetic resonance spectroscopy (MRS) of the brain in a baby with CD. In principle, MRS is a technique that provides information about the chemical composition of the brain when added to the MRI scan. It determines the relative concentrations of the molecules/resonance frequency of chemical elements. Among the metabolites detectable by applying this technique are lactate, choline, and *N*-acetyl aspartate (NAA). We here demonstrate a strong increase in the NAA peak (indicated by the bold arrow on the graph) in a baby with CD

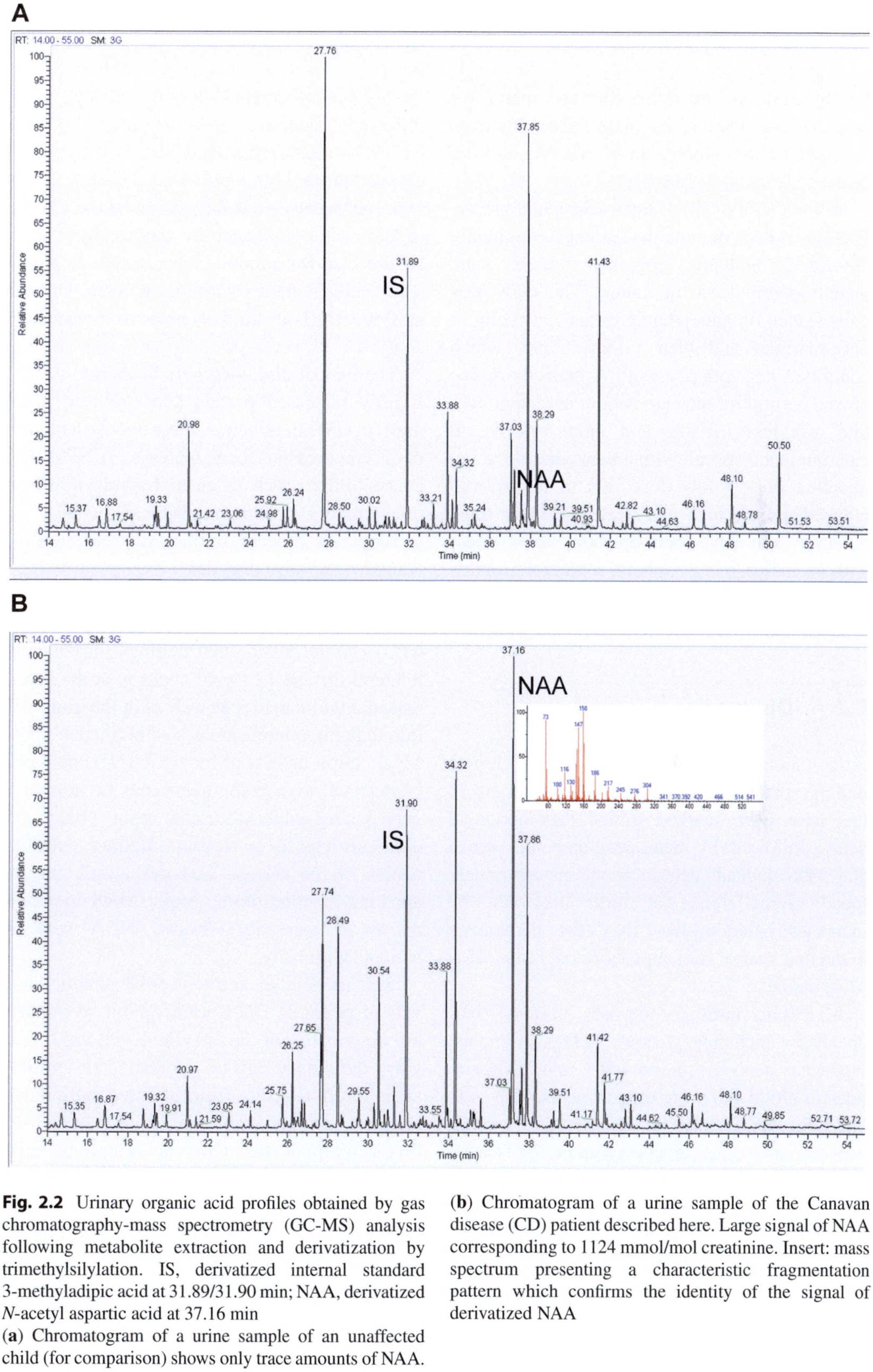

Fig. 2.2 Urinary organic acid profiles obtained by gas chromatography-mass spectrometry (GC-MS) analysis following metabolite extraction and derivatization by trimethylsilylation. IS, derivatized internal standard 3-methyladipic acid at 31.89/31.90 min; NAA, derivatized *N*-acetyl aspartic acid at 37.16 min

(**a**) Chromatogram of a urine sample of an unaffected child (for comparison) shows only trace amounts of NAA.

(**b**) Chromatogram of a urine sample of the Canavan disease (CD) patient described here. Large signal of NAA corresponding to 1124 mmol/mol creatinine. Insert: mass spectrum presenting a characteristic fragmentation pattern which confirms the identity of the signal of derivatized NAA

lymphocytes immortalized by Epstein-Barr virus (EBV). Molecular genetic analysis, on the DNA level including sequence analysis of the *ASPA*-coding sequence and splice sites and search for deletions and duplications by multiplex ligation-dependent probe amplification (MLPA), did not detect a mutation in this case.

Although her initial electroencephalography (EEG) had been normal, the patient subsequently developed epileptic seizures which were predominantly tonic in nature. The baby was commenced on antiepileptic drugs consisting of phenobarbital and then valproic acid, which controlled her epilepsy well. A benzodiazepine proved helpful for management of her irritability. She was less irritable and more reactive to environmental stimuli with a benzodiazepine but tolerated only a low dose due to her overall reduced muscle tone and head lag. Over time, however, she developed episodes of stiffness with an increase in peripheral tone and progressive spasticity. By this point, she required tube feeding.

2.2 Diagnosis

From a metabolic point of view, the key clinical findings that had warranted further work-up in this case were marked global developmental delay, followed by neurodevelopmental arrest and neurological deterioration, macrocephaly (head circumference at birth may not be increased; enlarging head may cross percentiles in the first year of life), and abnormal brain MRI/MRS studies.

Metabolic testing comprises baseline biochemistry, including glucose, lactate, ammonia, blood gas analysis, amino acids, and acylcarnitines in blood and urine organic acids. This is to check for inborn metabolic disorders such as aminoacidopathies (e.g., maple syrup urine disease); organic acidurias, including "cerebral" organic acidurias that predominantly involve the brain (e.g., CD, glutaric aciduria type 1); and mitochondrial disorders, particularly respiratory chain defects. Urinary glycosaminoglycans and oligosaccharides together with plasma/white cell lysosomal enzymes are important tools in checking for lysosomal disorders, including Krabbe disease (galactocerebrosidase deficiency), in the differential diagnosis. However, only CD (OMIM #27190) is characterized biochemically by grossly elevated urinary NAA excretion. This key metabolite can be detected in the patient's urine, plasma, or CSF, while no diagnostic abnormalities are to be expected from routine laboratory tests, amino acid analysis, or acylcarnitine profiles. Urine is generally the body fluid of choice for organic acid analysis.

The patient discussed here furthermore had a slightly increased protein concentration (>450 mg/l) in CSF. In principle, there may be a number of different reasons for an increase in CSF protein concentration, such as an underlying infection, inflammatory or immune response, hemorrhage, or neoplastic process. In this case, the cause is a neurodegenerative disorder.

Brain MRI demonstrates very abnormal symmetric white matter changes in patients with CD. In more detail, such findings may include bilateral diffuse T2 signal changes in the supratentorial white matter as well as in the globi pallidi, thalami, pontine areas, and brainstem tracts. Myelination defects of the optic tracts may also be detected. Eye examination may be normal in very young patients despite poor vision, but optic atrophy is a frequent finding over the course of the disease, and nystagmus may be apparent. Furthermore, brain MRS typically reveals a markedly elevated NAA peak in patients with CD.

Enzyme studies and molecular genetic analysis of genomic DNA are available to confirm the diagnosis. On the enzyme level, aspartoacylase deficiency can be confirmed in cultured skin fibroblast cells from affected individuals and in lymphoblastoid cells prepared from blood lymphocytes. Clinical symptoms arise from the enzyme deficiency in the brain, specifically in the oligodendrocytes which are the cells that create the myelin layers and thereby compose the white matter of the brain. CD belongs to a group of genetic disorders which are designated as "leukodystrophies." The diagnostic hallmarks for these conditions are distinctive white matter findings which are usually symmetric and progressive, reflecting abnormal development and destruction of the brain myelin. The deterioration of white matter findings can correlate with the neurological decline.

Postmortem examination in patients with CD can reveal a soft-gelatinous and spongy brain appearance. CD is named after the neuropathologist Dr Myrtelle Canavan. Her index case was a male infant who had initially presented with macrocephaly, nystagmus, and failure to thrive and who died at 16 months of age (Canavan 1931). Neuropathological examination of the brains from deceased CD patients revealed the distinctive widespread vacuolation in both white and gray matters (van Bogaert and Bertrand 1949). Vacuoles that have been found within protoplasmic astrocytes have been interpreted as the main cause for the vacuolated appearance of the white matter (Adachi et al. 1972).

In the vast majority, but not in all patients with CD, the diagnosis can also be confirmed at a molecular genetic level by identifying pathogenic mutations in the *ASPA* gene. Prenatal diagnosis is possible in families with a known index case with CD (Lienhard and Sass 2011). However, enzyme assays in amniocytes have been considered non-reliable, and prenatal diagnosis may be achieved by molecular genetic analysis of the *ASPA* gene (if the index case has two confirmed pathogenic mutations in the *ASPA* gene) together with organic acid analysis of amniotic fluid to check for NAA concentrations.

2.3 Biochemical Perspectives

Aspartoacylase (EC 3.5.15) is a dimeric metalloenzyme that requires Zn^{2+}; it is also known as aminoacylase 2. It is highly expressed in different organs and tissues, including the brain (oligodendrocytes) (Sommer and Sass 2012). Aspartoacylase catalyzes specifically the conversion of NAA to aspartate and acetate (Fig. 2.3), while the hydrolysis of a wide range of other *N*-acylated amino acids is catalyzed by aminoacylase 1 (EC 3.5.1.14). In CD patients, the hydrolysis of NAA is blocked, and NAA accumulates predominantly in the white matter of the brain, followed by progressive neurological decline. However, the underlying pathophysiology is still enigmatic.

Several factors have been discussed to explain the pathophysiological mechanisms in CD. In more detail, it was discussed that high concentrations of *N*-acetylaspartylglutamate in the brain, secondary to high concentrations of NAA, could have detrimental effects by disturbing *N*-methyl-D-aspartate (NMDA) receptor-dependent processes or by causing accumulation of glutamate (Burlina et al. 1999). It was also suggested that NAA functions as a molecular water pump in myelinated neurons and that NAA accumulation leads to osmotic dysregulation in the brain which can subsequently lead to dysmyelination and subcortical vacuolation observed in CD patients (Baslow 2003). It has also been presumed that aspartoacylase may be involved in the epigenetic regulation of genes relevant to myelin and genes responsible for the differentiation of oligodendrocytes, cells that

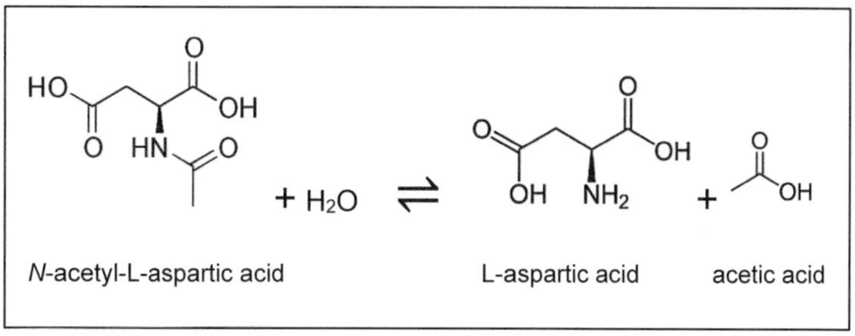

Fig. 2.3 Reaction catalyzed by aspartoacylase (aminoacylase 2), the enzyme deficient in patients with Canavan disease (CD): *N*-acetyl-L-aspartic acid is hydrolyzed to L-aspartic acid and acetic acid

express much aspartoacylase (Kumar et al. 2009). Another hypothesis has attributed the pathology observed in CD to the inability to generate free acetate from NAA due to the lack of aspartoacylase, leading to impaired myelin synthesis during postnatal axonal myelination. However, it was demonstrated in a CD mouse model that acetate derived from NAA is not essential for myelin synthesis, thereby refuting the latter hypothesis and further highlighting the pathophysiological role of NAA accumulation in the pathogenesis of CD (Guo et al. 2015; Maier et al. 2015).

The clinical course may vary between patients, and it has been suggested that there may be a correlation between residual enzyme activity, genotype, and clinical course in CD patients (Mendes et al. 2017).

2.4 Genetics

The human *ASPA* gene maps to chromosome 17 (17p13.3). It includes six exons and five introns. While mutations in the *ASPA* gene encoding the aspartoacylase enzyme have first been reported in CD patients of Ashkenazi descent (Kaul et al. 1993) where CD is highly prevalent, CD is now considered a panethnic disease, with the missense mutation A305E representing about half of the alleles in non-Jewish patients of European origin (Sistermans et al. 2000). CD is an autosomal recessive genetic disorder. Notably, as reflected in our case, conventional DNA sequencing, i.e., Sanger sequencing, does not always reveal the underlying disease-causing mutations as it does not allow identification of deep intronic mutations, changes in regulatory regions, and large deletions or insertions as well as chromosomal rearrangements. In the patient presented here, MLPA analysis was added but did not reveal any deletion or duplication in the ASPA gene. This highlights the importance of diagnostic enzyme assays in confirming a diagnosis of CD which is based on clinical, neuroradiological, and metabolite findings.

2.5 Therapy

No curative therapy is currently available for patients with CD. Treatment is therefore focussed on supporting the patient's general health, well-being, and quality of life, including adequate nutrition and controlling epilepsy. Novel treatment options have been explored and are currently under investigation, including gene therapy trials with a viral vector carrying the *ASPA* gene; however, no curative agent has emerged. Data obtained from a CD mouse model suggests that reducing the excessive amount of NAA could potentially slow down disease progression in patients with CD (Guo et al. 2015; Maier et al. 2015; Sohn et al. 2017). However, there is also evidence that complete inhibition of NAA synthase may have "unforeseeable risks for patients" (Maier et al. 2015). It has been described in a limited number of patients that lithium can reduce NAA concentrations in affected CD patients; however, no improvement in gross motor function of affected children has been found (Assadi et al. 2010), and its therapeutic efficacy remains in question (Kölker et al. 2014).

2.6 Summary

CD is a rare, autosomal recessive neurometabolic disorder with often devastating clinical consequences. It typically presents in early infancy. Clinical symptoms and MRI-based findings are chronic neurological deterioration with irritability and severe global developmental delay, loss of early milestones, macrocephaly, and seizures together with white matter abnormalities in the brain ("leukodystrophy"). CD is biochemically characterized by a deficiency of aspartoacylase along with the accumulation of NAA in the brain and in body fluids, particularly in urine. In the majority of CD patients, the disease leads to early death. As there is no cure for this condition, management of affected children and their families is symptom-orientated and supportive and includes a multidisciplinary medical team.

End-of-Chapter Questions

1. What are the typical clinical symptoms in patients with CD?
2. Can you describe the most important diagnostic tests to establish and confirm the diagnosis of CD?
3. How can you explain the observation that not all patients with CD have an identifiable mutation in the coding region of the *ASPA* gene?
4. How would you manage a patient with confirmed CD?
5. What do you know about the pathomechanisms underlying CD?

References

Adachi M, Torii J, Schneck L, Volk BW (1972) Electron microscopic and enzyme histochemical studies of the cerebellum in spongy degeneration (van Bogaert and Bertrand type). Acta Neuropathol 20:22–31

Assadi M, Janson C, Wang DJ, Goldfarb O, Suri N, Bilaniuk L, Leone P (2010) Lithium citrate reduces excessive intra-cerebral *N*-acetyl aspartate in Canavan disease. Eur J Paediatr Neurol 14:354–359

Baslow MH (2003) *N*-acetylaspartate in the vertebrate brain: metabolism and function. Neurochem Res 28:941–953

Burlina AP, Ferrari V, Divry P, Gradowska W, Jakobs C, Bennett MJ, Sewell AC, Dionisi-Vici C, Burlina AB (1999) *N*-acetylaspartylglutamate in Canavan disease: an adverse effector? Eur J Pediatr 158:406–409

Canavan MM (1931) Schilder's encephalitis periaxialis diffusa: report of a case in a child aged sixteen and one-half months. Arch Neur Psych 25:299–308

Guo F, Bannerman P, Mills Ko E, Miers L, Xu J, Burns T, Li S, Freeman E, McDonough JA, Pleasure D (2015) Ablating *N*-acetylaspartate prevents leukodystrophy in a Canavan disease model. Ann Neurol 77:884–888

Kaul R, Gao GP, Balamurugan K, Matalon R (1993) Cloning of the human aspartoacylase cDNA and a common missense mutation in Canavan disease. Nat Genet 5:118–123

Kölker S, Struys EA, van der Knaap M, Jacobs C (2014) Cerebral organic Acidurias. In: Blau N et al (eds) Physician's guide to the diagnosis, treatment and follow-up of inherited metabolic diseases. Springer, Berlin, pp 143–156

Kumar S, Biancotti JC, Matalon R, de Vellis J (2009) Lack of aspartoacylase activity disrupts survival and differentiation of neural progenitors and oligodendrocytes in a mouse model of Canavan disease. J Neurosci Res 87:3415–3427

Lienhard U, Sass JO (2011) Canavan disease: a neurometabolic disease caused by aspartoacylase deficiency. J Pediatr Sci 3(1):e7

Maier H, Wang-Eckhardt L, Hartmann D, Gieselmann V, Eckhardt M (2015) *N*-Acetylaspartate synthase deficiency corrects the myelin phenotype in a canavan disease mouse model but does not affect survival time. J Neurosci 35:14501–14516

Mendes MI, Smith DE, Pop A, Lennertz P, Ojeda MR, Kanhai WA, van Dooren SJ, Anikster Y, Barić I, Boelen C, Campistol J, de Boer L, Kariminejad A, Kayserili H, Roubertie A, Verbruggen KT, Vianey-Saban C, Williams M, Salomons GS (2017) Clinically distinct phenotypes of Canavan disease correlate with residual aspartoacylase enzyme activity. Hum Mutat 38:524–531

Sistermans EA, de Coo RF, van Beerendonk HM, Poll-The BT, Kleijer WJ, van Oost BA (2000) Mutation detection in the aspartoacylase gene in 17 patients with Canavan disease: four new mutations in the non-Jewish population. Eur J Hum Genet 8:557–560

Sohn J, Bannerman P, Guo F, Burns T, Miers L, Croteau C, Singhal NK, McDonough JA, Pleasure D (2017) Suppressing *N*-Acetyl-L-Aspartate synthesis prevents loss of neurons in a murine model of Canavan Leukodystrophy. J Neurosci 37:413–421

Sommer A, Sass JO (2012) Expression of aspartoacylase (ASPA) and Canavan disease. Gene 505:206–210

van Bogaert L, Bertrand I (1949) Sur une idiotie familiale avec dégénérescenece spongieuse de neurone (note preliminaire). Acta Neurol Belg 49:572–587

Prominent Insulin Resistance in Congenital Generalized Lipoatrophy

Jun Wada and Yashpal S. Kanwar

Keywords
Lipodystrophy · Fatty liver · Insulin resistance · Leptin · Diabetes

3.1 Case Report

A 35-year-old female was admitted to the division of endocrinology and metabolism for generalized lipoatrophy and hyperglycemia. She was markedly deficient of adipose tissue since birth and developed diabetes and hypertension at the age of 20 years. She began receiving premixed biosynthetic human insulin with a ratio of 30% regular and 70% NPH (neutral protamine Hagedorn) insulin at the age of 27 years. The glycemic control remained poor despite the increase in her insulin dose to 60–80 units/day. Her hemoglobin A1c levels remained high, and they varied between 8 and 11%. Current admission revealed a female with 161 cm height, body weight 53.0 kg, body mass index (BMI) 20.4 kg/m^2, and systemic blood pressure of 146/86 mm Hg. She had muscular stature with generalized loss of subdermal fatty tissue. She also had umbilical hernia and acromegalic features with slight enlargement of her hands, feet, and mandible (Fig. 3.1a). She had difficulty in controlling her excessive appetite.

Her fasting plasma glucose was 222 mg/dl and HbA1c 8.4%. Serum c-peptide was 4.1 ng/mL and daily urinary excretion 85.2 μg/day, indicating her insulin secretion was not impaired. Her serum leptin level was extremely low, i.e., 0.8 ng/ml (normal range 1.7–14.5 ng/ml). Urinalysis revealed glucosuria and microalbuminuria of 261.3 mg/day. She had mild elevation of liver enzymes: aspartate transaminase (AST, 41 IU/l), alanine aminotransferase (ALT, 49 IU/l), alkaline phosphatase (ALP, 409 IU/l), and γ-glutamyl transferase (γ-GT, 53 IU/l) (Table 3.1). Computed tomography revealed absence of subdermal and visceral adipose tissues but notable fatty liver (Fig. 3.1b and c). She also had dyslipidemia with low HDL-cholesterol (35 mg/dl) and high triglyceride (257 mg/dl) levels. Euglycemic hyperinsulinemic clamp study targeting plasma glucose concentration of 95 mg/dl with insulin infusion rate of 1.25 mU/kg/min suggested a prominent insulin resistance since glucose infusion rate of 2.5 mg/kg/min (normal range 8.0–12.0 mg/kg/min) was required to maintain target glucose levels.

J. Wada (✉)
Department of Nephrology, Rheumatology, Endocrinology and Metabolism, Okayama University Graduate School of Medicine, Dentistry and Pharmaceutical Sciences, Okayama, Japan
e-mail: junwada@okayama-u.ac.jp

Y. S. Kanwar
Department of Pathology, Feinberg School of Medicine, Northwestern University, Chicago, IL, USA
e-mail: y-kanwar@northwestern.edu

T. Oohashi et al. (eds.), *Human Pathobiochemistry*, https://doi.org/10.1007/978-981-13-2977-7_3

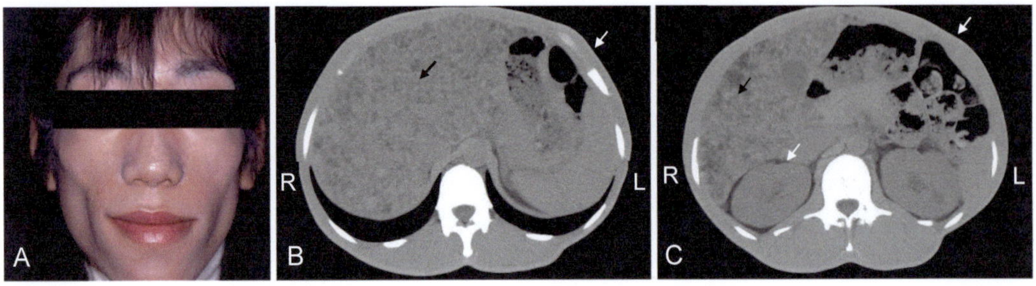

Fig. 3.1 The presented case with congenital lipoatrophy. (**a**) Photograph of the patient. (**b**) and (**c**) Computed tomography of the abdomen. Prominent fatty liver is noted, and both subdermal and abdominal adipose tissues are absent. Black arrows indicate prominent fatty liver and white arrows absence of subdermal and retroperitoneal adipose tissues

Daily leptin replacement therapy was started with recombinant methionyl human leptin (metreleptin; 0.04 mg/kg BW), and after 1 month the dose was increased to 0.08 mg/kg BW. Subsequently, insulin therapy was discontinued, and metabolic profile, such as HbA1c and serum triglycerides, returned to within normal limits.

3.2 Diagnosis

This is a characteristic case of generalized lipoatrophy, diabetes with prominent insulin resistance, hypertension, dyslipidemia, and severe fatty liver. The lipodystrophic disorders are classified into two main categories, familial/congenital versus acquired lipodystrophy (Table 3.2). Both categories include total (generalized) and partial lipoatrophy. Genome sequencing analysis revealed that the patient had a mutation in the *BSCL2* (seipin lipid droplet biogenesis associated) gene, which confirmed the diagnosis (Table 3.3). Although the mutation of *BSCL2* gene was found later in life, the patient had generalized lipoatrophy since birth, and she was labeled with the diagnosis having congenital generalized lipodystrophy type 2 (Berardinelli-Seip syndrome).

3.3 Biochemical and Molecular Perspectives

BSCL2 gene encodes multi-pass transmembrane protein seipin, which localizes to the endoplasmic reticulum. It is essential for adipocyte differentiation, lipid storage, and lipid droplet maintenance. Upon adipogenic differentiation, induced pluripotent stem (iPS) cells derived from patients with BSCL2 deficiency exhibit marked reduction of lipid droplet formation. In addition, seipin interacts with adipose differentiation-related protein (ADRP), and normal cytoplasmic punctate localization of ADRP has been reported to be lost in patient-derived iPS cells (Mori et al. 2016). Thus, the failure to form lipid droplets is linked to lipoatrophic phenotypes of such patients (Table 3.4).

The loss of adipose tissues in patients with Berardinelli-Seip syndrome can be linked to the reduction of secreted factors known as adipokines, such as adiponectin, leptin, and others. Leptin, a 167 amino acid adipokine, is mainly expressed in white adipose tissue (WAT) and plays an important role in energy homeostasis (Fig. 3.2). Leptin was initially discovered as a mutation in the leptin gene resulting in massive obesity and type 2 diabetes in human as well as in rodents. The adipose tissue mass positively correlates with serum concentration of leptin, and it

Table 3.1 Laboratory data in present case

Urinalysis and biochemistry	(Normal range in female)
Protein	(−)
Sugar	4+
pH	5.5
Ketone	(−)
Occult blood	(−)
Creatinine clearance	83.6 mL/min/1.73m^2 (80–140)
Daily albumin excretion	261.3 mg/day (−30)
Peripheral blood cell count	
White blood cells	3400/μL (3500–9000)
Red blood cells	370 × 10^4/μL (350–500)
Hemoglobin	11.4 g/dL (12.0–16.0)
Platelets	19.4 × 10^4/μL (15–35)
Serum biochemistry	
Total protein	8.1 g/dL (6.5–8.0)
Albumin	4.4 g/dL (4.0–5.0)
Aspartate transaminase	41 IU/L (10–35)
Alanine aminotransferase	49 IU/L (5–30)
Lactate dehydrogenase	194 IU/L (120–220)
Alkaline phosphatase	409 IU/L (100–350)
γ-glutamyl transferase	53 IU/L (10–30)
Creatinine	0.62 mg/dL (0.4–0.8)
Uric acid	5.4 mg/dL (2.5–6.0)
Urea nitrogen	13.2 mg/dL (8–20)
Sodium	137 mmol/L (138–145)
Potassium	4.5 mmol/L (3.6–4.8)
Chloride	98 mmol/L (101–108)
Total cholesterol	190 mg/dL (130–220)
Triglyceride	226 mg/dL (30–150)
High-density lipoprotein-cholesterol	38 mg/dL (40–100)
Low-density lipoprotein-cholesterol	97 mg/dL (65–163)
Glucose metabolism	
Hemoglobin A1c	8.4% (4.6–6.2)
Fasting plasma glucose	222 mg/dL (80–110)
Serum C-peptide	4.1 ng/mL (1.1–4.4)
Daily urinary C-peptide excretion	95.2 μg/day (50–100)
Leptin	0.8 ng/mL (1.7–14.5)
Immunoreactive insulin (IRI)*	191.7μU/mL (1.7–10.4)

*IRI was measured under daily injection of insulin

Table 3.2 Classification of lipodystrophies (Huang-Doran et al. 2010)

Congenital lipodystrophy
Congenital generalized lipoatrophy (CGL)
Familial partial lipoatrophy (FPLD)
Acquired lipodystrophy
Acquired generalized lipoatrophy
Autoimmune disorders-associated (juvenile dermatomyositis, SLE, autoimmune hemolytic anemia, autoimmune hepatitis, low C4 complement levels)
Acquired partial lipoatrophy
HIV-associated
C3 nephritic factor-associated

Table 3.3 Classification of congenital generalized and partial lipodystrophy and causal genes

Congenital generalized lipoatrophy (CGL)	Official gene symbol
CGL type 1 (CGL1)	*AGPAT2*
CGL type 2 (CGL2)	*BSCL2*
CGL type 3 (CGL3)	*CAV1*
CGL type 4 (CGL4)	*PTRF*
Familial partial lipodystrophy (FPLD)	**Official gene symbol**
FPLD type 1 (Kobberling's syndrome)	Unknown
FPLD type 2 (Dunnigan's syndrome)	*LMNA*
FPLD type 3	*PPARG*
FPLD type 4	*AKT2*
FPLD type 5	*PLIN1*

Table 3.4 Major clinical features of lipodystrophies (Brown et al. 2016)

Generalized or regional absence of body fat
Severe acanthosis nigricans
Prominent muscles and veins
Cushingoid, acromegaloid, and progeroid appearance
Diabetes mellitus with high insulin requirements
Severe hypertriglyceridemia
Nonalcoholic steatohepatitis in a nonobese individual
Early-onset cardiomyopathy
Polycystic ovarian syndrome (PCOS)
Renal dysfunction (proteinuria, diabetic nephropathy)
Significant hyperphagia

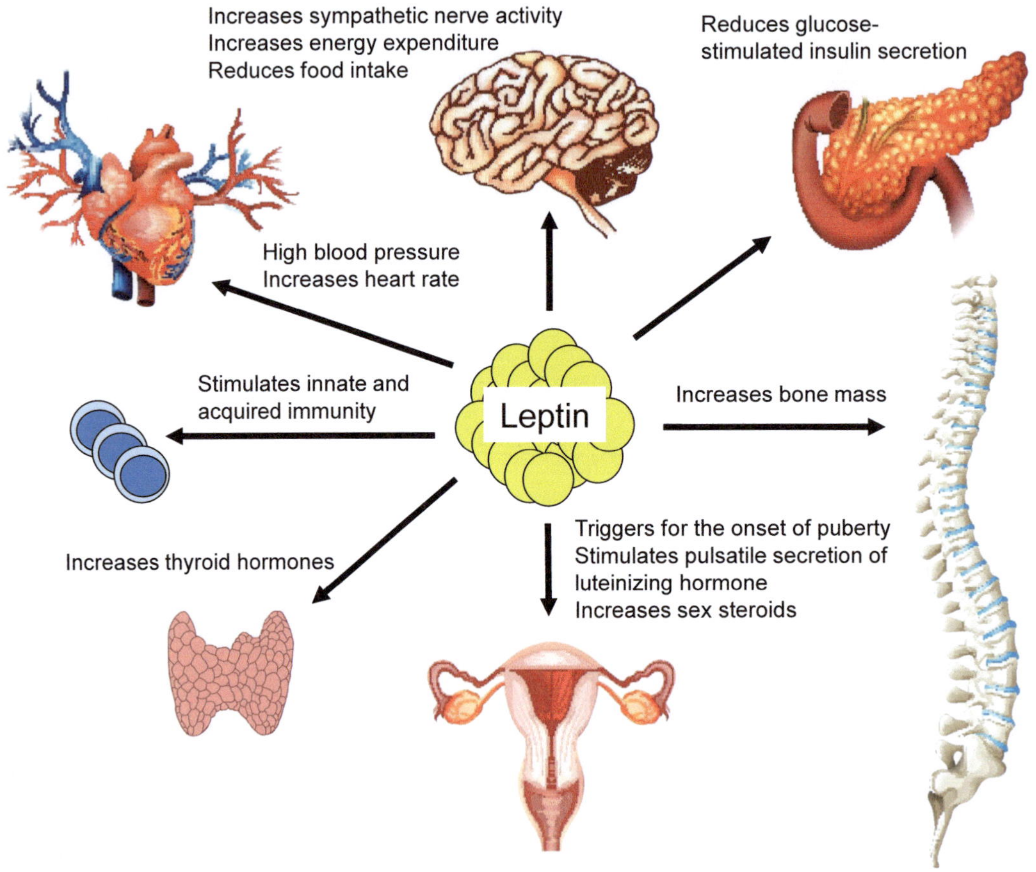

Fig. 3.2 Diverse physiological actions of leptin

is a marker of acute changes in energy intake. Upon increased energy intake and subsequent increase in adipose tissue mass, leptin is secreted by the adipocytes (Fig. 3.3).

Once in blood circulation, it crosses the blood-brain barrier, and binds to leptin receptor in the hypothalamus, including supraoptic, paraventricular, periventricular, and arcuate nuclei and lateral hypothalamus, following which it activates Janus kinase (JAK)-signal transducer and activator of transcription 3 (STAT3) signal pathway. Specifically, leptin activates anorexigenic pro-opiomelanocortin (POMC) and cocaine- and amphetamine-regulated transcript (CART) neurons, while it suppresses orexigenic agouti-related peptide (AgRP) and neuropeptide Y (NPY) neurons. These series of events result in the inhibition of feeding behavior and increased energy expenditure to maintain whole-body adipose tissue mass (Fig. 3.4). Leptin also regulates hypothalamus-pituitary-endocrine organs axis, and low leptin levels are associated with reduced levels of gonadotropins, thyroid hormones, and growth hormone secretion in obese patients with congenital leptin deficiency(Farooqi et al. 1999). In such young patients, thyroid and luteinizing hormones (LH) with pulsatile secretion, sex steroid levels can be restored with the supplementation of leptin, and the rising leptin levels can be interpreted as triggers for the onset of normal puberty (Mantzoros et al. 1997) (Fig. 3.2).

In addition to central action, leptin also exerts various effects on peripheral organs, such as the adipose tissue, muscle, liver, pancreas, cardiovas-

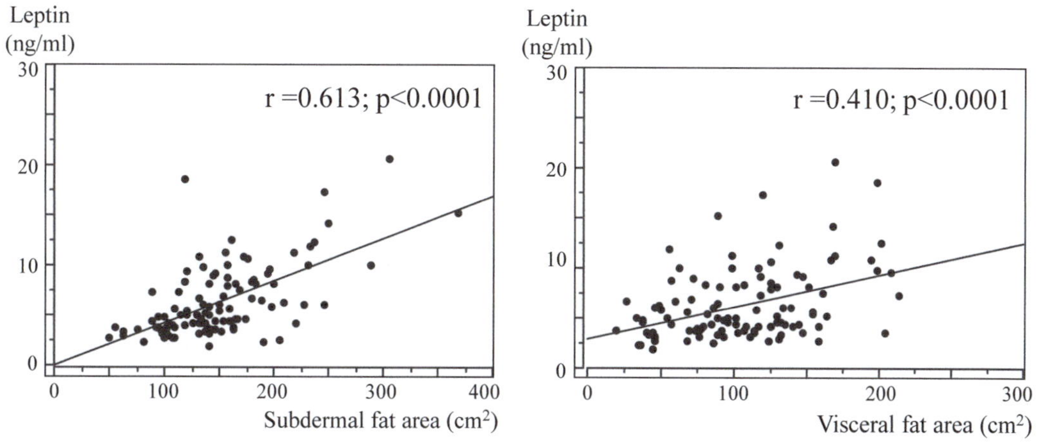

Fig. 3.3 The correlation of serum leptin levels with subdermal and visceral adipose tissues in Japanese men (Kunitomi et al. 2002)

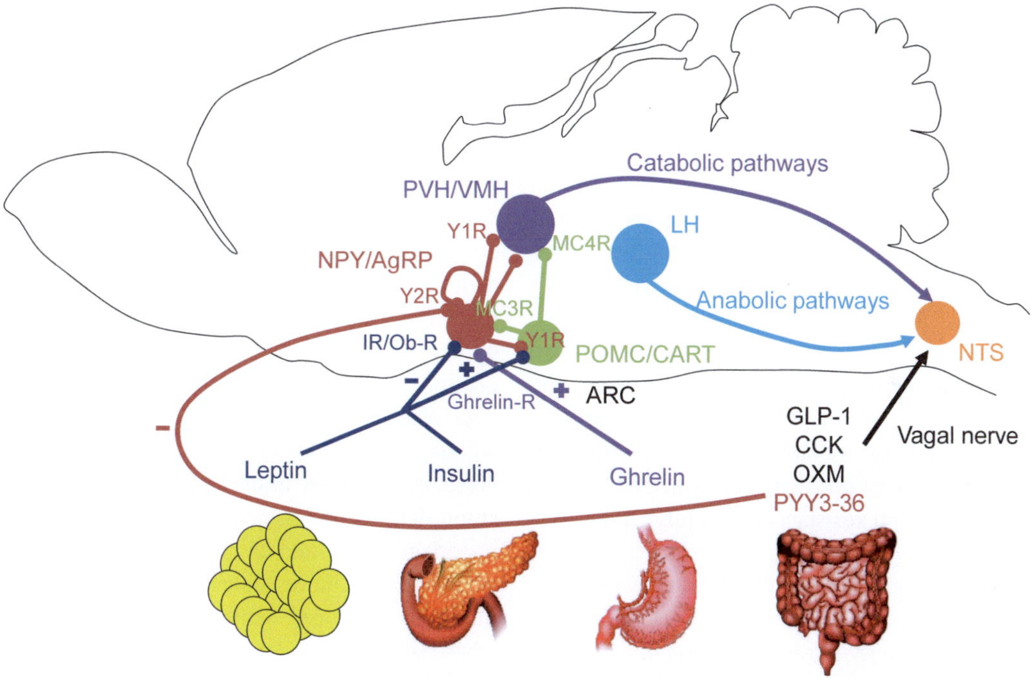

Fig. 3.4 Peripheral and central control of appetite and satiety. *ARC* arcuate nucleus, *CCK* cholecystokinin, *GLP-1* glucagon-like peptide-1, *Ghrelin-R* ghrelin receptor, *IR* insulin receptor, *MC3R* melanocortin 3 receptor, *MC4R* melanopcortin 4 receptor, *NPY/AgRP* neuropeptide Y/agouti-related protein neurons, *NTS* nucleus of the solitary tract, *Ob-R* leptin receptor, *OXM* oxyntomodulin, *PVH/VMH* paraventricular hypothalamus/ventromedial hypothalamus, *PYY3-36* peptide YY3-36, *POMC/CART* proopiomelanocortin/cocaine- and amphetamine-regulated transcript, *Y1R* peptide YY Y1 receptor, *Y2R* peptide YY Y2 receptor

cular system, and bone. Leptin directly acts on brown (BAT) and WAT, increases glucose utilization, and upregulates rate of lipolysis, as demonstrated in Zucker *fa/fa* rats with loss of function in the mutated leptin gene (Siegrist-Kaiser et al. 1997). In leptin-deficient *ob/ob* mice, accumulation of triacylglycerol (TAG) in skeletal muscle contributes to insulin resistance, and it should be noted that normally leptin decreases glycogen synthesis and oleate incorporation into TAG and increased oleate oxidation (Muoio et al. 1999). Acute leptin infusion stimulates hepatic fatty acid oxidation and reduces TAG, and these metabolic processes are dependent of phosphoinositol-3-kinase (PI3K) activity (Huang et al. 2006). In addition, leptin treatment reverses hyperglycemia in animal models of poorly controlled type 1 diabetes by reducing hepatic gluconeogenesis (Perry et al. 2014). One of the important peripheral actions of leptin is inhibition of insulin biosynthesis and secretion by pancreatic β-cells. In turn, insulin stimulates leptin secretion from adipose tissue; and this hormonal negative feedback system, an adipo-insular axis, plays an important role in the maintenance of body weight and glucose metabolism. In fact, the disruption of leptin receptor in pancreatic β-cells in mice results in obesity, fasting hyperinsulinemia, impaired glucose-stimulated insulin release, and glucose intolerance without changes in food intake and satiety responses to leptin (Covey et al. 2006).

Despite marked obesity associated with components of metabolic syndrome, such as visceral obesity, insulin resistance, hyperglycemia, dyslipidemia in patients with leptin, or leptin receptor deficiency, they are not hypertensive or have increased sympathetic nerve system (SNS) activities. Interestingly, chronic injection of leptin into lean rodents results in gradual elevation of blood pressure, suggesting slow-acting mechanism, such as modest renal SNS activation to increase sodium reabsorption rather than massive systemic SNS activation associated with peripheral vasoconstriction (Foo et al. 2014). Also the consensus is that chronic administration may lead to "selective" resistance to appetite suppressing actions of leptin, whereas cardiovascular effects of leptin remain operative or even get enhanced in the control of blood pressure.

The generalized decrease in systemic adipose tissue (lipoatrophy) and localized loss of adipose mass is associated with hypoleptinemia. This is commonly categorized under lipodystrophic disorders, consisting of both familial/congenital as well acquired lipodystrophy (Table 3.2).

Although congenital lipodystrophic disorders are rare, they are more commonly associated with the patients with human immunodeficiency virus (HIV) and use of highly active antiretroviral therapy (HAART), and this is referred to as HIV-associated lipodystrophy. By the use of HAART, viremia in HIV patients is well-controlled, and they live longer; however, they acquire significant metabolic risks including development of lipodystrophy, hypoleptinemia, insulin resistance, and atherosclerosis (Tsoukas et al. 2015). These patients usually manifest with hyperinsulinemia, prominent insulin resistance, diabetes, hepatic steatosis, and dyslipidemia with notable elevation of TAG. In such a state, inadequate and reduced capacity for lipid accumulation in adipose tissue results in ectopic lipid deposition and lipotoxicity in various organs, e.g., liver, pancreas, and heart. These lipodystrophic patients suffer from pseudoathletic muscular hypertrophy, hypertrophic and dilated cardiomyopathy, acute pancreatitis, steatohepatitis, and liver cancers. Moreover, in obese patients, the adipocyte overload with lipid causes fatty acid efflux from adipose tissue and ectopic lipid deposition. The obese subjects usually demonstrate hyperleptinemia, and the leptin resistance causes prominent insulin resistance with increased food intake and reduced satiety response to leptin. Obese subjects also develop metabolic syndrome manifesting with prominent insulin resistance, hyperglycemia, hypertension, dyslipidemia, and fatty liver disease. Although the physique characteristics and serum concentrations of leptin are totally different in patients with lipodystrophy and obesity, they have similar metabolic defects and common target organ complications (Fig. 3.5).

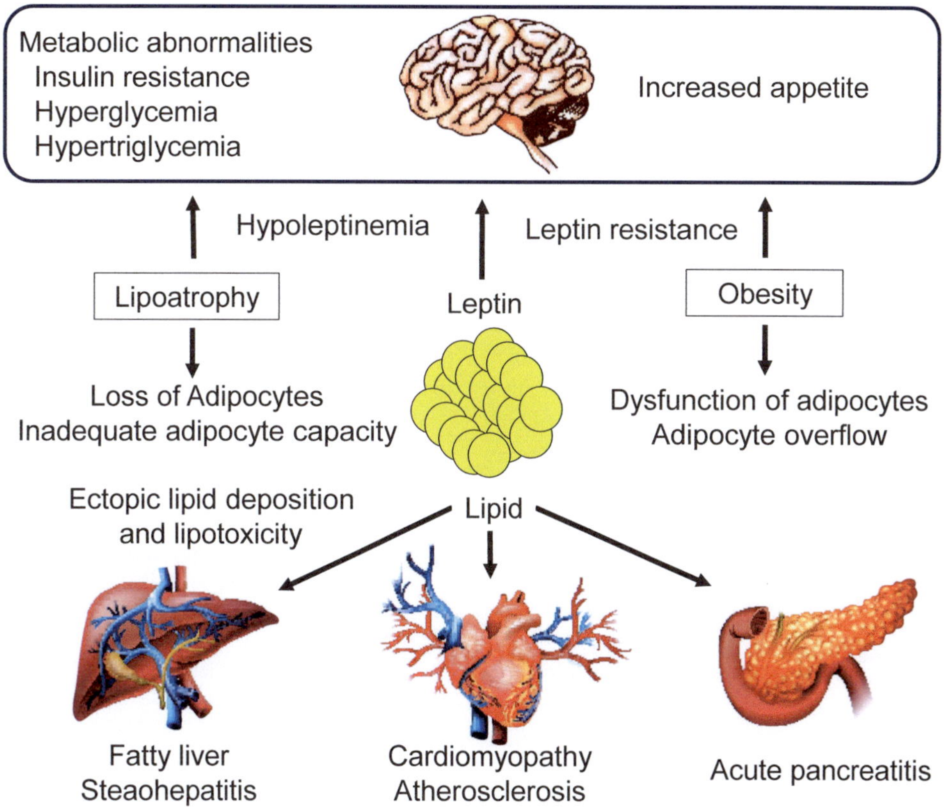

Fig. 3.5 The analogies in the pathogenesis of lipoatrophy and obesity

3.4 Therapy

Severe lipodystrophy is associated with leptin deficiency, insulin resistance, hypertriglyceridemia, and hepatic steatosis; and leptin replacement therapy ameliorates these metabolic disturbances (Oral et al. 2002). Currently available therapy in humans is the administration of metreleptin. The US Food and Drug Administration (USFDA) and Pharmaceuticals and Medical Devices Agency (PMDA) in Japan have approved the use of metreleptin to treat complications of leptin deficiency in patients with congenital or acquired generalized lipodystrophy as a replacement therapy in addition to diet therapy. In certain patients undergoing replacement therapy develop anti-metreleptin antibodies with neutralizing activity and severe infection and begin to show signs indicative of the loss of metreleptin efficacy.

Due to the beneficial effects of leptin, the use of metreleptin has been also evaluated in other disease states, such as obesity, type 1 and 2 diabetes, HIV-associated lipodystrophy, and hypothalamic amenorrhea (HA). In general, metreleptin has not been shown to be effective in state of obesity without congenital leptin deficiency since the development of neutralizing anti-metreleptin antibodies has been reported in such patients. In obesity, at times there is prominent leptin resistance, and metreleptin effectiveness would be compromised, in such cases a combination of pramlintide/metreleptin has been evaluated in clinical trials (Ravussin et al. 2009). It is worth mentioning here that energy homeostasis and control of body weight involve complex communications between the central nervous system and peripheral neurohormonal signals from various tissues that integrate long-term adiposity signals of leptin (metreleptin) and short-term satiation signals of amylin (pramlintide). Thus, combined use of these polypeptide hormones would be warranted. Unfortunately, the combined treatment with pramlintide and metreleptin although resulted in a significant weight loss, the development of anti-metreleptin antibodies precluded their use in patients with obesity.

Fundamental defect in type 1 diabetes is impaired insulin secretion by pancreatic β-cells due to autoimmune-mediated loss, and thus insulin replacement is regarded as the choice of therapy in such patients. However, the involvement of other hormones, such as glucagon and leptin, should be also taken into account in the treatment of diabetes. The absence of glucagon receptor prevents the elevation of blood glucose, fasting and non-fasting free fatty acid, and β-hydroxy butyrate levels in mouse model treated with a double dose of streptozotocin to maximize β-cell destruction (Lee et al. 2011). Leptin treatment also reverses hyperglycemia in rat models of poorly controlled type 1 diabetes by correcting hyperglycemia and hepatic gluconeogenesis (Perry et al. 2014). Reduced leptin levels increase hypothalamic pituitary adrenal (HPA) axis activity, elevate CRF, ACTH and corticosterone levels, and promote fasting hyperglycemia and diabetic ketoacidosis in rat models of poorly controlled type 1 diabetes, as a result, a markedly higher rate of lipolysis and hepatic gluconeogenesis would be anticipated (Fig. 3.6). In addition, leptin replacement has been shown to decrease central fat mass and improve insulin sensitivity, dyslipidemia, and glucose levels in lipodystrophy commonly associated with HIV (Tsoukas et al. 2015; Lee et al. 2011). In view of the above, it may be worth investigating thoroughly long-term

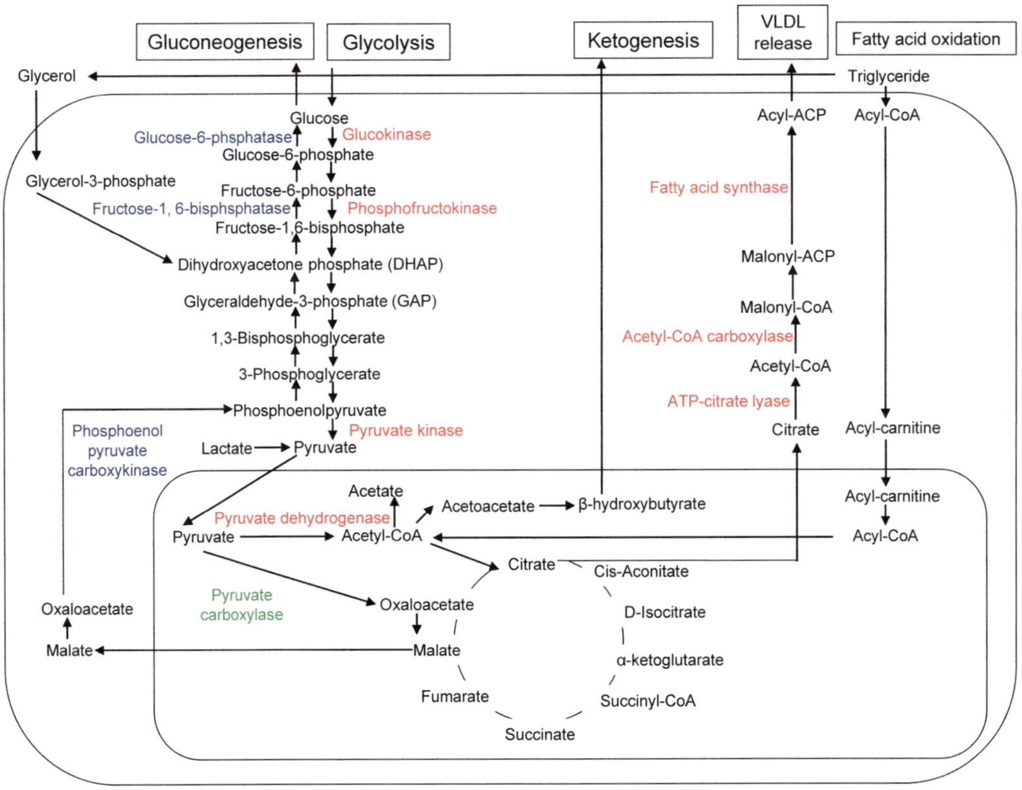

Fig. 3.6 Lipid and glucose metabolic pathways in liver cells. Glycolytic and fatty acid synthetic enzymes are stimulated by insulin and inhibited by cortisol (red). The enzymes involved in gluconeogenesis is simulated by cortisol and inhibited by insulin (blue). The activity of pyruvate carboxylase is enhanced by acetyl-CoA and inhibited by insulin (green). In diabetic ketoacidosis, hypoleptinemia results in enhanced CRF/ACTH/corticosterone axis. It results in higher rate of lipolysis, conversion of glycerol to glucose, and conversion of pyruvate to glucose through greater hepatic acetyl-CoA allosteric activation of pyruvate carboxykinase flux

benefits of leptin replacement whether the administration of leptin protects the patients with HIV-associated lipodystrophy and shields them from the development of severe atherosclerosis and thus improves their life span.

Finally, it should be noted that young women with ample physical activity, reduced food intake, and too much stress, such as in athletes, are known to develop relative leptin deficiency and HA. In athletes, HA is often accompanied with eating disorders, infertility, and premature osteoporosis. The leptin replacement therapy usually restores the rhythmicity of menstrual cycles, and levels of estrogens, thyroid hormones, and IGF-1, and also increases bone formation (Foo et al. 2014).

End-of-Chapter Questions

1. Which symptoms does the leptin resistance provoke in the patients with obesity?
2. Which symptoms does hypoleptinemia induce in the patients with lipoatrophy?
3. Which symptoms does the congenital leptin deficiency and leptin receptor deficiency induce in human?
4. What are the differences in effectiveness of leptin administration in the patients with obesity, lipoatrophy, congenital leptin deficiency, and leptin receptor deficiency?

References

Brown RJ, Araujo-Vilar D, Cheung PT et al (2016) The diagnosis and Management of Lipodystrophy Syndromes: a multi-society practice guideline. J Clin Endocrinol Metab 101(12):4500–4511. https://doi.org/10.1210/jc.2016-2466 published Online First: 2016/10/07

Covey SD, Wideman RD, McDonald C et al (2006) The pancreatic beta cell is a key site for mediating the effects of leptin on glucose homeostasis. Cell Metab 4(4):291–302. https://doi.org/10.1016/j.cmet.2006.09.005

Farooqi IS, Jebb SA, Langmack G et al (1999) Effects of recombinant leptin therapy in a child with congenital leptin deficiency. N Engl J Med 341(12):879–884. https://doi.org/10.1056/NEJM199909163411204

Foo JP, Polyzos SA, Anastasilakis AD et al (2014) The effect of leptin replacement on parathyroid hormone, RANKL-osteoprotegerin axis, and Wnt inhibitors in young women with hypothalamic amenorrhea. J Clin Endocrinol Metab 99(11):E2252–E2258. https://doi.org/10.1210/jc.2014-2491

Huang W, Dedousis N, Bandi A et al (2006) Liver triglyceride secretion and lipid oxidative metabolism are rapidly altered by leptin in vivo. Endocrinology 147(3):1480–1487. https://doi.org/10.1210/en.2005-0731

Huang-Doran I, Sleigh A, Rochford JJ et al (2010) Lipodystrophy: metabolic insights from a rare disorder. J Endocrinol 207(3):245–255. https://doi.org/10.1677/JOE-10-0272 published Online First: 2010/09/28

Kunitomi M, Wada J, Takahashi K et al (2002) Relationship between reduced serum IGF-I levels and accumulation of visceral fat in Japanese men. Int J Obes Relat Metab Disord 26(3):361–369. https://doi.org/10.1038/sj.ijo.0801899

Lee Y, Wang MY, Du XQ et al (2011) Glucagon receptor knockout prevents insulin-deficient type 1 diabetes in mice. Diabetes 60(2):391–397. https://doi.org/10.2337/db10-0426

Mantzoros CS, Flier JS, Rogol AD (1997) A longitudinal assessment of hormonal and physical alterations during normal puberty in boys. V. Rising leptin levels may signal the onset of puberty. J Clin Endocrinol Metab 82(4):1066–1070. https://doi.org/10.1210/jcem.82.4.3878

Mori E, Fujikura J, Noguchi M et al (2016) Impaired adipogenic capacity in induced pluripotent stem cells from lipodystrophic patients with BSCL2 mutations. Metabolism 65(4):543–556. https://doi.org/10.1016/j.metabol.2015.12.015

Muoio DM, Dohm GL, Tapscott EB et al (1999) Leptin opposes insulin's effects on fatty acid partitioning in muscles isolated from obese ob/ob mice. Am J Phys 276(5 Pt 1):E913–E921

Oral EA, Simha V, Ruiz E et al (2002) Leptin-replacement therapy for lipodystrophy. N Engl J Med 346(8):570–578. https://doi.org/10.1056/NEJMoa012437

Perry RJ, Zhang XM, Zhang D et al (2014) Leptin reverses diabetes by suppression of the hypothalamic-pituitary-adrenal axis. Nat Med 20(7):759–763. https://doi.org/10.1038/nm.3579

Ravussin E, Smith SR, Mitchell JA et al (2009) Enhanced weight loss with pramlintide/metreleptin: an integrated neurohormonal approach to obesity pharmacotherapy. Obesity (Silver Spring) 17(9):1736–1743. https://doi.org/10.1038/oby.2009.184

Siegrist-Kaiser CA, Pauli V, Juge-Aubry CE et al (1997) Direct effects of leptin on brown and white adipose tissue. J Clin Invest 100(11):2858–2864. https://doi.org/10.1172/JCI119834

Tsoukas MA, Farr OM, Mantzoros CS (2015) Leptin in congenital and HIV-associated lipodystrophy. Metabolism 64(1):47–59. https://doi.org/10.1016/j.metabol.2014.07.017

Fabry Disease

4

Ken Sakurai and Toya Ohashi

Keywords
Fabry disease · α-Galactosidase A · Globotriaosylceramide · Enzyme replacement therapy · Substrate reduction therapy · Chaperone · Gene therapy

4.1 Case Report

A 41-year-old male visited our hospital with proteinuria and hearing loss. The patient is Japanese, had never experienced pain attacks in the extremities, and sweats normally. He did not present with chest pains and palpitations and often had diarrhea without abdominal pain. Proteinuria was diagnosed when he was 25 years old, and he was then diagnosed with chronic nephritis. At 30 years of age, he experienced recurrent sudden-onset deafness, sometimes with tinnitus. He had hypertension, hypercholesterolemia, and hyperuricemia. His electrocardiogram revealed arrhythmia, and he had a panic disorder. He gave up smoking when he was 30 years old. His mother had cardiac involvement with Fabry disease but no history of kidney diseases. She also had a c.465 T > A (amino acid: D155E) heterozygous mutation in *α-galactosidase A* of white blood cells. Physical examination revealed his body mass index was 22.5 kg/m^2 and had regular heartbeats, but no hepatosplenomegaly, no angiokeratoma, no rash, and no any neurological sign. In his blood counts, there were normal white blood cells and platelet but no anemia. His serum creatinine 0.97 mg/dL (normal; 0.65–1.07), estimated glomerular filtration rate (eGFR) 69 mL/min/1.73 m^2 (normal; >90) and cystatin C 0.90 mg/L (normal; 0.63–0.95), LDL-cholesterol 98 mg/dL (65–139), and triglyceride 188 mg/dL (normal; 40–149) were almost normal, but brain natriuretic peptide 55.3 pg/mL(normal; <18.5) was slightly elevated. He exhibited proteinuria (3+) and quantitative urine protein/creatinine ratio 3.8 g/g Cr (normal; <0.3) but not hematuria. Serum α-galactosidase A activity 2.82 nmol/hr/mg protein in his white blood cells was extremely low (normal control: 155.85–222.59). Thin-layer chromatography of his urinary lipids indicated globotriaosylceramide (Gb3) positive (Fig. 4.1). Gene analysis identified c.465 T > A (amino acid: D155E) hemizygous mutation in *α-galactosidase A* of his white blood cells.

K. Sakurai (✉)
Department of Pediatrics, The Jikei University School of Medicine, Minato-ku, Tokyo, Japan
e-mail: kenken@jikei.ac.jp

T. Ohashi
Department of Pediatrics, The Jikei University School of Medicine, Minato-ku, Tokyo, Japan

Division of Gene Therapy, Research Center for Medical Sciences, The Jikei University School of Medicine, Minato-ku, Tokyo, Japan

T. Oohashi et al. (eds.), *Human Pathobiochemistry*, https://doi.org/10.1007/978-981-13-2977-7_4

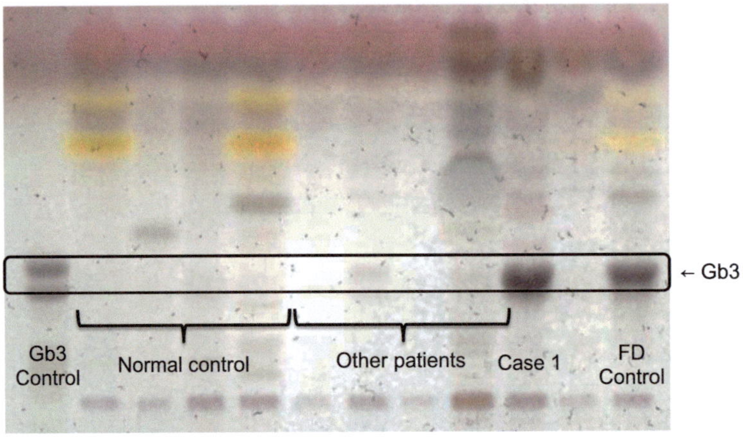

Fig. 4.1 Analysis of patient urine samples for presence of Gb3 by thin-layer chromatography.
The band from urine sample of case 1 patient via thin-layer chromatography indicates the presence of Gb3, as well as Gb3 control marker (far left lane) and Fabry disease (FD) patient's positive control (far right lane), whereas the Gb3 band is not detected in urine samples of normal healthy controls and patients with other diseases. *FD* Fabry disease, *Gb3* globotriaosylceramide

4.2 Diagnosis

Fabry disease (FD) is an inherited lysosomal disorder caused by a genetic deficiency in α-galactosidase A (*GLA*). A deficiency in GLA leads to the accumulation of globotriaosylceramide (Gb3) mainly in the kidney, heart, and brain (Desnick et al. 2017). Usually, FD is divided into three phenotypes: classical type, late-onset type, and heterozygote female type. The symptoms of classical-type FD include pain in the extremities and hypohidrosis that begin when the patient is of school age as well as kidney disease, heart disease, and cerebrovascular disease at middle age (Figs. 4.2 and 4.3).

The average age of onset is 9 and 13 years for males and females, respectively. The average age of diagnosis is 23 and 32 years for males and females, respectively. The diagnostic delays are due to non-specific symptoms. The main symptoms of classical-type FD in males are neurological pain (62%), skin symptoms including hypohidrosis (31%), gastroenterological symptoms (19%), and renal signs (17%); in females, this includes neurological pain (41%), gastroenterological symptoms (13%), and skin symptoms (12%) (Eng et al. 2007).

Because patients with FD exhibit pain in their fingers and toes, they are often misdiagnosed with collagen diseases, fibromyalgia, or mental disorders. Some patients with FD are also misdiagnosed with other cardiomyopathies by echocardiogram and electrocardiogram findings, or with chronic nephritis, such as in the case report described above. Specific techniques used in magnetic resonance imaging (MRI), such as T2-weighted and fluid-attenuated inversion recovery (FLAIR) sequences, often show multiple high intensities of the brain in FD patients (Fig. 4.3c). This finding may lead to a misdiagnosis of multiple sclerosis. Many patients with FD may not have been correctly diagnosed at the onset of initial symptoms. The most important point to be able to make a precise diagnosis is to suspect FD from unspecific findings such as finger pain, hypohidrosis, and cardiomyopathy.

To make a definitive diagnosis, male patients should be evaluated for their GLA enzyme activity in plasma or white blood cells, and the amount of Gb3 in their urine. Many female patients have normal blood GLA activity in plasma or white blood cells and normal amounts of Gb3 in urine, thus requiring a genetic analysis of the *GLA* locus. Another pitfall is the existence of pseudo-deficiency alleles, which have a mutation that may alter or reduce protein function, but do not result in disease. Thus, even in male patients, a genetic analysis is required if GLA

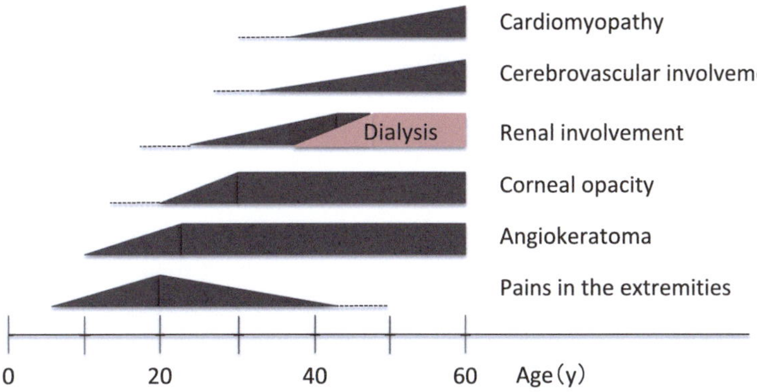

Fig. 4.2 Clinical course of classical type of Fabry disease.
Classical type of Fabry disease (FD) presents clinical symptoms with advancing age. First, patients with FD experience pains in extremities at roughly 4 years old and then angiokeratoma, corneal opacity, and proteinuria at young adulthood, cardiomyopathy and cerebrovascular involvement at middle age, and finally renal pathology requiring hemodialysis

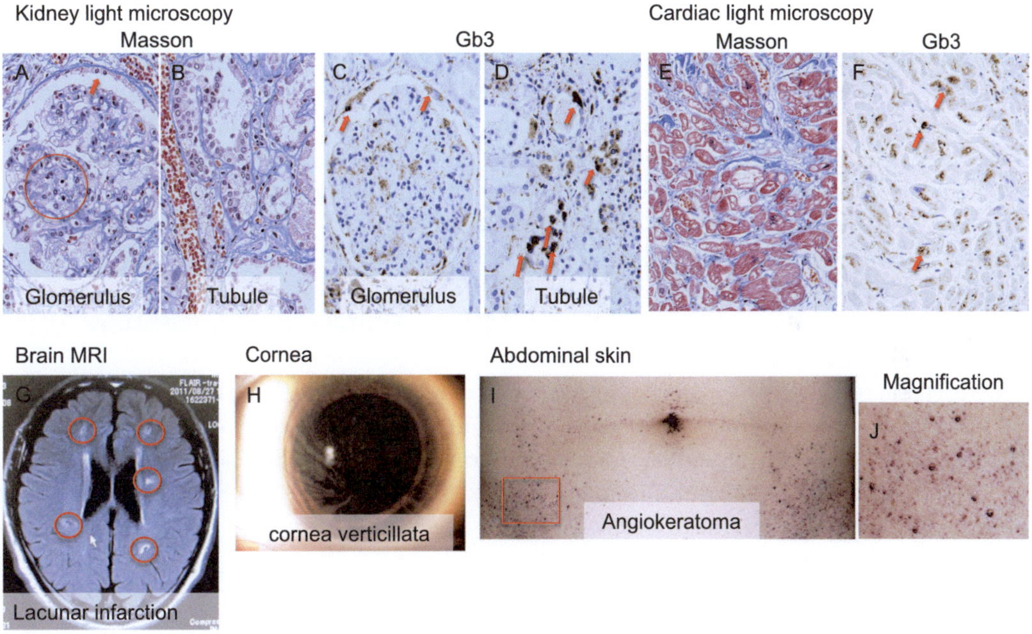

Fig. 4.3 Specific examinations for Fabry disease.
The kidneys from FD male patient at 66 years old examined by light microscopy show glomerulus (**a**, **c**), tubules (**b**, **d**), and cardiomyocytes (**e**, **f**). He had received ERT for 6 years, but he had had still Gb3 accumulation of tissues, because of acquiring high titer of immunoglobulin G against GLA. The histological findings (**a**–**f**) are a typical FD. Masson staining of glomerulus (**a**) shows enlarged epithelial cells and Bowman's capsules (red arrows) and an increased number of stromal cells (red circle). Tubules show almost normal structure by Masson staining (**b**). Gb3 staining shows the accumulation, brown spots (red arrows) mainly in epithelial cells of the glomerulus (**c**) and in distal and collecting tubules (**d**). Light microscopy of cardiac tissue (**e**, **f**), from the patient, shows almost all enlarged myocardial fibers (**e**), and light brown staining of cardiomyocytes (red arrows) indicates the accumulation of Gb3 (**f**). Brain magnetic resonance imaging (MRI) that consisted of fluid-attenuated inversion recovery (FLAIR) sequence shows several small high-intensity spots in the deep brain parenchyma (red circles), which indicate lacunar infarctions (**g**). Slit lamp examination shows the corneal verticillata (**h**). The red spots on the skin indicate angiokeratoma from 40s FD male patient (**i**). The magnification is from red square of the left picture (**j**). *Gb3* globotriaosylceramide

activity is moderately decreased. The E66Q mutation in the *GLA* gene is thought to be a functional polymorphism, and there is no abnormal accumulation of Gb3 in tissues, even if the E66Q mutation reduces the enzymatic activity of GLA (Kobayashi et al. 2012).

4.3 Biochemical and Molecular Perspectives

FD is an inherited metabolic disorder caused by gene mutations in *GLA*, which is one of over 40 lysosomal enzymes. Decreased GLA activity leads to the accumulation of Gb3 in lysosomes of various tissues including the kidney and heart, and Gb3 accumulation causes lacunar infarctions of the brain, cornea verticillata, and autonomic nerve (Fig. 4.3).

GLA is located on Xq22.1, spans 13 kb, includes 7 exons and 6 introns, and encodes 398 amino acids. More than 600 gene mutations in *GLA* have been described, including splicing mutations, nonsense mutations, missense mutations, and deletions. Genotype and phenotype correlations of FD are unclear, resulting in the same mutation from the same family with different phenotypes. The main substrate of GLA is Gb3, which consists of ceramide with three monosaccharides (Fig. 4.4). Other substrates include galabiosylceramide and blood group B substances. GLA degrades the terminal galactose of Gb3 in an acidic lysosome. Mutant GLAs are unstable biochemically, and huge deletions in *GLA* also cause unstable RNA transcripts (Fig. 4.5a). Furthermore, mutant GLAs can result in the accumulation of Gb3 in lysosomes (Fig. 4.5b) and can affect various tissues. Skewed X-inactivation might explain the variable clinical presentation in female patients heterozygous for FD (Gubler et al. 1978). Random inactivation occurs on one X chromosome per cell. If X-inactivation occurs on the X chromosome carrying the wild type *GLA* allele, the mutant *GLA* allele remains active and can lead to very low GLA activity. In contrast, if the X chromosome carrying the mutant *GLA* allele is inactivated, GLA activity is normal. Each organ has a mosaicism of cells with normal or mutant GLA protein (Fig. 4.5c).

Table shows the clinical characterization of FD. Pain in the extremities, caused by Gb3 accumulation in dorsal root ganglions, begins from childhood and regresses spontaneously after middle age. Figure 4.3i and j shows typical angiokeratoma on a patient's skin. Gb3 accumulation also occurs in cardiac myocytes and affects the conducting system of the heart, resulting in cardiac ventricular hypertrophy and arrhythmia (Fig. 4.3e, f). Cardiac-type FD is documented in 3% of patients with left ventricular hypertrophy (LVH) with unknown etiology. The kidney is the main organ affected in FD. Gb3 accumulation occurs in various kidney cells (Fig. 4.3a–d). Proteinuria is a common finding in young adults, and eGFR progressively decreases with age. Untreated patients will require renal transplantation. Specific findings in the kidney include Gb3 accumulation in various cell types such as podocytes, mesangial cells, and renal tubular cells (Fig. 4.3a–d). Recently, fused foot processes in podocytes have been reported (Kanai et al. 2011) (Table 4.1).

4.4 Therapy and Prevention

It is very important for the clinical management of FD to relieve the excruciating pain and prevent the progression of the disease in the kidney, heart, and brain.

4.4.1 Pain Relief

Carbamazepine and phenytoin are more effective than other drugs for the pain experienced by patients with FD, which has been evidenced in a systematic review (Schuller et al. 2016).

4.4.2 Enzyme Replacement Therapy

To remove the accumulated Gb3, replacements with normal enzymes have been attempted in vitro and in vivo. Lysosomal enzymes are

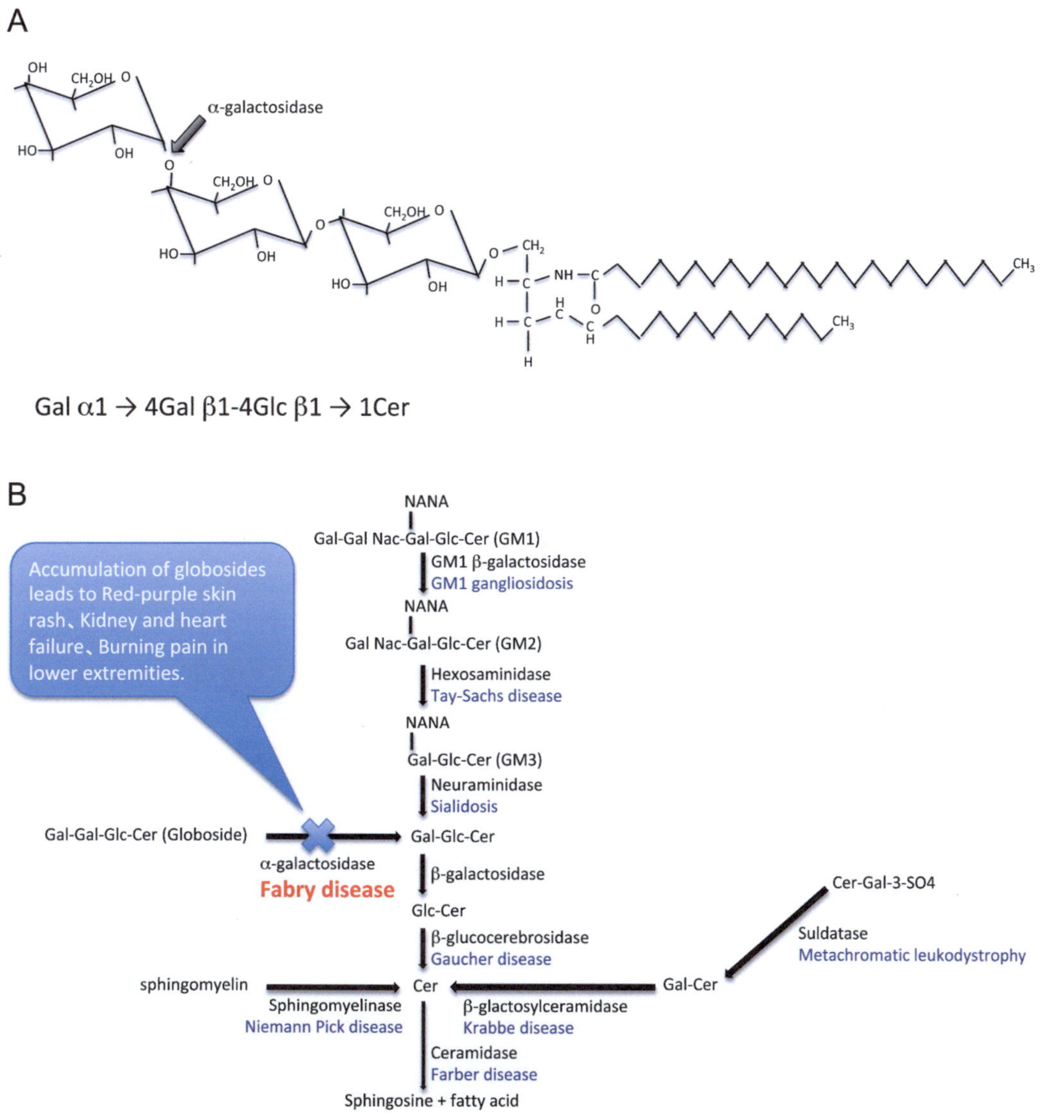

Fig. 4.4 Structure of Gb3 and GLA cleavage site and metabolic pathway of glycolipids.
Globotriaosylceramide (Gb3) has one glucose and two galactose molecules bound to ceramide (**a**) In normal cells, Gb3 is degraded by normal GLA at the terminal galactose, and the metabolic pathway of Gb3 is continuously active. Patients with Fabry disease cannot degrade the terminal galactose, resulting in an accumulation of Gb3. Gb3 is present in many cell types, including endothelial cells and thus can affect the kidney, heart, and brain. The map (**b**) shows the metabolic pathway of glycolipids and enzyme-specific lysosomal diseases. The process to metabolize Gb3 to sphingosine depends on several lysosomal enzymes to degrade glycolipids. *GLA* α-galactosidase A, *Gal* galactose, *Glc* glucose, *Cer* ceramide

taken up into cells and can function in the lysosomes. This is called "cross correction." In 1973, a partially purified enzyme from human placental tissue was intravenously administered to patients with FD (Brady et al. 1973). The substrates in their plasma decreased to approximately 50% in 45 min. Subsequently, independent Phase III clinical trials with recombinant human GLA have been performed. Currently, two recombinant human GLAs for the treatment of FD are available. One is produced in Chinese hamster ovary

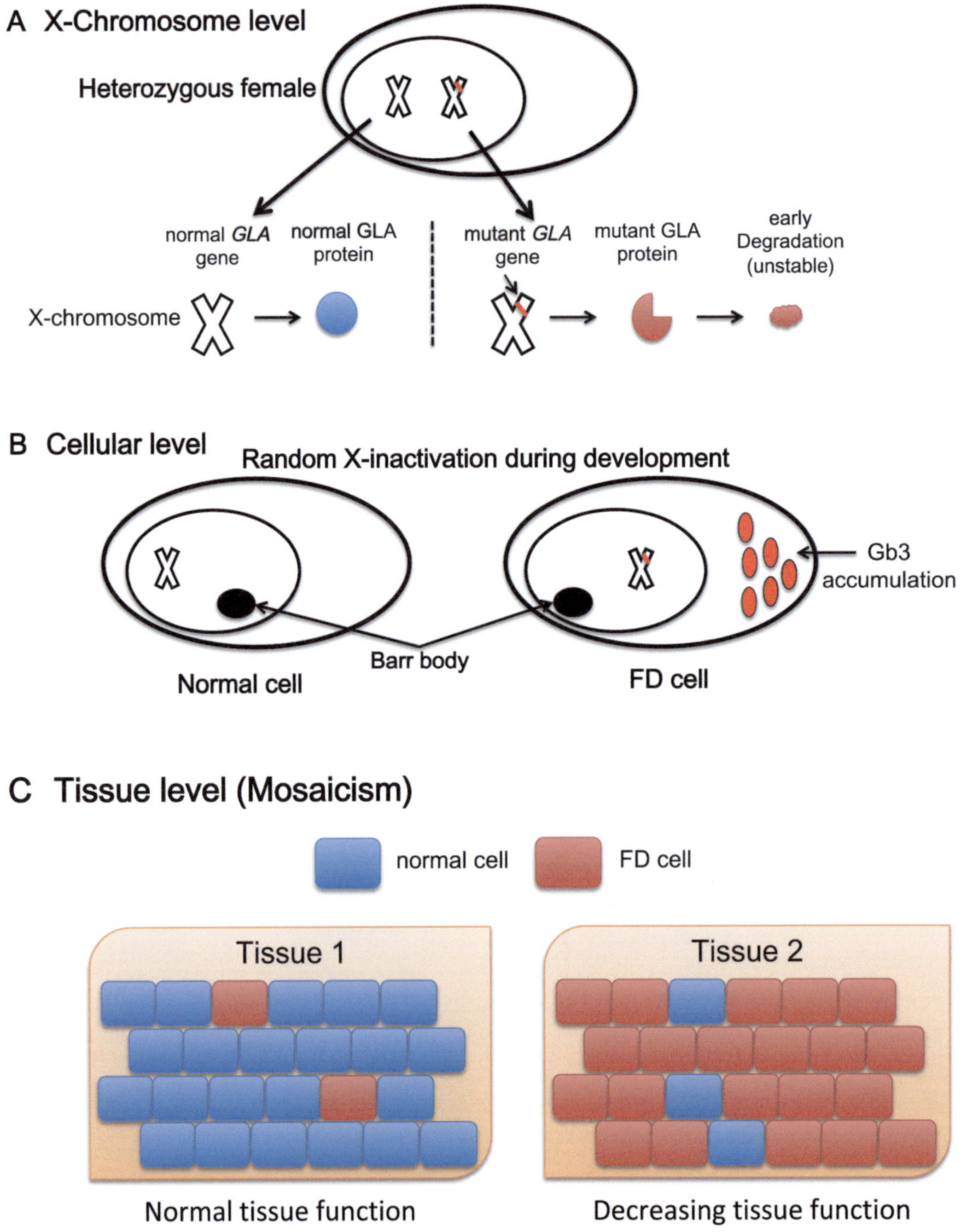

Fig. 4.5 Pathogenesis in females heterozygous for Fabry disease as suggested by Lyon's hypothesis.
Normal GLA protein is produced only by the normal *GLA* gene (left). Mutant *GLA* gene produces unstable GLA protein, which is degraded prematurely (right) (**a**). Lyon's hypothesis states that in a heterozygous female, X-inactivation causes either normal or mutant GLA to be produced per cell (**b**) In certain tissues in females heterozygous for Fabry disease, random X-inactivation results in chimeras consisting of normal cells producing normal GLA protein and mutant cells producing mutant GLA protein. The ratio of these chimeras results in various phenotypes for female patients with Fabry disease (C). *GLA* α-galactosidase A, *FD* Fabry disease

Table 4.1 Clinical characterization of Fabry disease

Classification	Classical type	Renal variant	Cardiac variant	Cerebrovascular variant
Onset age (y)	4~8	25	40 <	20~55
Death age	41	?	60 <	?
Hypohidrosis	++	±	–	+
Pain in the extremities	++	±	–	+
Angiokeratoma	++	±	–	±
Corneal clouding	++	±	–	?
Cardiac involvement	LVH	LVH	LVH	±
	MI		MI	
Cerebrovascular involvement	Brain infarction	?	–	Brain infarction
Renal disturbance	RF	RF	Proteinuria	+
Residual GLA activity	< 1%	< 5%	< 10%	?

Table shows clinical characterization of Fabry disease (FD) depending on classification. The ages in table are approximate ages. *LVH* left ventricular hypertrophy, *MI* myocardial infarction, *RF* renal failure, *GLA* α-galactosidase A

(CHO) cells (agalsidase beta) and the other in human fibroblasts (agalsidase alpha).

These clinical trials with open-label extensions demonstrated the frequency of immunoglobulin G (IgG) anti-GLA antibody formation among patients with enzyme replacement therapy; i.e., 89.7% of the patients with FD were treated with agalsidase beta (Wilcox et al. 2004) and 56.0% of patients with agalsidase alpha (Schiffmann et al. 2006). An experimental study also demonstrated that the serum, from patients with FD included IgG against a therapeutic enzyme, reduced GLA activity in Fabry fibroblast cells and Fabry mice tissues (Ohashi et al. 2008). Another study described immunologic tolerance for the therapeutic protein observed in some patients. Female patients with FD were significantly more tolerant than men; i.e., 58% female patients versus 11% male. Patients with FD that have nonsense mutations in *GLA* were more likely to develop anti-GLA antibodies than were patients with missense mutations. The frequency of infusion-associated reactions for enzyme replacement therapy in males was more than that in females (26% vs 11%, respectively) (Wilcox et al. 2012).

4.4.3 Pharmacological Chaperones

Certain missense mutations produced active mutant enzymatic proteins that are unstable and easily degraded in the endoplasmic reticulum (ER). However, these mutant proteins can be stabilized by small molecules. Migalastat, a pharmacological chaperone of GLA, binds to the active site of GLA and stabilizes certain mutant enzymes. Such mutant enzymes are then translocated to the lysosome, escaping degradation in the ER, and eventually degrade accumulated substrates in the lysosome. Clinical trials for migalastat demonstrated that eGFR was stabilized and LVH was improved in patients with amenable mutations. Moreover, the severity of diarrhea, reflexes, and indigestion was also decreased in treated patients. Based on these observations, migalastat has been approved for use in the European Union. Each patient's mutation is tested by HEK (HEK: human embryonic kidney) assay, that the mutation is transfected into HEK cells, to determine whether the patient has amenable mutations, before migalastat treatment. Although migalastat is only effective in patients who carry amenable mutations the advantage of migalastat is that it can be administered orally, without any serious adverse event (Germain et al. 2016).

4.4.4 Substrate Reduction Therapy

In substrate reduction therapy, the oral administration of Genz-682452, an antagonist of glucosylceramide synthase, results in selective inhibition of glucosylceramide synthase and reduced tissue levels of Gb3 and lyso-Gb3 in

Fabry mice while also delaying the loss of the thermal nociceptive response. Moreover, Genz-682452, but not recombinant enzymes, can cross the blood-brain barrier and reduce the levels of accumulated glycosphingolipids in treated murine brains. A combination therapy of substrate reduction and enzyme replacement may give additive therapeutic merits to patients with FD (Ashe et al. 2015).

4.4.5 Gene Therapy

Several experimental gene therapies for lysosomal storage diseases have been performed. NOD/SCID-Fabry mice were generated, and ex vivo gene therapies were examined in these mice. Human CD34+ cells, a marker for hematopoietic stem and progenitor cells, that were transduced with lentiviral vectors carrying human *GLA*/IRES-human CD25 were transplanted into the model mice. This transplant experiment demonstrated that plasma GLA activity was significantly elevated and Gb3 accumulation in various tissues was significantly reduced (Pacienza et al. 2012). Phase I clinical trial of gene therapy for FD has started at the University of Calgary in Canada (Press release, University of Calgary).

End-of-Chapter Questions

(1) A young male patient presents in your clinic with pain in the extremities as well as hypohidrosis. What should you look for to make a definitive diagnosis for Fabry disease?
(2) Is genetic analysis of *GLA* required for definitive diagnosis of females with symptoms of Fabry disease?
(3) Fabry disease is X-linked. What is the mechanism by which female patients develop Fabry disease?
(4) If only one treatment can improve Fabry patients completely, what would be that treatment?

Answers

1. While some symptoms or combination of symptoms can lead to a diagnosis of Fabry disease, most of the symptoms are not definitive, such as pain in the extremities. For definitive diagnoses, in vitro functional assays should be done on whole blood (or white blood cells) patient samples for GLA activity, and urine should be tested for presence of Gb3.
2. Yes, because GLA activity from a female patient with Fabry disease cannot be distinguished from a normal control individual. A female, heterozygous at the *GLA* locus, will likely still produce functional enzyme in an in vitro assay from half the cells in the sample. Only an analysis at the gene sequence level can a possible Fabry disease diagnosis be confirmed for a female.
3. Females will likely have tissue mosaicism, giving a variable disease presentation and progression from patient to patient. As random X-inactivation occurs during development, some cells will inactivate the normal allele, resulting in Fabry disease cells that will constitute some of the cells in the resulting tissue. The degree of symptoms will depend on the degree of lyonization.
4. Theoretically, a one-time treatment using gene therapy can improve patients with Fabry disease by addition of a healthy exogenous allele of the *GLA* gene. Successful gene therapy would result in production of functional enzyme by all cells treated ex vivo and would reduce excess Gb3. A limitation would be that only cells treated ex vivo would produce functional enzyme, not all cells in the patient. Therefore, there may be residual symptomology.

References

Ashe KM, Budman E, Bangari D et al (2015) Efficacy of enzyme and substrate reduction therapy with a novel antagonist of glucosylceramide synthase for Fabry disease. Mol Med 21:389–399

Brady RO, Tallman JF, Johnson WG et al (1973) Replacement therapy for inherited enzyme deficiency-use of purified Ceramidetrihexosidase in Fabry's disease. N Engl J Med 289:9–14

Desnick RJ, Ioannou YA, Eng CM (2017) 150: α-Galactosidase a deficiency: Fabry disease in the online metabolic and molecular bases of inherited disease. http://ommbid.mhmedical.com/content.aspx?sectionid=62644837&bookid=971&Resultclick=2

Eng CM, Fletcher J, Wilcox WR et al (2007) Fabry disease: baseline medical characteristics of a cohort of 1765 males and females in the Fabry registry. J Inherit Metab Dis 30:184–192

Germain DP, Hughes DA, Nicholls K et al (2016) Treatment of Fabry's disease with the pharmacologic chaperone Migalastat. New Engl J Med 375:545–555

Gubler MC, Lenoir G, Grunfeld JP et al (1978) Early renal changes in hemizygous and heterozygous patients with Fabry's disease. Kidney Int 13(3):223–235

Kanai T, Yamagata T, Ito T et al (2011) Foot process effacement with normal urinalysis in classic Fabry disease. JIMD Rep 1:39–42

Kobayashi M, Ohashi T, Fukuda T et al (2012) No accumulation of globotriaosylceramide in the heart of a patient with the E66Q mutation in the α-galactosidase A gene. Mol Genet Metab 107(4):711–715

Ohashi T, Iizuka S, Ida H et al (2008) Reduced a-Gal A enzyme activity in Fabry fibroblast cells and Fabry mice tissues induced by serum from antibody positive patients with Fabry disease. Mol Genet Metab 94:313–318

Pacienza N, Yoshimitsu M, Mizue N et al (2012) Lentivector transduction improves outcomes over transplantation of human HSCs alone in NOD/SCID/Fabry mice. Mol Ther 20(7):1454–1461

Press release (April 8th, 2016) Phase I Gene Therapy Study for Fabry Disease, University of Calgary, Canada

Schiffmann R, Ries M, Timmons M et al (2006) Long-term therapy with agalsidase alpha for Fabry disease: safety and effects on renal function in a home infusion setting. Nephrol Dial Transplant 21:345–354

Schuller Y, Linthorst GE, Hollak CEM et al (2016) Pain management strategies for neuropathic pain in Fabry disease – a systematic review. BMC Neurol 16:25

Wilcox WR, Banikazemi M, Guffon N et al (2004) International Fabry disease study group, long-term safety and efficacy of enzyme replacement therapy for Fabry disease. Am J Hum Genet 75:65–74

Wilcox WR, Linthorst GE, Germain DP et al (2012) Anti-α-galactosidase A antibody response to agalsidase beta treatment: data from the Fabry registry. Mol Genet Metab 105:443–449

Familial Hypercholesterolemia 5

Masa-aki Kawashiri and Daniel J. Rader

Keywords
Atherosclerosis · Cholesterol · Familial hypercholesterolemia · Low-density lipoprotein · Low-density lipoprotein receptor

5.1 Case Report

A 40-year-old woman with a history of cutaneous and tendon xanthoma (abnormal accumulation of lipids in macrophages of several tissues) and hypercholesterolemia attended an outpatient clinic of the Kanazawa University Hospital with progressive chest pain lasting for 1 month. The patient was diagnosed with severe hypercholesterolemia (total cholesterol, 800 mg/dL) at 3 years of age by the dermatologist examining her cutaneous xanthoma. The cholesterol biosynthesis inhibitor pravastatin was administered when she was 20 years of age; however, she discontinued the medication because it was ineffective in reducing her cholesterol levels. At an approximately 30 years of age, she experienced chest pain induced by exerting effort or cold atmosphere. Therefore, she restricted her daily activity to minimize chest pain. At 40 years of age, she visited her general physician complaining of progressive chest pain, and she was found to have severe hypercholesterolemia {total cholesterol, 450 mg/dL; normal range < 220 mg/dL, low-density lipoprotein cholesterol (LDL-C), 360 mg/dL; normal range < 140 mg/dL} in laboratory test and ST depression in electrocardiogram, suggesting myocardial ischemia.

The patient was born from a non-consanguinity marriage. Her father was diagnosed with hypercholesterolemia. Her mother died from colon cancer, and no blood cholesterol data was obtained. However, her mother's elder brother had xanthelasma on his eyelids, suggesting the existence of family history of maternal hypercholesterolemia. The patient's son and daughter were also diagnosed with hypercholesterolemia; however, their cholesterol levels were milder than that of the proband. She had no coronary risk factors, such as hypertension, impaired glucose tolerance, and smoking habit, other than hypercholesterolemia. She had xanthelasma on her eyelids, xanthoma on her bilateral third and fourth extensor tendons of the hands, tuberculous skin xanthoma on her bilateral elbows, and Achilles tendon xanthomas, which were 18 mm thick (Fig. 5.1). She also had surgical scars from resection of skin xanthoma on her wrists. She had a grade 4/6 late systolic murmur (abnormal sound heard during auscultation of the heart) on the

M.-a. Kawashiri (✉)
Department of Cardiovascular and Internal Medicine, Kanazawa University Graduate School of Medical Science, Kanazawa, Japan
e-mail: mk@med.kanazawa-u.ac.jp

D. J. Rader
Departments of Genetics, Perelman School of Medicine, University of Pennsylvania, Philadelphia, PA, USA
e-mail: rader@mail.med.upenn.edu

T. Oohashi et al. (eds.), *Human Pathobiochemistry*, https://doi.org/10.1007/978-981-13-2977-7_5

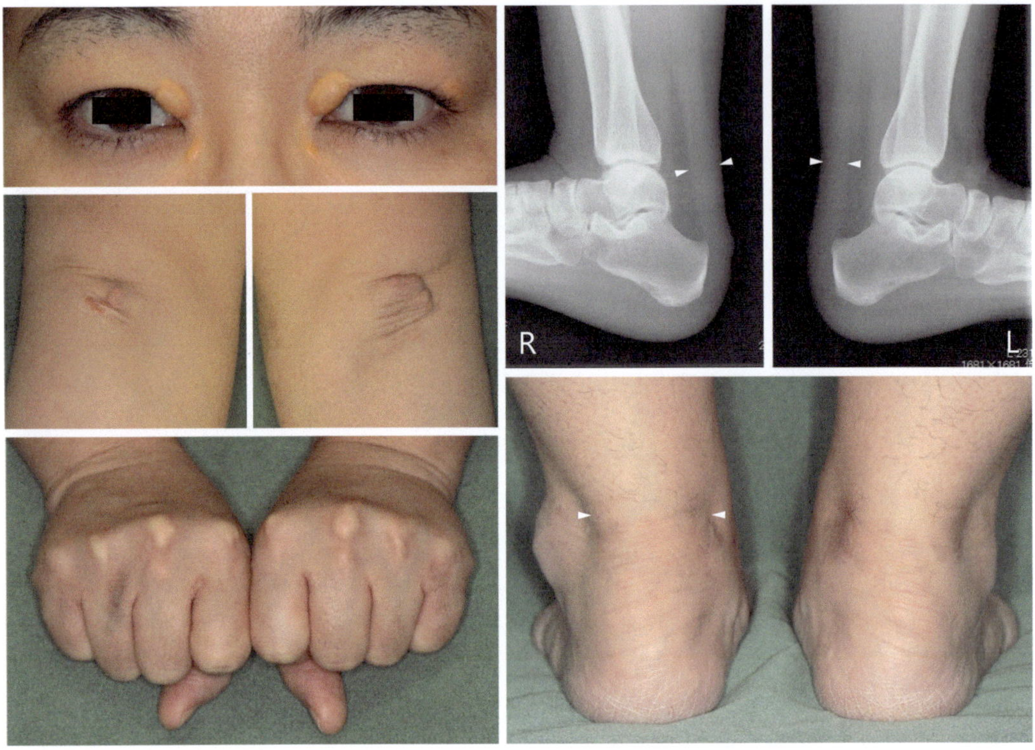

Fig. 5.1 Physical findings of a patient with compound heterozygous familial hypercholesterolemia
The presented case had xanthelasma on her eyelids (top left), tuberculous skin xanthoma on her bilateral elbows (middle left), xanthoma on her bilateral third and fourth extensor tendons of the hands (bottom left), and Achilles tendon xanthoma (18 mm thick in side view of roentgenograph; top and bottom right). The thickness of Achilles tendon on the roentgenograph is to be measured perpendicularly at its thickest part (arrow head)

second right sternal border radiating to her neck and bilateral carotid bruits (abnormal sound heard during auscultation of the vessels suggesting its narrowing). A cardiac echocardiography revealed shrinkage of the aortic root and narrowing of the ascending aorta just above the valve (supravalvular aortic stenosis), and Doppler and planimetric analysis suggested severe aortic stenosis (blockage), resulting in a significant pressure gradient. Carotid ultrasonography revealed approximately 50% and 75% stenosis in the right and left common carotid artery, respectively; cardiac catheterization revealed moderate coronary atherosclerosis. Her chest pain resolved successfully following aortic valve replacement surgery.

Her serum levels of total cholesterol, LDL-C, and high-density lipoprotein (HDL) cholesterol (normal range; ≥ 40 mg/dL) were 523, 429, and 43 mg/dL, respectively, and her triglyceride level (normal range; < 150 mg/dL) was 122 mg/dL. Her serum level of apolipoprotein B (apoB) (normal range; < 100 mg/dL) was 269 mg/dL. Genetic analysis revealed three mutations in the LDL receptor gene: two (c.904 T > G and c.907C > T) were identified in her father, both of which were changed amino acid sequences. A large deletion in exon 6 of the same gene was found in her maternal family member including uncle. Finally, she was diagnosed with compound heterozygous familial hypercholesterolemia [(FH) HeFH].

She was treated with a low-fat, low-cholesterol, restricted caloric diet and 20 mg daily of rosuvastatin (maximum doses in Japan), a cholesterol biosynthesis inhibitor stronger than old generation statin, such as pravastatin. Her serum levels of LDL-C and apoB reached to 257 mg/dL and 165 mg/dL, respectively. The selective cholesterol absorb inhibitor ezetimibe was also introduced. Following treatment, her LDL-C and

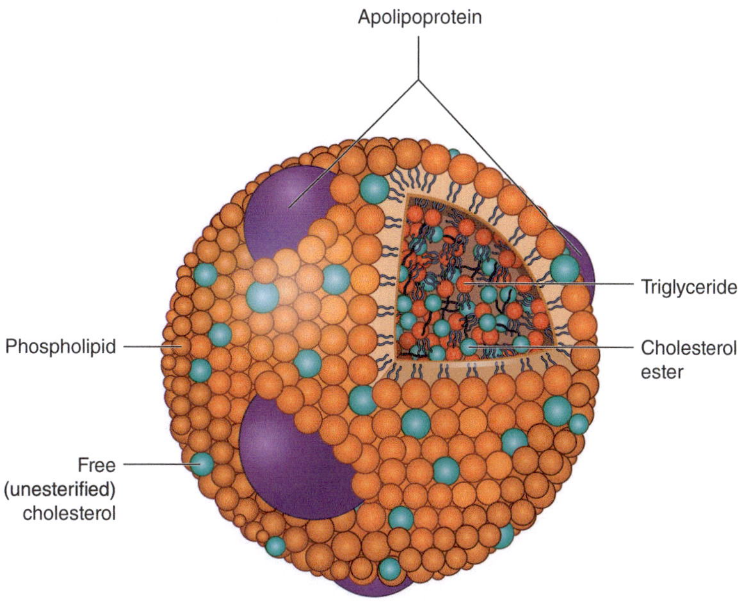

Fig. 5.2 Basic structure of lipoprotein Lipoproteins are composed of a core neutral lipid, mainly cholesteryl ester and triglycerides, surrounded by free cholesterol, phospholipids, and apolipoproteins

apoB levels decreased to approximately 240 and 155 mg/dL, respectively. Because these medicines were only partially effective, LDL-apheresis therapy, which is extracorporeal selective adsorption of apoB-containing lipoproteins, was introduced along with the anti-proprotein convertase subtilisin/kexin type 9 (PCSK9) antibody evolocumab, which inhibits intracellular degradation of LDL receptor. Finally, her serum levels just before LDL-apheresis of LDL-C and apoB reached approximately 170 and 130, respectively. However, even after these intensive cholesterol-lowering therapies, her serum LDL-C levels did not reach to the target level for secondary prevention of coronary artery disease (100 mg/dL): therefore, microsomal triglyceride transfer protein inhibitor is the next treatment option.

5.2 Diagnosis

5.2.1 Compound Heterozygous Familial Hypercholesterolemia

Lipids including cholesterol are hydrophobic; thus, they are bound with apolipoproteins to create lipoproteins that circulate in the blood stream. Lipoproteins are composed of a core neutral lipid, mainly cholesteryl ester and triglycerides, surrounded by free cholesterol, phospholipids, and apolipoproteins (Fig. 5.2).

Cells require cholesterol as an essential material for its membrane. The liver synthesizes bile acid from cholesterol, which is required for uptake lipid from the intestine (exogenous pathway). In some endocrine organs, such as adrenal gland, testis, and ovary, cholesterol is also used as a substrate for steroids, such as testosterone, androsterone, estradiol, and progesterone. Cells can synthesize cholesterol from acetyl-CoA; however, the quantity is not sufficient for the cells, except for hepatocytes. Thus, cholesterol is transported from the liver to the other cells mainly by LDL via the LDL receptor (endogenous pathway). Cholesterol is returned from the peripheral cells including macrophages in atherosclerotic lesion to the liver by HDL (reverse cholesterol transport pathway) (Fig. 5.3).

FH is the most severe form of monogenic hypercholesterolemia and is mainly caused by mutations in the LDL receptor gene (Rader et al. 2003, Mabuchi 2017). Endogenous cholesterol is secreted from hepatocytes (liver cells) into circulation as very low-density lipoprotein (VLDL), dilipidated by lipoprotein and hepatic lipases, and consequently metabolized to LDL through intermediate-density lipoprotein (IDL) (Fig. 5.3).

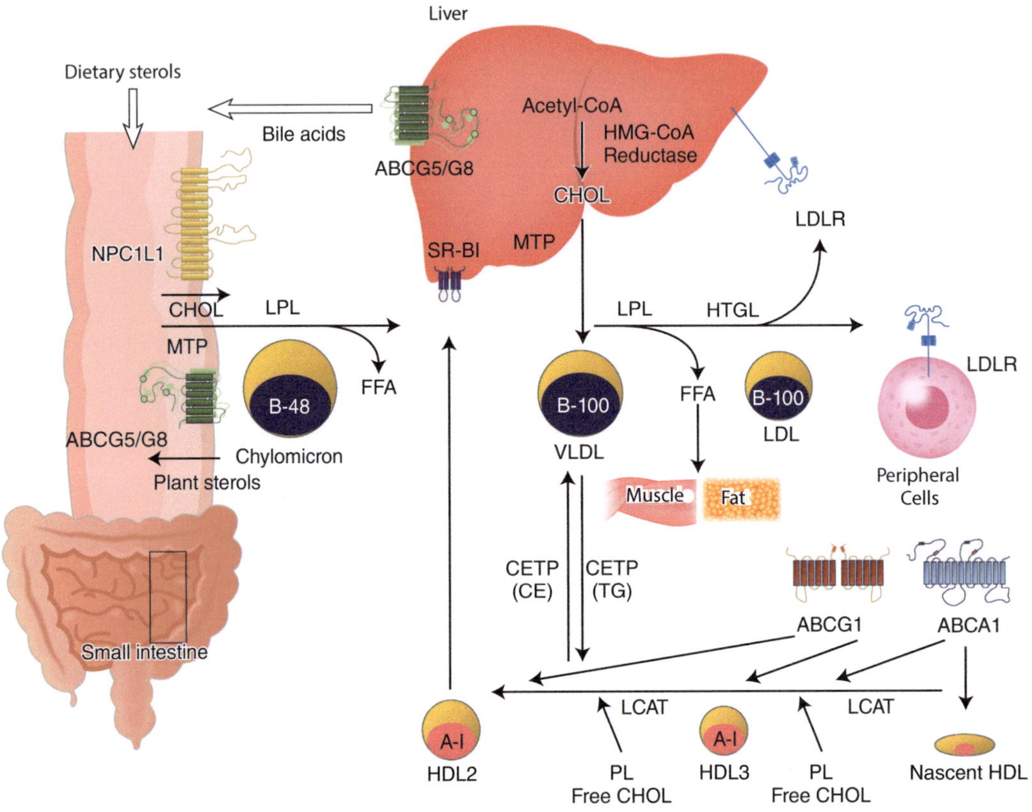

Fig. 5.3 Lipoprotein metabolism
Human lipoprotein metabolism is divided into three categories; the exogenous pathway (chylomicron), the endogenous pathway (mainly LDL), and the reverse cholesterol transport pathway (HDL). LDL receptor, the causal protein of FH, is the main regulator of human cholesterol levels
ABCA1 adenosine triphosphate-binding cassette sub-family A member 1, *ABCG1* adenosine triphosphate-binding cassette sub-family G member 1, *ABCG5G8* adenosine triphosphate-binding cassette sub-family G member 5/sub-family G member 8, *CE* cholesteryl ester, *CETP* cholesteryl ester transfer protein, *CHOL* cholesterol, *FFA* free fatty acid, *HDL* high-density lipoprotein, *HTGL* hepatic triglyceride lipase, *HMG-CoA* hydroxymethylglutaryl-CoA, *LCAT* Lecithin-cholesterol acyltransferase, *LDL* low-density lipoprotein, *LDLR* low-density lipoprotein receptor, *LPL* lipoprotein lipase, *MTP* microsomal triglyceride transfer protein, *NPC1L1* Niemann-Pick C1-like 1, *PL* phospholipid, *SR-BI* scavenger receptor class B type 1, *TG* triglyceride

LDL is the major transport particle of cholesterol to peripheral cells via the LDL receptor, and the remaining LDL particles in circulation are cleared by the LDL receptor on hepatocytes. LDL particles of FH without functional LDL receptors are congested in circulation because of the inability of the LDL receptor to clear the particles (Fig. 5.3).

FH is a common autosomal dominant hereditary disorder, and there are two forms of FH: individuals with two mutated LDL receptor alleles (homozygous FH: HoFH) and those with a single mutated LDL receptor allele (heterozygous FH: HeFH). The blood total cholesterol levels of HoFH, HeFH with LDL receptor gene mutation, and their normal siblings are trimodal as follows: 713 ± 122, 332 ± 60, and 179 ± 26 mg/dL, respectively (Fig. 5.4) (Mabuchi 2017). Initially, the frequency of HeFH was reported as 1 in 500 general populations; recent advances in genetic analysis have expanded the frequency to 1 in 200–300 general populations (Mabuchi 2017, Nordestgaard et al. 2013). Accordingly, the frequency of HoFH is estimated as 1 in

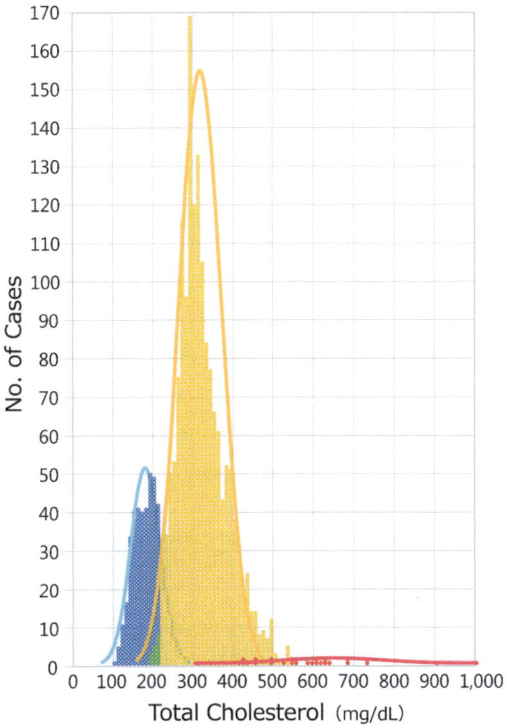

Fig. 5.4 Total cholesterol levels of HoFH, HeFH, and their normal siblings
The blood total cholesterol levels of HoFH, HeFH with LDL receptor gene mutation, and their normal siblings are trimodal as follows: 713 ± 122, 332 ± 60, and 179 ± 26 mg/dL, respectively (Mabuchi 2017)

160,000–360,000 general populations. In certain populations, such as Afrikaners in South Africa, French Canadians, and the Lebanese, the frequency of HeFH is substantially higher than in other regions because of the founder gene effect, whereby these population groups originated from a small number of individuals with a specific LDL receptor gene mutation at a higher frequency than that of the general population.

HoFHs cause the development of systemic cutaneous xanthoma and atherosclerosis from childhood; cutaneous xanthomas between the hand fingers and buttocks are specific physical findings for HoFH. Interestingly, as in the case presented here, the manifestation of early atherosclerosis in HoFH is often due to the formation of atheroma along with aortic root, aortic valve, and coronary ostia, which extends to coronary and systemic atherosclerosis thereafter. Supravalvular and aortic valve stenosis can cause acute heart failure or sudden cardiac death (Mabuchi 2017). The development of coronary artery disease (CAD) is decided by the cumulative values of LDL-C levels (Nordestgaard et al. 2013). Thus, HoFH could cause the development of premature CAD before 20 years of age along with extremely high LDL-C levels from the birth if effective cholesterol-lowering therapy is not administered. Thus, early surgical intervention including aortic valve replacement or a Bentall procedure, which is the composite graft replacement of aortic valve, aortic root, and ascending aorta, and coronary artery bypass surgery may be needed. Although it is also important to optimize the other cardiovascular risk factors, such as blood pressure and diabetes, these have minimal impact on the clinical course of HoFH.

HeFHs also initially present as tendon xanthomas in the backs of hands and Achilles' tendons, and cutaneous tuberculous xanthomas occur during and after the second decade of life. Arcus cornealis and xanthelasma on eyelids are characteristics of FH; however, they are not specific or suitable for the diagnosis of FH. Coronary atherosclerosis detected by coronary computed tomography starts to develop at 23 and 33 years in male and female HeFH patients, respectively (Tada et al. 2015b). Clinically significant CAD usually manifests after 30 years of age for men and 40 years for women.

There are substantial interindividual differences in LDL-C levels among patients with HoFH and HeFH. HoFH is classically divided into two categories: receptor-negative FH with no or minimum LDL receptor activity and receptor-defective FH with substantial residual LDL receptor activity. Although LDL receptor activity can be measured in vitro using skin fibroblasts, it is not routinely performed in clinical practice. The clinical course of receptor-negative FH is usually worse compared with that of receptor-defective FH because of substantially higher cumulative LDL-C levels. Standard medical cholesterol-lowering therapies, which eventually increase hepatic LDL receptor, do not effectively reduce LDL-C levels in patients with receptor-negative HoFH but can substantially reduce LDL-C levels in patients with receptor-defective HoFH.

The diagnosis of HeFH is made by combining elevated LDL-C levels, clinical manifestations (particularly xanthomas), and family history of hypercholesterolemia and/or premature CAD. Although several different clinical diagnostic criteria of FH have been proposed, the existence of tendon or skin xanthomas, which are the reflection of continuous hypercholesterolemia from birth, is essential for the clinical diagnosis of FH. Of those, the following three diagnostic criteria are major: (a) Simon Broome Register Diagnostic Criteria for FH, (b) Make Early Diagnosis to Prevent Early Death (MedPed) Program Diagnostic Criteria for FH, and (c) Dutch Lipid Clinic Network Diagnostic Criteria for FH (Table 5.1a, 5.1b, and 5.1c). The diagnosis of HoFH is made based on markedly elevated LDL-C levels and a family history of parental HeFH on both sides, except for very rare recessive hereditary conditions.

A genetic diagnosis is considered to be definitive; however, the annotation of the specific mutation is not always easy. Very recently, comprehensive genetic analysis of known causal genes (*LDL receptor, apoB, PCSK9, and LDL receptor adaptor protein 1*) has become readily available. Thus, it can be speculated that genetic analysis will play a greater role in future diagnostic criteria of FH. Regarding the patient in the case report, three different mutations in the LDL receptor gene were detected, two (c.904 T > G and c.907C > T) were of a paternal origin and the other (a large deletion in exon 6) was of a maternal origin. Thus, the final diagnosis of the case was compound heterozygous FH, which has a similar significance as HoFH in the clinical setting.

Table 5.1a Simon Broome Register Diagnostic Criteria for FH

Definite FH is defined as	
(a)	Total cholesterol > 260 mg/dL or LDL-cholesterol above 155 mg/dL in a child < 16 years
	Total cholesterol > 290 mg/dL or LDL-cholesterol above 190 mg/dL in an adult
	(levels either pretreatment or highest on treatment)
	plus
(b)	Tendon xanthoma in patient
	or in 1st-degree relative (parent, sibling, child)
	or in 2nd-degree relative (grandparent, uncle, aunt)
	or
(c)	DNA-based evidence of an LDL receptor mutation or familial defective apoB-100
Possible FH is defined a	
(a)	Total cholesterol > 260 mg/dL or LDL-cholesterol above 155 mg/dL in a child < 16 years
	Total cholesterol > 290 mg/dL or LDL-cholesterol above 190 mg/dL in an adult
	(Levels either pretreatment or highest on treatment)
	plus one of (d) or (e)
(d)	Family history of myocardial infarction
	below age of 50 in 2nd-degree relative
	below age of 60 in 1st-degree relative
(e)	Family history of raised cholesterols
	> 290 mg/dL in adult in 1st- or 2nd-degree relative
	> 260 mg/dL in child or sibling under 16

Table 5.1b Make Early Diagnosis to Prevent Early Death (MedPed) Program Diagnostic Criteria for FH

Age (years)	First-degree relative with FH	Second-degree relative with FH	Third-degree relative with FH	General Population
<20	220	230	240	270
20–29	240	250	260	290
30–39	270	280	290	340
≥40	290	300	310	360

FH is diagnosed if total cholesterol exceeds these cut points in mg/dL
The total cholesterol cut points for FH is dependent upon the confirmed cases of FH in the family
If FH is not diagnosed in the family, then the cut point for diagnosis is as per "general population"

5.3 Differential Diagnosis

1. Familial defective apolipoprotein B-100 (FDB or FH2)
2. Autosomal dominant hypercholesterolemia due to gain-of-function mutations in pro-protein convertase subtilisin/kexin type 9 (PCSK9) gene (FH3)

Table 5.1c Dutch Lipid Clinic Network Diagnostic Criteria for FH

Criteria		Points
Family history		
	First-degree relative with known premature* coronary and vascular disease, OR	1
	First-degree relative with known LDL-C level above the 95th percentile	
	First-degree relative with tendinous xanthomata and/or arcus cornealis, OR	2
	Children aged less than 18 years with LDL-C level above the 95th percentile	
Clinical history		
	Patient with premature* coronary artery disease	2
	Patient with premature* cerebral or peripheral vascular disease	1
Physical examination		
	Tendinous xanthomata	6
	Arcus cornealis prior to age 45 years	4
Cholesterol levels (mg/dL)		
	LDL-C ≥ 330 mg/dL	8
	LDL-C 250 – 329 mg/dL	5
	LDL-C 190 – 249 mg/dL	3
	LDL-C 155 – 189 mg/dL	1
DNA analysis		
	Functional mutation in the LDL receptor, apoB or PCSK9 gene	8
Diagnosis (diagnosis is based on the total number of points obtained)		
	Definite FH	≥ 8
	Probable FH	6 - 8
	Possible FH	3 - 5
	Unlikely FH	< 3

*Premature means < 55 years in men; < 60 years in women

3. Autosomal recessive hypercholesterolemia (ARH)
4. Sitosterolemia
5. Secondary hyperlipidemia, such as hypothyroidism

5.4 Biochemical and Molecular Perspectives

Brown and Goldstein received the Nobel Prize in Physiology or Medicine in 1985 for discovering the mechanisms of receptor-mediated endocytosis and the disruption of the LDL receptor as the cause of FH (Brown and Goldstein 1983). To date, dysfunctions of the LDL receptor pathway are recognized as the causes of FH, including FH2, FH3, and ARH.

The size and densities of the different classes of lipoproteins are determined by their composition; larger lipoproteins are buoyant and have more triglycerides and less apolipoproteins, while smaller lipoproteins are dense and have less triglycerides and more apolipoproteins (Table 5.2). Surface charge revealed by agarose gel electrophoresis varies between lipoproteins according to the amount of charged lipids and the conformations of their apolipoproteins. Hence, lipoproteins can be isolated by numerous techniques including size exclusion chromatography, nuclear magnetic resonance, ultracentrifugation, agarose gel electrophoresis, and polyacrylamide

Table 5.2 Properties of major lipoprotein classes

Lipoproteins	Diameter (nm)	Density (g/mL)	Electrophoretic mobility	Composition (%)					Major apolipoproteins
				Core		Surface			
				CE	TG	FC	PL	Protein	
Chylomicrons	80–500	< 0.93	Origin	3	86	2	7	2	B-48, E, A-I, A-II, A-IV, C
VLDL	30–80	0.95–1.006	Pre-beta	12	55	7	18	8	B-100, C-I, C-II, C-III, E
IDL	25–35	1.006–1.019	Slow pre-beta	29	23	9	19	19	B-100, E
LDL	21.6	1.019–1.063	Beta	42	6	8	22	22	B-100
HDL_2	10	1.063–1.125	Alpha	17	5	5	33	40	A-I, A-II
HDL_3	7.5	1.125–1.210	Alpha	13	3	4	25	55	A-I, A-II
Lp(a)	30	1.055–1.085	Slow pre-beta	33	3	9	22	33	B-100, apo(a)

CE cholesteryl ester, *FC* free cholesterol, *PL* phospholipid, *Lp(a)* lipoprotein (a)

gel electrophoresis. Specific lipoproteins are associated with specific apolipoproteins; however, there is some overlaps (Table 5.2). For example, apoB-100 is associated with all lipoproteins except for chylomicron and HDL, and apoA-I is mainly associated with HDL. LDL is the most abundant lipoproteins in humans.

The LDL receptor is a cell surface glycoprotein, which mediates the removal of LDLs and remnant lipoproteins from circulation by binding to apoB and apoE and plays a major role in blood cholesterol level regulation. The extracellular component of the LDL receptor comprises the following domains: a ligand-binding domain; an epidermal growth factor (EGF) precursor homology domain, which contains a six-bladed beta-propeller flanked by cysteine-rich EGF domain; and an O-liked sugar-rich domain (Fig. 5.5) (Gidding et al. 2015). LDL receptors are synthesized in the endoplasmic reticulum (ER) and glycosylated in the Golgi apparatus and then transported to clathrin-coated pits on the cell surface. After LDL binds to the ligand-binding domain on the LDL receptor, the LDL-LDL receptor complex is internalized and delivered to endosomes. In the acidic environment of the endosome, the LDL particle is dissociated from the receptor, which is recycled back to the cell surface, and the particle is degenerated for the storage of intracellular cholesterol (Fig. 5.6).

Numerous mutations (more than 1100) at the LDL receptor locus have been described as the cause of FH (http://www.ucl.ac.uk/fh). Interindividual phenotypic variation among patients with HeFH and HoFH is at least partially explained by LDL receptor activity and their causal mutations. The LDL receptor gene contains 18 exons and 17 introns and encodes 860 amino acids (Fig. 5.5). Mutations in the LDL receptor gene can be divided into five different functional classes; Class 1 mutations (null alleles) result in LDL receptor synthesis alteration. These mutations include nonsense mutations with a premature stop codon occurring early in the protein, large rearrangements including many bases, and insertions or deletions causing frameshift. Class 2 mutations cause an alteration in receptor transport to the Golgi apparatus or to the plasma membrane (completely, Class 2A; partially, Class 2B). Class 3 mutations result in an alteration of binding to apoB-100 despite normal transport to the plasma membrane. Class 4 mutations result in an endocytosis alteration despite normal binding to LDL, and class 5 mutations cause an alteration in the recycle mechanism (Fig. 5.6). Class 1 and class 2A are usually associated with null or less than 2% LDL receptor activity in cultured fibroblasts (receptor negative), while other classes are associated with reduced LDL receptor activity (receptor defective). Although the LDL receptor activity of the present case was not determined in vitro, it can be speculated that her LDL receptor function was preserved because cholesterol-lowering medication, including statin, was partially effective.

ApoB-100 is the ligand for the LDL receptor, and initially a missense mutation in the LDL

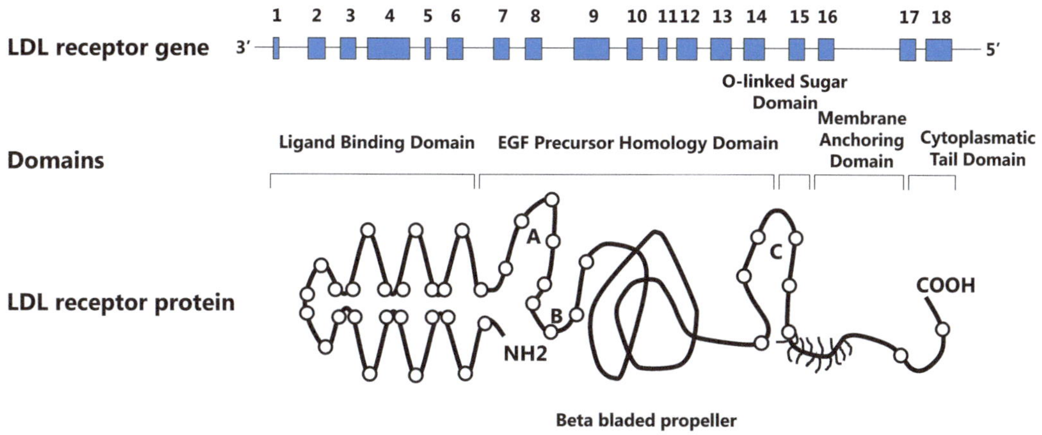

Fig. 5.5 Protein and domains of the LDL receptor
The different domains in LDL receptor protein are encoded by specific regions in the LDL receptor gene
EGF endothelial growth factor

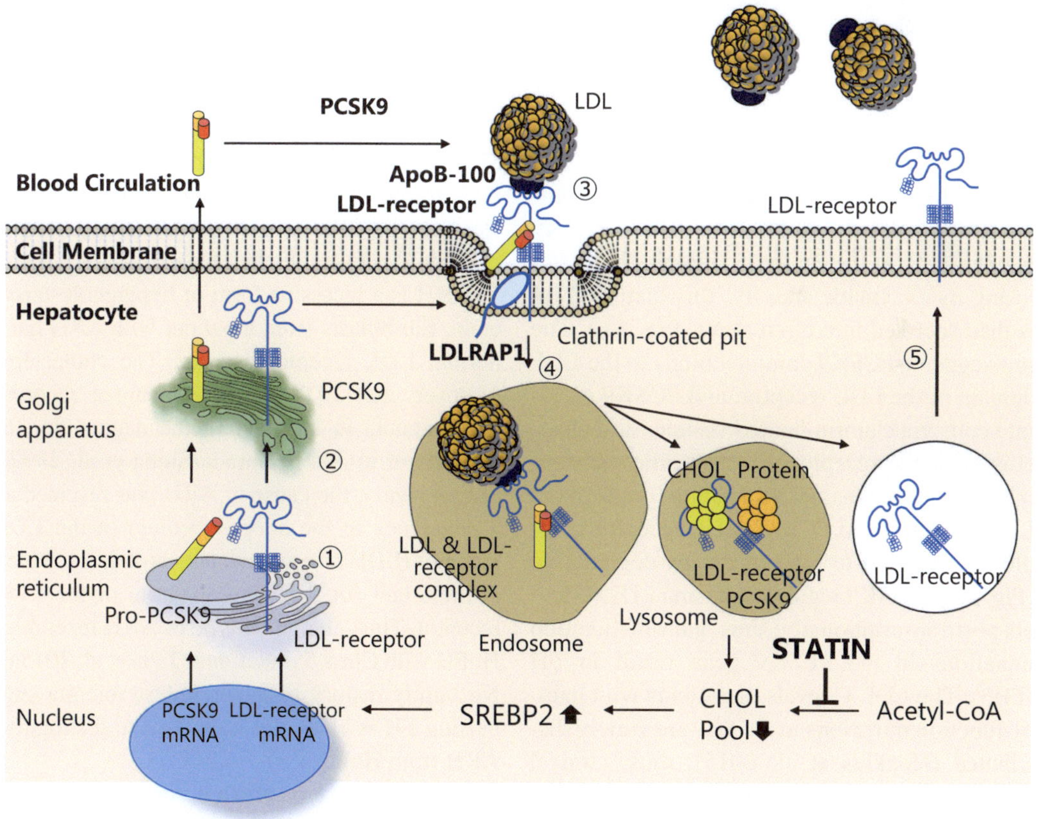

Fig. 5.6 Regulation of the LDL receptor
Both the LDL receptor and proprotein convertase subtilisin/kexin type 9 (PCSK9) are regulated by sterol regulatory element-binding protein 2 (SREBP2). The numbers indicate the sites of functional classes of LDL receptor gene mutations.
LDLRAP1 LDL receptor adaptor protein 1

receptor-binding domain of apoB-100 (Arg3500Gln) was described, followed by another mutation in the same codon (Arg3500Trp), as a cause of familial defective apolipoprotein B-100 (FDB) or FH2 (Soria et al. 1989). Both mutations are thought to reduce binding activity with the LDL receptor; eventually LDL particles in circulation are not cleared by the receptor. There is a clear localization of FDB; the frequency in Western countries is 1 in 1000 in the general population, and no cases of FDB have been reported in Japan so far (Nohara et al. 1995). Although clinical manifestations of FDB are usually milder than those of FH with the LDL receptor gene mutation (FH1), it is impossible to clinically distinguish FDB from FH1 without genetic analysis.

The proprotein convertase subtilisin/kexin type 9 (PCSK9) is the 9th member of the proprotein convertase family, which was the third cause of FH (Abifadel et al. 2003). After PCSK9 is synthesized as pre-pro PCSK9 in the ER, signal peptide (aa 1-30) is cleaved, followed by cleavage of the pro-domain (aa 31-152) by its own enzymatic activity. The pro-domain non-covalently binds to mature PCSK9 (aa 153-692) at the catalytic subunit, thereafter losing its enzymatic activity; this mature form is then secreted into circulation (Fig. 5.6). The pro-segment-PCSK9 complex binds to the EGF domain of the LDL receptor, and PCSK9 enters into cells via clathrin-coated vesicles as a chaperon with the receptor. In the acidic environment of endosomes, PCSK9 tightly binds to the LDL receptor, and PCSK9-LDL receptor complex is escorted to lysosomes for degradation (Fig. 5.6). PCSK9 regulates plasma LDL-C levels posttranscriptionally; thus, gain-of-function mutations in the *PCSK9* gene result in FH (FH3). The LDL-C levels of patients with gain-of-function mutations in *PCSK9* are widely distributed (Hopkins et al. 2015); thus, clinical distinguishing FH1 from FH3 is also difficult without measuring PCSK9 activity or genetic analysis. If PCSK9 is absent or inactivated, the LDL receptor effectively returns to the cell surface for recycling, eventually causing hypocholesterolemia.

Interestingly, the LDL receptor and PCSK9 are regulated by the same transcriptional factor, sterol regulatory element-binding protein 2 (SREBP2). SREBPs are inactive proteins bound to the ER. In the ER, SREBP binds to the SREBP cleavage-activating protein (SCAP), which is a sensor of intracellular sterols. When cellular cholesterol content is low, the SCAP-SREBP complex is able to move to the Golgi apparatus, where SREBP is sequentially cleaved by two proteases, S1P and S2P, and the NH2 terminal of SREBP is released to the cytoplasm. SREBP then translocates to the nucleus, where it activates transcription by binding to SRE-1 in the promoter regions of many genes encoding proteins involved in the homeostasis of cholesterol and other lipids. Statin, which is an inhibitor of cholesterol biosynthesis, upregulates SREBP2, increases the transcriptions of LDL receptors, and then increases PCSK9, which degrades LDL receptors. In contrast, when the cellular cholesterol content is in excess, SCAP undergoes conformational changes, which inhibits the transport of the SCAP-SREBP complex to the Golgi apparatus. As a result, the NH2 terminal of SREBP is not cleaved, and SREBP does not translocate to the nucleus to activate the target genes.

ARH is a recessive form of hypercholesterolemia. Fibroblasts from a patient with ARH have normal LDL receptor activity. The cholesterol turnover after LDL-apheresis using a rebound curve results in a defect in the degradation of cholesterol in ARH (Harada-Shiba et al. 1992). Subsequently, the cause of ARH was revealed as a deficiency in the adopter protein of the LDL receptor (LDL receptor adaptor protein 1), which is important for the internalization of the LDL receptor. Thus, the phenotype of ARH resembles HoFH with Class 5 mutations (Tada et al. 2015a). No family history of hypercholesterolemia suggesting FH is clinically important to distinguish ARH from HoFH.

Sitosterol is a type of plant sterol, which enters an enterocyte via the Niemann-Pick C1-like 1 (NPC1L1) protein together with cholesterol, and is excreted from the adenosine triphosphate-binding cassette sub-family G member 5/sub-family G member 8 (ABCG5/G8) (Fig. 5.3).

Sitosterolemia is a rare autosomal recessive hyperlipidemia caused by dysfunctional mutations in ABCG5 or ABCG8. Sitosterolemia can exhibit severe hypercholesterolemia as well as tendon and skin xanthoma, resembling HoFH. In particular, infantile sitosterolemic patients with severe hypercholesterolemia sometimes exhibit systemic intertriginous xanthomas, which diminish quickly when weaned from breastfeeding (Tada et al. 2015c). Detecting hypersitosterolemia and determining the absence of family history of severe hypercholesterolemia are important to distinguish sitosterolemia from HoFH.

It is known that double heterozygous FH, which has two different mutations in FH-causing genes (i.e., LDL receptor and apoB or PCSK9), resembles HoFH but with somewhat milder clinical manifestations.

5.5 Therapy

The primary goal for HeFH, and even for HoFH, is to have the same life expectancy and the same frequency of cardiovascular events as normal subjects. Most of hyperlipidemia other than FH develops after the second decade of life; thus, cumulative LDL-C from the birth is higher in FH than in other hyperlipidemia, even when the LDL-C levels after adulthood were the same. Therefore, intense LDL-C-lowering therapy for FH from an adequate age is crucial. A diet rich in saturated fat and cholesterol is associated with an increased flux of fats and cholesterol from the intestine. An increased cellular cholesterol content results in the downregulation of hepatic LDL receptor expression. Therefore, diet therapy consists of an adequate amount of total calories and restriction of carbohydrate, saturated fat, and cholesterol; maintaining adequate body weight is essential.

Although diet therapy is important, it is insufficient in reducing LDL-C levels in patients with FH to adequate levels. Thus, all FH patients ultimately become candidates for medical LDL-C-lowering therapy; most LDL-C-lowering medication upregulates LDL receptor activity by reducing the intracellular cholesterol content. HeFH patients, whereby half of the LDL receptors are intact, can usually be treated with ordinal cholesterol-lowering medication to acceptable levels. On the other hand, HoFH patients, who mostly lack the LDL receptor, are highly resistant to standard medical cholesterol-lowering therapy.

Statins inhibit 5-hydroxyl-3-methylglutaryl-coenzyme A (HMG-CoA) reductase, a rate limiting enzyme of cholesterol biosynthesis in the mevalonate pathway, and reduce plasma LDL-C levels by increasing hepatic expression of LDL receptors. Statin is the most widely used LDL-C-lowering medication and is proved to reduce not only cardiovascular events but also all-cause mortality of patients at high risk of cardiovascular events. Moreover, the reduction rate of cardiovascular events and achieved LDL-C levels from several randomized-controlled studies using statin were linearly and significantly associated. Ezetimibe is the inhibitor of NPC1L1, the transmembrane protein that plays a key role in the cholesterol absorption by facilitating its uptake via vesicular endocytosis. Thus, ezetimibe also reduces the cholesterol content in hepatocytes and increase hepatic expression of LDL receptors. Ezetimibe combined with simvastatin therapy significantly reduced cardiovascular events in patients with acute coronary syndrome (Cannon et al. 2015). Bile acid sequestrants (resins) interrupt the enterohepatic circulation of bile acids, resulting in hepatic upregulation of LDL receptors. Combination of these three drugs, all of which increases hepatic LDL receptors, could decrease LDL-C levels to one third of pretreatment levels in patients with HeFH (Kawashiri et al. 2012).

PCSK9 regulates plasma LDL-C levels by directing hepatic cell surface LDL receptors to lysosomal degradation, resulting in an accumulation of LDLs in circulation because of reduced clearance. Evolocumab and alirocumab are fully human monoclonal antibody against human PCSK9, which can robustly reduce plasma LDL-C levels by more than 50% with monotherapy, and can be effectively combined with the other cholesterol-lowering drugs, such as statins and ezetimibe. Evolocumab combined with statin therapy significantly reduced cardiovascular events and ischemic strokes in patients with atherosclerotic cardiovascular diseases compared with statin therapy alone (Sabatine et al. 2017).

Microsomal triglyceride transfer protein (MTP) is a key protein in the assembly and secretion of apoB-containing lipoproteins in the liver (VLDL) and intestine (chylomicron) (Fig. 5.3). Lomitapide, a small molecule that inhibits MTP, does not require LDL receptor to decrease apoB-containing lipoproteins; thus, it is approved in several countries as an orphan drug for HoFH. While lomitapide decreases plasma LDL-C levels by 50% in HoFH, special attention should be paid for hepatic steatosis (Cuchel et al. 2013). Mipomersen is an antisense oligonucleotide, which is a small strand of nucleic acid that binds to messenger RNA to inhibit translation, for apoB. It also does not require the LDL receptor to decrease apoB-containing lipoproteins in HoFH (Raal et al. 2010). LDL-apheresis is a procedure that removes all apoB-containing lipoproteins (VLDL, IDL, LDL, and Lp(a)), the surface of which are exceptionally charged positive, using negatively charged dextran sulfate. Although LDL-apheresis is safe and does not remove HDL nor other negatively charged proteins with the exception of apoB-containing lipoproteins, it should be repeated weekly or biweekly because of rapid reaccumulation of LDL. Furthermore, it is an invasive and time-consuming procedure only available at specialized medical centers. Liver transplantation is an alternative therapy for HoFH and effectively reduces LDL-C; however, it is highly invasive and lifelong immunosuppressive therapy is required.

Under these conditions, innovative therapies are needed for HoFH. Among future treatment options, regenerative medicine using induced pluripotent stem (iPS) cells, which can introduce auto-hepatocyte with functional LDL receptors, is a theoretically attractive approach. Liver-directed gene therapy is another attractive option, but long-term stable expression of de novo gene by the host is difficult because of its immunologic exclusion. Antisense oligonucleotide therapy is also of interest for rare inherited disorders such as HoFH; antisense oligonucleotides against Lp(a) and PCSK9 are currently under development.

End-of-Chapter Questions

1. FH is thought to be the most frequent monogenic hereditary disorder. Why do you think FH is so frequent worldwide?
2. Why is tendon xanthoma very specific for FH? Why does hyperlipidemia other than FH not induce the development of tendon xanthoma even when the plasma LDL-C levels are the same with those of FH after adulthood?
3. What are the molecular causes of FH? What are the clinical characteristics and lipid levels of both HeFH and HoFH? Please explain what the compound heterozygous FH or double heterozygous FH.
4. What is the mechanism of statins, ezetimibe, and PCSK9 inhibitors to reduce LDL-C? What are the specific and novel therapies for HoFH and how do they reduce LDL-C?

Additional Materials

Which of the following lipoproteins are *not* associated with apolipoprotein B-100?

1. Chylomicron
2. VLDL
3. LDL
4. Lp(a)

Answer 1

Which of the following combinations between lipoproteins and their role is correct?

1. Chylomicron & Endogenous cholesterol carrier
2. VLDL & Provider of free fatty acids
3. LDL & Reverse cholesterol transporter
4. HDL & Exogenous cholesterol carrier

Answer 2

Which of the following physical findings is specific for FH?

1. Achilles tendon xanthoma
2. Arcus cornealis
3. Eruptive xanthoma on the skin
4. Xanthelasma on eyelids

Answer 1

Which of the following disorders are *not* be seen more frequently in FH than in normal subjects?

1. Abdominal aortic aneurysm
2. Cerebral infarction
3. Myocardial infarction
4. Peripheral artery disease

Answer 2

Which of the following combinations between drug and target molecule is *not* correct?

1. Evolocumab & PCSK9
2. Ezetimibe & NPC1L1
3. Lomitapide & apolipoprotein B-100
4. Statins & HMG-CoA reductase

Answer 3

References

Abifadel M, Varret M, Rabès JP, Allard D, Ouguerram K, Devillers M, Cruaud C, Benjannet S, Wickham L, Erlich D, Derré A, Villéger L, Farnier M, Beucler I, Bruckert E, Chambaz J, Chanu B, Lecerf JM, Luc G, Moulin P, Weissenbach J, Prat A, Krempf M, Junien C, Seidah NG, Boileau C (2003) Mutations in PCSK9 cause autosomal dominant hypercholesterolemia. Nat Genet 34:154–156

Brown MS, Goldstein JL (1983) Lipoprotein receptors in the liver. Control signals for plasma cholesterol traffic. J Clin Invest 72:743–747

Cannon CP, Blazing MA, Giugliano RP, McCagg A, White JA, Theroux P, Darius H, Lewis BS, Ophuis TO, Jukema JW, De Ferrari GM, Ruzyllo W, De Lucca P, Im K, Bohula EA, Reist C, Wiviott SD, Tershakovec AM, Musliner TA, Braunwald E, Califf RM, Investigators IMPROVE-IT (2015) Ezetimibe added to statin therapy after acute coronary syndromes. N Engl J Med 372:2387–2397

Cuchel M, Meagher EA, du Toit Theron H, Blom DJ, Marais AD, Hegele RA, Averna MR, Sirtori CR, Shah PK, Gaudet D, Stefanutti C, Vigna GB, Du Plessis AM, Propert KJ, Sasiela WJ, Bloedon LT, Rader DJ, Phase 3 HoFH Lomitapide Study investigators (2013) Efficacy and safety of a microsomal triglyceride transfer protein inhibitor in patients with homozygous familial hypercholesterolaemia: a single-arm, open-label, phase 3 study. Lancet 381:40–46

Gidding SS, Champagne MA, de Ferranti SD, Defesche J, Ito MK, Knowles JW, McCrindle B, Raal F, Rader D, Santos RD, Lopes-Virella M, Watts GF, Wierzbicki AS (2015) American heart association atherosclerosis, hypertension, and obesity in young committee of council on cardiovascular disease in young, council on cardiovascular and stroke nursing, council on functional genomics and translational biology, and council on lifestyle and cardiometabolic health. The agenda for familial hypercholesterolemia: a scientific statement from the the American Heart Association Circulation 132:2167–2192

Harada-Shiba M, Tajima S, Yokoyama S, Miyake Y, Kojima S, Tsushima M, Kawakami M, Yamamoto A (1992) Siblings with normal LDL receptor activity and severe hypercholesterolemia. Arterioscler Thromb 12:1071–1078

Hopkins PN, Defesche J, Fouchier SW, Bruckert E, Luc G, Cariou B, Sjouke B, Leren TP, Harada-Shiba M, Mabuchi H, Rabès JP, Carrié A, van Heyningen C, Carreau V, Farnier M, Teoh YP, Bourbon M, Kawashiri MA, Nohara A, Soran H, Marais AD, Tada H, Abifadel M, Boileau C, Chanu B, Katsuda S, Kishimoto I, Lambert G, Makino H, Miyamoto Y, Pichelin M, Yagi K, Yamagishi M, Zair Y, Mellis S, Yancopoulos GD, Stahl N, Mendoza J, Du Y, Hamon S, Krempf M, Swergold GD (2015) Characterization of autosomal dominant hypercholesterolemia caused by PCSK9 gain of function mutations and its specific treatment with Alirocumab, a PCSK9 monoclonal antibody. Circ Cardiovasc Genet 8:823–831

Kawashiri MA, Nohara A, Noguchi T, Tada H, Nakanishi C, Mori M, Konno T, Hayashi K, Fujino N, Inazu A, Kobayashi J, Mabuchi H, Yamagishi M (2012) Efficacy and safety of coadministration of rosuvastatin, ezetimibe, and colestimide in heterozygous familial hypercholesterolemia. Am J Cardiol 109:364–369

Mabuchi H (2017) Half a century tales of familial hypercholesterolemia in Japan. J Atheroscler Thromb 24:189–207

Nohara A, Yagi K, Inazu A, Kajinami K, Koizumi J, Mabuchi H (1995) Absence of familial defective apolipoprotein B-100 in Japanese patients with familial hypercholesterolaemia. Lancet 345:1438

Nordestgaard BG, Chapman MJ, Humphries SE, Ginsberg HN, Masana L, Descamps OS, Wiklund O, Hegele RA, Raal FJ, Defesche JC, Wiegman A, Santos RD, Watts GF, Parhofer KG, Hovingh GK, Kovanen PT, Boileau C, Averna M, Borén J, Bruckert E, Catapano AL, Kuivenhoven JA, Pajukanta P, Ray K, Stalenhoef AF, Stroes E, Taskinen MR, Tybjærg-Hansen A, European Atherosclerosis Society Consensus Panel (2013) Familial hypercholesterolaemia is underdiagnosed and undertreated in the general population: guidance for clinicians to prevent coronary heart disease: consensus statement of the European atherosclerosis society. Eur Heart J 34:3478–3490a

Raal FJ, Santos RD, Blom DJ, Marais AD, Charng MJ, Cromwell WC, Lachmann RH, Gaudet D, Tan JL, Chasan-Taber S, Tribble DL, Flaim JD, Crooke ST (2010) Mipomersen, an apolipoprotein B synthesis inhibitor, for lowering of LDL cholesterol concentrations in patients with homozygous familial hypercholesterolaemia: a randomised, double-blind, placebo-controlled trial. Lancet 375:998–1006

Rader DJ, Cohen J, Hobbs HH (2003) Monogenic hypercholesterolemia: new insights in pathogenesis and treatment. J Clin Invest 111:1795–1803

Sabatine MS, Giugliano RP, Keech AC, Honarpour N, Wiviott SD, Murphy SA, Kuder JF, Wang H, Liu T, Wasserman SM, Sever PS, Pedersen TR, FOURIER Steering Committee and Investigators (2017) Evolocumab and clinical outcomes in patients with cardiovascular disease. N Engl J Med 376:1713–1722

Soria LF, Ludwig EH, Clarke HR, Vega GL, Grundy SM, McCarthy BJ (1989) Association between a specific apolipoprotein B mutation and familial defective apolipoprotein B-100. Proc Natl Acad Sci U S A 86:587–591

Tada H, Kawashiri MA, Nohara A, Inazu A, Kobayashi J, Mabuchi H, Yamagishi M (2015a) Autosomal recessive hypercholesterolemia: a mild phenotype of familial hypercholesterolemia: insight from the kinetic study using stable isotope and animal studies. J Atheroscler Thromb 22:1–9

Tada H, Kawashiri MA, Okada H, Teramoto R, Konno T, Yoshimuta T, Sakata K, Nohara A, Inazu A, Kobayashi J, Mabuchi H, Yamagishi M, Hayashi K (2015b) Assessment of coronary atherosclerosis in patients with familial hypercholesterolemia by coronary computed tomography angiography. Am J Cardiol 115:724–729

Tada H, Kawashiri MA, Takata M, Matsunami K, Imamura A, Matsuyama M, Sawada H, Nunoi H, Konno T, Hayashi K, Nohara A, Inazu A, Kobayashi J, Mabuchi H, Yamagishi M (2015c) Infantile cases of Sitosterolemia with novel mutations in the ABCG5 gene: extreme Hypercholesterolaemia is exacerbated by breastfeeding. JIMD Rep 21:115–122

Gaucher Disease

6

Hiroyuki Ida

Keywords
Gaucher disease · Genotype/phenotype correlation · Enzyme replacement therapy · Pharmacological chaperone therapy · Substrate reduction therapy

6.1 Case Report

A 7-month-old infant was admitted to the hospital because of afebrile convulsion. The patient was born with a normal pregnancy at full term. The delivery was uncomplicated. He developed normally until 3 months of age. However, he manifested stridor and poor feeding at 4 months old and exhibited retroflexion of the neck and strabismus. Feeding problems and difficulty in handling salivary secretion appeared at 6 months old. He manifested tonic-clonic convulsion without fever. Since convulsion continued for more than 30 min, he was transferred to the emergency room.

His body weight, height, and head circumference were 6.2 kg (−2.2SD), 67.0 cm (−0.8SD), and 41.1 cm (3 percentile), respectively. He could control his head, but he could not roll over or sit up. Facial appearance was normal. Severe stridor and retraction were noted upon chest exam. No heart murmur was audible. The abdomen was swollen and moderate splenomegaly was pointed out. The patient's liver was palpable at 3 cm under the right costal margin. Additionally, the tip of spleen was palpable between the level of the umbilicus and the pelvic rim without extension to the right side of the abdomen. On neurological examination, he showed tonic posture, trismus, and spasticity.

Hemoglobin level, white blood cell, and platelet counts were 10.8 g/dl, 6700 /μl, and 9.0×10^4/μl, respectively. On blood biochemistry, aspartate aminotransferase (AST; normal range is 25–80 U/l), alanine aminotransferase (ALT; normal range is 10–60 U/l), lactate dehydrogenase (LDH; normal range is 370–820 U/l), and total bilirubin (normal range is 0.2–1.4 mg/dl) were 32 U/l, 26 U/l, 380 U/l, and 0.2 mg/dl, respectively. The levels of acid phosphatase (ACP) and angiotensin-converting enzyme (ACE) were 88 IU/l/ and 98 IU/l, respectively. Normal ranges of ACP and ACE are less than 14 IU/l and 7–30 IU/l, respectively.

6.2 Diagnosis

6.2.1 Clinical Diagnosis

A clinical hallmark of Gaucher disease (MIM #230800) is hepatosplenomegaly. Based on the

H. Ida (✉)
Department of Pediatrics, The Jikei University School of Medicine, Tokyo, Japan
e-mail: hiroy@jikei.ac.jp

T. Oohashi et al. (eds.), *Human Pathobiochemistry*, https://doi.org/10.1007/978-981-13-2977-7_6

Table 6.1 Clinical phenotypes of Gaucher disease

	Type 1 non-neuronopathic form	Type 2 acute neuronopathic form	Type 3 subacute neuronopathic form
		Neuronopathic Gaucher disease	
Onset	Child ~ adult	Neonate/infant	Child ~ adolescence
Neurological manifestation	-	+++	+~++
Hepatosplenomegaly	- ~ +++	+	+ ~ +++
Bone involvement	- ~ +++	-	- ~ +++
Prognosis	Good	Poor	Various

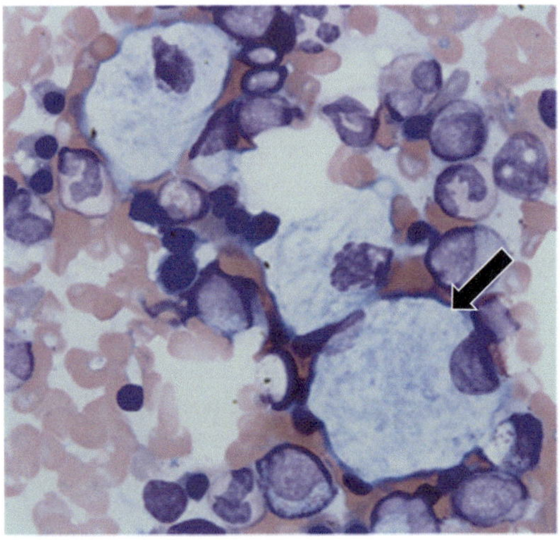

Fig. 6.1 Gaucher cell
The size of abnormal cell (arrow) is approximately 80 μm. The cytoplasm has "wrinkled tissue paper" appearance and the nucleus is eccentric

presence and rate of progression of neurological symptoms, Gaucher disease is classified into three clinical phenotypes (Table 6.1). In type 1 Gaucher disease, patients have no neurological manifestations and chronic course. A clinical feature of type 1 Gaucher disease is variability in age of onset, severity, and progression. Type 2 Gaucher disease is characterized by severe and progressive neurological deterioration and is either fatal at birth or within 2–3 years. Patients with type 3 Gaucher disease have neurologic symptoms with later onset and a more chronic course than that observed in type 2 disease.

Thrombocytopenia is the common peripheral blood abnormality. Anemia is usually mild. Leukopenia also occurs in some patients. These hematologic manifestations are probably due to a combination of increased splenic sequestration and decreased production because of replacement of the bone marrow by Gaucher cells (Fig. 6.1). Since Gaucher cells contain ACP and macrophages produce ACE, these enzymes are elevated in patients with Gaucher disease. The skeletal manifestation of Gaucher disease can be totally debilitating. The "Erlenmeyer flask deformity," which is the expanded cortex of the distal femur, is a common radiographic finding. "Bone crises," pathologic fracture of the long bone, vertebral collapse, and avascular necrosis of the femoral neck can be seen in type 1 patients with a severe phenotype. Involvement of the central nervous system can be seen in patients with type 2 and type 3 Gaucher disease (neuronopathic Gaucher disease). Oculomotor abnormalities are often the first manifestations with the appearance of bilateral, fixed strabismus or oculomotor apraxia. Patients may also show hypertonia of the neck muscles with extreme retroflexions of the neck; bulbar signs, limb rigidity, and seizures occur (Fig. 6.2).

When you see a patient with hepatosplenomegaly, Niemann-Pick disease and cholesteryl ester storage disease should be ruled out. Other possible differential diagnoses are leukemia, osteomyelitis, and Perthes disease when a patient manifests skeletal manifestations.

6.2.2 Biochemical Diagnosis

Gaucher disease is a lysosomal storage disorder caused by the deficiency of glucocerebrosidase (GBA) activity and consequent storage of gluco-

Fig. 6.2 Diagnostic flow for Gaucher disease Physicians suspect Gaucher disease based on clinical findings, and then first step examinations will be done. Definite diagnosis can be made by enzyme assay, and less than 10% value of normal enzyme activity leads the diagnosis of Gaucher disease. Mutation analysis of GBA gene is confirmative

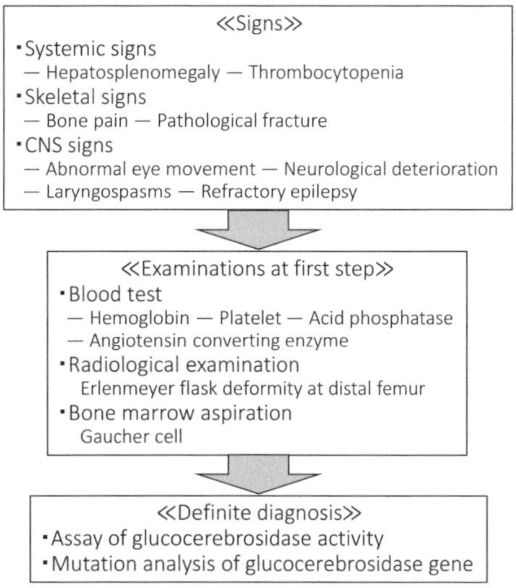

Physicians suspect Gaucher disease based on clinical findings, and then first step examinations will be done. Definite diagnosis can be made by enzyme assay and less than 10% value of normal enzyme activity leads the diagnosis of Gaucher disease. Mutation analysis of GBA gene is confirmative.

cerebroside in the cells of the monocyte/macrophage lineage (Figs. 6.3 and 6.4). Artificial, water-soluble substrates for the β-glucosidase enzyme are very useful in the diagnosis of Gaucher disease. 4-Methylumbelliferyl-β-glucoside (4MU-β-glucoside) can be used as a substrate to make the diagnosis of Gaucher disease. However, since the activity of the β-glucosidase in monocytes, lymphocytes, and granulocyte is different, it is better to separate these cells for enzyme assays. The enzyme activity in cultured skin fibroblasts is stable and reliable. There is the overlap of enzyme activity between the normal subjects and heterozygotes, and enzyme-based diagnosis is not useful in differentiating neuronopathic and non-neuronopathic Gaucher disease.

6.2.3 Pathological Diagnosis

The pathological characteristics of Gaucher disease are the presence in various tissues of lipid-engorged cells, referred to as a Gaucher cell, due to their hallmark appearance. A typical Gaucher cell contains one or more nuclei and cytoplasm with a striated, fibrillary, or tubular pattern, which is so-called wrinkled tissue paper or crumpled skin (Fig. 6.1). Gaucher cells are distributed systemically; however, they are mainly found in the spleen, sinusoids of the liver, bone marrow, and parenchyma of the lymph nodes. Their distribution and origin are important in the pathophysiology of Gaucher disease. However, very similar cells, pseudo-Gaucher cells, are found in other disorders such as chronic granulocytic leukemia, thalassemia, multiple myeloma, Hodgkin disease, and plasmacytoid lymphomas. Because biochemical and molecular diagnosis are specific and less invasive, pathologic diagnosis of Gaucher disease is supportive.

6.2.4 Molecular Diagnosis

There have been over 300 reported causative mutations in the GBA gene. The N370S mutation and c.84–85insG mutation are common in Jewish patient population (Fig. 6.5). Mutations N370S, L444P, c.84–85insG, and IVS2 + 1G → A account for more than 95% of the mutated alleles in Ashkenazi Jewish patients. The N370S mutation is a G to A transition at nucleotide 5841 of the GBA gene, substituting serine for asparagine. The c.84–85insG mutation is an insertion of an extra G at nucleotide 84 of the cDNA. Due to the insertion of an extra nucleotide, shift of the protein translation frame occurs, resulting in prema-

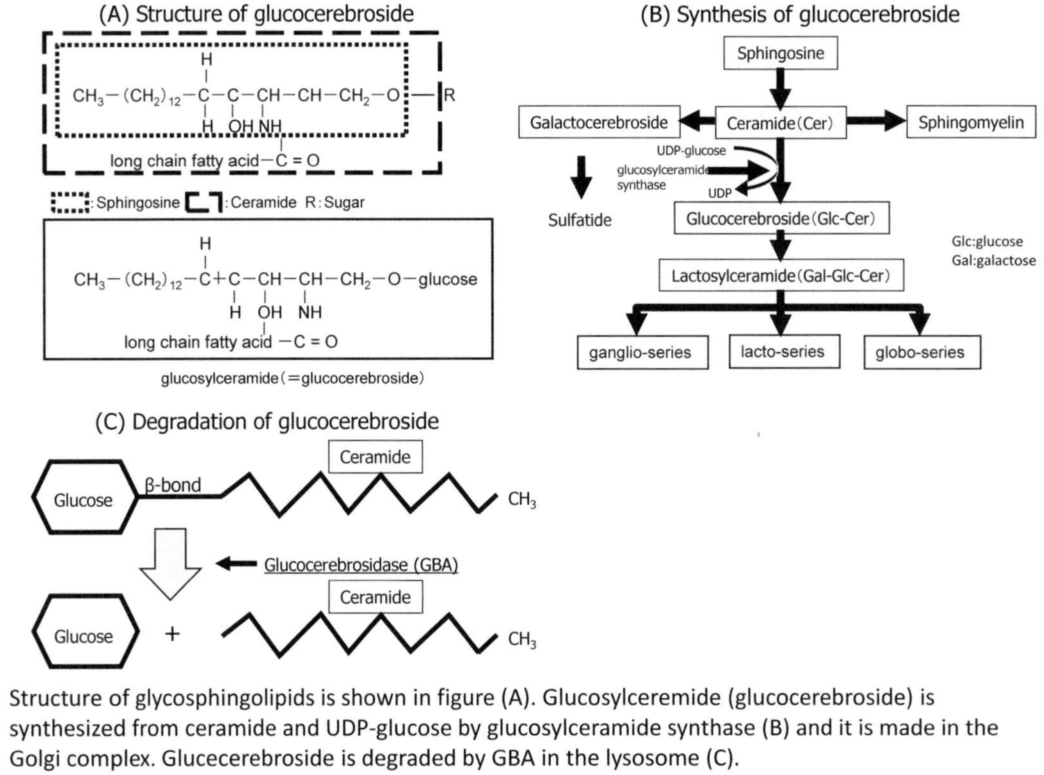

Fig. 6.3 Structure, synthesis, and degradation of glucocerebroside The structure of glycosphingolipids is shown in figure (**a**). Glucosylceramide (glucocerebroside) is synthesized from ceramide and UDP-glucose by glucosylceramide synthase (**b**), and it is made in the Golgi complex. Glucocerebroside is degraded by GBA in the lysosome (**c**)

Fig. 6.4 Pathophysiology of Gaucher disease GBA gene mutation causes the loss of enzyme activity, resulting in accumulation of glucocerebroside and glucosylsphingosine. These biochemical abnormalities lead the clinical manifestations

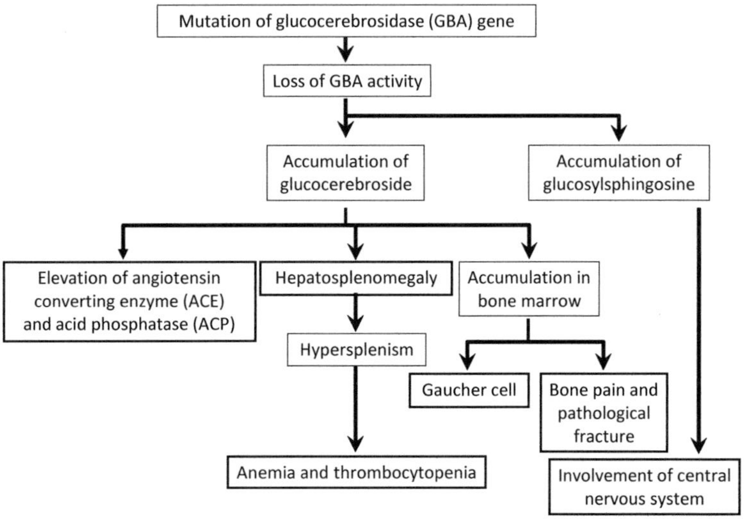

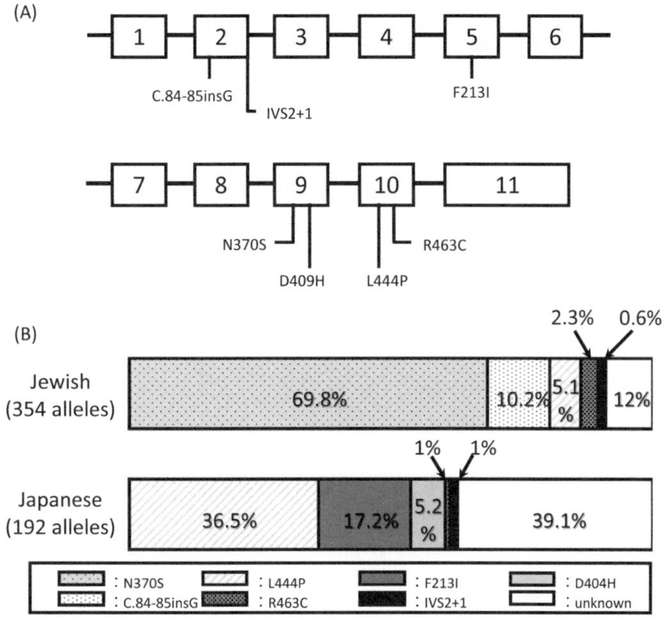

Fig. 6.5 Structure of GBA gene and location of common mutations (**a**) and mutation prevalence in Jewish and Japanese patients with Gaucher disease (**b**)

ture termination. If Gaucher disease is suspected clinically in a Jewish patient, molecular diagnosis is useful to confirm the diagnosis because of characteristic mutation prevalence described above. However, since the prevalence of mutations is different among various ethnicities, particularly in the Japanese population, a DNA-based diagnosis is less helpful (Fig. 6.5). Unlike an enzyme-based diagnosis, mutation analysis can distinguish a heterozygote from a normal individual and offers the possibility of predicting the clinical phenotype. The presence of the N370S mutation can exclude the possibility of neuronopathic Gaucher disease, and it is highly associated with type 1 patients with mild phenotypic expression. Homozygotes for the L444P mutation all have severe visceral involvement and usually have neurological manifestations.

6.2.5 Diagnosis of This Case

This case is an infant with neurological involvement, splenomegaly, thrombocytopenia, and elevation of ACP and ACE. Bone marrow aspiration was carried out, and characteristic Gaucher cells were found (Fig. 6.1). To confirm the diagnosis, an enzyme assay using cultured skin fibroblast was performed, and GBA activity was determined to be 56 nmol/hour/mg protein. This value was 5.1% of control fibroblasts, suggesting that the diagnosis of this patient is Gaucher disease. Mutation analysis using white blood cells was done, and his genotype was L444P/RecNciI (refer to the Molecular basis of GBA gene in the following section). A diagnostic flow chart is shown in Fig. 6.2. Based on the data described above, this patient can be diagnosed as type 2 Gaucher disease.

6.3 Biochemical and Molecular Perspectives

6.3.1 Biochemistry

Glucocerebroside is synthesized from ceramide and UDP-glucose by glucosylceramide synthase, and it is degraded to ceramide and glucose by GBA (Fig. 6.3). GBA is one of the lysosomal enzymes and is a glycoprotein. It has a molecular weight of about 65KDa. There are five consensus sequences for glycosylation sites. Terminal sugars are important to target the monocytes/macrophage in which glucocerebroside accumulates (see the enzyme replacement therapy (ERT) in the

Treatment section). GBA gene mutations that lead to the loss of enzyme activity result in the accumulation of glucocerebroside in the lysosomes of macrophages. This biochemical abnormality causes clinical manifestations such as hepatosplenomegaly, anemia, thrombocytopenia, and involvement of bone tissue. As a result of accumulation of glucosylsphingosine, a lysoderivative of glucocerebroside, patients with neuronopathic Gaucher disease manifest neurological symptoms. The pathophysiology of Gaucher disease progression is represented in Fig. 6.4.

6.3.2 Molecular Basis of GBA Gene

The functional GBA gene is about 7Kb in length and consists of 11 exons. The active site of the GBA enzyme is encoded by exons 9 and 10. Seven common mutations (c.84–85insG, IVS2 + 1G → A, F213I, N370S, D409H, L444P, R463C) are well known (Fig. 6.5). It has been reported that the mutation prevalence among various races is different. Mutations N370S, L444P, c.84–85insG, and IVS2 + 1G → A account for more than 95% of the mutated alleles in Ashkenazi Jewish patients, although they constitute less than 75% of the mutated alleles in non-Jews (Horowitz 1993). Among Japanese patients with Gaucher disease, neither mutation N370S nor c.84–85insG has not been identified (Ida et al. 1995) (Fig. 6.5). In the Portuguese Gaucher population, the N370S mutation accounts for 63% of the mutated alleles, and two other rare mutations, G377S and N396 T, are common (Amaral et al. 1999).

Interestingly, a nearly identical pseudogene is located 16 kb downstream from the functional gene. Sequencing of cDNA clones from patients with Gaucher disease showed complex alleles with several mutations. One complex allele had three point mutations, L444P, A456P, and V460 V, and has been designated as RecNciI. This complex allele could have been generated by a phenomenon where the pseudogene causes the transfer of mutations into the active gene via unequal, homologous recombination or gene conversion. The RecNciI allele is always associated with moderate to severe disease. The patients carrying L444P/RecNciI genotype always present with type 2 disease.

6.3.3 Genotype/Phenotype Correlation

The N370S mutation is highly linked with non-neuronopathic Gaucher disease (type 1) and has a very mild phenotypic expression. The existence of the N370S mutation implies that the patient has type 1 Gaucher disease. In addition, nearly 90% of the patients, carrying the N370S/N370S genotype, have a mild clinical course of the disease, and many of them are asymptomatic (Sibille et al. 1993). GBA protein carrying the N370S mutation is stable, has the same affinity toward the artificial substrate (4MU-β-glucoside) as the normal enzyme, but is not stable enough to catalyze its hydrolysis normally. Based on the characteristics of mutation prevalence in the Japanese population, incidence of patients with neuronopathic Gaucher disease is higher than that of Jewish patients, and Japanese type 1 disease tends to be more severe and progressive than the Jewish patients.

As shown in Fig. 6.6, the majority of Jewish patients are type 1. In contrast, approximately 50% of Japanese patients are classified into neuronopathic Gaucher disease (Tajima et al. 2009). Over 60% of Japanese patients have onset at less than 5 years of age. Among Japanese patients, 64% are splenectomized, and another 46% develop severe bone involvement. Over mean interval ranging from 4.9 to 6.9 years, mean relative height and weight, severity score index, and platelet counts all worsen to a highly significant degree (Ida et al. 1998) (Table 6.2).

6.4 Treatment

6.4.1 Enzyme Replacement Therapy (ERT)

Since ERT is safe and effective, it is the common treatment strategy for Gaucher disease and

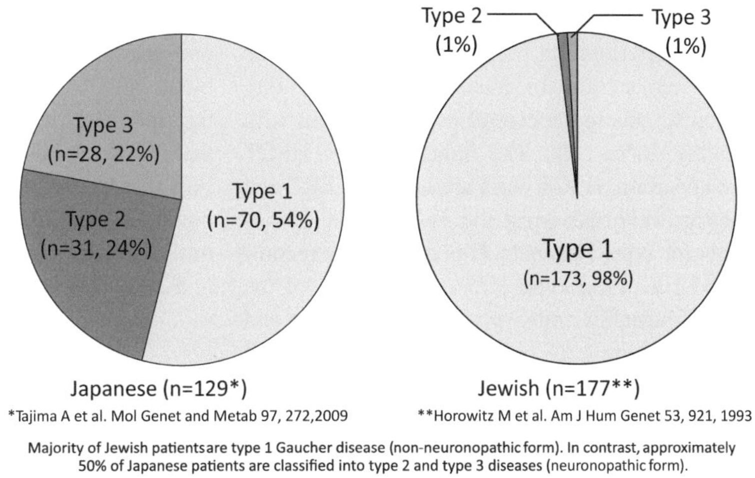

Fig. 6.6 Prevalence of clinical phenotypes in Gaucher disease Majority of Jewish patients are type 1 Gaucher disease (non-neuronopathic form). In contrast, approximately 50% of Japanese patients are classified into type 2 and type 3 diseases (neuronopathic form)

Table 6.2 Natural history of Japanese Gaucher disease type 1

	At baseline	At evaluation	Follow-up (year)	p-value#
Hemoglobin value (g/dl)	9.9±2.5	9.4±1.6	4.9	*
Platelet count (×10^4/mm^3)	11.5±7.9	7.7±5.5	4.9	**
Severity score index (SSI)	8.1±2.7	12.0±4.5	6.9	***
Body weight (SD)	−0.7±1.1	−1.2±0.8	5.0	**
Height (SD)	−1.9±1.0	−2.7±1.3	5.1	**

The values are expressed as mean values±SD, respectively
#In statistical analysis, paired t-tests were used to analyze changes in relative physical growth, and Wilcoxon signed-rank tests, to analyze changes in SSI and hematological measurements
*:p>0.05, **:p<0.01, ***:p<0.001

widely used. By reduction of the amount of accumulated glucocerebroside, hepatosplenomegaly, hematological abnormalities, and bone manifestations improve. The therapeutic enzyme is produced by DNA recombinant techniques using Chinese hamster ovary cells or genetically engineered human cultured skin fibroblasts. The terminal sugar on the glycosyl chains added on to the enzyme at the sites of glycosylation are modified to high-mannose (two N-acetylglucosamines with multiple mannose residues) molecule to target the enzyme to macrophages, via the macrophage mannose receptor. Enzyme therapy is administered by drip infusion every 2 weeks. Exogenous enzyme incorporates into monocytes/macrophages and is subsequently transported to lysosomes, ultimately degrading accumulated glucocerebroside.

One report of ERT between 2 and 5 years of treatment in 1028 adult patients with type 1 Gaucher disease demonstrated that ERT was effective for hematologic, visceral, and skeletal involvement (Weinreb et al. 2002). In another study, within 8 years of initiating ERT for 884 children with Gaucher disease type 1, most clinical parameters became normal or nearly normal (Andersson et al. 2008). However, neurological involvement does not respond to ERT since the administered enzyme does not cross the blood-brain barrier.

6.4.2 Human Stem Cell Transplantation (HSCT)

HSCT has the goal of replacing glucocerebrosidase activity in Gaucher patients. This is a consequence of the transplanted monocyte/macrophage system from donor cells expressing normal glucocerebrosidase. A small group of patients with type 3 disease (Norrbottnian type) has been successfully treated by HSCT (Ringden et al. 1995).

Interestingly, these patients had no further neurological deterioration posttreatment, perhaps due to the replacement of macrophages surrounding blood vessels in the central nervous system with healthy donor cells. The indications for HSCT are uncertain. HSCT costs are less than ERT and may prevent and arrest the neurological symptoms for type 3 patients. It is difficult to recommend HSCT for type 1 disease because of the 10% mortality rate posttransplantation and adverse long-term effects on growth and development.

6.4.3 Pharmacological Chaperone Therapy (PCT)

PCT is a new treatment strategy for Gaucher disease. The basic mechanism of PCT is to correct the folding and stabilize mutated GBA proteins, resulting in an increase in enzyme activity as well as increased concentration of enzyme (Parenti 2009). The mechanism of PCT is shown in Fig. 6.7. Orally administered chaperones enter the cell and bind to less stable, misfolded enzyme. The stabilized enzyme becomes properly folded and enters the lysosomes, where it continues to break down substrate. Since pharmacological chaperones are small molecules and designed to cross the blood-brain barrier, it is one of the candidates for the treatment of neuronopathic Gaucher disease that is not responsive to ERT. On the other hand, PCT is effective only for patients carrying specific mutations that affect enzyme stability. A clinical trial of PCT for neuronopathic Gaucher disease is underway (Narita et al. 2016).

6.4.4 Substrate Reduction Therapy (SRT)

SRT is an approach involving partial inhibition of glucosylceramide synthase, the enzyme that transfers glucose to ceramide to form glucosylceramide (glucocerebroside), to more evenly balance the rate of synthesis with the impaired rate of catabolism (Fig. 6.3). Since the SRT is an oral drug, patients do not have to visit a hospital every 2 weeks for administration by drip infusion. SRT for untreated adults with Gaucher disease type 1 resulted in significant improvement in spleen and liver volume, hemoglobin level, and platelet count. In addition, SRT maintained hematological and organ volume stability in adults with type

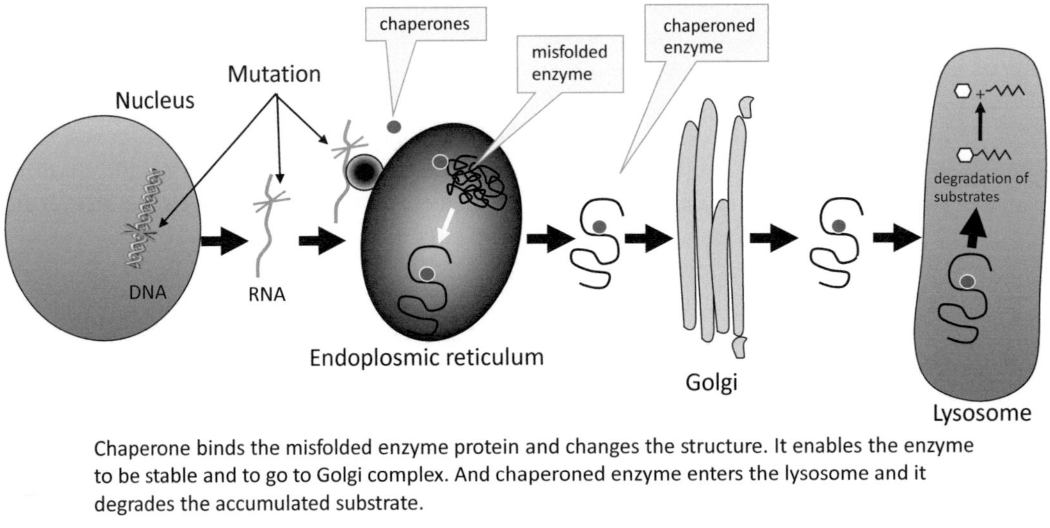

Fig. 6.7 Proposed mechanism of pharmacological chaperone therapy (PCT)
Chaperone binds the misfolded enzyme protein and changes the structure. It enables the enzyme to be stable and to go to Golgi complex. And the chaperoned enzyme enters the lysosome, and it degrades the accumulated substrate

1 Gaucher disease already controlled by ERT and could be useful as maintenance treatment (Cox et al. 2015). Major problems of SRT are adverse gastrointestinal effects and interactions with medications related to CYP2D6 metabolism (e.g., classes 1 and 3 antiarrhythmic drugs, selective serotonin reuptake inhibitors, tricyclic antidepressants, and so on).

End of Chapter Questions

1. Explain pathophysiology of Gaucher disease.
2. Explain the clinical manifestations of Gaucher disease.
3. Explain the treatment for Gaucher disease. And explain advantage and disadvantage of each treatment.

References

Amaral O, Lacerda L, Marcao A, Pinto E, Tamagnini G, Miranda MCS (1999) Homozygosity for two mild glucocerebrosidase mutations of probable Iberian origin. Clin Genet 56:100–102

Andersson H, Kaplan P, Kacena K, Yee J (2008) Eight-year clinical outcomes of long-term enzyme replacement therapy for 884 children with Gaucher disease type 1. Pediatrics 122:1182–1190

Cox TM, Drelichman G, Cravo R, Balwani M, Burrow TA, Martins AM, Lukina E, Rosenbloom B, Ross L, Angell J, Puga AC (2015) Eliglustat compared with imiglucerase in patients with Gaucher's disease type 1 stabilized on enzyme replacement therapy: a phase 3, randomized, open-label, non-inferiority trial. Lancet 385:2355–2362

Horowitz M, Tzuri G, Eyal N, Berebi A, Kolodony EH, Brady RO, Barton NW, Abrahamov A, Zimran A (1993) Prevalence of nine mutations among Jewish and non-Jewish Gaucher disease patients. Am J Hum Genet 53:921–930

Ida H, Iwasawa K, Kawame H, Rennert OM, Maekawa K, Eto Y (1995) Characteristics of gene mutations among 32 unrelated Japanese Gaucher disease patients: absence of the common Jewish 84GG and 1226G mutations. Hum Genet 95:717–720

Ida H, Rennert OM, Ito T, Maekawa K, Eto Y (1998) Type 1 Gaucher disease: phenotypic expression and natural history in Japanese patients. Blood Cells, Mol Dis 24:73–81

Narita A, Higaki K, Maegaki Y, Ohno K, Suzuki Y (2016) Ambroxol chaperone therapy for neuronopathic Gaucher disease: a pilot study. Ann Clin Translat Neurol 3:200–215

Parenti G (2009) Treating lysosomal storage diseases with pharmacological chaperones: from concept to clinics. EMBO Mol Med 1:268–279

Ringden O, Groth CG, Erikson A, Granqvist S, Mansson JE, Sparrelid E (1995) Ten years's experience of bone marrow transplantation for Gaucher disease. Transplantation 59:864–870

Sibille A, Eng CM, Kim SJ, Pastores G, Grabowski GA (1993) Phenotype/genotype correlations in Gaucher disease type 1: clinical and therapeutic implications. Am J Hum Genet 52:1094–1101

Tajima A, Yokoi T, Ariga M, Ito T, Kaneshiro E, Eto Y, Ida H (2009) Clinical and genetic study of Japanese patients with type 3 Gaucher disease. Mol Genet Metab 97:272–277

Weinreb NJ, Charrow J, Andersson HC, Laplan P, Kolodony EH, Mistry P, Pastores G, Rosenbloom BE, Scott CR, Wappner RS, Zimran A (2002) Effectiveness of ERT in 1028 patients with type 1 Gaucher disease after 2 to 5 years treatment. Am J Med 113:112–119

Heme Oxygenase-1 Deficiency

7

Akihiro Yachie

Keywords
Heme oxygenase-1 · Oxidative stress · Asplenia · Hemolytic anemia · Endothelial injury

7.1 Case Report

A 26-month-old boy was admitted because of recurrent fever, generalized erythematous rash, and joint pain. The fever persisted, and the joint pain waxed and waned. The combination of fever, erythematous rash, and joint pain suggested that the patient suffered from a childhood chronic inflammatory illness such as systemic juvenile idiopathic arthritis (sJIA) or chronic infantile neurological cutaneous and articular syndrome (CINCA, alternatively called neonatal-onset multisystem inflammatory disease or NOMID). However, neither of the conditions could explain the unique clinical features and the extraordinary laboratory findings seen in this patient. The patient had characteristic facial features such as flat nasal bridge, frontal bossing, and prominent edema of the eyelids. Marked hepatomegaly was noted, but the spleen was absent. Unlike most of patients with asplenia, this patient did not have any form of congenital heart disease. His brother and sister were healthy. The mother had experienced two intrauterine fetal deaths.

The white cell count was 51,600/μL (normal range; 3300 to 8800), and the platelet count was 226 x 10^4/μL (normal range; 13–35). The marked leukocytosis and thrombocytosis both persisted throughout the course of the illness. He had significant anemia with erythrocyte count 1.48 x 10^6/μL (normal range; 4.30–5.50) and hemoglobin concentration 4.9 g/dL (normal range; 13.5–17.0). Peripheral blood smear showed numerous fragmented erythrocytes and erythroblasts (Fig. 7.1). Serum iron concentration was 64 μg/dL (normal range; 50–170). Lactate dehydrogenase (LDH) was 17,470 IU/L (normal range; 196–355), and aspartate aminotransferase (AST) was 442 IU/L (normal range; 9–42), but alanine aminotransferase (ALT) was within normal limits. Serum ferritin was 780 ng/mL (normal range; 26–280). Marked abnormalities were noted in parameters of both coagulation and fibrinolysis system, although he did not show apparent bleeding tendency or signs of accelerated coagulation. Fibrinogen was 109 mg/dL (normal range; 196–356), fibrin degradation product (FDP) was 300.1 μg/mL (normal range; <5), d-dimer was 186.1 μg/mL (normal range; <2.5), thrombin-

A. Yachie (✉)
Department of Pediatrics, School of Medicine, Graduate School of Medical Sciences, Kanazawa University, Kanazawa, Japan
e-mail: yachie@staff.kanazawa-u.ac.jp

T. Oohashi et al. (eds.), *Human Pathobiochemistry*, https://doi.org/10.1007/978-981-13-2977-7_7

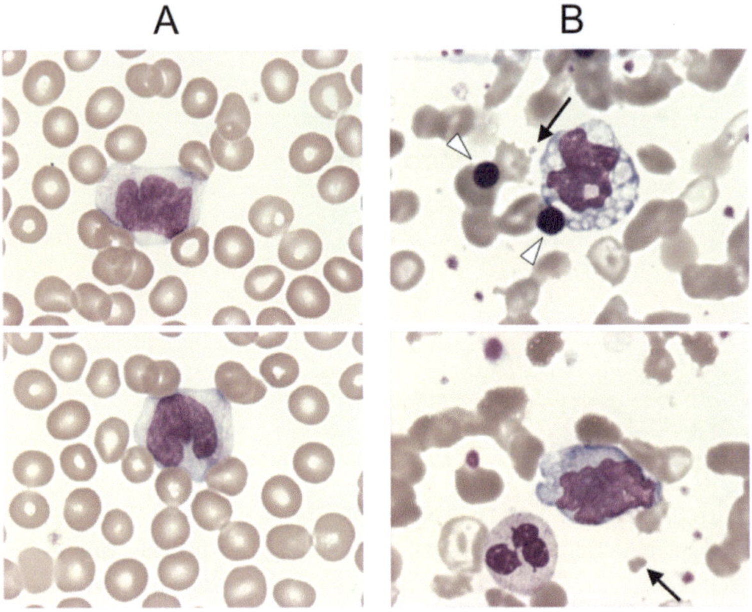

Fig. 7.1 May-Grünwald Giemsa staining of the control (**a**) and the patient (**b**) peripheral blood. Normal control smear shows intact erythrocytes and normal appearing monocytes, whereas the patient's smear shows numerous fragmented erythrocytes (arrows), erythroblasts (arrow heads), and monocytes with numerous vacuoles and basophilic cytoplasm

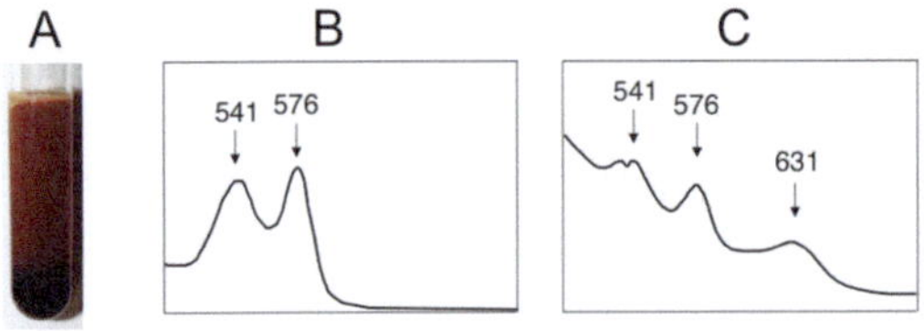

Fig. 7.2 Appearance of the centrifuged patient blood was highly specific. The hematocrit was significantly low, and the serum was turbid with dark brownish tint (**a**). Spectrophotometer analysis of the fresh hemolysate of the patient erythrocytes showed two distinct peaks of OxyHb at 541 nm and 576 nm (**b**). The serum analysis showed an additional peak of MetHb at 631 nm (**c**)

antithrombin complex was 202.2 μg/L (normal range; <3), and plasmin-α_2 plasmin inhibitor complex was 22.3 μg/mL (normal range; <0.8). Thrombomodulin was 12 ng/mL (normal range; <3.5) and von Willebrand factor was 580% (normal range; 60–170). Hyperlipidemia was another prominent finding, with triglyceride at 638 mg/dL (normal range; 32–115) and total cholesterol at 552 mg/dL (normal range; 32–115) with predominance of low-density lipoprotein cholesterol.

The appearance of his serum was peculiar. Freshly separated serum was always turbid with brownish tint (Fig. 7.2a). Hemolysate of the patient's erythrocytes showed two distinct peaks of oxyhemoglobin (OxyHb) at 541 nm and 576 nm (Fig. 7.2b). In addition to the two peaks of OxyHb, the patient's serum showed a third unique peak at 631 nm, which corresponded to methemoglobin (MetHb) (Fig. 7.2c). Both gross appearance of the serum and the absorption spectrum analysis suggested that heme in the form of OxyHb and MetHb was markedly increased in the patient's serum. These results indicated that either massive hemolysis is constantly taking place in vivo or that Hb accumulated in the serum due to defects in Hb catabolic pathway. Haptoglobin (Hp) concentration was extremely elevated at 800–1200 mg/dL (normal range; 19–170), and a large amount of the Hb-Hp complex was detected in the patient's urine sample. Repeated measurement of serum bilirubin concentration was always low at 0.1–0.3 mg/dL (normal range; 0.2–1.3). Hemopexin was undetectable by immunoelectrophoresis. Serum heme concentration was extremely high at 490 μM (normal range; <1 μM). Both direct and indirect Coombs' tests were negative on repeated occasions. Putting all these data together, it was highly suggested that there is a certain abnormality in the process of hepatic uptake of Hb-Hp complex or in heme degradation pathway (Fig. 7.3).

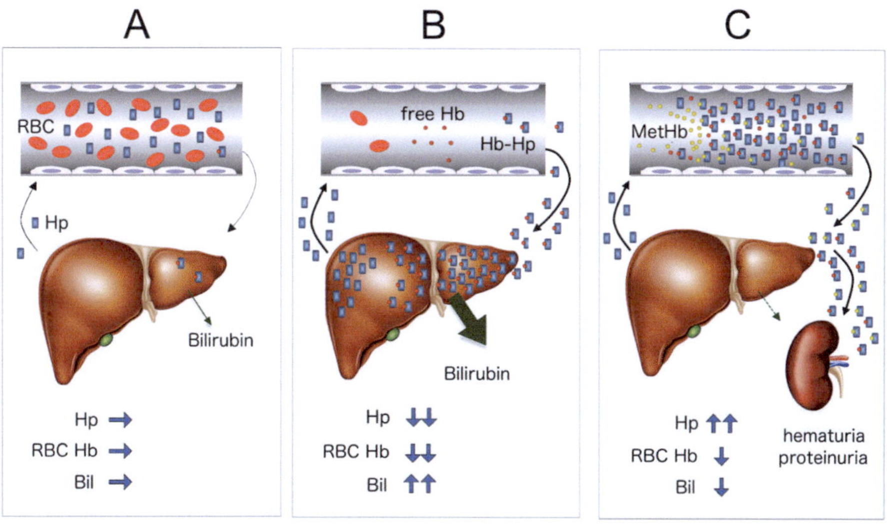

Fig. 7.3 Hepatic uptake of Hb-Hp complex and heme degradation into bilirubin. In normal condition, small amount of Hb derived from hemolysis is rapidly bound by circulating Hp, forming Hb-Hp complex, which is rapidly uptaken by the liver through specific receptors (**a**). When massive hemolysis occurs, as in the case of hemolytic anemia, large amount of free Hb is bound by serum Hp and transferred to the liver (**b**). The results are significant reduction of serum Hp and increased of indirect bilirubin derived from the heme degradation pathway. In the HO-1-deficient patient, there was accumulation of massive levels of both OxyHb and MetHb bound to Hp within the serum. Despite the apparent intravascular hemolysis, total Hp content was significantly increased, serum bilirubin level remained low, and there was overflow of Hb-Hp complex within the urine (**c**)

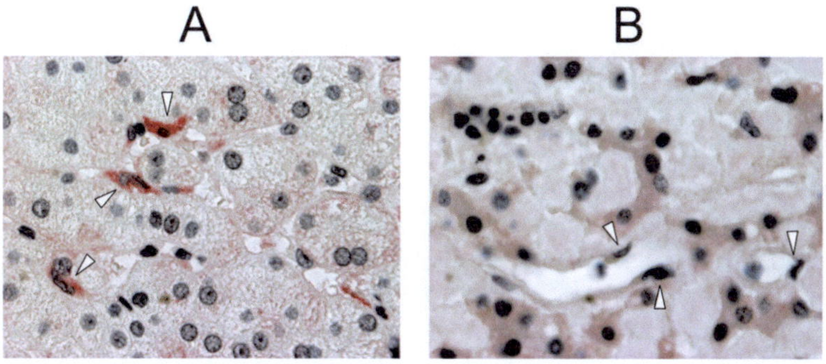

Fig. 7.4 HO-1 production by hepatic Kupffer cells. HO-1 immunostaining was performed using anti-HO-1 rabbit antiserum and alkaline phosphatase-conjugated goat anti-rabbit antibody. Alkaline phosphatase activity was visualized using Fast Red TR salt. Kupffer cells of the control liver produced significant levels of HO-1 (A, arrow heads), whereas HO-1 was not detectable within Kupffer cells of the patient liver (B, arrow heads)

Heme oxygenase (HO) is an enzyme which plays a key role in heme degradation to biliverdin. Three isoforms of HO are known, including HO-1, HO-2, and HO-3 (Maines 1988). Among these isoforms, HO-1 is peculiar in that it is rapidly induced in response to various oxidative stresses (Nath 2006). In contrast, HO-2 is a constitutive isoform that is expressed under homeostatic conditions. HO-3 is not catalytically active, but is thought to work in oxygen sensing. Immunohistochemical analysis of the liver biopsy specimen showed that Kupffer cells did not produce HO-1 in the patient's liver (Fig. 7.4). Exposure of Epstein-Barr virus-transformed lymphoblastoid cell line (LCL) derived from the patient to oxidative stress such as hemin, cad-

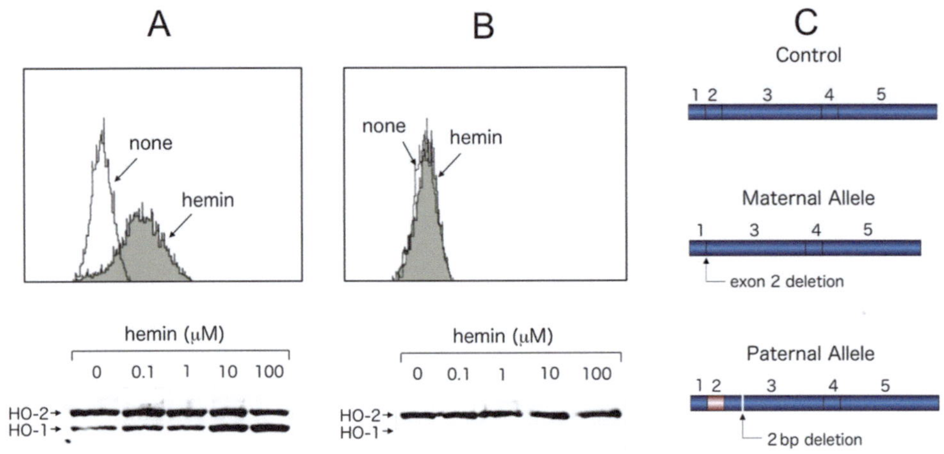

Fig. 7.5 Hemin-induced HO-1 production by LCL. LCL from control and the patient were stimulated with hemin and HO-1 production was compared by flow cytometry and immunoblotting. Large amount of HO-1 was produced dose-dependently upon hemin exposure of normal control LCL (**a**), whereas it was not induced in the patient LCL (**b**). HO-2 was constitutively produced in both control and the patient LCLs. Mutation analysis of the patient and the parents revealed that the patient had compound heterozygote mutations of HO-1 gene. Exon 2 deletion in the maternal allele and 2 bp deletion within exon 3 of the paternal allele were detected (**c**)

mium, or sodium arsenite did not induce HO-1 protein (Fig. 7.5a and b). HO-2 was constitutively expressed in both controls and the patient LCL. HO-1 gene analysis revealed that the patient had compound heterozygotes of HO-1 gene mutations (Yachie et al. 1999). The maternal allele lacked the second exon, and the paternal allele showed two base pair deletions within the third exon (Fig. 7.5c).

Pathological examination of the first case of HO-1 deficiency revealed characteristic tissue injury. Notably, cellular injury was confined to selected organs and cell types, including the kidney, the liver, circulating monocytes, and vascular endothelial cells. In the kidney, mild mesangial proliferation and thickening of the capillary loop were observed within the glomeruli (Fig. 7.6a). Electron microscopy revealed marked swelling of the endothelial cells and their detachment throughout the glomerular capillary (Fig. 7.6b). In addition to the glomerular damage, tubulointerstitial injury with tubular atrophy was significant. The liver was massively enlarged, and there was a significant amyloid accumulation resulting in marked atrophy of hepatocytes (Fig. 7.6c, d). Scattered foci of iron deposits were observed in both the kidney and the liver (Kawashima et al. 2002). Cytoplasm of the circulating monocytes was vacuolated, and monocyte surface antigens were significantly different from normal profiles, as described later.

After we reported the first case of human HO-1 deficiency, only five additional cases have been confirmed by HO-1 gene analysis (Radhakrishnan 2011a, b, and personal communication). Fever, absence of the spleen, hemolytic anemia, hematuria/proteinuria, and the absence of jaundice seem to be the common denominators of the disease. The common clinical and laboratory features are summarized in Tables 7.1 and 7.2, respectively. All cases from India (from the second to the sixth cases) had identical homozygous mutation. However, the disease onset and durations vary significantly depending on the case. Laboratory findings were characterized by marked increase of platelet. Levels of hepatic enzymes, such as LDH, AST, and ALT, were significantly elevated. Low to normal serum bilirubin and high Hp concentration in the presence of hemolysis seems to be the hallmark of the illness.

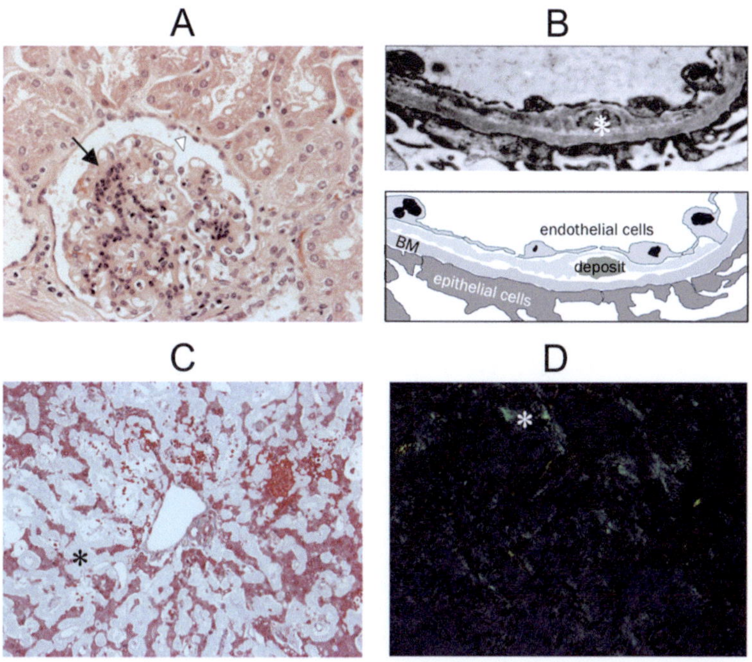

Fig. 7.6 Renal and hepatic pathology of the HO-1-deficient patient. Hematoxylin and eosin stain of the renal biopsy specimen showed mild mesangial proliferation and thickening of the capillary loop were observed within the glomeruli (**a**). Electron microscopy revealed marked swelling of the endothelial cells and their detachment throughout the glomerular capillary (**b**). The liver was massively enlarged, and there was a significant amyloid accumulation. Azan staining shows marked atrophy of hepatocytes, but fibrosis was minimum (**c**). Congo red stain shows apple-green birefringence under polarized light (**d**)

7.2 Diagnosis

Because only limited numbers of patients with HO-1 deficiency (OMIM#614034) are known, there are no established diagnostic criteria for this very rare disease. Combinations of (1) absence or hypoplasia of the spleen, (2) extremely elevated LDH and ferritin, (3) leukocytosis and thrombocytosis, and (4) hemolytic anemia without jaundice are highly suggestive of this disease. Although the age of onset among the six HO-1 deficiency cases varied, ranging from infancy to 15 years of age, laboratory data and clinical profiles were surprisingly uniform. Fever, hemolytic anemia, and hematuria/proteinuria were consistent findings. In all cases, bilirubin remained within normal range, while serum ferritin and LDH values were invariably high. Although the absence or hypoplasia of the spleen seems to be the hallmark of the illness, its functional significance in patients with HO-1 deficiency has not been determined.

Differential diagnosis includes various childhood inflammatory illnesses including sJIA and other autoinflammatory diseases such as CINCA or NOMID.

7.3 Biochemical Perspectives or Molecular Perspectives

Heme is a major component of Hb, the product of erythrocyte destruction. Heme is constantly produced in vivo and it is extremely toxic to cells (Balla et al. 2007). Therefore, constitutive mechanisms exist to cancel the toxic effect of heme. Serum Hp binds free Hb efficiently, and the Hb-Hp complex is promptly taken up by phagocytes and hepatocytes which express the recep-

Table 7.1 Clinical characteristics of 6 cases

Findings	1st case (our patient)	2nd case	3nd case	4th case	5th case	6th case
Sex	Male	Female	Male	Male	Female	Male
Onset/diagnosis	2 year/5 year	15 year/16 year	6 month/2 year	10 year/10 year	7 year/9 year	13 month/20 month
Death at	6 year	16 year	3 year	Alive	Alive	2 year
Fever	+	+	+	+	+	+
Arthralgia	+	–	–	–	–	–
Hemolytic anemia	+++	++	++	++	++	++
Jaundice	–	–	–	–	–	–
Hematuria/proteinuria	+/+	+/+	+/+	+/+	+/+	+/+
Asplenia	+	+	+	+	+	+
Prominent forehead	+	+	+	+	+	+
Hypertension	+	+	+	+	+	–
Cerebral bleeding	+	+	+	–	–	–
Growth delay	+	–	+	+	+	±

Table 7.2 Mutations and laboratory data

Findings	1st case (our patient)	2nd case	3rd case	4th case	5th case	6th case
Mutation (s)	exon2 del	R44X	R44X	R44X	R44X	R44X
HO-1 gene	2bp del (exon3)	(homo)	(homo)	(homo)	(homo)	(homo)
Mother	exon2 del	Not done	Not done	R44X/ wild	R44X/wild	R44X/ wild
Father	exon3;2bp del	Not done	Not done	R44X/ wild	R44X/wild	R44X/ wild
CRP (mg/dL; <0.4)	6.7	30.8	5.3	Normal	24.0	4.8
FDP-DD (mg/mL; <2.5)	186.1	>8	Not done	Normal	Not done	Not done
WBC ($\times 10^3/\mu$L; 3.3–8.8)	51.6	18.5	38.0	39.6	Not available	43.2
Plt ($\times 10^4/\mu$L; 13–35)	226	137	109	117	100	123
Ferritin (ng/mL; <280)	780	4912	15,530	15,358	2500	>2000
LDH (IU/L; <229)	17,470	9462	12,858	16,000	4000	21,400
AST/ALT (IU/L; <33/27)	448/74	982/149	1080/283	689/68	300/80	652/133
Bilirubin (mg/dL; <1.3)	0.2	0.64	1.2	0.3	0.4	Low

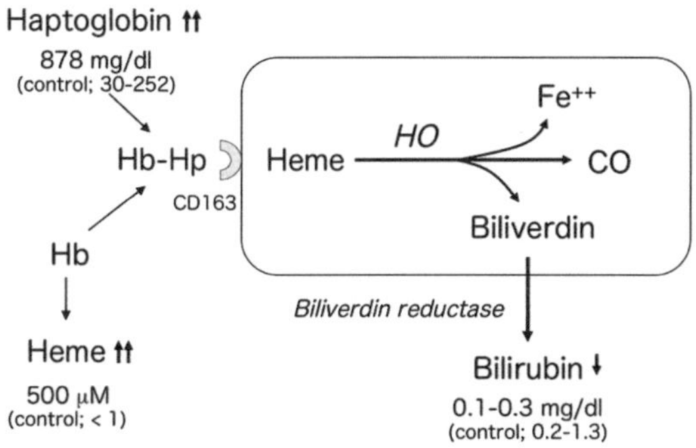

Fig. 7.7 Catabolic pathway of heme within the liver. Hb-Hp complex is uptaken by hepatocytes via its receptor, CD163. Heme is then catabolized to Fe^{++}, CO, and biliverdin by HO. All three molecules are known to exert potent antioxidative action

tors for the complex, now known as CD163 (Madsen et al. 2004).

Heme catabolic pathway within the liver is shown in Fig. 7.7. Of particular importance is the fact that HO-1 not only plays role as a catabolic enzyme of heme degradation pathway; it also induces molecules, including carbon monoxide (CO), which exerts potent antioxidative functions (Nakahira et al. 2006). HO-1 deficiency will lead to significant reduction of these molecules and subsequent cellular and tissue injury. However, there are numerous questions to be answered regarding the pathogenesis and the biochemical and molecular perspectives of HO-1 deficiency. Some of these questions are shown below.

7.3.1 Question 1: Why Only Certain Organs and Cell Types Are Damaged in HO-1 Deficiency?

Of particular interest is the fact that only selected cells or organs are damaged in HO-1 deficiency. Several mechanisms may explain this. First of all, these susceptible cells seem to be the targets of constant oxidative stress exposure, both for anatomical and functional reasons. Vascular endothelial cells are the target of shear stress and are exposed to multiple oxidative stresses, including hemolysis, pH changes, and hypoxemia for obvious reasons. In the HO-1-deficient patient, secondary accumulation of heme proteins, cholesterol, and fragmented erythrocytes further aggravates the oxidative stresses. Renal tubular cells are constantly exposed to hematuria, proteinuria, and various other excretory substances in urine. Tissue macrophages and circulating monocytes are frequently turned on to exert their scavenger functions. In addition, they function as one of the central players of innate and acquired immunity. Second, while these cells may be particularly sensitive to oxidative injury, they may serve as a high-quality sensor of oxidative stress for the susceptible organs. Upon exposure to oxi-

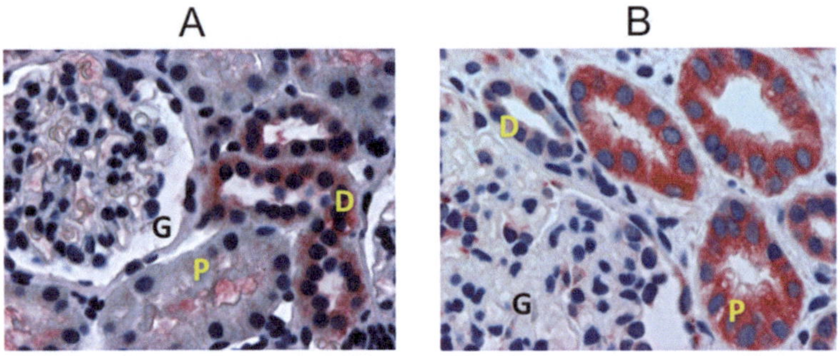

Fig. 7.8 HO-1 production by renal tubular epithelium. In control kidney (**a**), HO-1 is not detectable within the glomerulus (g) or proximal tubular epithelial cells (p). In contrast, distal tubular epithelial cells (d) constantly express significant level of HO-1. In patients with hemolytic uremic syndrome (**b**), hematuria induced high levels of HO-1 within proximal tubular epithelial cells (p), whereas glomerulus (g) produced little HO-1, and distal tubular epithelial cells (d) produced only modest level of HO-1

dative insults, these sensor cells rapidly induce antioxidative molecules. As a result, the organs are saved from critical damage. HO-1 may participate as a central molecule to neutralize the oxidative stress.

In normal rat and human kidneys, renal tubular epithelial cells express HO-1 selectively. Distal tubules express significantly more HO-1 than the proximal tubules in a normal kidney (Fig. 7.8a). However, significant level of HO-1 expression is induced within proximal tubules in patients with hematuria or proteinuria (Fig. 7.8b). Proximal tubular epithelial cells also produce significant levels of HO-1 upon various oxidative stimuli in vitro, indicating that these cells are inherently capable of responding to exogenous noxious insults (Yang et al. 2003).

7.3.2 Question 2: How Does the Hypoplasia of the Spleen Occur and What Is Its Functional Significance?

We do not know exactly why all the patients with HO-1 deficiency lacked functional spleen. One hypothesis is that HO-1 deficiency and subsequent vascular endothelial injury results in early (most likely in utero) vascular damage involving the splenic artery. Due to accumulating injured blood cells, the patients' spleens are overworked. The continuing splenic overwork may lead to the enhanced oxidative damage to the splenic vasculatures which are already vulnerable to oxidative stress. It is intriguing in this respect that hypomorphic mutation of HO-1 in mice resulted in the progressive atrophy of the spleen due to fibrotic changes in the splenic artery (Kovtunovych et al. 2010). These assumptions will remain as such until cases of HO-1 deficiency with intact splenic function are discovered.

Regardless of the mechanism leading to splenic hypoplasia, splenic dysfunction helps the survival of damaged, but still functioning, blood cells and prevents the fatal outcome. In contrast, the presence of functionally intact spleen may accelerate the destruction of injured blood cells thereby leading to fatal hemolysis and thrombocytopenia.

7.3.3 Question 3: What Is the Mechanism of Accelerated Cell Injury Seen in HO-1 Deficiency?

Direct consequences of HO-1 deficiency include extensive cell injury due to lack of the enzyme and induction of cell dysfunction, in particular, scavenger functions of macrophages. As we cannot extrapolate that these are the universal features observed in every HO-1-deficient patient, we analyzed the functional significance of HO-1

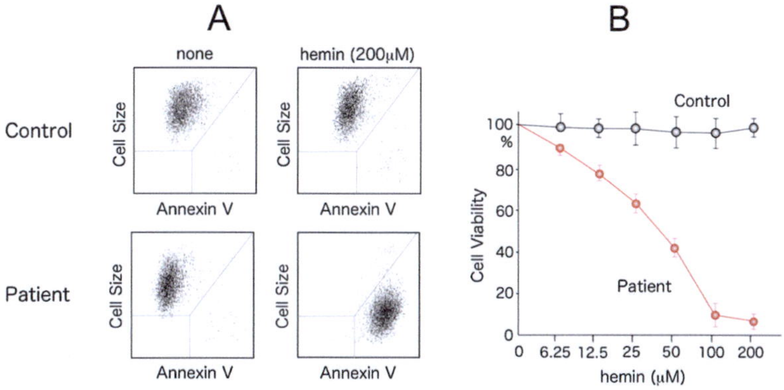

Fig. 7.9 Hemin-induced cell injury in HO-1 deficiency. LCLs from control and the patients were induced with different concentrations of heme. Apoptosis or injury of LCL was evaluated by morphological examination and flow cytometry. Apoptotic cells are identified by low cell size and high annexin V binding. Few control LCL, but most of the patient LCL underwent apoptosis with 200 μM hemin (**a**). Hemin induced rapid apoptosis or cell injury in HO-1-deficient patient in a dose-dependent fashion, whereas the cells remained intact in controls (**b**)

deficiency using the LCL derived from the patient.

LCL of the HO-1-deficient patient (HO-1-deficient LCL) was extremely sensitive to hemin-induced cellular injury (Fig.7.9). Extreme sensitivity of HO-1-deficient LCL to hemin stimulation could not be reversed by the addition of apoferritin or bilirubin. Furthermore, ferritin production by HO-1-deficient LCL was comparable with that by control LCL with or without the addition of hemin. Although ferritin and bilirubin may act as antioxidant in certain situations, they do not contribute much to the protection of cells from hemin-induced cell injury, at high hemin concentrations. LCL transfected with HO-1 gene was significantly less sensitive to hemin-induced cellular injury (Yachie et al. 1999). These results support the notion that degradation of heme by HO-1 is directly responsible for the reversal of cellular injury.

7.3.4 Question 4: What Are the Functional Defects of Macrophages?

In addition to cellular injury to the resident macrophages, the patient exhibited peculiar findings indicating a disturbance of macrophage scavenger function. Asplenia was certainly a significant contribution to the reduced scavenging function in the patient, resulting in the overload by other reticuloendothelial systems, including circulating macrophages and the hepatic Kupffer cells. The patient's peripheral blood monocytes exhibited some morphological characteristics, including prominent vacuolation and basophilic cytoplasm (Fig. 7.1b). These changes may reflect a persistent, systemic inflammatory reaction. In addition to morphological changes, the surface antigen expression by these monocytes was abnormal, i.e., expressions of HLA-DR and CD36 were significantly reduced as compared with normal monocytes (Yachie et al. 2002). CD14 expression was comparable to that in the control. The reduction in the antigen expression was constantly observed after repeated examinations, indicating that these changes reflect abnormal monocyte functions in the patient.

Consistent with the changes in the surface molecules, monocyte phagocytic functions were also impaired dramatically. Although phagocytosis of fluorescence-labeled latex beads did not change significantly, that of opsonized erythrocytes was almost completely abolished. These results indicate that reduction of surface molecules are directly related to the abolishment of receptor-mediated phagocytosis of opsonized erythrocytes by monocytes. Various hematological and biochemical abnormalities seen in the

HO-1-deficient patient, including increased Hp concentration, abundance of fragmented erythrocytes and thrombocytosis, and hyperlipidemia, may all be explained by the reduced scavenging functions of phagocytes.

7.3.5 Question 5: What Is the Role of Damaged Endothelial Cells in the Regulation of the Coagulation/Fibrinolytic System?

Another distinct feature of the first human HO-1 deficiency case was the defective endothelial function, as represented by extremely abnormal parameters of coagulation/fibrinolysis. Unlike in cases of other hematological illnesses associated with disseminated intravascular coagulation, our patient exhibited extraordinarily elevated values for thrombin-antithrombin complex, FDP, and plasmin-α_2 plasmin inhibitor complex. Paradoxically, the platelet numbers constantly increased. The data indicated that HO-1 or HO-1 products, such as CO, may be associated with the regulation of the coagulation/fibrinolytic system.

We recently demonstrated in in vitro cultures that a CO-releasing molecule suppressed TNF-α-induced upregulation of tissue factor and plasminogen activator inhibitor type 1 by human umbilical vein endothelial cells. It also suppressed mitogen-activated protein kinases and NK-κB signaling pathway activation by TNF-α. Lipopolysaccharide (LPS)-induced TNF-α production by circulating mononuclear cells was also significantly inhibited by the CO-releasing molecules (Maruyama et al. 2012). These results may explain the characteristic findings seen in the HO-1-deficient patient. At the same time, the data support the view that CO-releasing molecules may constitute a novel anticoagulative and anti-inflammatory therapy.

A summary of the macrophage activation and endothelial cell dysfunction is shown in Fig. 7.10. Lack of HO-1 resulted in unregulated activation of macrophages with excess inflammatory cytokine release. At the same time, HO-1 deficiency resulted in overproduction of tissue factor by endothelial cells and macrophages, leading to the abnormal activation of the coagulation/fibrinolysis system. The figure illustrates the role of HO-1 as an inhibitor of cytokine overproduction and endothelial cell dysfunction. It was intriguing that the HO-1-deficient patient did not run an acute catastrophic course. On the contrary, prolonged and sustained activation of macrophages, platelets, and endothelial cells led to exhaustion and dysfunction of these cells.

7.3.6 Question 6: What Is the Impact of HO-1 Deficiency on Macrophage Function?

Monocytes/macrophages are composed of at least two functionally distinct subsets, M1 and M2 (Geissmann et al. 2003). The different subsets of the monocyte/macrophage lineage differentiate in response to environmental stimuli. M1 macrophages are the "classical" macrophages, and they comprise the pro-inflammatory subset, whereas M2 macrophages are "alternatively" activated macrophages. They resolve inflammatory responses, perform scavenger functions, and promote tissue remodeling and repair. Interferon (IFN)-γ is the key cytokine driving the M1 pathway, whereas IL-4, IL-10, and steroids promote monocyte differentiation into M2 macrophages.

We reported previously that circulating monocytes produce significant levels of HO-1 during Kawasaki disease and infectious diseases, suggesting its anti-inflammatory role during these illnesses (Yachie et al. 2003). Furthermore, we investigated the profiles of cytokine mRNA expression in two subsets of circulating monocytes (Mizuno et al. 2005). In this study, freshly isolated $CD16^{high}/CCR2^{negative}$ monocytes expressed significant levels of HO-1 mRNA in vivo. They produced little IL-10 upon stimulation with LPS. In contrast, the major subset of $CD16^{low}/CCR2^{positive}$ monocytes did not express HO-1 mRNA in vivo, whereas they responded significantly to LPS and produced IL-10. The fractions of $CD16^{high}/CCR2^{negative}$ monocytes

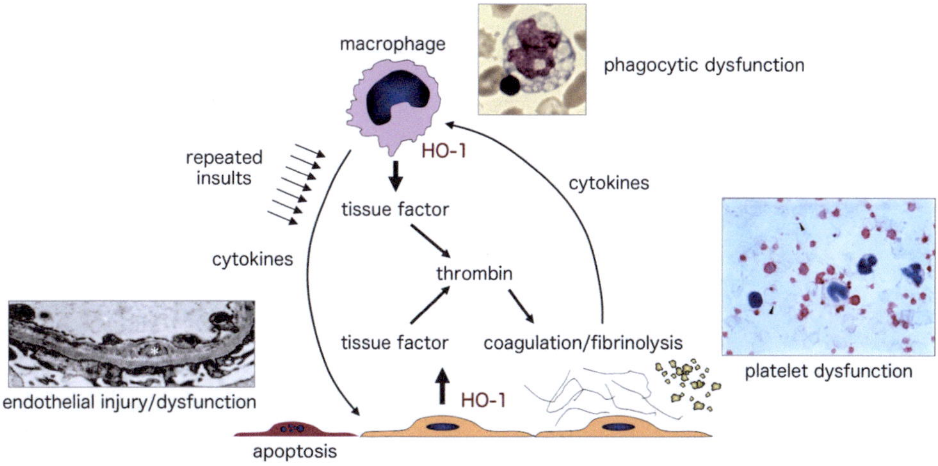

Fig. 7.10 Summary of the macrophage activation and endothelial cell dysfunction. Lack of HO-1 resulted in unregulated activation of macrophages with excess inflammatory cytokine release. At the same time, HO-1 deficiency resulted in overproduction of tissue factor by endothelial cells, leading to the abnormal activation of the coagulation/fibrinolysis system. Prolonged and sustained activation of monocytes, platelets, and endothelial cells leads to exhaustion and dysfunction of these cells.

increased during various acute inflammatory diseases, such as Kawasaki disease and influenza virus infection, suggesting that monocytes play anti-inflammatory roles through HO-1 production.

In macrophages and dendritic cells, CO reduces pro-inflammatory and increases anti-inflammatory cytokine secretion in response to LPS (Lee and Chau 2002). HO-1-mediated anti-inflammatory effects may therefore be closely linked to anti-inflammatory mechanisms, such as the suppression of the immune and inflammatory responses in macrophages via diminished antigen-presenting capacity and cytokine synthesis (Listopad et al. 2007). This is consistent with our finding in the HO-1-deficient patient, in whom the lack of HO-1 resulted in a marked rise in circulating heme and subsequent oxidative vascular and tissue injury, anemia, and chronic inflammation. Schaer et al. reported that macrophages express upregulated levels of CD163 in sepsis-induced hemophagocytic syndrome (Schaer et al. 2006). HO-1 is induced by CD163-mediated Hb-Hp complex uptake (Fig. 7.11) (Yamazaki et al. 2007). These macrophages expressed significant levels of HO-1, suggesting their role as a negative regulator of inflammation.

7.4 Therapy

No effective therapy is known. However, careful avoidance of various types of external stress, including infections, physical stresses, and medical interventions, may at least prolong the onset of the catastrophic episodes. There is no evidence to support the roles of anti-inflammatory drugs or immunosuppressive agents. Experiences with only limited numbers of HO-1-deficient patients indicate that once the onset of the inflammatory processes is triggered, it is extremely difficult to prevent the progression of the subsequent organ damage.

End-of-Chapter Questions

1. What are the common clinical findings of human HO-1 deficiency?
2. What are the common laboratory features of human HO-1 deficiency?
3. What could be the differential diagnosis of patients with HO-1 deficiency?
4. The ages of onset for each HO-1-deficient patient are significantly different. What are the possible reasons for this?

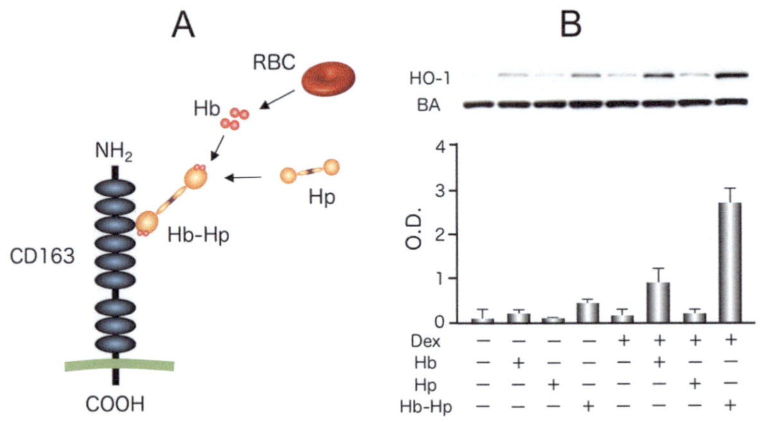

Fig. 7.11 HO-1 induction by CD163-mediated Hb-Hp complex uptake by monocytes. Hb-Hp complex binds to CD163 on monocyte surface (**a**). Binding of Hb-Hp complex efficiently induces HO-1 production by monocytes. Peripheral blood mononuclear cells were pretreated with or without dexamethasone (Dex) for 24 h. Cells were then treated with Hb, Hp, or Hb-Hp complex for 24 h. In controls, cells were cultured with growth medium for the same period. Cell lysates were prepared, and HO-1 and β-actin (BA) expression were examined by Western blotting. Intensity of each band was quantified by densitometry, and optical density (O.D.) is shown (**b**). $*p < 0.1$. $**p < 0.01$. Bars indicate standard deviations

References

Balla J, Vercellotti GM, Jeney V, Yachie A, Varga Z, Jacob HS, Eaton JW, Balla G (2007) Heme, heme oxygenase, and ferritin: how the vascular endothelium survives (and dies) in an iron-rich environment. Antioxid Redox Signal 9:2119–2137

Geissmann F, Jung S, Littman DR (2003) Blood monocytes consist of two principal subsets with distinct migratory properties. Immunity 19:71–82

Kawashima A, Oda Y, Yachie A, Koizumi S, Nakanishi I (2002) Heme oxygenase-1 deficiency: the first autopsy case. Hum Pathol 33:125–130

Kovtunovych G, Eckhaus MA, Ghosh MC, Ollivierre-Wilson H, Rouault TA (2010) Dysfunction of the heme recycling system in heme oxygenase 1-deficient mice: effects on macrophage viability and tissue iron distribution. Blood 116:6054–6062

Lee T-S, Chau L-Y (2002) Heme oxygenase-1 mediates the anti-inflammatory effect of interleukin-10 in mice. Nat Med 8:240–246

Listopad J, Asadullah K, Sievers C, Ritter T, Meisel C, Sabat R, Döcke WD (2007) Heme oxygenase-1 inhibits T cell-dependent skin inflammation and differentiation and function of antigen-presenting cells. Exp Dermatol 16:661–670

Madsen M, Møller HJ, Nielsen MJ, Jacobsen C, Graversen JH, van den Berg T, Moestrup SK (2004) Molecular characterization of the haptoglobin-hemoglobin receptor CD163. J Biol Chem 279:51561–51567

Maines MD (1988) Heme oxygenase: function, multiplicity, regulatory mechanisms, and clinical application. FASEB J 2:2557–2568

Maruyama K, Morishita E, Yuno T, Sekiya A, Asakura H, Ohtake S, Yachie A (2012) Carbon monoxide (CO)-releasing molecule-derived CO regulates tissue factor and plasminogen activator inhibitor type 1 in human endothelial cells. Throm Res 130:e188–e193

Mizuno K, Toma T, Tsukiji H, Okamoto H, Yamazaki H, Ohta K, Ohta K, Kasahara Y, Koizumi S, Yachie A (2005) Selective expansion of $CD16^{high}$ $CCR2^{-}$ subpopulation of circulating monocytes with preferential production of haem oxygenase (HO)-1 in response to acute inflammation. Clin Exp Immunol 142:461–470

Nakahira K, Kim HP, Geng XH, Nakao A, Wang X, Murase N, Drain PF, Wang X, Sasidhar M, Nabel EG, Takahashi T, Lukacs NW, Ryter SW, Morita K, Choi AM (2006) Carbon monoxide differentially inhibits TLR signaling by regulating ROS-induced trafficking of TLRs to lipid rafts. J Exp Med 203:2377–2389

Nath KA (2006) Heme oxygenase-1: a provenance for cytoprotective pathways in the kidney and other tissues. Kidney Int 70:432–443

Radhakrishnan N, Yadav SP, Sachdeva A, Pruthi PK, Sawhney S, Piplani T, Wada T, Yachie A (2011a) Human heme oxygenase-1 deficiency presenting with hemolysis, nephritis, and asplenia. J Pediatr Hematol Oncol 33:74–78

Radhakrishnan N, Yadav SP, Sachdeva A, Wada T, Yachie A (2011b) An interesting tetrad of asplenia, inflammation, hemolysis, and nephritis. Pediatr Hematol Oncol 28:723–726

Schaer DJ, Schaer CA, Schoedon G, Imhof A, Kurrer MO (2006) Hemophagocytic macrophages constitute a major compartment of heme oxygenase expression in sepsis. Eur J Haematol 77:432–436

Yachie A, Niida Y, Wada T, Igarashi N, Kaneda H, Toma T, Ohta K, Kasahara Y, Koizumi S (1999) Oxidative stress causes enhanced endothelial cell injury in human heme oxygenase-1 deficiency. J Clin Invest 103:129–135

Yachie A, Toma T, Shimura S, Yue L, Morimoto K, Maruhashi K, Niida Y, Ohta K, Kasahara Y, Saikawa Y, Koizumi S (2002) Human HO-1 deficiency and the oxidative injury of vascular endothelial cells. In: Abraham NG (ed) Heme oxygenase in biology and medicine. Kluwer Academic/Plenum Publishers, New York

Yachie A, Toma T, Mizuno K, Okamoto H, Shimura S, Ohta K, Kasahara Y, Koizumi S (2003) Heme oxygenase-1 production by peripheral blood monocytes during acute inflammatory illnesses of children. Exp Biol Med 228:550–556

Yamazaki H, Ohta K, Tsukiji H, Toma T, Hashida Y, Ishizaki A, Saito T, Arai S, Koizumi S, Yachie A (2007) Corticosteroid enhances heme oxygenase-1 production by circulating monocytes by up-regulating hemoglobin scavenger receptor and amplifying the receptor-mediated uptake of hemoglobin-haptoglobin complex. Biochem Biophys Res Commun 358:506–512

Yang Y, Ohta K, Shimizu M, Morimoto K, Goto C, Nakai A, Toma T, Kasahara Y, Yachie A, Seki H, Koizumi S (2003) Selective protection of renal tubular epithelial cells by heme oxygenase (HO)-1 during stress-induced injury. Kidney Int 64:1302–1309

The Homocystinurias

8

John H. Walter and Henk J. Blom

Keywords
Homocystinuria · Homocystine · Homocysteine · Cystathionine beta-synthase deficiency · Hyperhomocysteinaemia

8.1 Case Report A

Patient A is found to have an increased blood methionine on newborn heel-prick screening of 250 μmol/L (ref range 20–50). He is seen by a specialist paediatrician at age 10 days. He appears well, is breastfeeding normally and is gaining weight. Physical examination is normal. Further blood and urine samples are collected. These confirm a raised blood methionine and in addition a blood total homocysteine (tHcy) of 190 μmol/L (ref range < 15 μmol/L). Homocystine is present in his urine. His parents have one other child, a boy age 5, whom they also bring with them to patient A's appointment. Patient A's brother is wearing glasses. He started school 1 year ago; his teacher has recently expressed concern to his parents about his slow progress. He has also been referred to an orthopaedic surgeon because of knock knees (genu valgum).

8.2 Case Report B

Patient B is a 26-year-old man who has presented, 2 years ago, with tenderness and swelling in his left calf. An ultrasound showed the presence of a deep vein thrombosis (DVT). This was thought to have been caused by immobility on a long-haul flight. He was treated with anticoagulants for a total of 3 months. He has now developed similar symptoms in the same leg but also has pain on breathing and shortness of breath. He is diagnosed as having had a further DVT but complicated by a pulmonary embolism. He is restarted on anticoagulant therapy and makes a good recovery. He has a past history of depression. Physical examination is unremarkable. He is of normal height. In view of the recurrence of the DVT, he has further investigations. He is found to have homocystine in his urine and a blood tHcy of 310 μmol/L.

J. H. Walter (✉)
Willink Biochemical Genetics Unit, Manchester Centre for Genomic Medicine, Royal Manchester Children's Hospital, St Mary's Hospital, University of Manchester, Central Manchester University Hospitals NHS Foundation Trust, Manchester, UK

H. J. Blom
Universitätsklinikum Freiburg, Laboratory of Clinical Biochemistry and Metabolism, University Medical Centre Freiburg, Freiburg, Germany

T. Oohashi et al. (eds.), *Human Pathobiochemistry*, https://doi.org/10.1007/978-981-13-2977-7_8

8.3 Case Report C

Patient C is a female born by normal vaginal delivery at term with a birth weight of 3.04 kg. She is the second child of healthy, non-consanguineous parents. She presented to her local hospital at 5 weeks of age with lethargy, reduced feeding and a low temperature. Treatment for sepsis with intravenous fluids and antibiotics was started but despite this she deteriorated further. A CT head showed posterior fatty areas of low attenuation but no acute changes. Blood tests revealed a raised homocysteine of 176 μmol/L (normal <15 μmol/L) and low methionine of <2 μmol/L (normal range 18–62 μmol/L). Urine organic acid analysis showed no increase in methylmalonic acid, and cerebrospinal fluid (CSF) folate (5-methyltetrahydrofolate) was undetectable. A skin biopsy was undertaken which confirmed the diagnosis of methylenetetrahydrofolate reductase (MTHFR) deficiency, with a severe reduction of fibroblast MTHFR activity to approximately 1% of the mean control value. Molecular genetic investigations showed her to be homozygous for a c.1781G > A mutation in the *MTHFR* gene.

8.4 Case Report D

Patient D, a 40-year-old woman, presented with a 3-month history of feeling increasingly tired and tingling in her hands and feet. More recently she had had difficulty in walking. On examination, she looked pale and had a decrease in lower limb vibration sense and proprioception. She had previously been well but over the past 15 years had chosen a strict vegan diet. On investigation she had a severe megaloblastic anaemia, a plasma vitamin B12 of only 28 pmol/L (reference range 155–700 pmol/L), blood homocysteine increased at 240 μmol/L (normal <15 μmol/L) and urine methylmalonic acid 2000 μmol/mmol of creatinine (normal <5 μmol/mmol creatinine). A diagnosis of subacute combined degeneration of the cord caused by severe dietary vitamin B12 deficiency was made. Treatment with vitamin B12 injections leads to a complete recovery over several months.

8.5 Diagnosis

Homocysteine is a sulphur-containing amino acid that results from the demethylation of the essential amino acid methionine. Hyperhomocysteinaemia refers to an increase in the concentration of homocysteine in plasma above normal and homocystinuria to the presence of homocystine (the disulphide form of homocysteine) in the urine. Homocystinuria only occurs when plasma levels of homocysteine are significantly raised. A number of inherited and acquired disorders may cause a high level of homocysteine. In this chapter we provide case reports that exemplify the wide phenotypic variability that can occur with these conditions and then discuss these and other disorders in more detail.

Case reports A to C document patients who have inherited disorders of sulphur amino acid metabolism. All have a raised blood homocysteine and the presence of homocysteine in their urine. Consequently, they can all be said to have hyperhomocysteinaemia and homocystinuria. Despite their disparate clinical phenotypes, the patients described in case studies A and B have a deficiency of the same enzyme, namely, cystathionine beta-synthase (CBS). The patient in case study C has a separate disorder – methylenetetrahydrofolate reductase (MTHFR) deficiency. Although both disorders are associated with an increase in homocysteine, plasma methionine is raised in CBS deficiency and low in MTHFR deficiency. Confirmation of diagnosis requires enzyme or DNA studies. Case report D demonstrates that acquired disease, such as severe dietary deficiency of vitamin B12 as in this case, can also cause neurological disease associated with hyperhomocysteinaemia. A diagnostic algorithm is shown in Fig. 8.1.

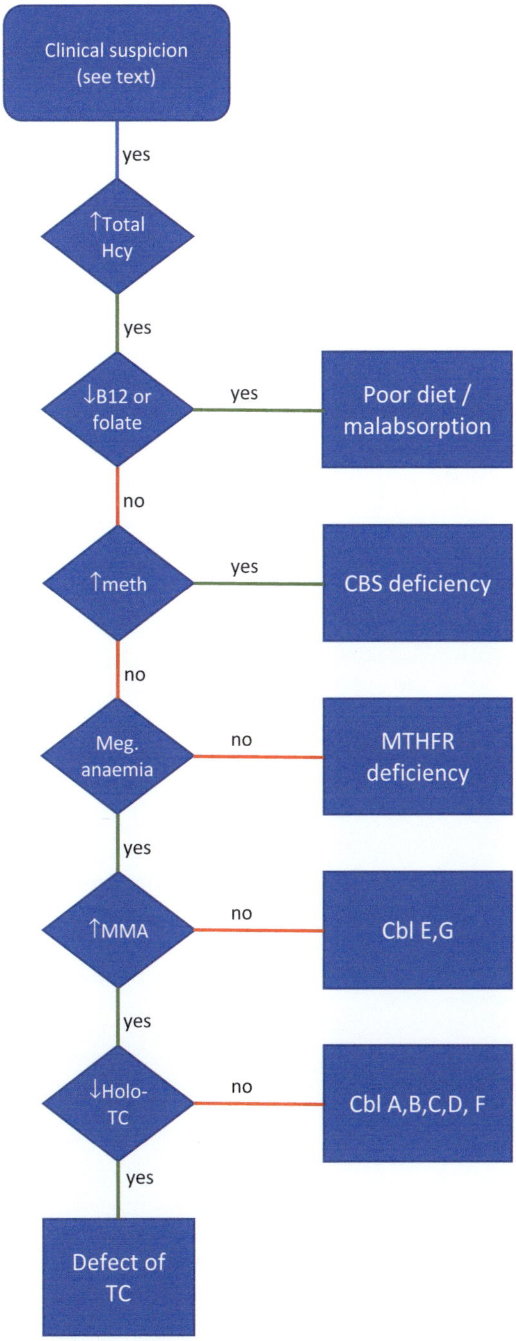

Fig. 8.1 Diagnostic algorithm. The arrows refer to changes in blood. *Hcy* homocysteine, *meth* methionine, *CBS* cystathionine beta-synthase, *meg* megaloblastic, *MMA* methylmalonic acid, *MTHFR* methylenetetrahydrofolate reductase, *Holo-TC* holo-transcobalamin, *Cbl* cobalamin

8.6 Biochemical and Molecular Perspectives

Homocysteine is a sulfur-containing amino acid which is formed from the demethylation of the essential amino acid methionine acting as a methyl donor for a number of biologically important substrates. Methionine is first activated by ATP to S-adenosylmethionine, which donates a methyl group in about 200 different methyltransferase reactions. The methylation of among others, DNA, RNA, proteins, lipids or small molecules, results in formation of S-adenosylhomocysteine, which in turn is hydrolysed to homocysteine.

Once formed homocysteine can either be remethylated to methionine or undergoes irreversible catabolism via the transsulfuration pathway. The methyl donor for remethylation is either 5-methyltetrahydrofolate or betaine. Transsulfuration of homocysteine occurs by cystathionine beta-synthase to form cystathionine and then by cystathionine gamma-lyase to form cysteine.

Homocysteine in plasma is mainly oxidized to sulphides either with itself or with cysteine or albumin (Fig. 8.2).

Whereas in the past homocystine (known as free homocysteine) and cysteine-homocysteine mixed disulphide were measured in plasma and urine by the use of amino acid analysers, over the last two decades, this has been superseded by an assay of tHcy that measures the sum of all forms of homocysteine in plasma. This latter test is both more sensitive and reliable and available in most clinical chemistry laboratories. In healthy individuals tHcy levels are less than 15 μmol/L. In children one should consider an inborn error of homocysteine metabolism if tHcy is above 50 μmol/L.

In order to consider the clinical and laboratory abnormalities that occur with the various causes of homocystinuria, it is necessary to have an understanding of the biochemical pathways involved (Fig. 8.3).

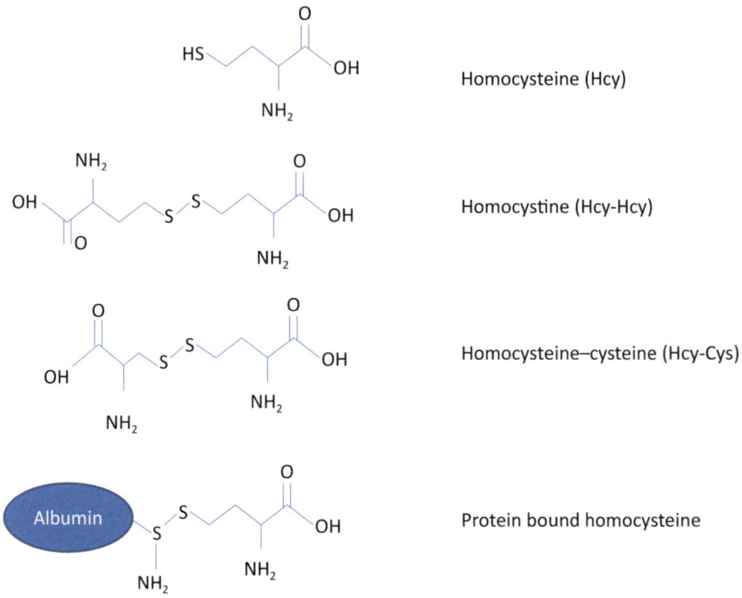

Fig. 8.2 The various forms of homocysteine in plasma. Collectively they form plasma total homocysteine (tHcy), but protein-bound homocysteine contributes approximately 80%

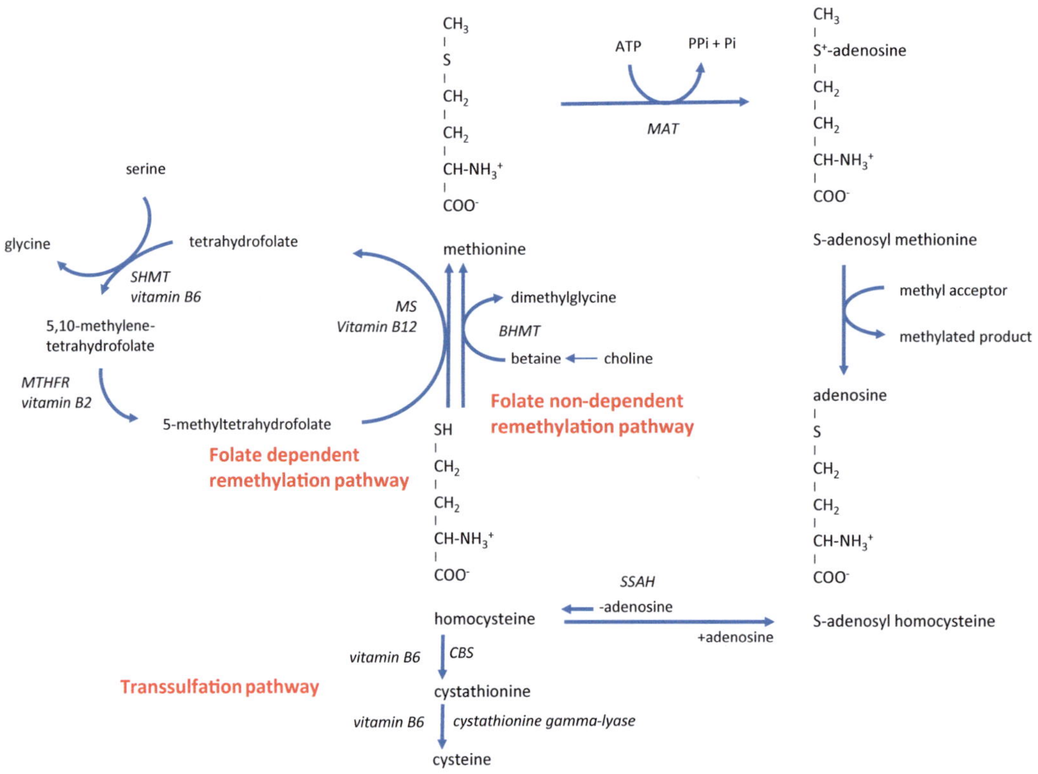

Fig. 8.3 Pathway of homocysteine and methionine metabolism. Methionine is converted to S-adenosylmethionine by methionine adenosyltransferase (MAT). S-adenosylmethionine is a methyl group donor in many methylation reactions and is converted to S-adenosylhomocysteine. S-adenosylhomocysteine is subsequently hydrolysed by S-adenosylhomocysteine hydrolase (SAHH) to homocysteine and adenosine. Homocysteine is catabolized by the transsulfuration or remethylated. *CBS* cystathionine beta-synthase, *MS* methionine synthase, *MTHFR* methylenetetrahydrofolate reductase, *SHMT* serine hydroxymethyltransferase, *BHMT* betaine-homocysteine S-methyltransferase

Table 8.1 Inherited and acquired disorders that may cause an increase in plasma total homocysteine (tHcy)

Disorder	Gene	tHcy	Plasma methionine	Urine or plasma methylmalonic acid	Megaloblastic anaemia
Dietary folate deficiency	–	↑	→	→	+
In combination with SNP MTHFR 677C>T	*MTHFR*	↑↑	→	→	–
Dietary B12 deficiency	–	↑ to ↑↑	→	↑ to ↑↑	+ to ++
Liver disease		↑	↑	→	–
Renal failure	–	↑ to ↑↑	→	↑	–
Pernicious anaemia (acquired intrinsic factor deficiency)	–	↑	→	↑ to ↑↑	++
Transcobalamin deficiency	*TCN2*	↑	→	↑	+
Transcobalamin receptor deficiency	*CD320*	↑	→	↑	
Cobalamin disorders with increased MMA	CblC: *MMACHC,* CblD: *MMADHC,* CblF: *LMBRD1,* CblJ: *ABCD4,* CblX: *HCFC1*	↑↑	→↑	↑↑	+
Cobalamin disorders with normal MMA	CblD, MMADHC CblE: *MTRR,* CblG: *MTR*	↑↑	→↓	→	++
CBS deficiency	*CBS*	↑↑↑	↑↑	→	–
MAT I/III deficiency	*MAT1A*	↑	↑↑↑	→	–
Severe MTHFR deficiency	*MTHFR*	↑↑	↓	→	–

tHcy total homocysteine, → normal, ↑ mildly increased, ↑↑ moderately increased, ↑↑↑ severely increased, – absent, + present, ++ moderate, +++ severe

An increase in tHcy occurs with both inherited and acquired disorders. These are listed in Table 8.1 together with the affected gene for inherited disorders and associated biochemical and haematological changes.

The most common inherited defect in the remethylation pathway is caused by CblC deficiency and in transsulfuration by CBS deficiency.

8.7 Disorders of Homocysteine Catabolism

8.7.1 Cystathionine Beta-Synthase Deficiency

CBS is a tetrameric cytosolic enzyme which is primarily expressed in the liver but also in the pancreas, kidney and brain. CBS requires the active form of vitamin B6 (pyridoxal 5′ phosphate) as a cofactor. CBS deficiency is a recessive disorder with a variable frequency in different populations ranging from 1 in 2,000 to 1:900,000. Over 160 mutations in the CBS gene have been described (see http://cbs.lf1.cuni.cz). Certain mutations are associated with disorders where treatment with pharmacological doses of pyridoxine (vitamin B6) leads to a marked improvement of the biochemical abnormalities (pyridoxine-responsive CBS deficiency). Approximately 50% of patients, however, have mutations which are associated with no such improvement with pyridoxine (non-responsive CBS deficiency) (Moat et al. 2004).

8.7.2 Clinical Findings

The severity of clinical manifestations is variable. Those with pyridoxine unresponsive disease generally present earlier and with more severe disease. The major systems affected include the central nervous system (developmental delay, seizures, psychiatric illness), eyes (early onset of myopia, followed by lens dislocation), skeleton (excessive growth, arachnodactyly, scoliosis, osteoporosis) and vascular (thromboembolism – particularly DVT, pulmonary embolism and sagittal sinus thrombosis). The clinical features that occur in untreated CBS deficiency resemble Marfan syndrome; hence patients are often described as having a Marfanoid appearance (Fig. 8.4). The pathophysiology is not fully understood, but damage to connective tissue appears to be related to increased homocysteine levels.

Guidelines for the diagnosis and treatment of CBS deficiency have recently been published (Morris et al. 2017).

8.7.3 Diagnosis

The diagnosis should be considered in all children with developmental delay or learning difficulties, particularly when associated with early-onset and progressive myopia and skeletal abnormalities such as genu valgum (knock knees) or pectus excavatum (pigeon chest). Those with milder variants may present in later childhood or adult life with thromboembolic disease.

The diagnosis is supported by finding a raised blood homocysteine (generally tHcy is >100 μmol/L, but lower values have been found), homocysteine in urine and a raised methionine. Further confirmation requires CBS enzyme analysis in cultured skin fibroblasts or mutation analysis.

Many countries have now introduced newborn screening for CBS deficiency based on finding a raised blood methionine. However, most patients with pyridoxine-responsive disease do not have a significant increase in methionine concentrations in the first weeks of life and as a consequence will be missed. Screening methods based on measuring tHcy or CBS mutation analysis will have a better sensitivity. For a review of newborn screening, see Huemer at et al. (2015).

8.7.4 Management

8.7.4.1 Pyridoxine

Patients who are responsive to pyridoxine are treated with oral pyridoxine at a dose of 10 mg/kg/day to a maximum of 500 mg/day. Higher doses (above 900 mg/day) are associated with the developmental of a peripheral neuropathy. Some patients are partially responsive to pyridoxine and require additional therapy (see below). Folate deficiency is often observed at diagnosis and needs to be corrected before pyridoxine responsiveness can be tested.

8.7.4.2 Betaine

Betaine acting as a methyl donor can remethylate homocysteine to form methionine by the hepatic enzyme betaine-homocysteine methyl transferase. Oral doses of betaine at a dose of 100–150 mg/kg/d given in two divided doses can significantly reduce tHcy levels. However, betaine used alone rarely reduces homocysteine sufficiently; it is most appropriately used

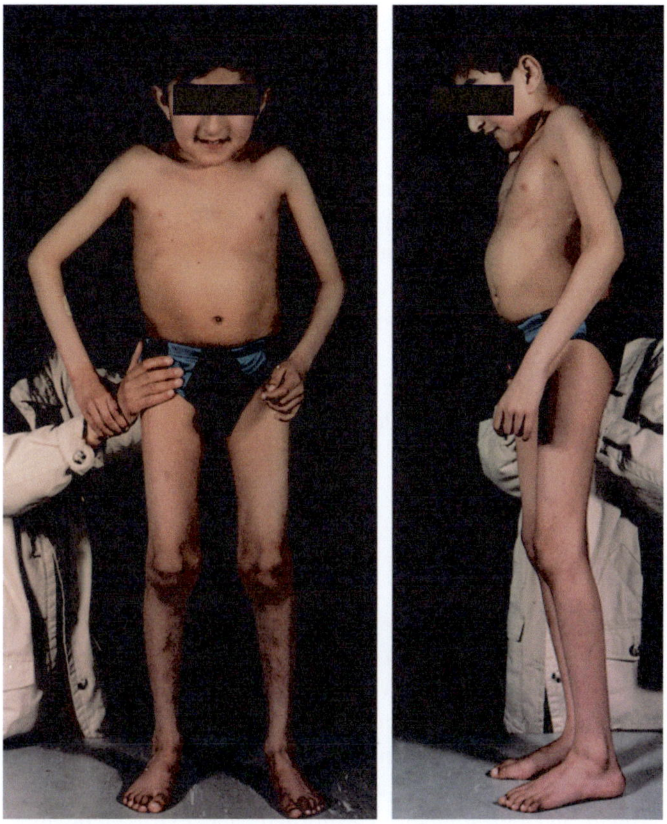

Fig. 8.4 A 7-year-old boy with untreated CBS deficiency. Note the skeletal abnormalities

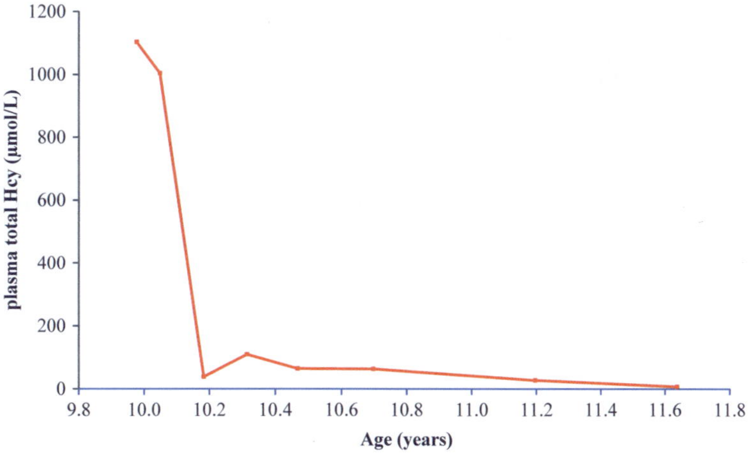

Fig. 8.5 The effect of starting dietary treatment on the plasma tHcy in a 10-year-old girl with newly diagnosed CBS deficiency. The girl was commenced on a commercially available methionine-free amino acid formula together with a severe restriction in her intake of protein-containing foods

as an adjunct to pyridoxine or to dietary therapy. Generally it is well tolerated but very rarely has been associated with cerebral oedema in those with very high plasma methionine levels. It may also cause a fishy odour in some patients.

8.7.4.3 Dietary Treatment

Restriction of dietary methionine can be highly effective in reducing homocysteine levels (Fig. 8.5). Such treatment requires a very severe restriction in foods contacting natural protein and, as a consequence, in order to be nutritionally complete, the use of commercial supplements of amino acids, vitamins and minerals. The diet is complex and requires management by a specialist metabolic dietitian and careful biochemical monitoring.

Dietary treatmet is most effective when started in early infancy, following a diagnosis made by newborn screening. Compliance is more difficult to achieve in children where the diagnosis is made later and who are used to eating normal foods.

Vitamin B12 and folic acid may be low in CBS deficiency, and additional supplements should be given.

8.7.5 Outcome

The outcome for patients with CBS deficiency is dependent upon a number of factors, including the severity of the disorder, age at diagnosis and compliance with treatment. Those diagnosed following newborn screening and treated with diet can avoid the complications of the disorder; growth and development are normal, and eye and vascular disease can be avoided. However, good compliance is essential, and treatment must be continued for life (Yap et al. 2001). Treatment for patients diagnosed later can also be effective in preventing progression of the disease but cannot reverse damage that may have already occurred.

In view of the increased risk of thrombosis, precautions need to be taken in patients requiring surgery and anaesthesia. Homocysteine levels should be well controlled, hydration maintained, and standard antithrombotic measures strictly applied. Fertility is not known to be affected in CBS deficiency, but thrombosis is a risk during pregnancy and delivery and in the post-partum period. It is recommended that anticoagulant treatment, in the form of low molecular weight heparin, is given from the 3rd trimester until at least 6 weeks post-partum. Oral contraception containing oestrogen should be avoided.

8.8 Disorders of Homocysteine Remethylation

8.8.1 Methylenetetrahydrofolate Reductase Deficiency

5-Methyltetrahydrofolate is required for the remethylation of homocysteine to methionine (Fig. 8.3) and is formed from 5,10-methylenetetrahydrofolate by the enzyme methylenetetrahydrofolate reductase (MTHFR). Pathogenic mutations cause severe enzyme deficiency lead to a wide spectrum of neurological symptoms, mainly encephalopathy, hypotonia, microcephaly, seizures, developmental delay and episodes of apnoea. Hydrocephalus is an additional rare but recognized complication. The condition generally presents in infancy and is associated with a high morbidity and mortality, although late-onset disease may also occur. Raised levels of homocysteine may also result in an increased risk of thrombosis.

The diagnosis is based on finding increased blood homocysteine and (in contrast to CBS deficiency) a low blood methionine. Confirmation requires enzyme assay in cultured skin fibroblasts and/or mutation analysis. Over 100 different mutations have been so far described (Froese et al. 2016).

In addition to the biochemical abnormalities in blood, total CSF folate levels are severely reduced. The cause of the neurological disease is not fully understood but may relate to a reduction in S-adenosylmethionine, an important methyl donor, required for the formation and maintenance of myelin in the brain.

8.8.2 Management

Treatment consists of betaine, hydroxocobalamin and folate (given as 5-methyltetrahydrofolate) which works in combination to reduce homocysteine, increase methionine and increase CSF folate levels. As in CBS deficiency, betaine lowers homocysteine by acting as a methyl donor for the remethylation of homocysteine to methionine, although the homocysteine levels remain clearly elevated. Methylcobalamin is derived from hydroxocobalamin and along with 5-MTHF is a cofactor in methionine production. In contrast to other forms of folic acid, 5-methyltetrahydrofolate appears to be most effective in increasing CSF folate levels. Guidelines have recently been published (Huemer et al. 2017).

8.8.3 Outcome

Some patients diagnosed and treated from early infancy have done well, but the prognosis is generally poor for the most severe forms.

8.9 Disorders of Cobalamin Metabolism

Vitamin B12, also known as cobalamin (Cbl), is the cofactor for two enzymes, methylmalonyl-CoA mutase and methionine synthase. Dietary vitamin B12, following absorption and transport to cells, undergoes a number of intracellular processes to form either adenosylcobalamin, the cofactor for methylmalonyl-CoA mutase, or methylcobalamin, the cofactor for methionine synthase.

The activities of these two enzymes can be affected, either separately or together, by acquired or inherited defects in vitamin B12 metabolism. Reduced activity of methylmalonyl-CoA mutase leads to an increase in blood and urine methylmalonic acid (MMA), while a reduction in methionine synthase activity disturbs remethylation of homocysteine leading to its accumulation and deficiency of methionine.

Acquired defects include poor dietary intake and disturbed absorption as a result of a lack of intrinsic factor that occurs in the autoimmune disorder, pernicious anaemia. Inherited defects of vitamin B12 metabolism include those affecting absorption (cubilin or megalin defects), transport (transcobalamin (TC) deficiency) and intracellular processing. Disorders of this intracellular processing are termed Cbl diseases and designated A to G, depending on the specific intracellular defect.

Defects, affecting only methylmalonyl-CoA mutase, such as inherited deficiency of the enzyme itself, and CblA and CblB disease, result in very high MMA levels but are not associated with an increase in tHcy. Both MMA and tHcy are increased in methionine synthase deficiency and in CblC, CblD and CblF disease.

Hyperhomocysteinaemia without any increase in MMA is found in defects affecting only methionine synthase, such as CblE and CblG disease, and in some CblD mutations, inhibiting the targeting of vitamin B12 to methionine synthase. For recent review see Hannibal et al. (Hannibal et al. 2016). These inherited disorders of vitamin B12 metabolism all are exceptionally rare apart from CblC disease which, in particular, is relatively prevalent in Mediterranean countries. Although CblC disease is associated with a wide variation in severity, most affected patients present in early infancy with a progressive neurological deterioration but may also have a multisystem disorder with anaemia, renal disease, retinopathy, optic atrophy, cardiomyopathy and interstitial pneumonia. Later-onset disease can occur but is less common. Treatment consists of intramuscular injections of high-dose hydroxocobalamin and oral betaine. As with MTHFR deficiency, the outcome is better in those diagnosed and treated as early as possible but unfortunately is generally poor, in particular regarding the eye and neurological problems. For a recent review of diagnosis and management, see Huemer et al. (Huemer et al. 2017).

8.10 Elevated Homocysteine as Risk Factor

Hyperhomocysteinaemia in the range of 15–25 μmol/L is often referred to as 'mild' hyperhomocysteinaemia and has been associated with increased risk for cardiovascular diseases, venous thrombosis, bone fractures, pregnancy complication and various neurological conditions such as Alzheimer disease. This form of hyperhomocysteinaemia is caused by nutritional, life style and genetic factors. Homozygosity for a common polymorphism in the *MTHFR* gene (c.677C > T) is a well-known example of genetically determined hyperhomocysteinaemia. Whether 'mild' hyperhomocysteinaemia is a cause, mediator or marker, continues to be a subject of strong debate. In particular the relation between hyperhomocysteinaemia and cardiovascular disease was a field of extensive research one to three decades ago (Hannibal and Blom 2017). However, large intervention studies have not demonstrated any reduction in risk of cardiovascular disease by folic acid, a very cheap vitamin that lowers homocysteine effectively (Marti-Carvajal et al. 2017). Intervention studies in the field of bone fractures or cognitive decline have also failed to show clinical beneficial effects. However, folic acid has proven to reduce the risk for neural tube defects dramatically if taken at least 4 weeks before conception.

End-of-Chapter Questions

1. In case report A, the patient is reported to have a number of problems. What (if any) investigations do you think should be undertaken on this boy? What result do you think is likely for each of these?
2. Not all countries advocate newborn screening for CBS deficiency. What are the arguments both for and against screening? How do these relate to published criteria for screening in general?
3. In case report B, the patient appears not to have had any symptoms in childhood. Why do you think this is? What treatment is necessary for this man, and how do you think he might respond?
4. Severe vitamin B12 deficiency is a known acquired cause of homocystinuria. What types of food contain this vitamin, and what groups of people might be particularly at risk from deficiency?
5. Over recent years there has been considerable controversy as to whether homozygosity for the common polymorphism (C677T) is a risk factor for certain disorders. What is the frequency of this variant in the general population? What conditions have been linked to it? What is the evidence that it does affect an individual's health?

References

Froese DS, Huemer M, Suormala T, Burda P, Coelho D, Gueant JL, Landolt MA, Kozich V, Fowler B, Baumgartner MR (2016) Mutation update and review of severe methylenetetrahydrofolate reductase deficiency. Hum Mutat 37:427–438

Hannibal L, Blom HJ (2017) Homocysteine and disease: causal associations or epiphenomenons? Mol Asp Med 53:36–42

Hannibal L, Lysne V, Bjorke-Monsen AL, Behringer S, Grunert SC, Spiekerkoetter U, Jacobsen DW, Blom HJ (2016) Biomarkers and algorithms for the diagnosis of vitamin B12 deficiency. Front Mol Biosci 3:27

Huemer M, Kozich V, Rinaldo P, Baumgartner MR, Merinero B, Pasquini E, Ribes A, Blom HJ (2015) Newborn screening for homocystinurias and methylation disorders: systematic review and proposed guidelines. J Inherit Metab Dis 38:1007–1019

Huemer M, Diodato D, Schwahn B, Schiff M, Bandeira A, Benoist JF, Burlina A, Cerone R, Couce ML, Garcia-Cazorla A, La Marca G, Pasquini E, Vilarinho L, Weisfeld-Adams JD, Kozich V, Blom H, Baumgartner MR, Dionisi-Vici C (2017) Guidelines for diagnosis and management of the cobalamin-related remethylation disorders cblC, cblD, cblE, cblF, cblG, cblJ and MTHFR deficiency. J Inherit Metab Dis 40:21–48

Martí-Carvajal AJ, Solà I, Lathyris D, Dayer M (2017) Homocysteine-lowering interventions for preventing cardiovascular events. Cochrane Database Syst Rev CD006612

Moat SJ, Bao L, Fowler B, Bonham JR, Walter JH, Kraus JP (2004) The molecular basis of cystathionine beta-synthase (CBS) deficiency in UK and US patients with homocystinuria. Hum Mutat 23:206

Morris AA, Kozich V, Santra S, Andria G, Ben-Omran TI, Chakrapani AB, Crushell E, Henderson MJ, Hochuli M, Huemer M, Janssen MC, Maillot F, Mayne PD, Mcnulty J, Morrison TM, Ogier H, O'Sullivan S, Pavlikova M, De Almeida IT, Terry A, Yap S, Blom HJ, Chapman KA (2017) Guidelines for the diagnosis and management of cystathionine beta-synthase deficiency. J Inherit Metab Dis 40:49–74

Yap S, Rushe H, Howard PM, Naughten ER (2001) The intellectual abilities of early-treated individuals with pyridoxine-nonresponsive homocystinuria due to cystathionine beta-synthase deficiency. J Inherit Metab Dis 24:437–447

9 Hypophosphatasia

Takeshi Taketani, Chigusa Oyama, Yasuaki Oda, and Lynne Murphy

Keywords
Hypophosphatasia · Mesenchymal stem cell · Bone-targeting enzyme replacement therapy · Congenital skeletal disease

9.1 Case Report

During the patient's fetal phase, hydramnios was present, and bone abnormality was detected. The male infant was born transvaginally at the gestational age of 41 weeks and 2 days. Respiratory impairment became worse soon after birth; therefore, mechanical ventilation was performed. Low titers of serum alkaline phosphatase (ALP) (9 IU/L; reference range, 530–1610 IU/L) and bone ALP (BAP) (0 μg/L; reference range, 3.7–20.4 μg/L) were found, and urinary phosphoethanolamine (PEA) (1195 μmol/L; normal range is non-detectable), one substrate of ALP, was detected. Hypomineralization, thin and short forearm bone, fluttering in the metaphyseal regions, and narrow thorax were revealed by X-ray (Fig. 9.1a, b). These indicated that the patient had developed perinatal lethal hypophosphatasia (HPP). Refractory convulsion developed 5 days after birth, which ceased with the administration of pyridoxine. Tracheobronchomalacia frequently appeared from 6 months after birth. To perform definite diagnosis, we extracted DNA from white blood cells and examined the gene analysis of the liver/bone/kidney alkaline phosphatase (*ALPL*) gene, the causative gene of HPP. As a result, the patient harbored a homozygous mutation in c.1559delT of the *ALPL* gene (Fig. 9.2) (Taketani T et al. 2015).

Allogeneic bone marrow transplantation (BMT) from the patient's mother (27 years old, HLA 1 locus mismatch) was performed at the age of 7 months with the parents' consent. The patient's mother had clinically normal bone structure and normal range of serum ALP titer. BM was engrafted 19 days after BMT, and mesenchymal stem cells (MSCs) were first transplanted 17 days after BMT. Severe acute gut graft versus host disease (GVHD) (grade 4) developed and did not improve with the administration of several immunosuppressive drugs, including cyclosporine A, tacrolimus, steroids, and infliximab. Interestingly, 1 week after the second mesenchymal stem cell transplantation (MSCT), which took place 51 days after BMT, intractable bloody diarrhea dramatically improved. Chronic

T. Taketani (✉) · C. Oyama · Y. Oda
Faculty of Medicine, Department of Pediatrics, Shimane University, Izumo, Shimane, Japan
e-mail: ttaketani@med.shimane-u.ac.jp; chigusa@med.shimane-u.ac.jp; y-oda@med.shimane-u.ac.jp

L. Murphy
Faculty of Medicine, Department of Medical English Education, Shimane University, Izumo, Shimane, Japan
e-mail: murphy@med.shimane-u.ac.jp

T. Oohashi et al. (eds.), *Human Pathobiochemistry*, https://doi.org/10.1007/978-981-13-2977-7_9

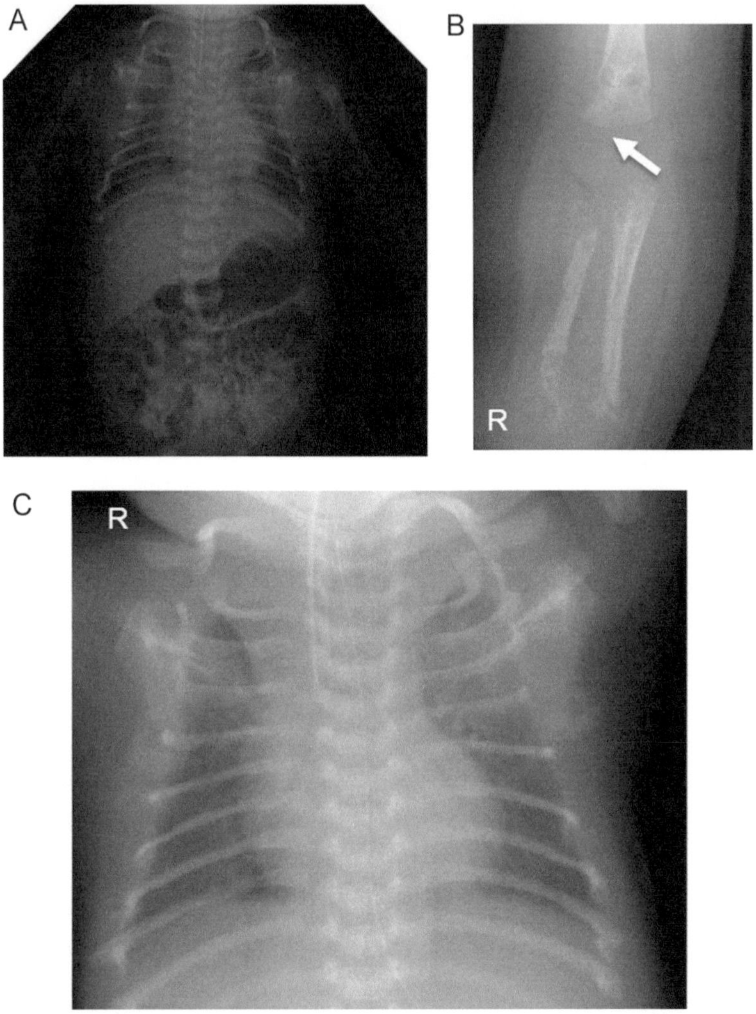

Fig. 9.1 Bone X-ray. (**a**) Whole body; (**b**) right upper extremity; (**c**) chest. Hypomineralization, thin and short forearm bone, flattering in the epiphyseal region (arrow), and narrow thorax are found. *R* right

liver GVHD, which occurred 3 months after BMT, was cured with the third MSCT and treatment with tacrolimus. As a result, we administered tacrolimus until the patient was 24 months old. However, respiratory disturbance and hypomineralization did not ameliorate. Therefore, we performed three additional MSCTs (fourth to sixth) until the patient was 17 months old. At the age of 18 months, his respiratory condition became stable, and tracheobronchomalacia resolved. Two additional MSCTs (seventh and eighth, at 21 and 24 months old, respectively) were performed because chronic hypoxemia and bone fragility persisted. When the patient was 29 months old, severe pneumonia occurred, and tracheobronchomalacia relapsed. Two months after the ninth MSCT (33 months old), the issue of tracheobronchomalacia was resolved. At the age of 3 years, the patient could sit on a chair, and his respiratory condition stabilized, although he breathed with the help of a ventilator. The number of MSCTs performed was nine (from 7 to 33 months old), and the average MSC number per transplantation was $1.8x10^6$ cells/kg.

Hypomineralization of both the long bone and the flat bone improved. In particular, mineralized areas of the cranial bone, which were lost before treatment, recovered dramatically (Fig. 9.3a, b).

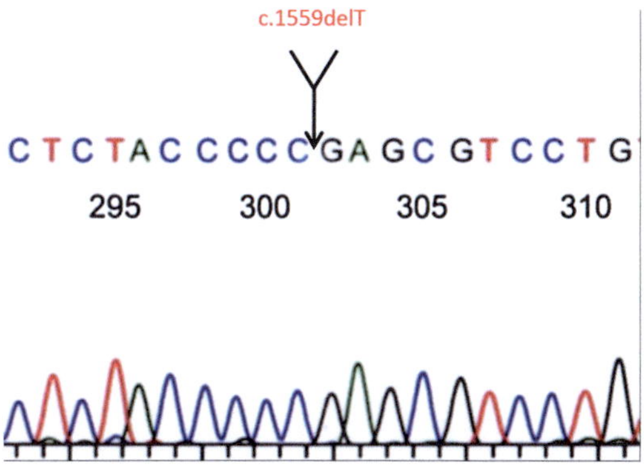

Fig. 9.2 *ALPL* mutation in this patient. We extracted DNA from white blood cells and analyzed the *ALPL* mutation by direct sequencing method. The patient harbored a homozygous mutation in c.1559delT of the *ALPL* gene

The length of the long bones also recovered gradually. Both bone density and muscle mass, measured by dual-energy X-ray absorptiometry (DXA), increased (Fig. 9.4). The titer of ALP started to increase from 11 months old but decreased from 20 months old. The ALP titer increased again at 30 months old, and urine PEA gradually decreased. Intriguingly, markers of bone formation (BAP, osteocalcin, undercarboxylated osteocalcin, and type I procollagen N-terminal propeptide) changed along with ALP and bone resorption markers (urine cross-linked N-telopeptide of type I collagen) decreased.

Some side effects of BMT occurred. Acute severe gut GVHD and chronic liver GVHD

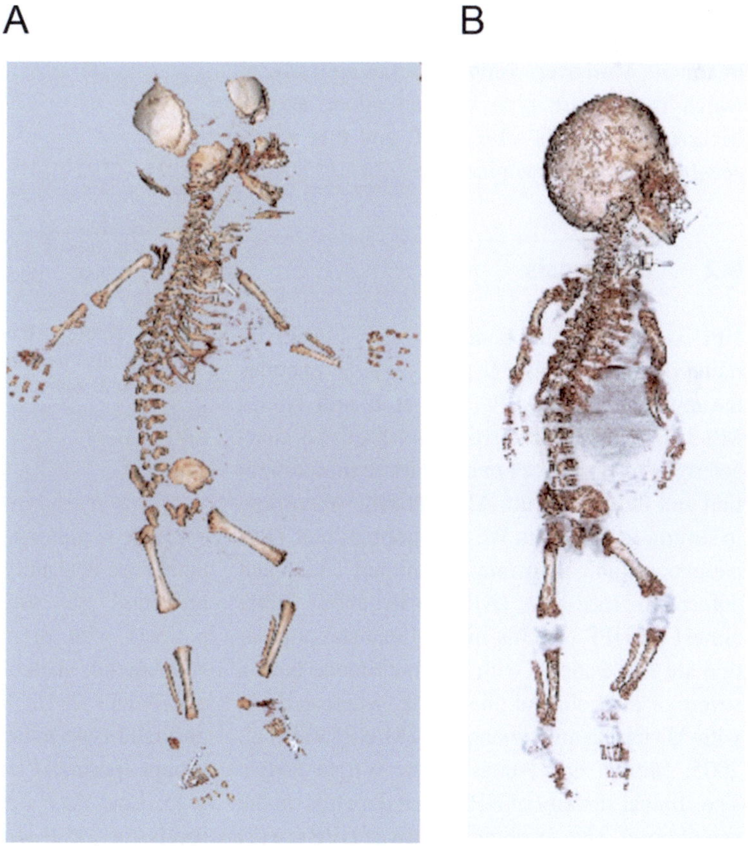

Fig. 9.3 Restoration of bone mineralization. Longitudinal bone computed tomography (CT) by three-dimensional reconstruction revealed that bone mineralization improved after mesenchymal stem cell transplantation with bone marrow transplantation. We evaluated a (**a**) 6-month-old and (**b**) 3-year-old

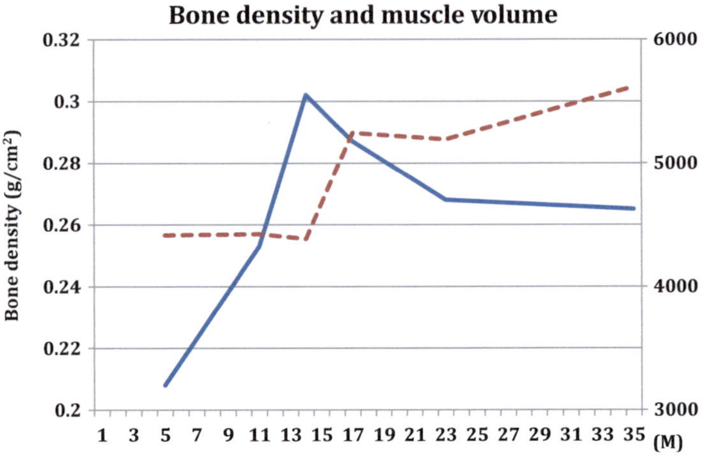

Fig. 9.4 Serial change of bone density and muscle volume. Bone density and muscle volume were measured by dual-energy X-ray absorptiometry (QDR Discovery (Hologic, Bedford, MA, USA). Blue line and red dash line indicate bone density and muscle volume, respectively. *M* month

improved with combination treatment with MSCs and immunosuppressant agents. Hypothyroidism developed 18 months after BMT. However, thyroid function recovered with thyroid hormone treatment. Moreover, complex partial epilepsy, of which the seizure type was an apnea attack, occurred 16 months after BMT, and this attack ceased with carbamazepine.

9.2 Diagnosis

HPP is a metabolic bone disorder caused by mutations in the *ALPL* gene, which encodes tissue-non-specific ALP (Mornet E and Nunes ME 2016; Whyte MP 2016). This disease is characterized by defective bone and tooth mineralization and reduced serum ALP activity. According to several reports from Western populations, HPP patients exhibit autosomal dominant (AD) and autosomal recessive (AR) inheritance, while almost all HPP patients in the Japanese population are AR. Patients with AR inheritance have a severe or mild clinical phenotype, whereas those with AD have a mild phenotype (Michigami et al. 2005; Mornet and Nunes 2016; Whyte 2016). The clinical severity of HPP often depends on the age of onset. The six clinical types of HPP are (1) perinatal type from fetal period to 1 month; (2) infantile from 1 to 6 months; (3) childhood type from the age of 6 months to 18 years; (4) odonto type, which is characterized by the premature loss of deciduous teeth by 5 years without apparent bone symptoms; and (5) adult type. The perinatal type is usually lethal because of a profound reduction in osteogenesis. However, some patients with prenatal onset have ameliorated spontaneous skeletal defects and survive (Mornet and Nunes 2016; Whyte 2016). So, nowadays, perinatal-type patients are divided into two subgroups, perinatal lethal type and prenatal benign type. Low ALP activity contributes to elevated levels of ALP substrates, that is, pyridoxal 5′-phosphate, PEA, and inorganic pyrophosphate. Less than 300 IU/L of ALP are found in almost all patients with HPP. Serum ALP level depends on severity of phenotype; the ALP titer in patients with severe HPP is very low (Fig. 9.5).

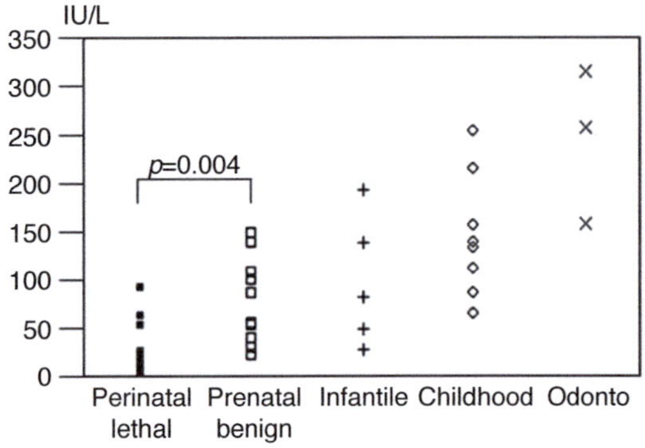

Fig. 9.5 Serum ALP titer by HPP phenotype. Serum ALP level depends on severity of phenotype; the ALP titer in patients with severe HPP is very low. Taketani T. et al. Arch Dis Child 2014. (Permission to post this figure by BMJ Journals)

Radiographic features revealed hypomineralization, deformity of long bones, flared metaphyses, hypolucent mid-metaphyses, osteochondral spurs, or narrow thorax.

Along with bone and tooth symptoms, many neurological symptoms, seizure, encephalopathy, intracranial hypertension, craniosynostosis, mental retardation (MR), deafness, and growth hormone deficiency (GHD), are frequently found in HPP patients (Taketani 2015; Whyte 2016). Seizure occurs in severe HPP types soon after birth and responds to pyridoxine but is an indicator of lethal prognosis. Encephalopathy with severe sequelae rarely presents in severe HPP types. Intracranial hypertension complicated in mild HPP types develops after the age of 1 year and sometimes needs neurosurgical intervention. MR, deafness, and GHD are more frequently found in Japanese HPP patients (Taketani 2015).

The six clinical features in detail were described as follows:

1. Perinatal Lethal Type and Prenatal Benign Type

Most patients with both types undergo the fetal ultrasonographic examination. Short extremities or deformed limbs and hydramnios are detected at 20–30 weeks of gestation. The main symptoms in both types are short extremities, deformed limbs, and bone fractures; respiratory failure including tracheobronchial spasm and pulmonary hypertension, vitamin B6-dependent convulsion, MR, deafness, and premature synostosis of the skull occur only in perinatal lethal type. The ALP level is significantly lower in perinatal lethal type compared with prenatal benign type. Radiographic examination shows that significant hypomineralization and narrow thorax are more frequent in perinatal lethal type than in prenatal benign type. In prenatal benign type, the ALP titer and urine PEA tend to be gradually increased and decreased, respectively, and hypomineralization also improves gradually. According to the prognosis, patients with perinatal lethal type have very poor prognosis within 1 year after birth. But, no one with prenatal benign type dies due to HPP.

2. Infantile Type

Most patients present with failure to thrive, hypercalcemia, hypomineralization, and flared metaphyses which resulted in rickets. For the follow-up period, premature loss of deciduous teeth is sometimes seen. A small number of patients have respiratory failure resulting in death. Living patients occasionally present with short stature, craniosynostosis, renal stone, and ambulation difficulty.

3. Childhood Type

The median age is 3 years at diagnosis. Bone pain, short extremities, short stature, and premature loss of deciduous teeth exist at diagnosis. MR and convulsion are also recorded. Long bone deformities and hypolucent mid-metaphyses are frequently found in radiographic examination. All patients have survived. Some patients with short stature are treated with GH.

4. Odonto Type

Premature loss of deciduous teeth, severe dental caries, and gingivitis are seen without bone lesion. Permanent teeth are not disturbed. In X-ray, alveolar bone wear, expansion of the pulp cavity, and root canal are found.

5. Adult Type

Patients with this type frequently present with bone pain due to stress fractures of metatarsals and false fractures of the femurs. A false fracture of the femur is seen as a fracture line called Looser zones in X-ray. Since middle age, cartilage calcification and osteoarthritis are sometimes merged. They have experienced rickets and premature loss of deciduous teeth in childhood.

9.3 Differential Diagnosis

The differential diagnosis of HPP depends on the age at which the diagnosis is considered (Mornet and Nunes 2016).

In utero, early prenatal ultrasound examination may lead to a consideration of osteogenesis imperfecta (OI), campomelic dysplasia, and chondrodysplasias with defects in bone mineralization, as well as HPP. After birth, OI, thanatophoric dysplasia, campomelic dysplasia, and chondrodysplasias with bone mineralization defects are outwardly difficult to distinguish from HPP. After infancy, rickets defines the physical and radiographic features of HPP. However, rickets due to nutritional and/or vitamin D deficiency, vitamin D resistance, or renal osteodystrophy is distinguished from HPP. In adulthood, we need to differentiate HPP and other bone diseases including osteoarthritis, pseudogout (secondary to calcium pyrophosphate dehydrate deposition), and osteopenia/osteoporosis.

9.4 Molecular Genetics

The *ALPL* gene is located on chromosome 1 p34–36, which consists of 12 exons with 524 amino acids. Since 17 amino acids at the N-terminus are signal peptides, which are removed by processing, the mature ALP protein is 507 amino acids. *ALPL* gene mutation has been found in most patients (95%) of HPP. More than 260 types of *ALPL* mutations have been identified in HPP patients, and 80% of these are missense mutations (Mornet and Nunes 2016; Whyte 2016). The phenotypes of HPP patients are closely related to the residual enzyme activity effects of *ALPL* mutations (Mornet and Nunes 2016; Whyte 2016; Taketani et al. 2014). A molecular-based estimation of the prevalence of HPP in the European population estimates the prevalence of severe and moderate HPP as 1/300,000 and 1/6370, respectively (Mornet et al. 2011). In the Japanese population, the prevalence of the 1559delT homozygous mutation in the *ALPL* gene, which is a common mutation that causes the perinatal lethal form, is estimated to be not less than 1/900,000 (Watanabe et al. 2011). This mutation is not found in other countries, suggesting that c.1559delT may be a founder mutation in the Japanese population (Michigami et al. 2005).

9.5 Biochemical and Molecular Perspectives

ALP hydrolyzes phosphate monoester bonds to phosphoric acid and alcohol at alkaline pH (pH 9–11). The substrate specificity is low. ALP is expressed in many tissues and cells. There are at least four isozymes in ALP including tissue-non-specific, intestinal, placental, and germ cell ALP. ALP is usually present as a membrane anchor-type enzyme on the cell surface. ALP structure is present as a glycosylphosphatidylinositol (GPI) anchor-type membrane protein forming a homodimer by noncovalent bond. When ALP-containing GPI-anchor proteins are treated with phosphatidyl inositol-specific phospholipase C or GPI-specific phospholipase D, ALP is released off the cell membrane. ALP does not lose its enzyme activity even if released from the cell membrane.

Bone mineralization begins in osteoblast and chondrocyte-derived matrix vesicles. Calcium and inorganic phosphate ions (Pi) contained in this vesicle are the constituents of hydroxyapatite crystals that are formed first. This crystal is released outside the cell, and the bone is formed by binding protein such as collagen, osteocalcin, osteopontin (OPN), osteonectin, and calcium phosphate. However, inorganic pyrophosphate (PPi) inhibits formation of hydroxyapatite crystals by antagonizing Pi. In other words, it is necessary that the balance of Pi and PPi is maintained in normal calcification. ALP, in addition to plasma cell membrane glycoprotein-1 (PC-1), ankylosis (ANK), and OPN are factors that regulate Pi and PPi (Whyte 2016, 2017). ALP exists in substrate vesicles and osteoblast membrane and hydrolyzes PPi to increase Pi (Fig. 9.6). PC-1 is present in substrate vesicles and osteoblast membrane and produces PPi from nucleotide triphosphate (Whyte 2016). ANK is a PPi transporter that exists in the osteoblast membrane and carries out PPi from within the cell to outside the cell (Whyte 2016). OPN inhibits the formation of hydroxyapatite crystals. Thus, ALP promotes mineralization, whereas PC-1, ANK, and OPN suppress calcification.

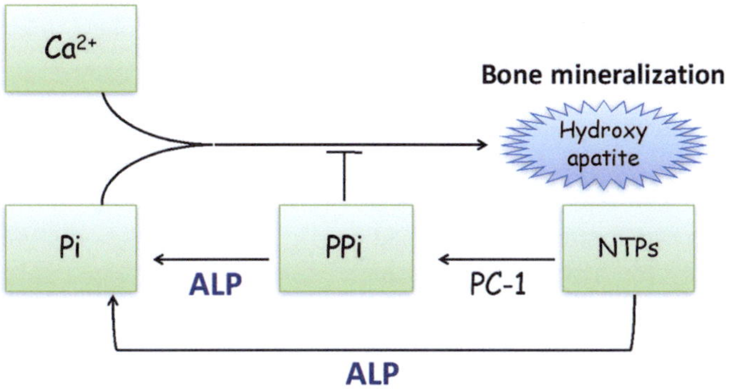

Fig. 9.6 Mechanism of ALP-related bone mineralization. Pi, inorganic phosphate; PPi, inorganic pyrophosphate; NTPs, nucleoside triphosphates; PC-1, plasma cell membrane glycoprotein-1; Pi is produced by ALP-induced dephosphorylation of PPi and NTPs. Reduction of ALP activity leads to decline of Pi and elevation of PPi, resulting in impaired production of hydroxyapatites, a principal component of bone and tooth

ALP, present in the neuronal membrane, also interacts with the synapses that are involved in neurotransmitter synthesis, synaptic stabilization, and myelin pattern formation (Taketani 2015). A study in mice also showed that ALP dysfunction compromises myelination and synaptogenesis in the brain (Taketani 2015). Thus, ALP plays a role in neural function such as developmental plasticity and activity-dependent cortical functions.

In the *ALPL* gene (*Akp2* gene in mouse) knockout mice, symptoms similar to severe forms of infantile HPP are presented. It causes growth disorder, accompanied by epileptic seizure and apnea, and dies before weaning. The hypomineralization of the skull and spinal cord is observed, the metaphyseal growth plate was short, and the differentiation of epiphyseal and growth plate cartilage is stopped (Whyte 2016, 2017).

Table 9.1 Supportive and preventive therapy in patients with HPP

Respiratory management	Prevention of respiratory infection, oxygen supply, mechanical respirator
Convulsion	Vitamin B6
Hypercalcemia	Low calcium milk, calcitonin
Intracranial hypertension	Craniotomy
Tooth management	Prevention of caries, denture
Bone pain	Non-steroid anti-inflammatory drugs
Bone fracture	Recombinant parathyroid hormone, orthopedic reduction

9.6 Therapy and Prevention

There is currently no well-established therapy for HPP. Previously, patients with HPP have taken supportive therapy (Table 9.1). But, currently, bone-targeting enzyme replacement therapy is performed as first-line therapy (Whyte et al. 2012). Cell therapy including BMT and MCST are under development for the treatment of HPP (Whyte et al. 2003; Cahill et al. 2007; Tadokoro et al. 2009; Taketani et al. 2015). In the future, gene-engineering therapy using normal *ALPL* gene insertion or gene editing of *ALPL* mutations is expected to be promising for curative treatment.

9.7 Bone-Targeting Enzyme Replacement Therapy

Plasma components derived from Paget's disease and placenta, of which plasma contains a high titer of ALP, were administered in the 1980s (Whyte et al. 2016). Bone mineralization did not improve although serum ALP level temporarily increased. The main problem considered was that the administered ALP enzyme does not reach the bone. Sekido reported that acidic oligopeptide sequences (many consecutive sequences of 2–9 acid peptides of aspartic acid (Asp) or glutamic acid) are commonly found in proteins, bone sialoprotein, and OPN that bind to hydroxyapatite (Sekido et al. 2001). The ALP enzyme tagged with a six or eight residue stretch of Asp had 30-fold higher affinity for hydroxyapatite than the untagged enzyme (Nishioka et al. 2006). In vitro mineralization assays with BM from a patient with HPP using the tagged enzyme provided evidence of bone mineralization (Nishioka et al. 2006). In 2008, a recombinant human ALP with a deca-Asp motif for hydroxyapatite targeting, asfotase alfa (AA), was produced (Millán et al. 2008). *Alp2* knockout mice, resulting in dying within 1 month, remained healthy when AA was administered subcutaneously (Millán et al. 2008).

Whyte et al. performed the clinical trial for patients with perinatal or infantile HPP of treatment with AA (Whyte et al. 2012). Muscle strength and skeletal mineralization improved substantially, sometimes rapidly, and were associated with better pulmonary, cognitive, and motor function. Now, after more than 5 years of treatment, no patient requires respiratory support, and all have made further gains (Whyte 2017). Whyte et al. also evaluated the efficacy and safety of AA administered to children of 6–12 years of age at baseline who were substantially impaired by HPP (Whyte et al. 2016). Improvement in skeletal disturbance and muscle strength was rapidly found. Resolution of bone pain and gait impairment was also recognized.

9.8 Mesenchymal Stem Cell Transplantation

MSCs reside in BM and other tissues such as fat, dental pulp, muscle, cartilage, umbilical cord, and amniotic fluid. MSCs possess self-renewal capacity and can differentiate into various mesenchymal lineages and several mesoderm lineages, including the osteogenic, chondrogenic,

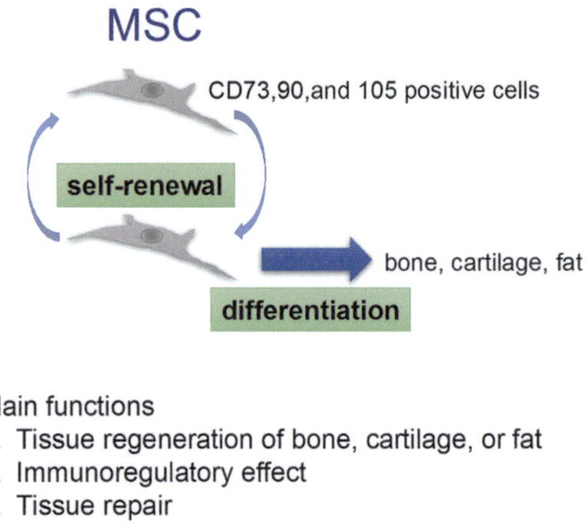

Fig. 9.7 General picture of mesenchymal stem cells. MSCs simultaneously express CD 73, 90, and 105 antigens in cell surface. Immunomodulatory effects: MSCs include suppression of T cell proliferation, induction of regulatory T cells, influencing dendritic cell maturation and function, suppression of B cell proliferation and terminal differentiation, and inhibition of NK cell function. Tissue repair: When tissue is damaged, inflammation occurs, and MSCs are mobilized to the site. MSCs differentiate into functional cells to replace damaged cells. MSCs and/or other cells stimulated by MSCs also produce trophic factors, cytokines, extracellular matrix, intracellular microvesicles, resulting in tissue repair through angiogenesis, remodeling of the extracellular matrix and the differentiation of tissue progenitor cells

and adipogenic lineages. MSCs function as a multipotent modality in immune regulation as well as the repair and regeneration of tissues and organs (Fig. 9.7). Their regenerative ability makes MSCs an ideal source for transplantation in patients with OI, inborn errors of metabolism, and ischemic heart diseases (Abdallah and Kaseem 2009). Furthermore, MSCs support the maintenance of hematopoietic stem cells, facilitate hematopoietic recovery, and reduce GVHD by modulating the immune functions of T cells, B cells, natural killer cells, monocytes, and dendritic cells.

Some studies have reported the use of BMT, BM stromal cells, bone fragment, and cultured osteoblasts in the treatment of HPP. However, the clinical efficacy of these procedures is limited (Whyte et al. 2003; Cahill et al. 2007). Tadokoro et al. previously performed allogeneic

BMT, MSCT, and osteogenic construct implantation for a patient with severe HPP. Symptoms of the patient improved to some extent, and new bone formation derived from cells of both the donor and the patient was detected. Nevertheless, the procedure was very complicated, and clinical features such as mineralization were not sufficient compared with healthy children, while the titer of serum ALP remained very low (Tadokoro et al. 2009).

To focus on the osteogenic effects of the MSCs, multiple infusions of ex vivo expanded allogeneic MSCs were performed for patients with severe HPP who had undergone BMT (Taketani et al. 2015). There were improvements in not only bone mineralization but also muscle mass, respiratory function, and mental development, resulting in the patients being alive at the age of 6. These data suggest that multiple MSC infusions, following BMT, are effective and brought about clinical benefits for patients with lethal HPP.

9.9 Gene Therapy

Gene therapy has proven to be a useful tool for the treatment of several inherited diseases. Many problems for gene therapy including oncogenic property and gene insertion efficacy have been resolved. In vivo research using Akp2 knockout mice is done. Adeno-associated virus 8 (AAV8)-mediated systemic gene therapy was an effective treatment for the rescue of Akp2 knockout mice (Matsumoto et al. 2011). They evaluated the therapeutic effects of AAV8 vectors including soluble ALP with deca-Asps (ALP-D10), bone-targeted ALP. BM cells transduced into ALP-D10 can also serve as a reservoir for stem cell-based enzyme replacement therapy to rescue the Akp2 knockout mice (Iijima et al. 2015). They evaluated the feasibility of neonatal ex vivo gene therapy in Akp2 knockout mice using lentivirally transduced BM cells expressing bone-targeted ALP, ALP-D10. The treated Akp2 knockout mice were normal in appearance and experienced no seizures. These suggest that gene therapy is a hopeful treatment for patients with HPP.

End-of-Chapter Questions

1. When you see an individual who has bone deformity, bone fracture, bone pain, refractory convulsion, premature loss of deciduous teeth, and short stature that is reminiscent of HPP, how do you diagnose the severity of HPP clinically and genetically?
2. What will you consider as the differential diagnosis of HPP in accordance with age?
3. What treatment will you recommend for patients with severe HPP?

References

Abdallah BM, Kassem M (2009) The use of mesenchymal (skeletal) stem cells for treatment of degenerative diseases: current status and future perspectives. J Cell Physiol 218:9–12

Cahill RA, Wenkert D, Perlman SA, Steele A, Coburn SP, McAlister WH, Mumm S, Whyte MP (2007) Infantile hypophosphatasia: transplantation therapy trial using bone fragments and cultured osteoblasts. J Clin Endocrinol Metab 92:2923–2930

Iijima O, Miyake K, Watanabe A, Miyake N, Igarashi T, Kanokoda C, Nakamura-Takahashi A, Kinoshita H, Noguchi T, Abe S, Narisawa S, Millán JL, Okada T, Shimada T (2015) Prevention of lethal murine Hypophosphatasia by neonatal ex vivo gene therapy using Lentivirally transduced bone marrow cells. Hum Gene Ther 26:801–812

Matsumoto T, Miyake K, Yamamoto S, Orimo H, Miyake N, Odagaki Y, Adachi K, Iijima O, Narisawa S, Millán JL, Fukunaga Y, Shimada T (2011) Rescue of severe infantile hypophosphatasia mice by AAV-mediated sustained expression of soluble alkaline phosphatase. Hum Gene Ther 22:1355–1364

Michigami T, Uchihashi T, Suzuki A, Tachikawa K, Nakajima S, Ozono K (2005) Common mutations F310L and T1559del in the tissue-nonspecific alkaline phosphatase gene are related to distinct phenotypes in Japanese patients with hypophosphatasia. Eur J Pediatr 164:277–282

Millán JL, Narisawa S, Lemire I, Loisel TP, Boileau G, Leonard P, Gramatikova S, Terkeltaub R, Camacho NP, McKee MD, Crine P, Whyte MP (2008) Enzyme replacement therapy for murine hypophosphatasia. J Bone Miner Res 23:777–787

Mornet E, Nunes ME (2007 Nov 20 [updated 2016 Feb 4]) Hypophosphatasia. In: Pagon RA, Adam MP, Ardinger HH, Wallace SE, Amemiya A, Bean LJH., Bird TD., Ledbetter N, Mefford HC, Smith

RJH, Stephens K (eds) GeneReviews® [Internet]. University of Washington, Seattle, Seattle: 1993–2017

Mornet E, Yvard A, Taillandier A, Fauvert D, Simon-Bouy B (2011) A molecular-based estimation of the prevalence of hypophosphatasia in the European population. Ann Hum Genet 75:439–445

Nishioka T, Tomatsu S, Gutierrez MA, Miyamoto K, Trandafirescu GG, Lopez PL, Grubb JH, Kanai R, Kobayashi H, Yamaguchi S, Gottesman GS, Cahill R, Noguchi A, Sly WS (2006) Enhancement of drug delivery to bone: characterization of human tissue-nonspecific alkaline phosphatase tagged with an acidic oligopeptide. Mol Genet Metab 88:244–255

Sekido T, Sakura N, Higashi Y, Miya K, Nitta Y, Nomura M, Sawanishi H, Morito K, Masamune Y, Kasugai S, Yokogawa K, Miyamoto K (2001) Novel drug delivery system to bone using acidic oligopeptide: pharmacokinetic characteristics and pharmacological potential. J Drug Target 9:111–121

Tadokoro M, Kanai R, Taketani T, Uchio Y, Yamaguchi S, Ohgushi H (2009) New bone formation by allogeneic mesenchymal stem cell transplantation in a patient with perinatal hypophosphatasia. J Pediatr 154:924–930

Taketani T (2015) Neurological symptoms of Hypophosphatasia. Subcell Biochem 76:309–322

Taketani T, Onigata K, Kobayashi H, Mushimoto Y, Fukuda S, Yamaguchi S (2014) Clinical and genetic aspects of hypophosphatasia in Japanese patients. Arch Dis Child 99:211–215

Taketani T, Oyama C, Mihara A, Tanabe Y, Abe M, Hirade T, Yamamoto S, Bo R, Kanai R, Tadenuma T, Michibata Y, Yamamoto S, Hattori M, Katsube Y, Ohnishi H, Sasao M, Oda Y, Hattori K, Yuba S, Ohgushi H, Yamaguchi S (2015) Ex vivo expanded allogeneic mesenchymal stem cells with bone marrow transplantation improved osteogenesis in infants with severe Hypophosphatasia. Cell Transplant 24:1931–1943

Watanabe A, Karasugi T, Sawai H, Naing BT, Ikegawa S, Orimo H, Shimada T (2011) Prevalence of c.1559delT in ALPL, a common mutation resulting in the perinatal (lethal) form of hypophosphatasia in Japanese and effects of the mutation on heterozygous carriers. J Hum Genet 56:166–168

Whyte MP (2016) Hypophosphatasia - aetiology, nosology, pathogenesis, diagnosis and treatment. Nat Rev Endocrinol 12:233–246

Whyte MP (2017) Hypophosphatasia: enzyme replacement therapy brings new opportunities and new challenges. J Bone Miner Res 32:667–675

Whyte MP, Kurtzberg J, McAlister WH, Mumm S, Podgornik MN, Coburn SP, Ryan LM, Miller CR, Gottesman GS, Smith AK, Douville J, Waters-Pick B, Armstrong RD, Martin PL (2003) Marrow cell transplantation for infantile hypophosphatasia. J Bone Miner Res 18:624–636

Whyte MP, Greenberg CR, Salman NJ, Bober MB, McAlister WH, Van Sickle B, Wenkert D, Edgar TS, Bauer ML, Hamdan M, Simmons JH, Bishop N, Lutz RE, McGinn M, Craig S, Moore JN, Taylor JW, Cleveland RH, Cranley WR, Lim R, Thacher TD, Mayhew JE, Downs M, Millan JL, Skrinar A, Crine P, Landy H (2012) Enzyme replacement therapy in life-threatening hypophosphatasia. N Engl J Med 366:904–913

Whyte MP, Madson KL, Phillips D, Reeves A, McAlister WH, Yakimoski A, Mack K, Hamilton K, Kagan K, Melian A, Thompson D, Moseley S, Odrljin T, Greenberg CR (2016) Asfotase alfa therapy for children with hypophosphatasia. JCI Insight 1:e85971 1–10

Phenylketonuria

10

Ikue Hata, Miori Yuasa, and Yuko Isozaki

Keywords
Phenylketonuria · Phenylalanine · Phenylalanine hydroxylase · Tetrahydrobiopterin · BH_4 deficiency

10.1 Case Reports

Case 1 was the third child of healthy and non-consanguineous parents without any family history of phenylketonuria (PKU). A newborn screening identified high level (>1200 µmol/L) of blood phenylalanine (Phe), and she was admitted to hospital at the age of 13 days. Physical examinations revealed no abnormalities, and clinical laboratory results were normal except for an elevated plasma Phe concentration of 3824 µmol/L (reference value, 42–108 µmol/L). The oral administration of tetrahydrobiopterin (BH_4), a cofactor for phenylalanine hydroxylase (PAH), did not affect the plasma Phe levels and thus excluded disordered BH_4 synthesis and recycling (BH_4 deficiency). Consequently, she was diagnosed with classical PKU and treated with a Phe-restricted diet. Breast milk was partially replaced with a Phe-free formula to lower the blood Phe levels to within the recommended range (120–360 µmol/L). After weaning, she continued on a Phe-restricted diet including an intake of low-protein foods, a Phe-free formula, and a low-Phe peptide powder supplement. The diet was well-tolerated, and blood Phe levels have remained mostly controlled and within the recommended range. Her growth and neurological development was normal to the point where she is currently 23 years of age and works as a nurse.

Case 2 was the first child of healthy and non-consanguineous parents with no family history of PKU. She was delivered at 31 weeks of gestation, with a birth weight of 1680 g. Her general health status was good at 18 days old, but a newborn screening revealed a high blood Phe level (2327 µmol/L). Further laboratory examination revealed a high plasma Phe level (1606 µmol/L), but no other abnormalities were detected. She was diagnosed with classical PKU and immediately placed on a Phe-restricted diet. However, she did not tolerate the diet therapy well, and the blood Phe levels remained above the recommended ranges. Currently, she is 30 years of age and has a mild intellectual disability.

Case 3 was the second child of healthy and non-consanguineous parents without a family history of PKU. A newborn screening revealed a mildly elevated level of blood Phe (455 µmol/L). At 14 days postdelivery, physical examinations revealed no abnormalities, and laboratory results were normal, with the exception of mildly ele-

I. Hata (✉) · M. Yuasa · Y. Isozaki
Department of Pediatrics, Faculty of Medical Sciences, University of Fukui, Fukui, Japan
e-mail: ikueh@u-fukui.ac.jp; miori@u-fukui.ac.jp; yuyu@u-fukui.ac.jp

T. Oohashi et al. (eds.), *Human Pathobiochemistry*, https://doi.org/10.1007/978-981-13-2977-7_10

vated plasma Phe (570 μmol/L). A single loading dose of BH_4 followed by BH_4 administration for 1 week decreased the plasma Phe by 20% and 35%, respectively. BH_4 deficiency was excluded by the results of analyses of blood and urinary pterins and dihydropteridine reductase activity. She was diagnosed with BH_4-responsive mild PKU, which was supported by *PAH* gene analysis. At 23 days of age, she was placed on BH_4 therapy without a Phe-restricted diet, and plasma Phe levels were mostly controlled under free feeding. However, plasma Phe levels exceeded the recommended range after weaning due to increased natural protein intake. Thus, in addition to the maximum dose of BH_4 (20 mg/kg/day), mild restriction of natural protein intake improved blood Phe levels to within the recommended range. She is presently 4 years of age, with normal growth and neurological development, and has remained free from adverse events related to BH_4 therapy to date.

10.2 Diagnosis

PKU is an amino acid metabolic disorder caused by the congenital deficiency of PAH which catalyzes the conversion of Phe to tyrosine (Fig. 10.1). An increased blood Phe concentration leads to severe neurological disorders. The catalytic activity of PAH requires BH_4 as a cofactor; therefore, BH_4 deficiency causes hyperphenylalaninemia as PKU. PKU and BH_4 deficiency are diagnosed through the following examinations.

1. Amino acid analysis

Blood Phe levels are measured through mass screening of newborns. Blood is taken 4–6 days after birth, dried on filter paper, and analyzed by tandem mass spectrometry. The cutoff value of Phe in newborn screening is approximately 120 μmol/L. If the Phe level at screening exceeds the cutoff level, plasma amino acid analysis must be performed by high-performance liquid chromatography. Hyperphenylalaninemia is defined as plasma Phe concentrations above 120 μmol/L.

Blood Phe levels vary with the residual activities of PAH. Patients with PKU are classified based on the blood Phe levels at diagnosis: classical PKU (blood Phe levels exceed 1200 μmol/L), mild PKU (blood Phe levels are 600 to 1200 μmol/L), and mild hyperphenylalaninemia (blood Phe levels are 120 to 600 μmol/L).

2. Pterin analysis

Analysis of blood and urinary pterins is necessary to distinguish BH_4 deficiency from PAH deficiency. The levels of neopterin and biopterin, which are intermediary metabolites of BH_4 synthesis, indicate profiles that are specific to each type of BH_4 deficiency. Furthermore, dihydropteridine reductase (DHPR) activity of dried blood spots should also be analyzed.

3. BH_4 loading test

Before the beginning of dietary Phe restriction, the BH_4 loading test should be performed. After administration of a BH_4 loading dose (20 mg/kg), some cases of mild PKU show a greater than 20% reduction in blood Phe levels, while most cases of classical PKU demonstrate no response. Cases that show a greater than 30% reduction in blood Phe levels after administration of BH_4 (20 mg/kg/day) for 1 week are diagnosed as BH_4-responsive mild PKU. Cases of BH_4 deficiency, excluding DHPR deficiency, demonstrate a prompt (4–8 hours after a dose of BH_4) reduction of blood Phe concentrations to within the normal range.

10.3 Biochemical Perspectives

Phe is an aromatic essential amino acid, which is indispensably obtained through the diet. Approximately three-fourths of dietary Phe is converted to tyrosine by PAH and is then utilized for energy and for generation of tyrosine-derived products, such as melanin and catecholamine (Fig. 10.1). The remainder of dietary Phe is converted to phenylpyruvate, phenylacetate, and phenyllactate, which are not metabolized further,

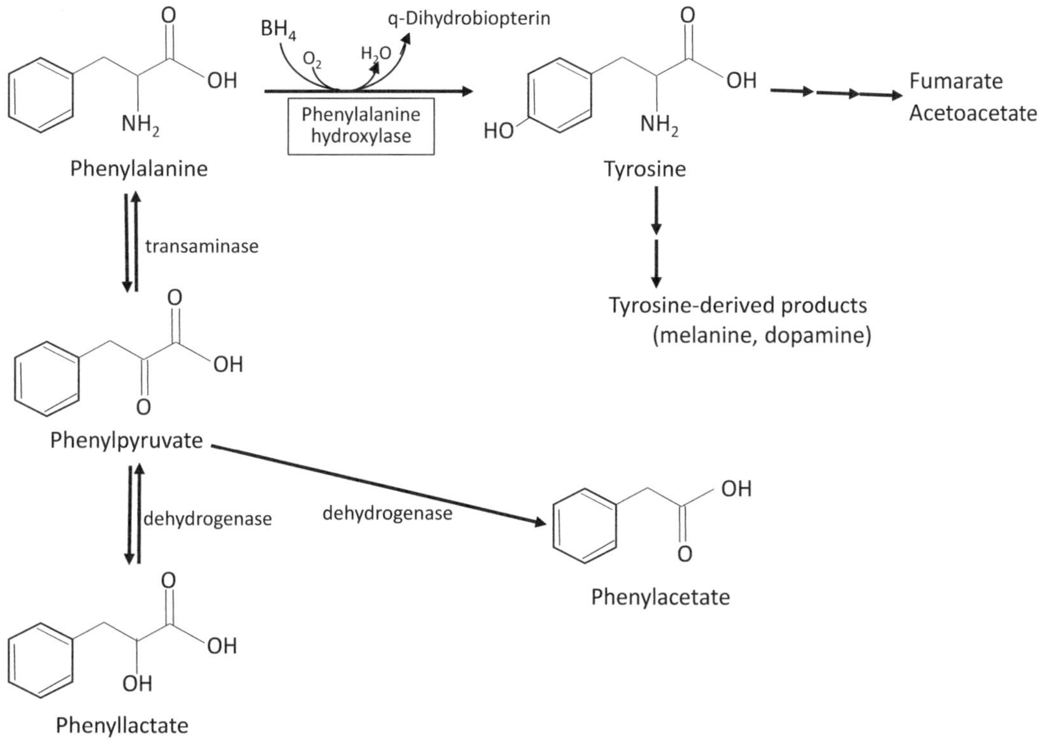

Fig. 10.1 Metabolism of phenylalanine. Dietary Phe is converted to tyrosine by PAH and is then utilized for energy and for generation of tyrosine-derived products. The remainder of dietary Phe is converted to phenylpyruvate, phenylacetate, and phenyllactate which are excreted via the urine

are accumulated in the blood and tissues, and are ultimately excreted via the urine.

PAH activity is most significant in the liver but is also high in the kidney and is about one-third of that in the liver (Heintz et al. 2012). The active form of PAH, which is an enzyme encoded by the *PAH* gene, exists as dimer or tetramer of subunits. Each subunit is about 51 kDa in size and contains a catalytic domain, a multimerization domain, and an N-terminal regulatory ACT domain that serves as a sensor that modulates an allosteric response to Phe (Fig. 10.2a) (Jaffe 2017). The activity of PAH requires Phe, molecular oxygen, non-heme iron, and BH_4 as a cofactor.

PAH assemblies exist as two structurally different types, which are called the resting-state PAH (RS-PAH) and the activated PAH (A-PAH) (Fig. 10.2b, c) (Jaffe 2017). The equilibrium between both types of assemblies fluctuates in response to Phe concentration. At low Phe concentration (<50 μmol/L), the predominant type is RS-PAH, which is characterized by a low affinity for Phe, and its protein structure consists of each regulatory domain existing separately. A rise in Phe concentrations is supposed to induce the rotation of these regulatory domains, and their interaction leads to the formation of A-PAH.

PAH is an allosteric enzyme that requires the binding of Phe for activation. It was previously demonstrated that Phe bound to some allosteric sites at the interface of the regulatory domains and served as a stabilizer in A-PAH (Zhang and Fitzpatrick 2016). BH_4 also acts to stabilize PAH as a "pharmacological chaperone" and works to restore the function of PAH in cases of BH_4-responsive mild PKU. BH_4 forms two hydrogen bonds with Ser23 of regulatory domains in RS-PAH in order to impede active site access, and this interaction is lost when Phe concentrations rise. This transient auto-inhibitory interaction is

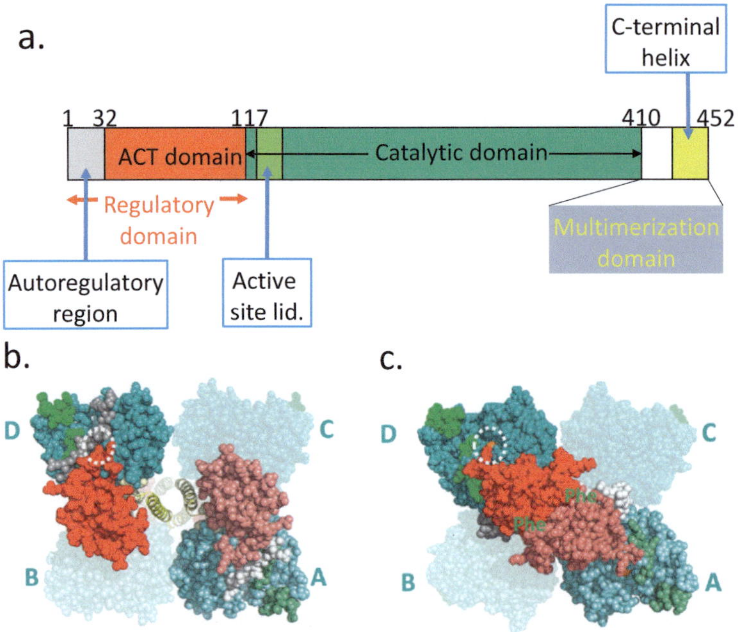

Fig. 10.2 The structure of PAH. (**a**) PAH protein consists of the regulatory, catalytic, and multimerization domains, which include subdomains. Numbers represent residues of the termini and hinges of domains. (**b**) The crystal structure of RS-PAH is illustrated. Each subunit is labeled in cyan near the catalytic domains, red near the regulatory domain, and yellow near C-terminal helix. The dotted white circle illustrates the autoregulatory region. (**c**) A composite homology model of the A-PAH tetramer contains close associated ACT domains, which form allosteric Phe binding site (green "Phe" label) (Reproduced and modified with permission from Mol Genet Metab, 2017; 121, 289–296)

necessary to maintain the equilibrium between RS-PAH and A-PAH (Jaffe 2017).

A deficiency of PAH activity causes the accumulation of Phe in the body. Untreated patients present with severe mental retardation, microcephalus, and seizures in early childhood. The accumulated Phe is converted to phenylpyruvate and other metabolites. These are excreted into urine and exude a peculiar "mousy" odor. In addition to the deficiency of tyrosine conversion from dietary Phe, the inhibition of tyrosinase by excessive Phe reduces the synthesis of melanin, resulting in a light pigmentation.

Several potential mechanisms of neurocognitive impairment, which is the main problem associated with PKU, have been demonstrated, although the specific details have not been proven sufficiently (de Groot et al. 2010). As Phe is a large neutral amino acid (LNAA), it competes with other LNAAs, such as arginine, histidine, isoleucine, leucine, lysine, methionine, threonine, tryptophan, tyrosine, and valine, for the large neutral amino acid type 1-transporter in the intestine and blood-brain barrier. High blood Phe concentrations in PKU patients lead to the uptake of large amounts of Phe into the central nervous system (CNS), and in competition with other LNAAs, this induces an amino acid imbalance. Insufficient levels of non-Phe LNAAs in CNS cause the reduction of protein synthesis that is essential for proper brain development and function. Myelin protein synthesis is inhibited by low LNAAs levels and/or high Phe concentrations (de Groot et al. 2010), which leads to white matter abnormalities and cognitive impairment in PKU patients. Reduced activity of 3-hydroxy-3-methylglutaryl coenzyme A reductase, the key enzyme for cholesterol synthesis, may cause a decrease of myelin lipid. In addition to the reduced supply of tyrosine and tryptophan, the suppression of tyrosine hydroxylase and tryptophan hydroxylase by excessive Phe levels results in deficiencies of monoamine neurotransmitters dopamine and serotonin. Other mechanisms, such as altered brain energy metabolism, impaired glutamatergic synaptic transmission, and increased oxidative stress, were also reported to be involved in the pathophysiology of brain damage from PKU. There have been some reports that Phe-reducing treatment corrected these alterations.

Because BH_4 is a common cofactor of PAH, tyrosine hydroxylase, and tryptophan hydroxylase, congenital disorders of BH_4 synthesis or recycling systems result in a deficiency of those enzymes. Dysfunction of PAH activity causes hyperphenylalaninemia, dysfunction of tyrosine hydroxylase causes catecholamine deficiency, and dysfunction of tryptophan hydroxylase causes serotonin deficiency. Deficiencies of these neurotransmitters lead to hypotonia of the trunk, lead-pipe rigidity, a high-pitched cry, seizures, and sleep disturbances during the infantile period. Untreated cases exhibit severe developmental delay and finally death. The BH_4 synthesis system consists of GTP cyclohydrolase I, 6-pyruvoyl-tetrahydropterin synthase, and sepiapterin reductase (SR). The BH_4 recycling system consists of pterin-4α-carbinolamine dehydratase and DHPR (Fig. 10.3) (Thöny and Martinez 2016). Patients with SR deficiency do not develop hyperphenylalaninemia because other enzymes substitute for SR in the liver.

10.4 Molecular Perspectives

The mode of inheritance of PAH deficiency is autosomal recessive. The *PAH* gene is mapped to human chromosome 12q24.1 with 13 exons and codes for 452 amino acids. More than 950 variants in the *PAH* gene have been identified in patients with PKU (Blau 2016). The reported *PAH* variants in the locus-specific database *PAH*vdb (http://www.biopku.org/pah/) contain missense variants (58.5%), deletions (15.9%), splice-site variants (13.7%), nonsense variants (6.0%), and insertions (3.1%). Although variants are identified in all exons, many of them are located in exons 6 (13.5%) and 7 (13.9%) of the catalytic domain and exon 3 (9.7%) of the regulatory domain (Wettstein et al. 2015). The frequency of *PAH* variants is specific for different world regions. In Japanese patients with PKU

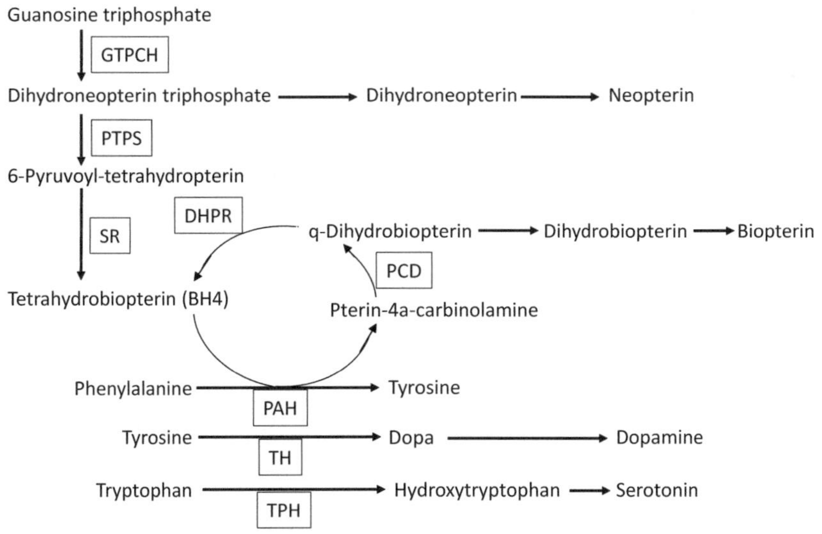

Fig. 10.3 The summary of biosynthesis and actions of BH_4. BH_4 is a common cofactor of phenylalanine hydroxylase (PAH), tyrosine hydroxylase (TH), and tryptophan hydroxylase (TPH). The BH_4 synthesis system consists of GTP cyclohydrolase I (GTPCH), 6-pyruvoyl-tetrahydropterin synthase (PTPS), and sepiapterin reductase (SR), and the BH_4 recycling system consists of pterin-4α-carbinolamine dehydratase (PCD) and dihydropteridine reductase (DHPR)

and hyperphenylalaninemia, R241C, R413P, IVS4-1g4>a, R111X, and R243Q variants are common (Okano et al. 2011).

Residual activities and association with phenotype of several mutated forms of PAH have been investigated. For example, p.R408W and p.R252W, which markedly reduce PAH activity, are frequently observed in classical PKU cases. Variants of p.E390G, p.Y414C, p.A300S, and p. R261Q show high residual PAH activities and often are observed in mild cases (Blau 2016). The correlation of the responsiveness to BH_4 with genotypes and phenotypes in PKU patients was also investigated. The analysis of data from the BIOPKU database (http://www.biopku.org/biopku/) revealed that most of patients with mild hyperphenylalaninemia (92.9%) were classified as BH_4 responders compared with mild PKU (63.9%) and classical PKU (15.5%) (Shen et al. 2016). The clear association of in vitro residual PAH activities of each variant with BH_4 responsiveness was also demonstrated (Blau 2016). Although the molecular basis of BH_4 responsiveness has not been sufficiently clarified, *PAH* variants in the multimerization domain were reported to be associated with BH_4 responsive phenotypes (77.4%) compared with other domains (Shen et al. 2016). As mentioned above, the severity of PKU phenotype and BH_4 responsiveness can be predicted by genotype to some extent.

However, inconsistencies are often observed in genotype-phenotype correlations. The principal cause of those inconsistencies is that most cases of PKU are compound heterozygotes. Co-expression of *PAH* variants *in vitro* revealed possible dominance effects by one allele, which could be both negative and positive. For example, when a null p.R408W variant co-expressed with other variants showing high residual activities, p.[I65T];[R408W] exhibited lower activity (15.0%) than expected (26.8%), and p.[P384S];[R408W] exhibited higher activity (56.1%) than expected (40.8%) (Shen et al. 2016). It was evaluated that BH_4 responsiveness could be accurately predicted by genotype alone in about 70% of PKU cases (Wettstein et al. 2015). While insufficient at present, molecular diagnosis is indispensable for the proper management for each PKU patients.

Five genes encoding the enzymes for BH_4 synthesis (*GCH1*, *PTS*, *SPR*) and recycling (*PCBD*, *QDPR*) have been identified. Detailed information of each enzyme is shown in Table 10.1. The modes of inheritance of each type of BH_4 deficiency are autosomal recessive, except for the autosomal dominant type of GTP cyclohydrolase I deficiency without hyperphenylalaninemia, called Segawa disease (Segawa 2000). Correlations between genotype and phenotype have not been recognized in BH_4 deficiency.

10.5 Treatment

10.5.1 PKU

1. Dietary therapy

Dietary therapy is the principal means of managing of PKU. Immediately after the diagnosis of PKU, a Phe-restricted diet should be started. The intake of natural proteins from meals needs to be restricted to reduce blood Phe levels below the

Table 10.1 Enzymes associated with the synthesis and the recycling of BH_4

Enzyme	OMIM number	No. of amino acids	Gene symbol	Cytogenetic location	No. of exons
GTP cyclohydrolase I (GTPCH)	600,225	250	*GCH1*	14q22.2	6
6-Pyruvoyl-tetrahydropterin synthase (PTPS)	612,719	145	*PTS*	11q23.1	6
Sepiapterin reductase (SR)	182,125	261	*SPR*	2p13.2	3
Pterin-4α-carbinolamine dehydratase 1 (PCBD1)	126,090	103	*PCBD1*	10q22.1	4
Dihydropteridine reductase (DHPR)	612,676	244	*QDPR*	4p15.32	7

Table 10.2 Recommended intake of Phe in PKU

	Phe intake	
Age	USA (mg/day)	Japan (mg/kg/day)
0–2 months	130–430	50–70
3–5 months	135–400	40–60
6–8 months	145–370	30–50
9–11 months	135–330	
1 year	200–320	20–40
2 years		20–35
3–4 years		15–35
4 years<	200–1100	

Table 10.3 Target blood Phe concentrations throughout life

	Phe concentrations (μmol/L)		
Age	Europe	USA	Japan
0–2 years	120–360	120–360	120–240
3–9 years			120–360
10–12 years			120–480
12 years<	120–600		120–600
Pregnancy	120–360		120–360

upper limit of the target range (Tables 10.2 and 10.3) (van Spronsen et al. 2017; Vockley et al. 2014; Kitagawa 2012). The required total protein intake is supplied by a Phe-free formula. However, insufficiency of Phe should be avoided because it is an essential amino acid. In infancy, the ratio of breast milk or ordinary infant formula, and a Phe-free formula, should be adjusted to maintain blood Phe levels within the recommended range. After weaning, the Phe-restricted diet must be continued with foods low in protein in addition to a Phe-free formula. Adolescent and adult patients can ingest a Phe-free L-amino acid supplement or low-Phe peptide powder. Total energy intake should be adjusted to the standard amounts for each patient's gender and age.

2. Pharmacotherapy

In patients diagnosed as BH_4-responsive mild PKU, BH_4, which is supplied as a biologically active synthetic form (sapropterin), may improve blood Phe levels to the recommended range without dietary therapy. If BH_4 alone is insufficient to control Phe levels, a diet slightly restricted in Phe should also be included. The reported adverse events of sapropterin therapy are trivial (e.g., headache, upper respiratory tract infections, diarrhea, and vomiting). The safety of sapropterin in children under 4 years of age (Shintaku and Ohura 2014) and pregnant women (Grange et al. 2014) is reported to be acceptable.

3. Enzyme replacement therapy

Phenylalanine ammonia lyase (PAL) is an enzyme derived from plants. It converts Phe to trans-cinnamic acid and then finally to hippuric acid which is excreted in urine. PAL given subcutaneously reduced blood Phe levels in a murine PKU model (Sarkissian et al. 2008). Recently, PAL conjugated with polyethylene glycol (rAvPAL-PEG), which reduces immunogenicity, has been studied in a clinical trial. It was observed that a single dose of rAvPAL-PEG injected subcutaneously reduced blood Phe levels in adult PKU patients. Frequently reported adverse events included injection-site reactions, dizziness, and rashes (Longo et al. 2014). The efficacy and safety of repeated doses of rAvPAL-PEG are being currently investigated.

4. Other therapies

As Phe competes for the large neutral amino acid type 1-transporter at the intestine and blood-brain barrier, large amount of non-Phe LNAAs inhibit the entry of Phe into the brain. In a murine PKU model, LNAAs supplementation restored brain monoamine concentrations as well as a severe Phe-restricted diet (van Vliet et al. 2017). Although the effects of LNAAs supplementation in human PKU patients have only been investigated in a limited number of studies, it may help to improve adherence to dietary therapy.

Glycomacropeptide, a glycophosphopeptide isolated from cheese whey, contains only a residual amount of Phe. Because it is a natural protein that can be easily obtained and used in the diet, it may improve adherence of PKU patients to a

Phe-restricted diet (Ney et al. 2016). However, the efficacy and nutritional changes associated with long-term glycomacropeptide use need to be clarified.

Gene therapy for PKU has been investigated with murine PKU models. Recombinant adeno-associated virus-mediated PAH, which was delivered to murine liver through the portal vein, led to a decrease in serum Phe concentration and restored neurodegenerative changes. Recombinant adeno-associated virus-mediated PAH injected into muscle could effectively decrease blood Phe concentration. In spite of the proven effects of gene therapy in animal models, clinical trials in humans have not been carried out (Gessler and Gao 2016).

5. Maternal PKU

In pregnant women with PKU, maternal blood Phe is transferred through the placenta, functions as a teratogenic factor, and affects the fetus, who is usually without PKU. Because blood Phe levels of the fetus are reported to be higher than that of the mother, women with mild hyperphenylalaninemia are also at risk of affecting their fetuses (Schoonheyt et al. 1994). In general, maternal blood Phe levels relate to the risks of fetal abnormalities that include intrauterine growth retardation, microcephalus, mental disabilities, heart anomalies, and peculiar facial structures similar to fetal alcohol syndrome. The rate of spontaneous abortions increases with Phe levels. To prevent these embryopathies, it is recommended that planned pregnancies should be preceded by dietary therapy to reduce blood Phe concentrations to lower levels (120–360 μmol/L) (van Spronsen et al. 2017). As maternal requirements for protein and Phe increase with fetal growth during pregnancy, proteins should be supplied via a Phe-restricted formula and Phe-free amino acid supplements or a low-Phe peptide powder. It was reported that BH_4 therapy is effective and safe even in pregnant women with BH_4-responsive mild PKU (Feillet et al. 2014).

6. Outcome

Patients with PKU who are treated early and appropriately can achieve normal intellectual ability. However, their average IQ scores are reported to be significantly lower than those of their unaffected siblings and controls (Gassio et al. 2005). Average blood Phe levels may correlate with IQ scores. A higher prevalence of behavior problems and psychological disturbance has also been reported in PKU individuals (Smith and Knowles 2000). High Phe concentration has been recognized to induce CNS damage leading to psychiatric disorders even among the adult patients (Brumm et al. 2010). Lifelong treatment is essential to prevent such problems.

10.5.2 BH_4 Deficiency

1. BH_4

After diagnosis of BH_4 deficiency, BH_4 administration should be started immediately. In this disorder, blood Phe levels can usually be maintained in the normal range without dietary therapy. However, some patients with DHPR deficiency require a larger dose of BH_4 or a Phe-restricted diet.

2. Neurotransmitter precursors

Because very little BH_4 enters the CNS, dopamine and serotonin deficiencies are not improved with BH_4 therapy, leading to the progression of CNS symptoms. Administration of L-dopa and 5-hydroxytryptophan, which are neurotransmitter precursors, is required to prevent these symptoms.

3. Folic acid

Patients with DHPR deficiency are also often deficient in folic acid. Increasing the supply of folic acid improves demyelination and calcification of the basal nuclei.

4. Outcome

Most patients with BH_4 deficiency show good response to treatment and show good outcomes. Some patients with DHPR deficiency develop neurological symptoms, such as intellectual disability and seizures.

Questions

1. If an infant shows hyperphenylalaninemia in newborn mass screening, what tests should be performed to diagnose PKU?
2. What symptoms do untreated PKU patients develop?
3. What is the difference between PKU and BH_4 deficiency in diagnosis and treatment?
4. What are the important components of the Phe-restricted diet?
5. How can pregnant women with PKU prevent fetal abnormalities?

References

Blau N (2016) Genetics of phenylketonuria: then and now. Hum Mutat 37:508–515

Brumm VL, Bilder D, Waisbren SE (2010) Psychiatric symptoms and disorders in phenylketonuria. Mol Genet Metab 99:S59–S63

de Groot MJ, Hoeksma M, Blau N et al (2010) Pathogenesis of cognitive dysfunction in phenylketonuria: review of hypotheses. Mol Genet Metab 99:S86–S89

Feillet F, Muntau AC, Debray FG et al (2014) Use of sapropterin dihydrochloride in maternal phenylketonuria. A European experience of eight cases. J Inherit Metab Dis 37:753–762

Gassio R, Fuste E, Lopez-Sala A et al (2005) School performance in early and continuously treated phenylketonuria. Pediatr Neurol 33:267–271

Gessler DJ, Gao G (2016) Gene therapy for the treatment of neurological disorders: metabolic disorders. Methods Mol Biol 1382:429–465

Grange DK, Hillman RE, Burton BK et al (2014) Sapropterin dihydrochloride use in pregnant women with phenylketonuria: an interim report of the PKU MOMS sub-registry. Mol Genet Metab 112:9–16

Heintz C, Troxler H, Martinez A et al (2012) Quantification of phenylalanine hydroxylase activity by isotope-dilution liquid chromatography-electrospray ionization tandem mass spectrometry. Mol Genet Metab 105:559–565

Jaffe EK (2017) New protein structures provide an updated understanding of phenylketonuria. Mol Genet Metab 121:289–296

Kitagawa T (2012) Newborn screening for inborn errors of metabolism in Japan. A history of the development of newborn screening. Pediatr Endocrinol Rev 10(Suppl 1):8–25

Longo N, Harding CO, Burton BK et al (2014) Single-dose, subcutaneous recombinant phenylalanine ammonia lyase conjugated with polyethylene glycol in adult patients with phenylketonuria: an open-label, multicentre, phase 1 dose-escalation trial. Lancet 384:37–44

Ney DM, Stroup BM, Clayton MK et al (2016) Glycomacropeptide for nutritional management of phenylketonuria: a randomized, controlled, crossover trial. Am J Clin Nutr 104:334–345

Okano Y, Kudo S, Nishi Y et al (2011) Molecular characterization of phenylketonuria and tetrahydrobiopterin-responsive phenylalanine hydroxylase deficiency in Japan. J Hum Genet 56:306–312

Sarkissian CN, Gamez A, Wang L et al (2008) Preclinical evaluation of multiple species of PEGylated recombinant phenylalanine ammonia lyase for the treatment of phenylketonuria. Proc Natl Acad Sci U S A 105:20894–20899

Schoonheyt WE, Clarke JT, Hanley WB et al (1994) Feto-maternal plasma phenylalanine concentration gradient from 19 weeks gestation to term. Clin Chim Acta 225:165–169

Segawa M (2000) Hereditary progressive dystonia with marked diurnal fluctuation. Brain Dev 22:S65–S80

Shen N, Heintz C, Thiel C et al (2016) Co-expression of phenylalanine hydroxylase variants and effects of interallelic complementation on *in vitro* enzyme activity and genotype-phenotype correlation. Mol Genet Metab 117:328–335

Shintaku H, Ohura T (2014) Sapropterin is safe and effective in patients less than 4-years-old with BH_4-responsive phenylalanine hydrolase deficiency. J Pediatr 165:1241–1244

Smith I, Knowles J (2000) Behaviour in early treated phenylketonuria: a systematic review. Eur J Pediatr 159(Suppl 2):S89–S93

Thöny B, Martinez A (2016) Pathobiochemistry of PKU: the phenylalanine hydroxylating system. In: Blau N (ed) Phenylketonuria and BH4 deficiencies, 3rd edn. UNI-MED, Bremen, pp 16–21

van Spronsen FJ, van Wegberg AM, Ahring K et al (2017) Key European guidelines for the diagnosis and management of patients with phenylketonuria. Lancet Diabetes Endocrinol 5:743–756 van Vliet D, van der Goot E, Bruinenberg VM et al (2017) Large neutral amino acid supplementation as an alternative to the phenylalanine-restricted diet in adults with phenylketonuria: evidence from adult *PAH-enu2* mice. J Nutr Biochem 53:20–27

Vockley J, Andersson HC, Antshel KM et al (2014) Phenylalanine hydroxylase deficiency: diagnosis and management guideline. Genet Med 16:188–200

Wettstein S, Underhaug J, Perez B et al (2015) Linking genotypes database with locus-specific database and genotype-phenotype correlation in phenylketonuria. Eur J Hum Genet 23:302–309

Zhang S, Fitzpatrick PF (2016) Identification of the allosteric site for phenylalanine in rat phenylalanine hydroxylase. J Biol Chem 291:7418–7425

11 Triglyceride Deposit Cardiomyovasculopathy

Ken-ichi Hirano, Ming Li, and Yoshihiko Ikeda

Keywords
Adipose triglyceride lipase · Heart failure · Long-chain fatty acids · Triglyceride deposit cardiomyovasculopathy · Triglyceride

Abbreviations

ATGL	Adipose triglyceride lipase
BMI	Body mass index
BMIPP	β-Methyl-p-[123I]-iodophenyl-pentadecanoic acid
CABG	Coronary artery bypass grafting
CTx	Cardiac transplantation
HF	Heart failure
LCFAs	Long-chain fatty acids
LV	Left ventricle
NLSD-I	Neutral lipid storage disease with ichthyosis
NLSD-M	Neutral lipid storage disease with myopathy
PCI	Percutaneous coronary intervention
SMCs	Smooth muscle cells
SPECT	Single-photon emission computed tomography
TG	Triglyceride
TGCV	Triglyceride deposit cardiomyovasculopathy
WOR	Washout rate

11.1 Case Reports

11.1.1 Case 1: Primary Triglyceride Deposit Cardiomyovasculopathy (TGCV) with Adipose Triglyceride Lipase (ATGL) Mutation

A 41-year-old man was admitted due to palpitation and ventricular tachycardia. His body measurements were as follows: height, 181 cm; body weight, 58.6 kg; and body mass index (BMI), 18 kg/m^2. Blood pressure was 106/70 mmHg. Physical examination showed no skin lesions but showed mild muscle weakness of his right arm. A peripheral blood smear showed vacuoles in polymorphonuclear leukocytes called Jordans' anomaly (Fig. 11.1A). Plasma triglyceride (TG) level

K.-i. Hirano (✉) · M. Li
Department of Cardiovascular Medicine, Laboratory of Cardiovascular Disease, Novel, Non-invasive, and Nutritional Therapeutics (CNT), Graduate School of Medicine, Osaka University, Suita, Osaka, Japan
e-mail: khirano@cnt-osaka.com; liming@cnt-osaka.com

Y. Ikeda
Department of Pathology, National Cerebral and Cardiovascular Center, Osaka, Japan
e-mail: yikeda@ncvc.go.jp

T. Oohashi et al. (eds.), *Human Pathobiochemistry*, https://doi.org/10.1007/978-981-13-2977-7_11

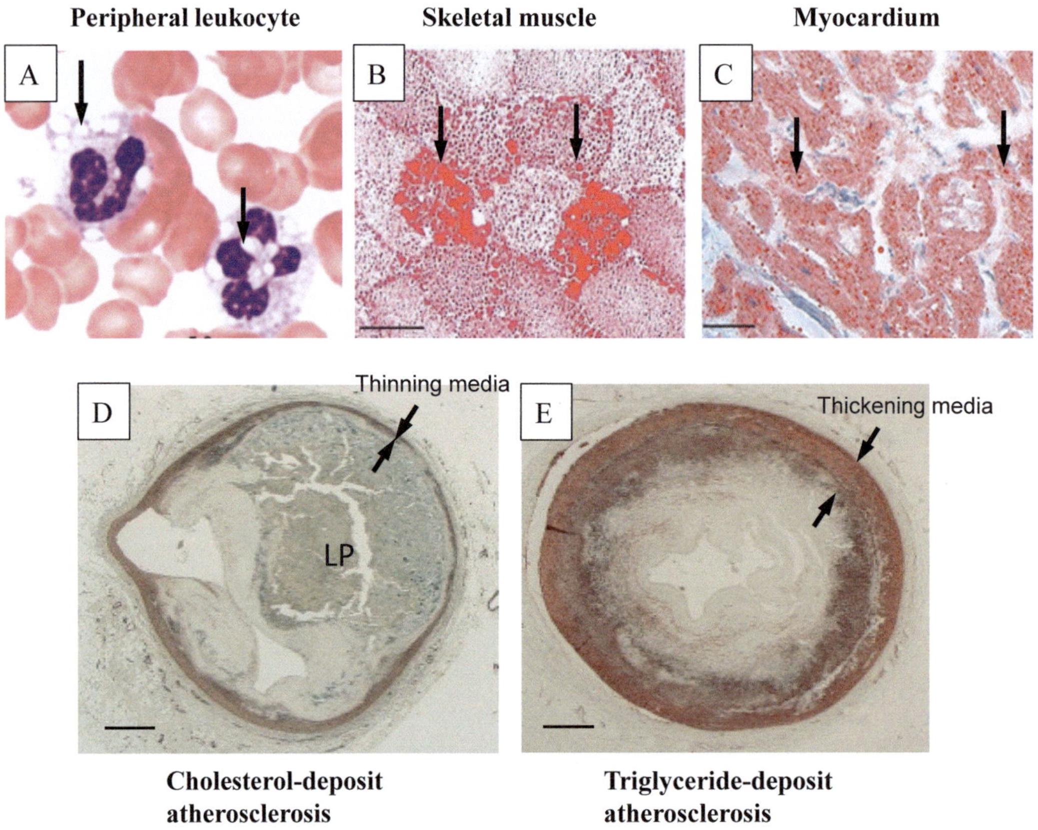

Fig. 11.1 Lipid deposition in peripheral leukocytes, skeletal muscle, myocardium, and coronary arteries in a patient (Case 1) with primary TGCV. Panel A shows peripheral blood smear with May-Giemsa staining. Arrows indicate vacuolar formation in neutrophils. Panel B shows oil red O staining of a biopsy specimen obtained from the patient's atrophied arm, with lipid droplets of predominantly type I fibers (arrows). All red dots denote lipid droplets. The scale bar represents 40 μm. Panel C shows oil red O-positive numerous vacuoles in the cytoplasm of cardiomyocytes (arrows) in the myocardium of the explanted heart after cardiac transplantation. The scale bar represents 50 μm. Panel D: cholesterol-deposit atherosclerosis had the following characteristics: focal and eccentric stenosis with large lipid pool (LP in the figure) paucity of SMCs in the fibrous cap and shoulders of the lesion and disrupted internal elastic lamina and thin medial layer (arrow) resulting in remodeling. In contrast, the patient with TG-deposit atherosclerosis showed diffuse concentric stenosis of the coronary arteries. The majority of foam cells were SMCs in thick intima and media (arrow in panel E)

was 137 mg/dl (normal value 30–150 mg/dl), and total cholesterol 160 mg/dl (normal value 51–219 mg/dl) was normal. Muscle biopsy revealed vacuolation positive for oil red O staining predominantly in type I fibers (Fig. 11.1B). Chest X-ray and echocardiography showed cardiomegaly with dilatation and reduced ejection fraction of the left ventricle (LV). Endomyocardial biopsy specimens showed massive vesicular vacuoles positive for oil red O in cardiomyocytes. The mitochondrial structure was normal. Myocardial scintigraphy with β-methyl-p-[123I]-iodophenyl-pentadecanoic acid (BMIPP) showed marked reduction of washout rate (WOR), suggesting abnormal metabolism of long-chain fatty acids (LCFAs), which are a major energy source for the normal heart.

Congestive heart failure (HF) developed progressively and became catecholamine-dependent within a couple of years. A left ventricular assist device was implanted, and the patient was registered as a heart transplant candidate on a waiting list. His passaged skin fibroblasts showed massive accumulation of TG and were used as a model for investigating abnormalities in cellular metabolism and elucidating the underlying mechanism. Eventually, the patient received cardiac transplantation (CTx). Because the biochemical analysis of explanted heart showed marked accumulation of TG in both myocardium and coronary arteries, we named this novel phenotype as TGCV (Hirano et al. 2008; Hirano 2009).

Genetic tests revealed that this patient was homozygous for the loss-of-function mutation in the *PNPLA2* gene encoding ATGL, which is the rate-limiting enzyme of intracellular hydrolysis of TG. The pathological analysis of explanted heart showed vacuoles observed in cardiomyocytes and stained positive with oil red O (Fig. 11.1C). Vacuoles positive for oil red O were also observed in the cytoplasm of endothelial cells and smooth muscle cells (SMCs) of the media of the coronary arteries. In contrast to usual cholesterol-deposit atherosclerosis with focal and eccentric stenosis (Fig. 11.1D), his coronary arteries with TG deposition showed diffuse and concentric stenosis (Fig. 11.1E) (Hirano 2009; Ikeda et al. 2014a, b).

11.1.2 Case 2: Idiopathic TGCV Without ATGL Mutation

A 69-year-old male was admitted to our hospital due to acute chest pain. His body measurements were as follows: height, 157 cm; body weight, 47.1 kg; and BMI, 19 kg/m^2. Blood pressure was 108/46 mm Hg. His medical history included diabetes mellitus, hypertension, and hyperlipidemia. His lipid profiles were as follows: TG level 57 mg/dl (normal value 30–150 mg/dl), high-density lipoprotein cholesterol 53 mg/dl (normal value 40–80 mg/dl), and low-density lipoprotein cholesterol 54 mg/dl (normal value <140 mg/dl). Coronary angiograms showed diffuse and narrowing arteries (Fig. 11.2A). He was diagnosed as having unstable angina pectoris and subsequently underwent percutaneous coronary intervention (PCI). Despite several rounds of PCI and coronary bypass surgery (CABG), his chest pain was intractable. A frontal chest X-ray showed enlargement of cardiothoracic ratio. Electrocardiogram showed a normal sinus rhythm (80 beats/min), complete right bundle branch block, and multisource premature ventricular contractions. Echocardiography showed akinesis of the basal-mid lateral-posterior segment of the left ventricular wall and the severe apical hypokinesis with enlarged cavity and reduced ejection fraction of LV. Myocardial scintigraphy with BMIPP showed a marked reduction of WOR (Fig. 11.2B), as similar to those observed in patients with primary TGCV.

Genetic test showed no mutations in the exons and introns of genes encoding ATGL or its coactivator CGI-58; however the ATGL activity was very low in his peripheral leukocytes (Takagi et al. 2018). Eventually, the patient died suddenly.

11.2 Diagnosis

11.2.1 Definition and Classification

TGCV is a novel disease concept that we found in Japanese CTx candidates in 2008. Probands carried mutations in the *PNPLA2* gene encoding ATGL. ATGL is the major enzyme that catalyzes the initial rate-limiting step of intracellular TG hydrolysis to release free nonesterified long-chain fatty acids (LCFAs), which are an essential energy source for the normal heart. Patients with TGCV show ectopic accumulation of triglycerides in cardiomyocytes and SMCs resulting from abnormal intracellular metabolism of triglycerides and LCFA. In TGCV, cardiomyovascular TG deposition is not related to BMI or body weight, which reflects the TG accumulation in adipose tissues in the body. Cases 1 and 2 were not obese, which was consistent with the diagnosis of TGCV.

TGCV is classified into primary and idiopathic TGCV with and without genetic ATGL deficiency, respectively. Both types of TGCV patients suffer from severe HF, arrhythmia, and coronary artery disease caused by lipotoxicity and energy failure at cellular levels (Hirano 2009).

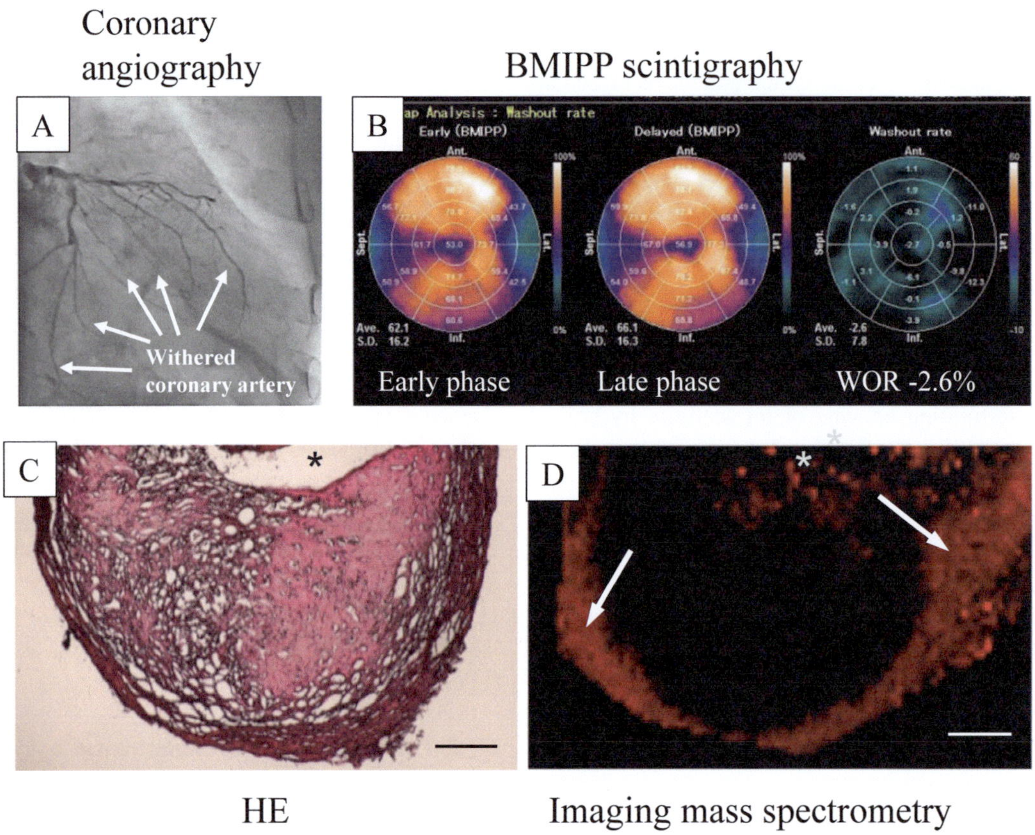

Fig. 11.2 Clinical imaging and pathological findings of an idiopathic TGCV patient (Case 2) without ATGL mutation. (**A**) Angiogram of left coronary artery shows diffuse narrowing like withered branches (arrow). (**B**) Bull's eye images of nuclear myocardial scintigraphy with BMIPP, a radioactive tracer for long-chain fatty acid (LCFA). Washout rate (WOR) was calculated as [BMIPP uptake (early phase)-remaining BMIPP (delay phase)/BMIPP uptake (early phase)] to evaluate the myocardial metabolism of LCFA. Orange colors in early and delayed phases denote the uptake of BMIPP. Blackout in the WOR image shows defective WOR (−2.4%, reference values >20%) (**C, D**). Spatial distribution of TG in coronary atherosclerotic lesions. Hematoxylin-eosin staining (**C**) and imaging mass spectrometry (**D**). Red dots (arrows) in **D** denote TG deposition, located in the fibrous cap and media of atherosclerotic lesions. Asterisks represent vascular lumen. The scale bars represent 500 μm

11.2.2 Clinical Signs and Symptoms

TGCV patients are healthy at birth but present with cardiac symptoms from their 20s to middle age. The symptoms of HF include palpitations, shortness of breath, dyspnea on exertion, fatigue, swelling (edema), and weight gain. As the disease progresses, more symptoms appear including difficulty breathing (dyspnea) at rest or when lying flat (orthopnea). Chest pain that radiates through the upper body occurs at rest and also at night, and does not respond to nitroglycerin or nitrites. Arrhythmia causes slow, irregular heartbeats or fainting (syncope). Many of patients have a history of cardiopulmonary arrest, and some have died of sudden cardiac death. In cases of primary TGCV, symptoms associated muscle weakness, and skeletal myopathy was frequently observed.

11.2.3 Laboratory and Imaging Data

In both primary and idiopathic TGCV, myocardial WOR of BMIPP calculated by early and delayed single-photon emission computed tomography (SPECT) images is markedly reduced, indicating the impaired metabolism of LCFA and TG. Early BMIPP-SPECT images can be used to detect the severity of myocardial damage and coronary artery lesions (Hirano et al. 2015). Coronary CT angiograms with TG imaging show diffuse narrowing stenosis with outside-in TG deposition of vascular wall (Higashi et al. 2017). As extra-cardiac signs, vacuolar formation in peripheral leukocytes called Jordans' anomaly can be observed particularly in primary TGCV. In idiopathic one, such vacuolar formation is minor in size but can sometimes be detected.

11.2.4 Postmortem Analysis

Postmortem analysis of cases with idiopathic TGCV showed the following characteristics (Ikeda et al. 2014a, b). (1) TGCV phenotype was frequently observed among autopsied diabetics who died of intractable cardiovascular diseases. (2) Their hearts were heavy in weight with frequent myocardial infarction and hypertrophy. Coronary arteries showed diffuse and concentric stenotic lesions at multivessels. (3) Immunoreactivities of ATGL were detectable in both myocardium and coronary arteries. (4) Imaging mass spectrometry showed that TG deposition was detected in the media and fibrous cap of plaques of coronary atherosclerotic lesions (Fig. 11.2C and red color in Fig. 11.2D). TG signals were overlapped with SMCs.

11.2.5 Genetic Testing

For the diagnosis of primary TGCV, genetic test including sequencing analysis of *PNPLA2* encoding ATGL is obviously useful. In Japan, five mutations have been reported so far. To investigate genetic causes or backgrounds for idiopathic TGCV is an interesting research focus.

11.2.6 Diagnostic Criteria for TGCV and Differential Diagnosis

The Japan TGCV study group provided the diagnostic criteria for TGCV, as shown in Table 11.1. The criteria include two major items (two points each) and two minor items (one point each). Four points or more and three points indicated definite and probable TGCV, respectively.

Here, we can provide the following clinical hints which may suspect TGCV: (1) heart failure that is poorly responsive to beta-blockers; (2) angina pectoris with chest pain that is resistant to nitrates (sometimes 4–5 sublingual tablets of nitroglycerin

Table 11.1 Diagnostic criteria for TGCV

1. Major items (two points)	
1.1 Myocardial TG deposition or impaired metabolism of LCFA	
At least one of the following:	
(1) Myocardial TG deposition by biopsy specimens[a]	
(2) Myocardial TG deposition by MR spectroscopy	
(3) Reduced washout rate (less than 10%) of BMIPP	
1.2 Diffuse narrowing coronary arteries demonstrated by CAG, CT angiography	
2. Minor items (one point)	
2.1 Jordans' anomaly (apparent vacuoles (about 1 μm in size) of polymorphonuclear leukocytes in peripheral blood smear)[b]	
2.2 Diabetes mellitus	
Diagnosis of TGCV	
(1) Four points or more	Definite TGCV
	Primary TGCV, if with ATGL gene mutations
	Idiopathic TGCV, if without ATGL gene mutations
(2) Three points	Probable TGCV
	Definite TGCV, if ATGL gene mutation confirmed

[a]For tissue TG content examination, frozen sections with osmium fixation, but not paraffin sections, should be used for prevention of lipid elution

[b]For difficult cases, May-Giemsa staining slides of the peripheral blood smear samples will be evaluated by the Japan TGCV study group

required for relief); and (3) worsening symptoms during fasting, such as in the morning.

It is important to differentiate TGCV from other cardiovascular diseases including dilated cardiomyopathy, hypertrophic cardiomyopathy, dilated-phase hypertrophic cardiomyopathy, and arrhythmogenic right ventricular cardiomyopathy. The differential diagnosis of TGCV also includes the following myocardial diseases, especially lipid storage metabolic diseases: (1) alcoholic heart disease, (2) neuromuscular myocardial disorders, (3) nutritional heart disease, and (4) metabolic myocardial disorders (e.g., Fabry disease, Pompe disease).

Furthermore, the conceptual difference between TGCV and other diabetic and metabolic heart diseases should be noted. Some known disease concepts need to be differentiated from TGCV. One is diabetic cardiomyopathy, which was originally defined as cardiomyopathy without significant stenosis in epicardial coronary arteries. Another concept is fat accumulation in epicardial adipose tissue, which is the over-deposition of TG in physiological tissue. TGCV is distinct from these two entities, because TGCV is characterized by the ectopic deposition of TG in the cardiomyovascular system with apparent involvement of epicardial coronary arteries.

11.2.7 Global Distribution of Patients with Genetic ATGL Deficiency

Here we review the literature (Hirano et al. 2008; Fischer et al. 2007; Pennisi et al. 2017; Kaneko et al. 2014; Chen et al. 2010; Reilich et al. 2011; Muggenthaler et al. 2016; Coassin et al. 2010) regarding patients with genetic ATGL deficiencies (homozygous mutation) published in recent decades. Of the 46 patients with genetic ATGL deficiencies (Fig. 11.3), there are 29 patients without cardiac symptoms and 17 patients with cardiac symptoms (TGCV phenotype) such as dilated or hypertrophic cardiomyopathy-like abnormalities and ventricular tachycardia.

Seven patients with primary TGCV have been so far identified in Japan. Four patients suffered from severe cardiac dysfunction that resulted in cardiac death, and two patients had to undergo heart transplantation due to dilated cardiomyopathy and heart failure (Higashi et al. 2015).

The concept of TGCV is relatively new, and physicians and medical staffs should pay more attention to this kind of disease when they encounter patients who suffer from congestive heart failure and/or coronary heart disease of unknown cause. A definitive diagnosis is essential for treatment and assessment of prognosis. Because cardiac involvement (TGCV phenotype) is critical for patients with ATGL deficiency, delayed diagnosis needs to be avoided. If all patients could be registered in the international registry (http://www.tgcv.org/r/home.html), physicians and medical staffs could share their experience with diagnosis and treatment to reduce patients' difficulties and disabling at earlier stages.

11.3 Biochemical Perspectives

TGCV is characterized by the ectopic accumulation of TG in cardiomyocytes and vascular SMCs that in turn causes pathological conditions such as severe heart failure, coronary artery disease, and arrhythmias (Fig. 11.4) (Hirano et al. 2008, 2014; Hirano 2009). Abnormal metabolism of cellular LCFAs resulting from ATGL deficiency induces both lipotoxicity and energy failure in cardiomyocytes, and these cause HF and coronary artery disease. ATGL is the enzyme that catalyzes the rate-limiting step of the hydrolysis of cellular (cytoplasmic) TG to release the first nonesterified LCFA (Zimmermann et al. 2004, 2009; Haemmerle et al. 2006).

Normal cardiomyocytes take up LCFAs through transporters such as CD36 and use them as a primary energy source. Once LCFAs enter into a cell, they follow one of two pathways: they are either transported directly to the mitochondria and undergo β-oxidation to produce ATP, or temporarily become TG, are promptly hydrolyzed by an enzyme such as ATGL, and are transported to the mitochondria to undergo β-oxidation (Fig. 11.5A). Patients with TGCV have abnormalities in ATGL and other components of the intracellular TG hydrolysis pathway that cause intracellular accumulation of triglycerides as lipid droplets (Hirano

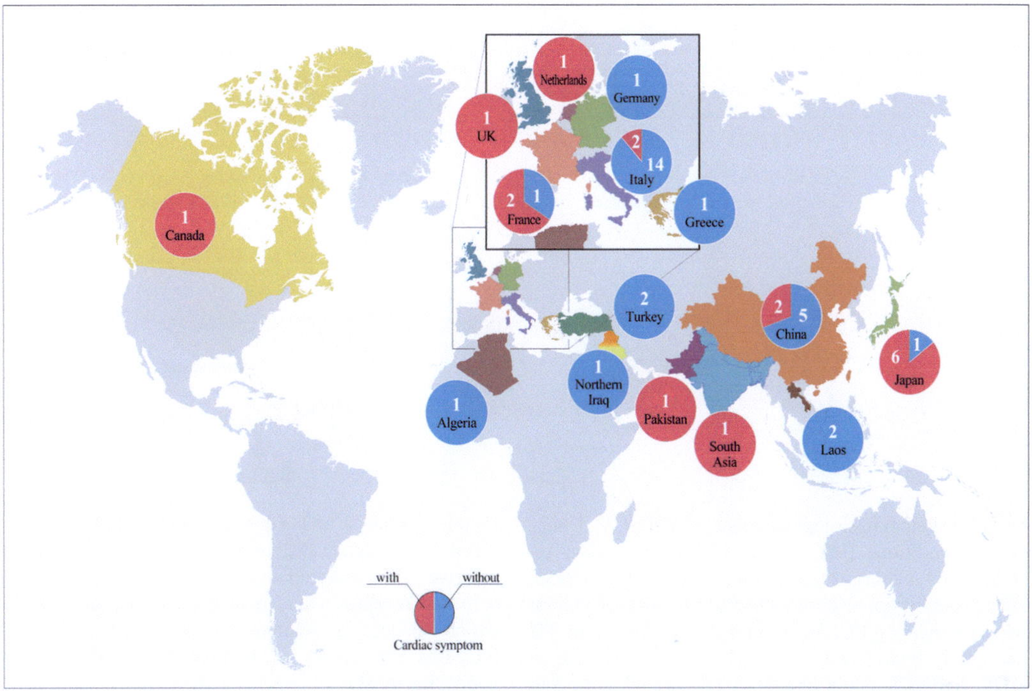

Fig. 11.3 Global distribution of patients with genetic ATGL deficiency Cases of genetic ATGL deficiency without cardiac symptoms are shown in blue, and cases with cardiac symptom (TGCV phenotype) are shown in red

et al. 2014, 2015). Lipotoxicity from intracellular TG accumulation and energy failure caused by the inability of cells to use this energy source are involved in the pathogenesis of TGCV (Fig. 11.5B).

11.4 Therapy

TGCV is resistant to standard medical treatments for HF, angina pectoris, or arrhythmia, including vasodilator, nitrates, and beta-blockers. Two patients with primary TGCV received CTx in Japan. The long-term follow-up indicates that the donor hearts can be maintained under the standard immunosuppression. However, it is noted that, even after CTx, skeletal myopathy develops gradually. Some patients with idiopathic TGCV suffer from slow-flow and no-reflow after PCI, requiring emergent CABG, as described in Case 2. The Japan TGCV study group has started physician-initiated clinical trials for primary and idiopathic TGCV patients to test efficacy and safety of nutritional therapies with medium-chain fatty acids (ClinicalTrial.gov: NCT02830763; NCT02502578).

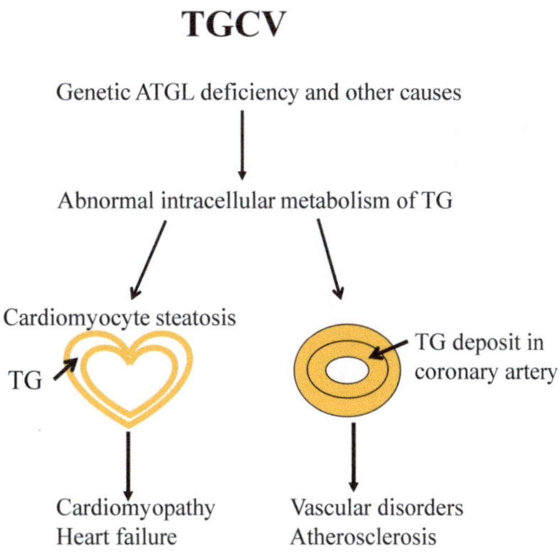

Fig. 11.4 Schematic presentation of the disease concept for TGCV. Genetic ATGL deficiency (primary TGCV) and other causes (idiopathic TGCV) induced cardiomyocyte steatosis and TG deposition of smooth muscle cells, leading to severe heart failure and TG-deposit atherosclerosis

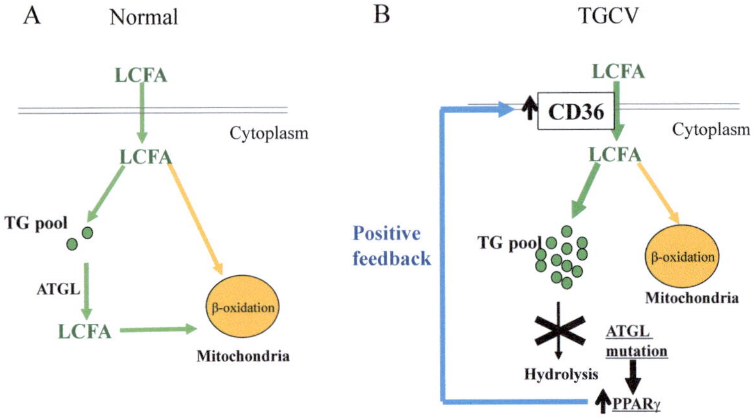

Fig. 11.5 Pathophysiological model for TGCV. Under normal conditions (left panel), LCFAs are taken up through LCFA transporters and receptors such as CD36. Some are transported to mitochondria for beta-oxidation and the remaining LCFAs are utilized as a source of TG and rapidly hydrolyzed by intracellular lipases such as ATGL. In TGCV (right panel), LCFAs are taken up and used to synthesize of TG that cannot be hydrolyzed due to ATGL insufficiency such as genetic mutations, leading to massive TG accumulation. Peroxisome proliferator-activated receptor γ and related genes are upregulated, which induces a positive feedback mechanism that increases the uptake of LCFAs despite the massive deposition of TG

Acknowledgments This was partially supported by research grants for rare and intractable diseases from the Japan Agency of Medical Research and Development (AMED) (Grant No. 17ek0109092h0003) and from the Ministry of Health, Labour, and Welfare of Japan.

Questions

1. Genetic ATGL deficiency is an extremely rare disease, and information is very limited. What is the best approach for summarizing information from case studies and educating healthcare providers?
2. How can TGCV be differentiated from other cardiovascular diseases?

References

Chen J, Hong D, Wang Z et al (2010) A novel PNPLA2 mutation causes neutral lipid storage disease with myopathy (NLSDM) presenting muscular dystrophic features with lipid storage and rimmed vacuoles. Clin Neuropathol 29:351–356

Coassin S, Schweiger M, Kloss-Brandstatter A et al (2010) Investigation and functional characterization of rare genetic variants in the adipose triglyceride lipase in a large healthy working population. PLoS Genet 6:e1001239

Fischer J, Lefevre C, Morava E et al (2007) The gene encoding adipose triglyceride lipase (PNPLA2) is mutated in neutral lipid storage disease with myopathy. Nat Genet 39:28–30

Haemmerle G, Lass A, Zimmermann R et al (2006) Defective lipolysis and altered energy metabolism in mice lacking adipose triglyceride lipase. Science 312:734–737

Higashi M, Hirano K, Kobayashi K et al (2015) Distinct cardiac phenotype between two homozygotes born in a village with accumulation of a genetic deficiency of adipose triglyceride lipase. Int J Cardiol 192:30–32

Higashi M, Ikeda Y, Miyauchi H, Zaima N, Suzuki A, Li M, Kobayashi K, Naito H, Hirano K (2017) Imaging modalities for triglyceride deposit cardiomyovasculopathy. Ann Nucl Cardiol 3:94–102

Hirano K (2009) A novel clinical entity: triglyceride deposit cardiomyovasculopathy. J Atheroscler Thromb 16:702–705

Hirano K, Ikeda Y, Zaima N et al (2008) Triglyceride deposit cardiomyovasculopathy. N Engl J Med 359:2396–2398

Hirano K, Tanaka T, Ikeda Y et al (2014) Genetic mutations in adipose triglyceride lipase and myocardial up-regulation of peroxisome proliferated activated receptor-gamma in patients with triglyceride deposit cardiomyovasculopathy. Biochem Biophys Res Commun 443:574–579

Hirano K, Ikeda Y, Sugimura K et al (2015) Cardiomyocyte steatosis and defective washout of iodine-123-beta-methyl iodophenyl-pentadecanoic acid in genetic deficiency of adipose triglyceride lipase. Eur Heart J 36:580

Ikeda Y, Hirano K, Fukushima N et al (2014a) A novel type of human spontaneous coronary atherosclerosis with triglyceride deposition. Eur Heart J 35:875

Ikeda Y, Zaima N, Hirano K et al (2014b) Coronary triglyceride deposition in contemporary advanced diabetics. Pathol Int 64:325–335

Kaneko K, Kuroda H, Izumi R et al (2014) A novel mutation in PNPLA2 causes neutral lipid storage disease with myopathy and triglyceride deposit cardiomyovasculopathy: a case report and literature review. Neuromuscul Disord 24:634–641

Muggenthaler M, Petropoulou E, Omer S et al (2016) Whole exome sequence analysis reveals a homozygous mutation in PNPLA2 as the cause of severe dilated cardiomyopathy secondary to neutral lipid storage disease. Int J Cardiol 210:41–44

Pennisi EM, Arca M, Bertini E et al (2017) Neutral lipid storage diseases: clinical/genetic features and natural history in a large cohort of Italian patients. Orphanet J Rare Dis 12:90

Reilich P, Horvath R, Krause S et al (2011) The phenotypic spectrum of neutral lipid storage myopathy due to mutations in the PNPLA2 gene. J Neurol 258:1987–1997

Takagi A, Ikeda Y, Kobayashi K et al (2018) Newly developed selective immunoinactivation assay revealed reduction in adipose triglyceride lipase activity in peripheral leucocytes from patients with idiopathic triglyceride deposit cardiomyovasculopathy. Biochem Biophys Res Commun 495(1):646–651

Zimmermann R, Strauss JG, Haemmerle G et al (2004) Fat mobilization in adipose tissue is promoted by adipose triglyceride lipase. Science 306:1383–1386

Zimmermann R, Lass A, Haemmerle G et al (2009) Fate of fat: the role of adipose triglyceride lipase in lipolysis. Biochim Biophys Acta 1791:494–500

Urea Cycle Disorders

12

Kimitoshi Nakamura, Jun Kido,
and Shirou Matsumoto

Keywords
Ammonia · Amino acids · Orotic acid · Neonatal onset · Late onset

12.1 Case Report

A female infant, weighing 2786 g, was born at 39 weeks gestational age with Apgar scores of 8 points at 1 min and 9 points at 5 min after birth. She had no family history of hereditary disorders. Although she had no medical problems at birth and consumed her mother's breast milk, she developed fever with increased white blood cell counts (WBC: 22,070/μL) and C-reactive protein level (CRP: 1.2 mg/dL) at 2 days after birth. Bacterial infection was suspected, and she received antibiotic therapy. Subsequently, she developed jaundice, convulsions, and respiratory failure and required respiratory support. As the doctors were concerned about a risk of neonatal herpes infection, she received γ-globulin and acyclovir treatments. In addition, she developed mild acidosis and hyperammonemia >400 μg/dL (normal range: 30–80 μg/dL). Urea cycle disorders were suspected, and she was treated with arginine, sodium benzoate, and high glucose infusion. Unfortunately, her plasma ammonia level increased to 2000 μg/dL at 3 days after birth (Fig. 12.1), and she was transferred to a neonatal intensive care unit for treatment with hemodialysis and possible liver transplantation. Her blood ammonia levels gradually improved and normalized at 21 days after birth. Her blood glutamine, glutamic acid, glycine, aspartic acid, citrulline, and arginine levels were 3262 μmol/L (control: 416–740 μmol/L), 265 μmol/L (12–83 μmol/L), 705 μmol/L (140–427 μmol/L), 29 μmol/L (trace–7.2 μmol/L), trace (18–48 μmol/L), and 69 μmol/L (32–150 μmol/L), respectively. The results of the neonatal tandem mass spectrometry screening test for inherited metabolic diseases and analysis test for urinary orotic acid were negative. She was definitively diagnosed with carbamoyl phosphate synthetase 1 deficiency (CPSD) by identification of a mutation of the carbamoyl phosphate synthetase 1 gene (c.2339G>A and c.2945G>A). She underwent a living donor liver transplantation with her father as the donor at the age of 6 months. The subject is presently 7 years and 5 months old, attends primary school, and lives a stable life.

K. Nakamura (✉) · J. Kido · S. Matsumoto
Department of Pediatrics, Kumamoto University Graduate School of Medical Sciences, Kumamoto, Japan
e-mail: nakamura@kumamoto-u.ac.jp

T. Oohashi et al. (eds.), *Human Pathobiochemistry*, https://doi.org/10.1007/978-981-13-2977-7_12

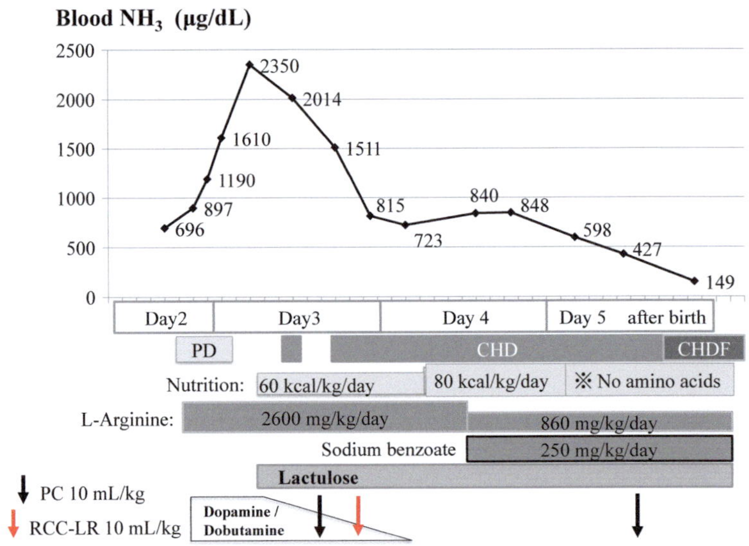

Fig. 12.1 Case report: female patient, 2 days old, blood ammonia concentrations at symptom onset and subsequent treatment. *CHD* continuous hemodialysis, *CHDF* continuous hemodiafiltration, *PC* platelet concentrates, *PD* peritoneal dialysis, *RCC-LR* red cell concentrates-leukocytes reduced

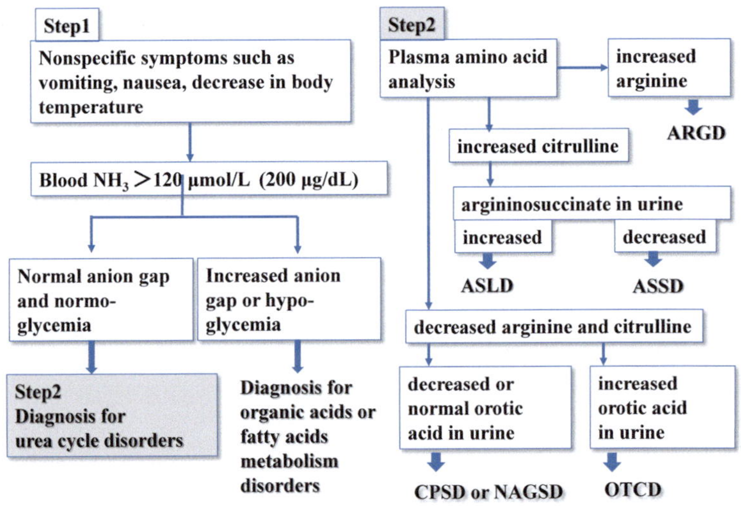

Fig. 12.2 Diagnostic flowchart of urea cycle disorders. *ARGD* arginase 1 deficiency, *ASLD* argininosuccinate lyase deficiency, *ASSD* argininosuccinate synthetase deficiency, *CPSD* carbamoyl phosphate synthetase 1 deficiency, *NAGSD* N-acetyl glutamate synthetase deficiency, *OTCD* ornithine transcarbamylase deficiency

12.2 Diagnosis

A medical doctor should correctly diagnose urea cycle disorders (UCDs) in patients developing hyperammonemia crises such as for the case described in the Case Report. A diagnostic flowchart for hyperammonemia is shown in Fig. 12.2. Encephalopathy associated with hyperammonemia in the neonatal period constitutes the most important clinical feature for diagnosis, although the late onset type can occur at any age. Other important diseases that present with hyperammonemia include mitochondrial disorders, fatty acid metabolism disorders, systemic sepsis, liver failure, and portosystemic shunt. Congenital metabolic disorders should be initially suspected in hyperammonemia without liver failure; however, cases with liver failure have also been reported, indicating the need for caution.

UCDs constitute diseases presenting with severe hyperammonemia, typically in the neonatal period, caused by congenital defects in the metabolic enzymes that comprise the urea cycle (urea synthesis pathway) (Nagata et al. 1991) (Table 12.1). UCDs are classified as rare diseases

Table 12.1 Characteristics of urea cycle disorders

Disease	Symptom	Elevated amino acids		Orotic acid	Inheritance gene		Enzyme
		Plasma	Urine				
CPSD	Hyperammonemia	Glutamine Glutamate		–	AR	CPS1	Carbamoyl phosphate synthetase 1
OTCD	Hyperammonemia	Glutamine Glutamate		++	XLR	OTC	Ornithine transcarbamylase
ASSD	Hyperammonemia	Citrulline		++	AR	ASS	Argininosuccinate synthetase
ASLD	Hyperammonemia Hepatomegaly Hair abnormality	Arginino- Succinate Citrulline	Arginino-succinate	+	AR	ASL	Argininosuccinate lyase
ARGD	Hyperammonemia Spastic paresis	Arginine	Arginine Lysine	++	AR	ARG1	Arginine syntase
NAGSD	Hyperammonemia	Glutamine Glutamate		–	AR	NAGS	N-acetyl glutamate synthetase

AR autosomal recessive, *XLR* X-linked
Nagata N et al. (1991) Am J Med Genet 39:228-229

and estimated to have a prevalence of 1/50,000 births (Nagata et al. 1991) in Japan and 1/35,000 in the EU/USA (Batshaw et al. 2014).

The urea cycle is a metabolic pathway that converts ammonia to urea in the liver and excretes endogenous or exogenous nitrogen sources out of the body. The urea cycle comprises five enzymes, two of which, carbamoyl phosphate synthetase 1 (CPS1: EC 6.3.5.5) and ornithine transcarbamylase (OTC: EC 2.1.3.3), are present in the mitochondrial matrix, whereas the remaining three (argininosuccinate synthetase (ASS: EC 6.3.4.5), argininosuccinate lyase (ASL: EC 4.3.2.1), and arginase 1 (ARG: EC 3.5.3.1)), are present in the cytoplasm. Furthermore, N-acetyl glutamate synthetase (NAGS: EC 2.3.1.1) and at least two other transporter proteins are essential to urea cycle function (Fig. 12.3).

In UCDs, ammonia cannot be converted to urea and excreted from the body because of dysfunction in any of these enzymes; therefore, patients with UCDs develop hyperammonemia and present with neurological abnormalities including vomiting, ataxia, confusion, and irritability. In the absence of appropriate intervention, the patients will develop seizures, become comatose, and die. The various conditions may be classified according to symptoms and age as follows:

- Presymptomatic, identified through family history analysis and screening tests
- Neonatal onset, presenting with severe hyperammonemia during the neonatal period
- Late onset, presenting with neurological symptoms or hyperammonemia after the neonatal period and deterioration of symptoms or severe hyperammonemia after a trigger such as infection or starvation

UCDs are further classified according to the affected enzyme/system as follows:

1. Primary urea cycle disorders
   Includes *N*-acetyl glutamate synthetase deficiency (NAGSD), carbamoyl phosphate synthetase deficiency (CPSD), ornithine transcarbamylase deficiency (OTCD), argininosuccinate synthetase (ASSD, type 1 citrullinemia), argininosuccinate lyase deficiency (ASLD), and arginase 1 deficiency (ARGD, argininemia)
2. Secondary urea cycle disorders
   Includes hyperornithinemia-hyperammonemia-homocitrullinuria (HHH) syndrome, lysinuric protein intolerance, citrin deficiency (type 2 citrullinemia), and ornithine aminotransferase deficiency (gyrate atrophy of choroid and retina)

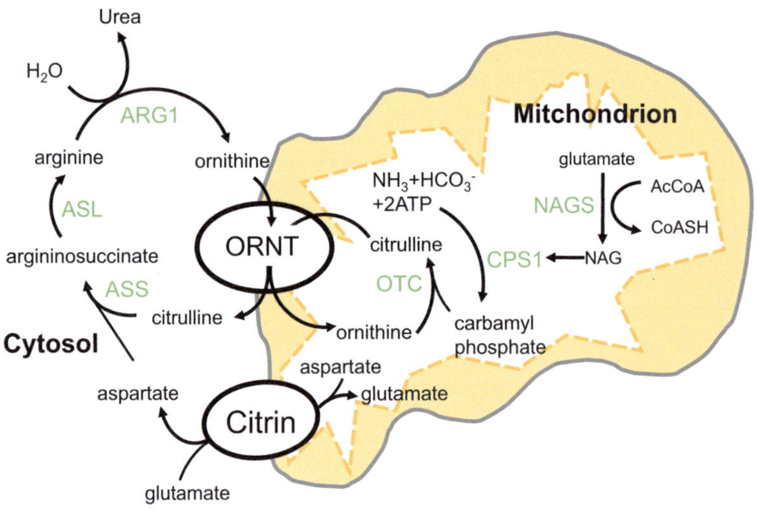

Fig. 12.3 Urea cycle. *AcCoA* acetyl-CoA, *ASL* argininosuccinate lyase, *ASS* argininosuccinate synthetase, *CPS1* carbamoyl phosphate synthetase 1, *CoASH* coenzyme Am, *NAGS* N-acetyl glutamate synthetase, *ORNT* mitochondrial ornithine/citrulline transporter, *OTC* ornithine transcarbamylase

12.3 Amino Acid Analysis

Amino acids profiling is useful for the diagnosis of UCDs, which allows the detection of an accumulation of precursor amino acids and a reduction in product amino acids resulting from and thereby specifying the defective enzyme. For example, in patients with OTCD, plasma glutamine, glutamate, and glycine levels should be increased, with a decrease in citrulline and arginine levels in the plasma and urine (Maestri et al. 1992; Berry and Steiner 2001). ARGD and ASLD can be differentiated by the concentrations of arginine and argininosuccinic acid, whereas hypercitrullinemia without elevated argininosuccinic acid is suggestive of ASSD. Decreased levels of citrulline and arginine with elevated urinary orotic acid are suggestive of OTCD, whereas reduced urinary orotic acid is suggestive of CPSD or NAGSD. HHH syndrome should be suspected with hyperammonemia associated with high blood ornithine and urinary homocitrulline. Accordingly, planning of pilot screening for UCDs in Japan using amino acids profiling is currently underway.

12.4 Urinary Orotic Acid

Orotic acid is effective for distinguishing between OTCD and CPSD (Fig. 12.4). Orotic acid is a heteroaromatic compound that was initially discovered in whey. Biochemically, orotic acid constitutes a biosynthetic intermediate of pyrimidine bases that is produced by dihydroorotate dehydrogenase from dihydroorotate and is converted to orotidine monophosphate by orotate phosphoribosyltransferase. Dihydroorotate is synthesized from carbamoyl phosphate and asparagine acid. This metabolic system for synthesis of pyrimidine bases is essential in all cells in the human body. Because the majority of OTC in human is expressed in the cytosol of hepatocytes with little in other cell types, orotic acid is considered to be synthesized consequent to the accumulation of carbamoyl phosphate in hepatocytes and excreted into the urine.

In UCDs, when the urinary orotic acid value is high, OTCD, ASSD, ASLD, and/or HHH syndrome should be suspected. In contrast, orotic acid is not elevated in CPSD. Notably, urinary orotic acid correlates with the clinical course and metabolic status in the body. In particular, urinary

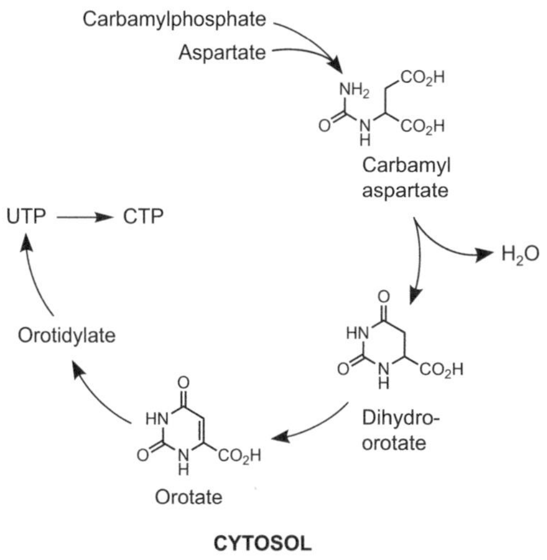

Fig. 12.4 Orotate synthesis system. *CTP* cytidine triphosphate, *UTP* uridine triphosphate

orotic acid excretion increases in allopurinol loading tests, which are performed to detect "non-symptomatic" female OTCD or select living donors for liver transplantation. However, the low sensitivity of these tests is a concerning issue of which clinicians should be aware.

12.5 Enzyme Activity

To directly interrogate enzyme function, hepatic tissue pathology is assessed, and the activity of urea cycle enzymes in the liver is measured. Specifically, the urea cycle is completed only in the liver with CPS, OTC, and NAGS expressed primarily in liver cells. Thus, it is essential to measure enzyme activity using liver tissue. Furthermore, because OTCD constitutes an X-linked genetic disorder, the enzyme activities in female patients with OTCD depend upon the location of the measured site in their liver, in accordance with the lyonization principle (i.e., will vary depending on the proportion of wild-type to mutant X chromosomes having undergone X-inactivation in the sample site).

12.6 Genetic Analysis

A definitive diagnosis is conducted through gene analysis using white blood cells. The sensitivity of genetic analysis in each country is approximately 80%, and genetic diagnosis in conjunction with prenatal diagnosis or other such means is essential. Although UCDs can often be classified with biochemical diagnostics, genetic analysis may be utilized for definitive diagnosis. For example, in the CPSD case discussed above, a compound heterozygous mutation in the *CPS1* gene was identified.

12.7 Biochemical and Molecular Perspectives

In 1932, Krebs et al. reported that ornithine activates urea synthesis and described the urea (ornithine) cycle. Specifically, the urea cycle removes ammonia generated via proteolysis. Previously, it was considered that the urea cycle was restricted to animals. However, it has been reported that in some diatoms, the urea cycle is used to generate ammonia from fixed nitrogen, a phenomenon that is attracting considerable attention. This metabolism state does not exist in the human body; instead, ammonia in the body is thought to be generated by the following four routes:

1. Ammonia production in the intestinal tract
   Ammonia is produced through catabolism of glutamine by glutaminase (mucosal epithelial cells), glutamate dehydrogenase (GDH; small intestine), and bacterial deaminase (large intestine). Urea is also broken down by bacterial urease (large intestine), generating ammonia.
2. Nitrogen metabolism associated with amino acid catabolism (deamination reactions)
3. Ammonia production by GDH
   In normal conditions, GDH functions to biosynthesize glutamate; however, the GDH reaction is reversible in catabolic states, such as during excessive exercise. In such states GDH

instead catabolizes glutamate to ammonia and α-ketoglutarate (brain, intestine, renal tubules).

4. Ammonia production by glutaminase
   The ammonia nitrogen produced in various tissues of the body is transported as blood glutamine or alanine into the liver, wherein these amino acids are converted into glutamate via transamination and ammonia is released through oxidative deamination. Glutamine is broken down into ammonia and glutamate by glutaminase.

The increase in ammonia with conditions such as a portosystemic shunt is thought to be due to the processes described in (1); moreover, if catabolism is accelerated, the reactions described in (2–4) increase, which leads to convulsions. The biosynthesis conducted in the urea cycle is shown in the following equation:

$$2NH_3 + CO_2 + 3ATP + 2H_2O \rightarrow NH_2CONH_2 + 2ADP + 2Pi + AMP + PPi$$

Thus, the conversion of two molecules of ammonia to urea requires 3ATP, which indicates that approximately 10% of the ATP produced in the mitochondria is consumed by the urea cycle. In this manner, our bodies actively excrete ammonia as fixed nitrogen in the form of urea.

Notably, the process of maturation in the human urea cycle is similar to that of urea cycle maturation during amphibian metamorphosis, consistent with the classic phrase that ontogeny recapitulates phylogeny. Among amphibians, aquatic larval tadpoles discharge ammonia into the water as nitrogen sources, similar to fish, but perform little urea synthesis (Fig. 12.5) (Mori 1991). However, during the transformation from tadpoles to frogs, the urea cycle enzymes are induced, allowing nitrogen to be excreted into urine as urea.

As also demonstrated in experiments in rat (Fig. 12.6), the expression of enzymes comprising the human urea cycle increases during the process of development from the fetal stage to infancy and then through childhood and adulthood.

12.8 Primary Urea Cycle Disorders

12.8.1 Ornithine Transcarbamylase Deficiency (OTCD)

OTCD is an X chromosome-linked, recessive genetic disease (Nagata et al. 1991). The condition is more severe in hemizygotic males than in heterozygotic females; thus, it was thought that heterozygote females either have a mild version of OTCD or are asymptomatic. However, Uchino et al. reported that female patients with late-onset OTCD in Japan did not necessarily represent mild cases (Uchino et al. 1998). The clinical symptoms observed in male neonates comprise severe hyperammonemia that typically manifests a few days after birth. Conversely, late-onset OTCD generally occurs in heterozygotic female patients and in some male patients. Late-onset OTCD is characterized by the episodic appearance of symptoms with acute hyperammonemia, interspersed with healthy periods, and it may occur at any age. These attacks often occur after consumption of a high-protein meal or after the individual has been in a catabolic state, such as during infection. Falling into a coma or death is possible during hyperammonemia, and mental developmental delays are highly possible after exposure to severe hyperammonemia. Laboratory findings during an episode of hyperammonemia include elevated plasma glutamine and alanine concentrations, as well as elevated urinary orotic acid. Definitive diagnosis can be achieved with genetic analysis or measurement of OTC enzyme activity in the liver. Prenatal diagnosis is performed via fetal liver biopsy or chorionic villi genetic analysis. Asymptomatic, heterozygote female carriers can sometimes be identified using an oral protein challenge (allopurinol loading test) that elevates plasma ammonia and urinary orotic acid concentrations. Notably, however, such "asymptomatic" carriers usually exhibit mild brain dysfunction compared to unaffected siblings.

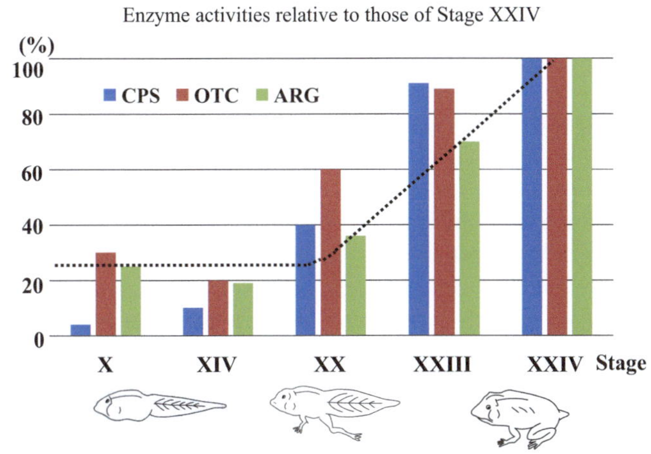

Fig. 12.5 Change of urea cycle enzymes activities under the growth process of the amphibian. (Citation from "Journey of nitrogen in the living body (Seitai no tisso no tabi)" by Mori 1991). The enzyme activities in the urea cycle are enhanced in the process through which tadpoles transform into a frog. *ARG* arginase 1, *CPS* carbamoyl phosphate synthetase 1, *OTC* ornithine transcarbamylase. Dashed line: Urea excreting ability

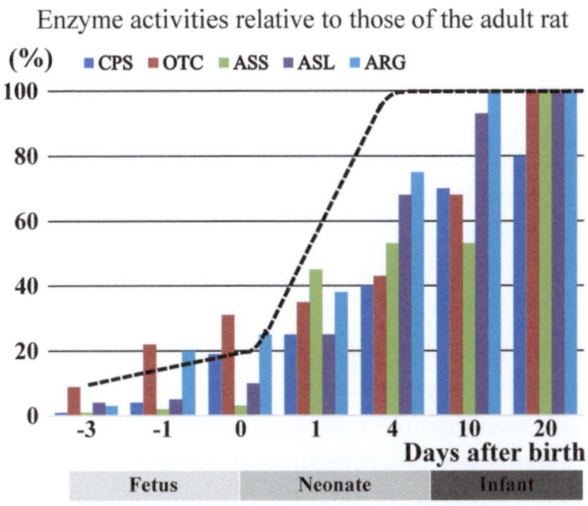

Fig. 12.6 Change of urea cycle enzymes activities under the growth process of the rat. (Modified for N.C.R. Räihä, in "The Urea Cycle" Grisolia S, et al. (ed.), 1976). The enzyme activities in the urea cycle are enhanced in the process through which the rat develops from a fetus to an infant. *ARG* arginase 1, *ASL* argininosuccinate lyase, *ASS* argininosuccinate synthetase, *CPS* carbamoyl phosphate synthetase 1, *OTC* ornithine transcarbamylase. Dashed line: Urea synthetic ability

12.8.2 Carbamoyl Phosphate Synthetase 1 Deficiency (CPSD)

The carbamoyl phosphate precursor, N-acetylglutamate (NAG), is synthesized by the enzyme NAG synthetase. NAG synthetase deficiency (NAGSD) and CPSD present with the same clinical and biochemical symptoms. CPSD exhibits more severe and varied symptoms, along with a broader range of onset age. Generally, this condition develops during the neonatal period a few days after birth, with the onset of hyperammonemia signs and symptoms, with cases often being severe and resulting in death or mental retardation. Laboratory findings show elevated plasma glutamine and alanine concentrations, whereas there is no increase in urinary orotic acid. Definitive diagnosis is achieved with genetic analysis or by measuring CPS1 enzyme activity in the liver; however, sophisticated technology is needed to differentiate this condition from NAGSD using enzyme activity measurements.

12.8.3 Argininosuccinate Synthetase Deficiency (ASSD)

ASSD presents with a considerable diversity of clinical and biochemical findings, with clinical symptoms ranging from severe to asymptomatic. The signs and symptoms during neonatal onset are identical with those observed in neonatal onset of CPSD and OTCD. The symptoms in late onset comprise either phased onset of stunted growth, frequent vomiting, developmental delay, and dry brittle hair or appear as acute episodes, as observed in OTCD. In some patients, the symptoms may not appear until the third decade of life. Laboratory findings include marked elevation of plasma citrulline concentration, elevation of plasma glutamine and alanine concentrations, and mild elevation of urinary orotic acid. Diagnosis is confirmed by genetic analysis and fibroblast enzyme activity measurement. Prenatal diagnosis is confirmed by amniotic cell enzyme activity measurement. Most patients carry two different recessive alleles and are therefore compound heterozygotes. The prognosis is extremely poor if the condition develops during the neonatal period, whereas a good prognosis is obtained for patients with mild disorders on a protein-restricted diet.

12.8.4 Argininosuccinate Lyase Deficiency (ASLD)

There is a substantively wide range in the severity of ASLD clinical and biochemical findings. In neonatal onset, severe hyperammonemia appears from a few days after birth, and the mortality rate is extremely high. Late onset is associated with mental retardation, stunted growth, and hepatomegaly. Hair abnormalities have diagnostic value specific to this disease. Furthermore, sequelae frequently observed in ASLD include continuous hepatomegaly associated with a bleeding tendency caused by mild elevation of deviant hepatic enzymes and coagulation factor abnormalities. Acute episodes of severe hyperammonemia occur when the patient is in a catabolic state, such as during infection. Laboratory findings include hyperammonemia, mild elevation of hepatic deviation enzymes, non-specific elevated plasma glutamine and alanine concentrations, moderate elevation of plasma citrulline concentration, and marked elevation of plasma argininosuccinate. Argininosuccinate can also be found in the urine and cerebrospinal fluid, with the concentrations being higher in the cerebrospinal fluid than in the plasma. ASL is present in red blood cells, the liver, and fibroblasts. Prenatal diagnosis is confirmed with amniotic cell enzyme activity measurements; specifically, there is an increased level of argininosuccinate in the amniotic fluid of affected fetuses.

12.8.5 Arginase 1 Deficiency (ARGD)

Humans carry two genetically different types of arginase, one of which is present in the cytoplasm and is expressed in the liver and red blood cells, whereas the other is found in kidney mitochondria. In ARGD, the cytoplasmic ARG enzyme is deficient. The clinical symptoms are completely different to those of other UCDs. At a few months after birth, infants are normally asymptomatic continuing for the first few years. Then, previously normal infants present with progressive spastic diplegia with scissors gait or choreiform movement and loss of developmental indices. Seizures occur frequently in this disease, and mental retardation is progressive. Hepatomegaly may also occur although severe hyperammonemia is infrequent. Laboratory findings include marked elevation of arginine in the plasma and cerebrospinal fluid and moderate elevation of urinary orotic acid. Plasma ammonia concentration is either normal or very slightly elevated. Excretion of arginine, lysine, cysteine, and ornithine in the urine may be elevated but is also sometimes normal. Urinary guanidino compounds (α-keto guanidino valerate and alginate) are significantly elevated. Plasma amino acid quantitative determination is extremely important. Diagnosis is determined based on red blood cell arginase measurement. Treatment includes a low-protein diet with absolutely no arginine along with frequent quantitative determination of plasma amino acid concentrations. The composition of the therapeutic diet and the daily consumption of protein must be carefully measured and controlled.

12.8.6 NAG Synthetase Deficiency (NAGSD)

NAGSD presents with hyperammonemia, similar to CPS1 deficiency, owing to deficiency in N-acetyl glutamate synthetase, which is essential for CPS activation. Administration of carbamylglutamate serves as an effective treatment.

12.9 Secondary Urea Cycle Disorders

In addition to the six aforementioned UCDs, urea cycle-related diseases include ornithine transaminase enzyme deficiency, citrin deficiensy, lysinuric protein intolerance, and hyperornithinemia-hyperammonemia-homocitrullinuria syndrome.

1. Lysinuric protein intolerance
   This is a functional abnormality of the dibasic amino acid transporter protein y + L. The amino acid transporter y + LAT-1 is mainly present in the epithelial basement membrane of the kidneys and small intestine. Hyperammonemia is caused by a secondary urea cycle disorder as a result of an absorption disorder of dibasic amino acids (lysine, arginine, and ornithine) in the small intestine epithelium owing to y + LAT-1 deficiency. Administration of citrulline is an effective therapy.
2. Hyperornithinemia-hyperammonemia-homocitrullinuria syndrome
   This disease constitutes an abnormality of the system for ornithine transport from the cytoplasm to the mitochondria, causing an accumulation of ornithine in the cytoplasm and deficiency of ornithine in the mitochondria. This in turn results in hyperornithinemia, homocitrullinuria, and hyperammonemia. Protein restriction and administration of ornithine are effective treatments.

12.10 Therapy

Following a definitive diagnosis, the treatment is implemented in accordance with the treatment guidelines for UCDs in each country. Each treatise or guideline (in the EU and USA) is available online. Collaboration with specialist doctors is essential; however, as transport is often difficult during acute attacks, it is preferable to receive treatment in a base hospital until the patient disease condition stabilizes while communicating with the specialist, rather than transferring the affected infant or child while they are in an unstable condition.

1. The main treatments in the acute phase are indicated as follows. Dialysis therapy is first instigated to rapidly reduce the plasma ammonia concentration.
   (a) Plasma ammonia reduction.
   Hemodialysis is one of the best treatments to immediately reduce the plasma ammonia concentration, as the faster the blood flow, the more rapidly the clearance will occur and symptoms improve. The dialysis method depends on the condition of the affected patient and the available equipment. Specific methods include blood filtration (either artery-vein, or vein-vein), hemodialysis, peritoneal dialysis, and continuous drainage peritoneal dialysis. Continuous hemodiafiltration is often used with neonates.
   (b) Pharmacotherapy to excrete surplus nitrogen via an alternative metabolic pathway. Inhibiting ammonia production can be achieved with intravenous administration of L-arginine hydrochloride and nitrogen-removing agents (e.g., sodium phenylbutyrate or/and sodium benzoate) (Brusilow 1991; Brusilow and Horwich 2001; Brusilow et al. 1979). After a loading dose, the patient receives a maintenance dose, which is initially performed using intravenous administration and then switched to oral administration once symptoms have stabilized. Moreover, oral lactulose contributes to decreasing blood ammonia by means of increasing the degree of acidity and lowering the production and absorption of ammonia in the intestinal tract.
   (c) Removal of surplus nitrogen in the diet.
   Excess dietary nitrogen can be eliminated through use of a protein-restricted diet or protein-free formula (Leonard 2001).

In addition, carbohydrates and fats can be used as calories during the acute phase, such as through intravenous administration of >10% glucose and fat emulsion or administration of protein-free milk via a gastric tube. It is considered best practice to transition from non-oral administration to enteral administration as early as possible. However, complete protein removal for more than 24 h to 48 h is not recommended because this induces catabolism owing to non-availability of essential amino acids. It is thus important to administer small amounts of protein (0.1–0.3 g/kg/day) from an early stage under the acute hyperammonemia state.

(d) Reduce the risk of neurological disorders.
It is vital to maintain the circulating plasma volume and blood pressure. However, excess fluid administration can promote cerebral edema; thus, vasopressors should be used appropriately, and the total body water and circulating volume must be controlled.

2. Management of the chronic phase.
   (a) Prevention of initial symptoms.
   Prevention of the recurrence of hyperammonemia is important. Treatment should include protein-restricted diet and formula.
   (b) Prevention of secondary infections.
   Every effort should be made to minimize the risk of respiratory infections and gastrointestinal infections in the household.
   (c) Prevent exposure to excessive stress.
   There are reports of hyperammonemia cases being induced by excessive internal bleeding after a major fracture or trauma as well as by steroid administration; therefore, considerable caution with regard to patients with UCDs under an enhanced state of metabolic stress is essential. Taking valproic acid, long-term hunger or starvation, intravenous steroids, and high intake of protein or amino acids can induce hyperammonemia and must be avoided.
   (d) Liver transplants have been performed in a large number of patients with UCDs, which has contributed to an improved prognosis (Morioka et al. 2005; Kasahara et al. 2010; Wakiya et al. 2011).
   (e) Treatments currently being researched are as follows:
   - Clinical trials of liver cell transplantation therapy are currently in progress in the USA and EU.
   - Clinical trials of liver stem cell transplantation therapy are currently in progress in Belgium and the USA.

As a result of the development of hemodialysis treatment and liver transplantation, patients with UCDs have a higher chance of survival compared to that before the turn of the century. The prognosis for neurological development has also improved (Uchino et al. 1998; Kido et al. 2012). Control of blood ammonia has become easier with the use of arginine, citrulline, benzoate, and sodium phenylbutyrate. However, without liver transplantation there is always the risk of developing hyperammonemia owing to metabolic stress, including the common cold or trauma. Recently, more facilities have begun to gain experience regarding the patients with UCDs being pregnant and giving birth. Meticulous care is essential for the delivery of patients with UCDs, and it is necessary to frequently check patient blood ammonia levels for approximately 2 weeks after the delivery (Langendonk et al. 2012; Kido et al. 2012). However, in relation to mental development, it is impossible to prevent an effect of raised glutamine and ammonia in the central nervous system even if the individual does not develop an intellectual disability. Therefore, it cannot accurately be stated that the long-term prognosis for patients with UCDs is favorable.

Questions

1. What should be considered when differentially diagnosing hyperammonemia? For UCDs, what tests should be performed subsequently?
2. What treatments are recommended for UCDs? Why is it essential to reduce elevated blood ammonia concentrations?
3. What are the most important developmental outcomes in the management of UCDs?
4. What kinds of treatments are expected in the future?

Answers

1. Important diseases that present with hyperammonemia include UCDs as well as mitochondrial disorders, fatty acid metabolism disorders, systemic sepsis, liver failure, and portosystemic shunt. Diagnostically, it is first important to measure blood ammonia. Specific diagnosis is possible with special tests including blood and urine amino acid analysis, orotic acid measurement, and genetic analysis.
2. Treatment includes hemodialysis and pharmacotherapy such as L-arginine hydrochloride, sodium phenylbutyrate, and sodium benzoate, in combination with a protein-restricted diet and protein-free milk. Furthermore, in the chronic stage, it is important to prevent secondary infection and exposure to excessive stress. Liver transplantation is sometimes performed as a radical treatment. The aim of treatment is to avoid brain and neuron disorders caused by hyperammonemia and to achieve comparatively normal neurodevelopment.
3. The most important outcomes are increasing normal height and weight and achieving normal neurodevelopment while avoiding hyperammonemia.
4. Prospective future treatments include hepatocyte stem cell transplantation and gene therapy.

References

Batshaw ML, Tuchman M, Summar M et al (2014) A longitudinal study of urea cycle disorders. Mol Genet Metab 113:127–130

Berry GT, Steiner RD (2001) Long-term management of patients with urea cycle disorders. J Pediatr 138:S56–S60

Brusilow SW (1991) Phenylacetylglutamine may replace urea as a vehicle for waste nitrogen excretion. Pediatr Res 29:147–150

Brusilow SW, Horwich AL (2001) Urea cycle enzymes. In: Scriver CR, Beaudet AL, Sly WS et al (eds) The metabolic and molecular basis of inherited disease, 8th ed. McGraw-Hill, New York, pp 1909–1963

Brusilow SW, Valle DL, Batshaw M (1979) New pathways of nitrogen excretion in inborn errors of urea synthesis. Lancet 2:452–454

Kasahara M, Sakamoto S, Shigeta T et al (2010) Living-donor liver transplantation for carbamoyl phosphate synthetase 1 deficiency. Pediatr Transplant 14:1036–1040

Kido J, Nakamura K, Mitsubuchi H et al (2012) Long-term outcome and intervention of urea cycle disorders in Japan. J Inherit Metab Dis 35:777–785

Langendonk JG, Roos JC, Angus L et al. (2012) A series of pregnancies in women with inherited metabolic disease. J Inherit Metab Dis 35:419–424

Leonard JV (2001) The nutritional management of urea cycle disorders. J Pediatr 138:S40–S44, discussion S44–S45

Maestri NE, McGowan KD, Brusilow SW (1992, August) Plasma glutamine concentration: a guide in the management of urea cycle disorders. J Pediatr 121(2):259–261

Mori M (1991) Urea cycle. In: Journey of nitrogen in the living body (Seitai no tisso no tabi). Kyoritsu Shuppan Co., Ltd., Tokyo, pp 48–49

Morioka D, Kasahara M, Takada Y et al (2005) Current role of liver transplantation for the treatment of urea cycle disorder: a review of the worldwide English literature and 13 cases at Kyoto University. Liver Transpl 11:1332–1342

Nagata N, Matsuda I, Oyanagi K (1991) Estimated frequency of urea cycle enzymopathies in Japan. Am J Med Genet 39:228–229

Uchino T, Endo F, Matsuda I (1998) Neurodevelopmental outcome of long-term therapy of urea cycle disorders in Japan. J Inherit Metab Dis 21:S151–S159

Wakiya T, Sanada Y, Mizuta K et al (2011) Living donor liver transplantation for ornithine transcarbamylase deficiency. Pediatr Transplant 15:390–395

Wilson Disease

13

Norikazu Shimizu and Tsugutoshi Aoki

Keywords
Wilson disease · ATP7B · Copper · Ceruloplasmin · Kayser-Fleischer ring

13.1 Case Reports

13.1.1 Case 1

An 8-year-old female visited our hospital because she was reportedly to have low serum ceruloplasmin levels. When she was 4 years old, she had a blood test to screen for the allergy condition at a different hospital. At that time, liver dysfunction, AST 193 IU/L (standard value, 24–38) and ALT 220 IU/L (standard value, 9–27), was pointed out. Despite the investigations, however, its cause remained unclear. She was treated with traditional Chinese medicine, but it did not decrease the serum transaminase levels. At the age of 8 years, she was introduced to our hospital for further evaluations.

At admission, her general condition was good. There were no Kayser-Fleischer rings, hepatosplenomegaly, or neurologic abnormalities. Routine laboratory examinations revealed elevated levels of AST, ALT, ALP, and γGTP (Table 13.1). Due to prolonged liver dysfunction and based on previous laboratory findings, we suspected Wilson disease. We measured the serum ceruloplasmin levels, serum copper levels, and urinary copper excretions (Table 13.1). Based on these results, the patient was diagnosed as having Wilson disease.

After the diagnosis was made, treatment with zinc acetate was initiated, and her liver dysfunction rapidly improved (Fig. 13.1). Although the urinary copper excretion level remained high, the serum AST and ALT levels were normalized 3 months later.

13.1.2 Case 2

A 13-year-old female presented with dysarthria and finger tremor. When she was 10 years old, microscopic hematuria was pointed out. However, further follow-ups were not done. Her mother and granduncle had Wilson disease.

Neurological examinations revealed that she had dysarthria, excessive salivation, finger tremors in both hands, writing disturbance due to finger tremor, and mild dystonia in the left arm. She also had hepatosplenomegaly. Furthermore, Kayser-Fleischer rings were present from ophthalmological assessments (Fig. 13.2).

Routine laboratory examinations were unremarkable except for the presence of hematuria (Table 13.2). Brain magnetic resonance imaging

N. Shimizu (✉) · T. Aoki
Department of Pediatrics, Toho University Ohashi Medical Center, Tokyo, Japan
e-mail: norikazu@med.toho-u.ac.jp

T. Oohashi et al. (eds.), *Human Pathobiochemistry*, https://doi.org/10.1007/978-981-13-2977-7_13

Table 13.1 Laboratory findings (Case 1)

Complete blood count	
RBC	459 × 10^4/μl
Hb	12.4 g/dl
Ht	38.5%
Plt	28.8 × 10^4/μl
WBC	5200/μl
Blood chemistry	
Total protein	7.6 g/dl
Albumin	4.3 g/dl
Total bilirubin	0.7 mg/dl
AST	280 IU/l
ALT	379 IU/l
LDH	968 IU/l
ALP	1012 IU/l
γGTP	45 IU/l
BUN	12 mg/dl
Creatinine	0.4 mg/dl
Na	142.0 mEq/l
K	3.8 mEq/l
Cl	106.0 mEq/l
CRP	0.1 mg/dl
Urinalysis	
pH	6.5
Glucose	(–)
Protein	(–)
Ketone	(–)
Occult blood	(–)
Sediment	
WBC	1–4/high-powered field
RBC	<1/high-powered field
Special laboratory findings	
Serum copper level	19.0 μg/dl
Serum ceruloplasmin level	2.3 mg/dl
Urinary copper excretion	176.2 μg/day

(MRI) showed high intensity areas in the basal ganglia and hypothalamus on the T2-weighted images (T2WI) (Fig. 13.3). Abdominal computerized tomography (CT) scan showed irregularities in the external contour and heterogeneous density of the liver and splenomegaly (Fig. 13.3). Special laboratory tests and an *ATP7B* gene analysis were subsequently performed. The results are shown in Table 13.2. Her serum ceruloplasmin and copper levels were low, but the urinary copper excretion levels were elevated. An *ATP7B* gene analysis showed that she had a p.Arg778Leu mutation in exon 8 and a p.Asn1270Ser mutation in exon 18. Based on these results, she was diagnosed as having Wilson disease. We also performed a gene analysis on her family (both parents and younger sister). Her mother (Wilson disease patient) had p.Arg778Leu mutation and p.Ala874Val mutation in exon 10. Her father was heterozygote of p.Asn1270Ser mutation, and her sister was heterozygote of p.Arg778Leu mutation (Fig. 13.4).

Subsequently the patient was started on trientine. This copper-chelating agent is reported to be effective especially for the neurologic symptoms of Wilson disease (Shimizu et al. 1999). Her neurologic symptoms improved gradually (Fig. 13.5) and 2 years 3 months later, all her neurologic symptoms disappeared.

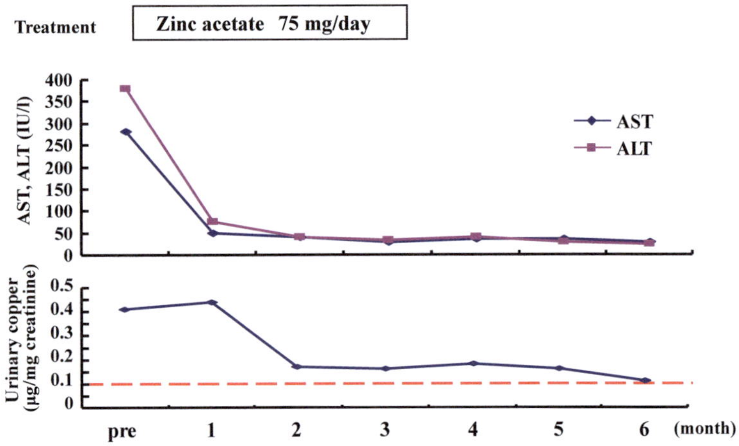

Fig. 13.1 Serum AST and ALT levels and urinary copper excretion in Case 1. Spot urinary copper excretions were measured and they were aiming at the level lower less than 0.1 μg/mg creatinine (red dotted line) for good control of copper balance with zinc treatment

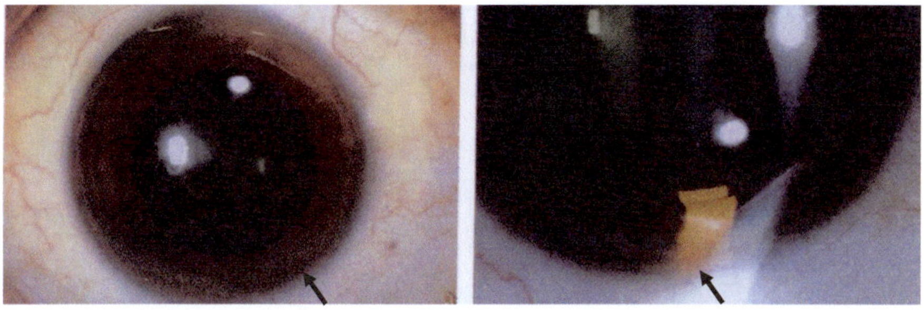

Slit lamp

Fig. 13.2 Ophthalmological findings in Case 2
Arrows indicate the Kayser-Fleischer ring

Table 13.2 Laboratory findings (Case 2)

Complete blood count	
RBC	514 × 10^4/μl
Hb	13.3 g/dl
Ht	40.1%
Plt	28.4 × 10^4/μl
WBC	14,700/μl
Blood chemistry	
Total protein	6.4 g/dl
Albumin	3.8 g/dl
Total bilirubin	0.5 mg/dl
AST	28 IU/l
ALT	6 IU/l
LDH	543 IU/l
ALP	492 IU/l
BUN	6 mg/dl
Creatinine	0.2 mg/dl
CPK	75.0 IU/l
Fe	72.0 μg/dl
Na	140.0 mEq/l
K	4.5 mEq/l
Cl	103.0 mEq/l
NH3	49 μg/dl
Urinalysis	
pH	7.5
Glucose	(–)
Protein	(–)
Ketone	(–)
Occult blood	(2+)
Sediment	
WBC	0–1/high-powered field
RBC	>100/high-powered field
Special laboratory findings	
Serum copper level	33.0 μg/dl
Serum ceruloplasmin level	2.8 mg/dl
Urinary copper excretion	71.5 μg/day (1.7 μg/kg/day)
ATP7B gene analysis	p.Arg778Leu/p.Asn1270Ser

13.2 Diagnosis

Wilson disease is an autosomal recessive disorder due to an inborn error in copper metabolism. Copper is primarily accumulated in various organs such as the liver, brain, cornea, and kidney in the patients. The incidence of Wilson disease is 1 in 35,000 to 45,000 in Japan (Aoki et al. 1996).

The principal clinical features of this disease are hepatic disorders, neurologic symptoms, and Kayser-Fleischer rings. The symptoms of hepatic disorders include fatigue, icterus, and edema in the legs. Many asymptomatic patients have elevated levels of hepatic enzymes. Neurologic symptoms are usually manifested as extrapyramidal symptoms among which dysarthria, gait disturbance, and frapping tremor are common and are the first to occur (Shimizu et al. 1996). The Kayser-Fleischer rings are specific for Wilson disease. Therefore, ophthalmological investigations should be performed for patients who are suspected of having Wilson disease. However, young children or infants and/or mild cases of this disease may not exhibit all of these symptoms. Despite the absence of Kayser-Fleischer rings, Wilson disease should still be considered in the differential diagnosis. Other symptoms, such as hematuria, proteinuria, and/or psychiatric symptoms (e.g., mental deterioration, depression, very labile mood, sexual exhibitionism, and frank psychosis), may also appear in patients with Wilson disease (Kitzberger et al. 2005). The screening criteria for Wilson disease are shown in Table 13.3.

Specific laboratory tests for the diagnosis of Wilson disease are measurements of serum ceru-

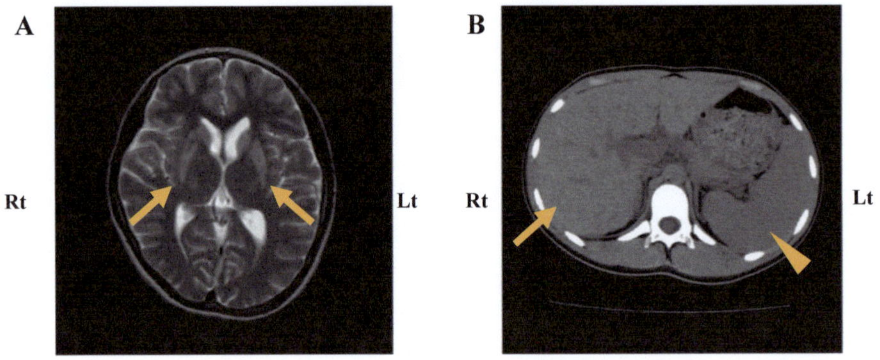

Fig. 13.3 Brain MRI and abdominal CT scan in Case 2. (**a**) T2-weighted image of brain MRI. Arrows indicate high intensity regions in the basal ganglia. (**b**) Abdominal CT scan. It shows irregularities of the external contour and heterogeneous density of the liver (arrow) and splenomegaly (arrowhead)

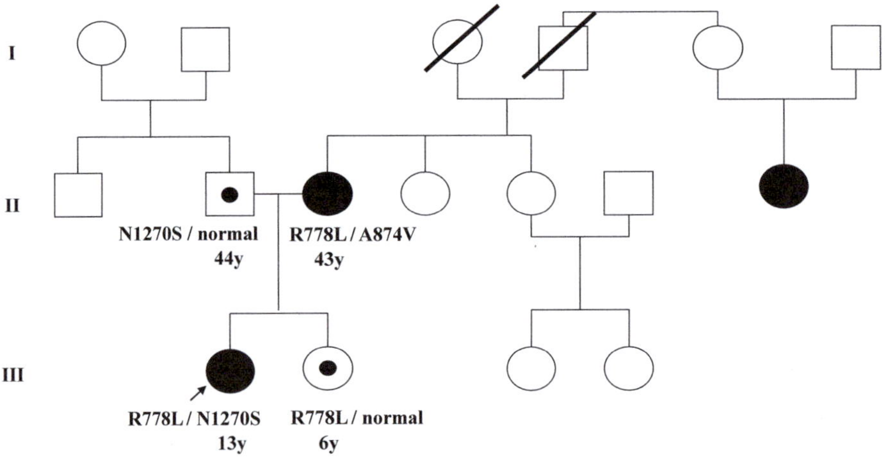

Fig. 13.4 Family tree of Case 2
Arrow indicates the patient (Case 2). R778L: p.Arg778Leu mutation, A874V: p.Ala874Val mutation, N1270S: p. Asn1270Ser mutation

Fig. 13.5 Clinical course of Case 2

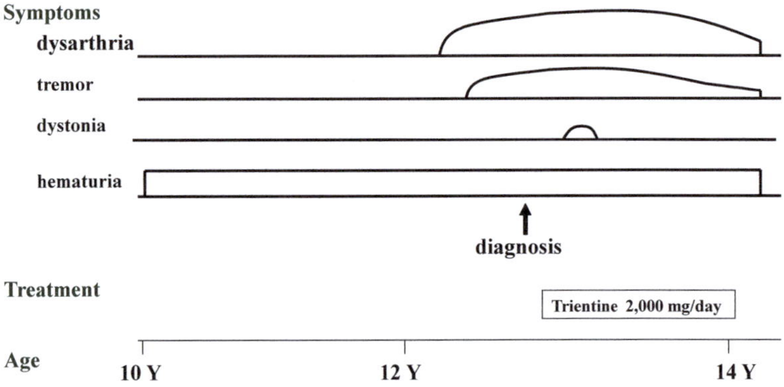

Table 13.3 Screening criteria for Wilson disease

Liver abnormalities except for infants
Hepatic disorders and/or hepatic failure with hemolysis
Neurologic symptoms, especially those with extrapyramidal signs, above 8 years of age
Hematuria/proteinuria with liver dysfunction except for infants
Psychiatric symptoms with liver dysfunction and/or neurologic symptoms after adolescence
Siblings of Wilson disease patients

Table 13.4 Diagnostic criteria for Wilson disease

Biochemical findings
1. Hepatic copper contents:
>200 μg/g wet tissue or > 250 μg/g dry tissue
2. Serum ceruloplasmin levels: <20 mg/dl
3. Urinary copper excretion:
1. >100 μg/day
2. >1.5 μg/kg/day (especially for children)
3. >0.2 μg/mg creatinine

Criteria: Wilson disease can be diagnosed if the patients have at least two of the three criteria. If the patients are younger than 3 years, criteria 1 and 2 or 2 and an ATP7B gene analysis are required

Reproduced with permission from Biomed Res Trace Elements 8: 13-21, 1997

loplasmin levels, serum copper levels, and urinary copper excretions. Further tests, such as the measurement of haptic copper contents and *ATP7B* gene analysis, should also be performed. More than 95% of patients with this disease have low serum ceruloplasmin levels. Serum copper levels are usually low; however, they are occasionally normal or increased. Urinary copper excretions are usually elevated after 4 years of age. Highly elevated hepatic copper contents (greater than 200 μg/g wet tissue or 250 μg/g dry tissue) are the most specific finding for the diagnosis of Wilson disease. The *ATP7B* gene analysis is also a useful tool. However, the mutation detection rate in Wilson disease patients is reported to be 85–90% (Nakamura et al. 2009). The diagnostic criteria of Wilson disease are shown in Table 13.4 (Fujii 1997).

13.3 Biochemical and Molecular Perspectives

The Wilson disease gene, *ATP7B*, is located on chromosome 13q14.3. More than 300 disease-specific mutations have been reported (Nakamura et al. 2009). The mutation spectrum of *ATP7B* displays a population-dependent distribution. Although the most frequent mutation in Caucasian patients (originating from European and North American countries) is the p. His1069Gln mutation and found in 10–40% of alleles (Ferenci 2006; Shah et al. 1997), it has not been detected in Japanese patients. In Japan, the most common mutation is the p.Arg778Leu mutation (Nakamura et al. 2009) and is also common in other Asian countries (Kim et al. 1998; Tsai et al. 1998). There is no definite correlation between genotype and phenotype in Wilson disease.

The pathogenesis of Wilson disease is due to copper accumulation in the liver, brain, cornea, and other organs. The ATP7B protein encodes a putative copper-transporting P-type ATPase (Fig. 13.6) (Shimizu 2017). This protein resides in hepatocytes in the trans-Golgi network and is responsible for the excretion of copper into the bile and its transportation into the secretory pathway for incorporation into apoceruloplasmin.

In Wilson disease, these two functions of copper metabolism are disturbed. As a result, copper accumulates in the hepatocytes and leads to liver damage (Fig. 13.7). The excess copper then flows out from the liver and into the blood stream where it accumulates in the brain, cornea, and kidney (Fig. 13.7). The accumulation of copper causes oxidative stress resulting in cell damage. Typically, the serum ceruloplasmin levels in patients with Wilson disease are markedly low, while the non-ceruloplasmin copper levels are high. This results in a net reduction of serum copper levels. The urinary copper excretion increases greatly, while the hepatic copper content also increases remarkably.

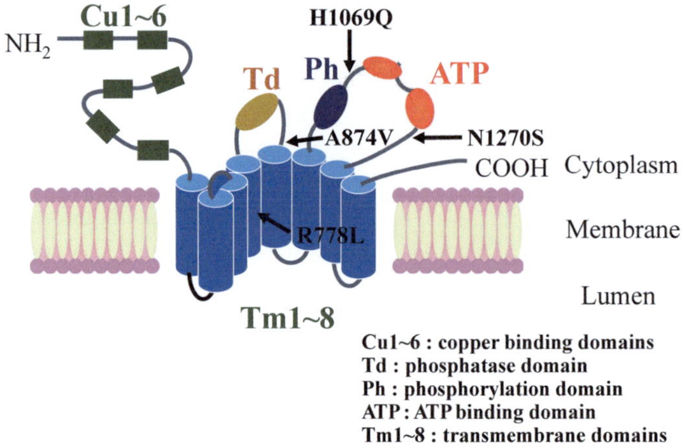

Fig. 13.6 The structure of ATP7B protein and position of mutations (Shimizu, 2016 modified). The ATP7B protein is a putative copper-transporting P-type ATPase, synthesized as a membrane protein. It has six copper-binding domains, phosphatase domain, phosphorylation domain, ATP-binding domain, and eight transmembrane domains. Arrows indicate the mutations of Wilson disease patients. The p.Arg778Leu is the most common in Japanese Wilson disease patients, and p.His1069Gln is the most common in Caucasian patients. R778L: p.Arg778Leu mutation, A874V: p.Ala874Val mutation, H1069Q: p.His1069Gln mutation, N1270S: p.Asn1270Ser mutation

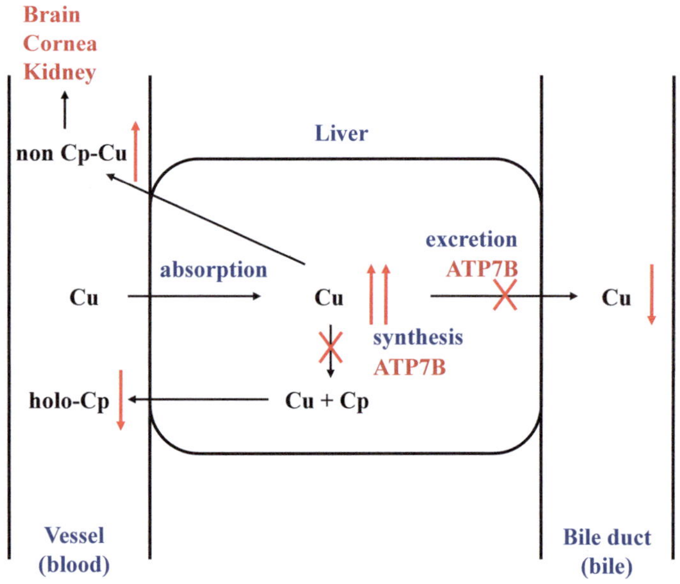

Fig. 13.7 Pathophysiology of Wilson disease. In Wilson disease, the excretion of copper into the bile and the incorporation of copper into apoceruloplasmin are disturbed. Copper accumulates in the hepatocytes, and excess copper flows into the blood and then accumulates in the brain, cornea, and kidney

13.4 Therapy and Prevention

Wilson disease is progressive and ultimately fatal if untreated. However, the prognosis can be improved dramatically by introducing chelating agents, such as penicillamine, for the removal of excess copper (Shimizu et al. 1999; Walshe 1956, 1973; Sternlieb 1990). Basic strategies for the treatment of Wilson disease are the administration of copper-chelating agents and/or zinc and having the patients follow a low-copper diet. Liver transplantation should be indicated for patients with severe liver failure.

1. D-penicillamine

This is the first drug of choice for patients with Wilson disease in Japan, and it has a strong chelating effect. The initial dose is usually 15 to 25 mg/kg/day (max, 1400 mg/day) given twice or thrice per day. For all patients, the urinary copper excretion should be monitored closely over 24 h, and the dose should be adjusted to achieve losses of over 2 mg/day in the early stages of treatment (Brewer and Askari 2005). D-penicillamine should be administered at least 1 h before or 2 to 3 h after meals. Otherwise, D-penicillamine may become bound with the copper in the diet and, as a result, cannot be absorbed into the intestine. When the basal copper excretion levels decrease to less than 70 μg/day, the administered dose of D-penicillamine can be decreased from 50% to 60% of the initial dose (i.e., 10 to 15 mg/kg/day).

Two important issues have been raised regarding the use of D-penicillamine. First, various side effects appear in approximately 20–25% of patients who receive it (Shimizu et al. 1999). Second, the symptoms of 30 to 50% of patients with neurologic symptoms may worsen in the early stages of treatment (Shimizu et al. 1999; Brewer and Askari 2005).

2. Trientine

Trientine is the second drug of choice for the treatment of Wilson disease in Japan. It is usually administered when D-penicillamine has to be withdrawn due to severe side effects or ineffectiveness. This agent has also shown good results in response to the neurologic symptoms of this disease (Shimizu et al. 1999; Suda et al. 1993). The initial dose is usually 40 to 50 mg/kg/day (max, 2500 mg/day) twice or thrice per day. Because the chelating effect of this drug is weaker than that of D-penicillamine, a higher dose must be used. However, it has few toxic effects. Trientine also should be administered 1 h before or 2–3 h after meals.

3. Zinc acetate

Zinc acetate blocks the intestinal copper absorption by inducing metallothionein synthesis in mucosal cells. It was approved in 2008 in Japan for maintenance therapy and for presymptomatic patients (Shimizu et al. 2010). The recommended dose of zinc acetate is 75 to 150 mg/day for adults (older than 16 years old) and 50 to 75 mg/day for children and should be administered twice (50 mg/day) or thrice per day. It should be administered 1 h before or 2 h after meals because some elements in the food, such as fiber and casein (Oelshlegel 1997; Pecoud et al. 1975), inhibit the absorption of zinc acetate. Zinc acetate has no severe side effects (Shimizu et al. 2010).

Zinc minimizes copper toxicity in the liver by promoting the nontoxic metallothionein-bound form. Furthermore, the therapeutic effect of zinc acetate for liver damage due to Wilson disease is relatively high. Thus, Brewer et al. recommend zinc acetate as the first-choice drug for mild to moderate hepatic Wilson disease patients since it can increase transaminase levels without producing symptoms related to compensated cirrhosis (Brewer and Askari 2005).

4. Combination therapy

Combination therapy of a copper-chelating agent and zinc has a strong copper exclusion effect. In this therapy, trientine is chosen over D-penicillamine because of its lower toxicity. The drugs must be separated from each other by at least an hour. We recommend the combination of trientine and zinc acetate for Wilson disease patients with decompensated cirrhosis from mild liver failure to severe neurologic symptoms.

5. Low-copper diet

A low-copper diet is also important. Dietary copper intake should be restricted to less than 1 mg/day or 0.5 mg/day (for infants or young children). When patients become well controlled and/or take zinc acetate, they can consume 1 to 1.5 mg/day of copper. The patients and/or their family should be aware of the types of foods that are copper-rich and thus avoid any excessive intake of copper-rich foods, such as shellfish, liver, mushrooms, broccoli, chocolate, and nuts.

6. Liver transplantation

Liver transplantation is an effective treatment for Wilson disease patients with acute liver failure, decompensated cirrhosis, or chronic liver failure that is unresponsive to medical therapy. Liver transplantation corrects not only hepatic copper metabolism but also improves extrahepatic copper metabolism.

7. Prevention

Wilson disease can be prevented in those who start treatment during the presymptomatic period. Various strategies such as mass screening and familial analysis are considered for the prevention of Wilson disease. Mass screening might be the best method to detect presymptomatic Wilson disease patients. However, a mass screening system for this disease is not currently available.

Familial analysis, especially for siblings, is useful to identify presymptomatic patients. When Wilson disease patients who have siblings are diagnosed, familial analysis for their siblings should be performed.

Questions

1. When you see a child (not an infant) with hepatic disorder, in what way do you perform the differential diagnosis?
2. When you see a patient who has family history of Wilson disease, what kinds of laboratory tests should be performed as screening of this disease?
3. How the difference is treatment strategy between Wilson disease patients with neurologic symptoms and without neurologic symptoms?
4. Wilson disease is a copper-accumulation disorder. However, serum copper levels are usually low. Why?
5. When you see a Wilson disease patient with low serum ceruloplasmin level and high serum copper level, what is happening to this patient?

References

Aoki T, Suzuki M, Fujioka Y, Shimizu N, Fuji H, Nakazono H, Kawase C, Yamaguchi Y, Arashima S, Matsuda I, Arima M (1996) Neonatal and Perinatal screening, the Asian pacific perspective. In: STS L, CCP P (eds) Nationwide survey of clinical features of Wilson's disease in Japan. The Chinese University Press, Hong Kong, pp 25–28

Brewer GJ, Askari FK (2005) Wilson's disease: clinical management and therapy. J Hepatol 42:S13–S21

Ferenci P (2006) Regional distribution of mutations of the ATP7B gene in patients with Wilson disease: impact on genetic testing. Hum Genet 120:151–159

Fujii H (1997) Ceruloplasmin and copper metabolism in presymptomatic patients with Wilson's disease, establishment of diagnostic criteria. Biomed Res Trace Elem 8:13–21

Kim EK, Yoo OJ, Song KY, Yoo HW, Choi SY, Cho SW, Hahn SH (1998) Identification of three novel mutations and high frequency of the Arg778Leu mutation in Korean patients with Wilson disease. Hum Mutat 11:275–278

Kitzberger R, Madl C, Ferenci P (2005) Wilson disease. Metab Brain Dis 20:295–302

Nakamura H, Hemmi H, Shimizu N (2009) Molecular diagnosis of Wilson disease in Japanese patients. Toho J Med 56:65–70

Oelshlegel FJ Jr, Brewer GJ (1997) Absorption of pharmacologic dose of zinc. Zinc metabolism: current aspects in health and disease. Alan R Liss, Inc, New York, pp p299–p316

Pecoud A, Donzel P, Schelling L (1975) Effect of foodstuffs on the absorption of zinc sulfate. Clin Pharmacol Ther 17:469–474

Shah AB, Chernov I, Zhang HT et al (1997) Identification and analysis of mutation in the Wilson disease gene (ATP7B): population frequencies, genotype-phenotype correlation, and functional analysis. Am J Hum Genet 61:317–328

Shimizu N (2017) Molecular diagnosis for Wilson disease in Japan. Kan Tan Sui 74:169–174 (in Japanese)

Shimizu N, Suzuki M, Yamaguchi Y, Aoki T, Matsuda I, Arima M (1996) A nation-wide survey for neurologic and hepato-neurologic type of Wilson disease: clinical features and hepatic copper content. No To Hattatsu 28:391–397 (in Japanese)

Shimizu N, Yamaguchi Y, Aoki T (1999) Treatment and management of Wilson's disease. Pediatr Int 41:419–422

Shimizu N, Fujiwara J, Ohnishi S, Sato M, Kodama H, Kohsaka T, Inui A, Fujisawa T, Tamai H, Ida S, Itoh

S, Ito M, Horiike N, Harada M, Yoshino M, Aoki T (2010) Effects of long-term zinc treatment in Japanese patients with Wilson disease: efficacy, stability, and copper metabolism. Transl Res 156:350–357

Sternlieb I (1990) Perspectives on Wilson's disease. Hepatology 12:1234–1239

Suda M, Kubota J, Yamaguchi Y, Fujioka Y, Saito Y, Aoki T (1993) A study of trienthine therapy in Wilson's disease with neurological symptoms. No To Hattatsu 25:429–434 (in Japanese)

Tsai CH, Tsai FJ, Wu JY, Chang JG, Lee CC, Lin SP, Yang CF, Jong YJ, Lo MC (1998) Mutation analysis of Wilson disease in Taiwan and description of six new mutations. Hum Mutat 12:370–376

Walshe JM (1956) Penicillamine a new oral therapy for Wilson's disease. Am J Med 21:487–495

Walshe JM (1973) Copper chelation in patients with Wilson's disease. A comparison of penicillamine and triethylene tetramine dihydrochloride. Q J Med 42:441–452

Part II

Genetics

14 Achondroplasia

Kosei Hasegawa, Hiroyuki Tanaka, and Yoshiki Seino

Keywords
Short stature · Skeletal disorder · Sleep apnea · Endochondral ossification · Chondrocyte

14.1 Case Report

A pregnant woman was referred to the obstetrics department at a hospital because her doctor considered that her baby might have some congenital skeletal disorder. Echography of the fetus revealed that the fetal femur growth stopped at 27 weeks of gestational age and that the femur length was very short at 33 weeks (−6.7 SD). The chest of the fetus was hypoplastic. In contrast to her baby's short limbs and narrow chest, its head was large (+1.9 SD). Cephalopelvic disproportion was suspected. Therefore, her doctor chose to conduct a cesarean section.

K. Hasegawa (✉)
Department of Pediatrics, Okayama University Hospital, Okayama, Japan
e-mail: haseyan@md.okayama-u.ac.jp

H. Tanaka
Department of Pediatrics, Okayama Saiseikai General Hospital, Okayama, Japan
e-mail: hrtanaka@saiseidr.jp

Y. Seino
Department of Pediatrics, JCHO Osaka Hospital, Osaka, Japan
e-mail: seino-yoshiki@osaka.jcho.go.jp

At 39 weeks of gestational age, a male baby was delivered by cesarean section. His Apgar scores were, respectively, 8 and 9 at 1 and 5 min. His mother (32 years old) and his father (39 years old) were non-consanguineous. Tachypnea and dyspnea were observed soon after birth. Therefore, mask bagging and oxygen supplementation were started. Physical examination showed a large head, depressed nasal bridge, and frontal bossing. His height was 46 cm (−1.4 SD). His weight was 2898 g (−0.23 SD). His head circumstance was 36.5 cm (2.3 SD), large for his stature. He had a narrow chest and short extremities. He also presented trident hand. Genitalia were of a normal male; testes were palpable in his scrotum.

Roentgenography at birth showed a small chest, hypoplasia of the iliac wing, and hypoplastic greater sciatic notches (Fig. 14.1a, b). Femurs, tibiae, humeri, ulnae, and radii were all short. The metaphyses of these bones were flared. Hydrocephalus was not identified by head CT analysis at 1 week after birth (Fig. 14.1c). The cranial base was hypoplastic (Fig. 14.1d). Foramen magnum was narrow, although compression of the cervical spinal cord was not observed. No abnormality was shown by blood examination or biochemical examination. Neonatal mass screening for congenital adrenal hyperplasia, congenital hypothyroidism, and congenital metabolic disorders all yielded negative results. Results of the physical and radiologi-

T. Oohashi et al. (eds.), *Human Pathobiochemistry*, https://doi.org/10.1007/978-981-13-2977-7_14

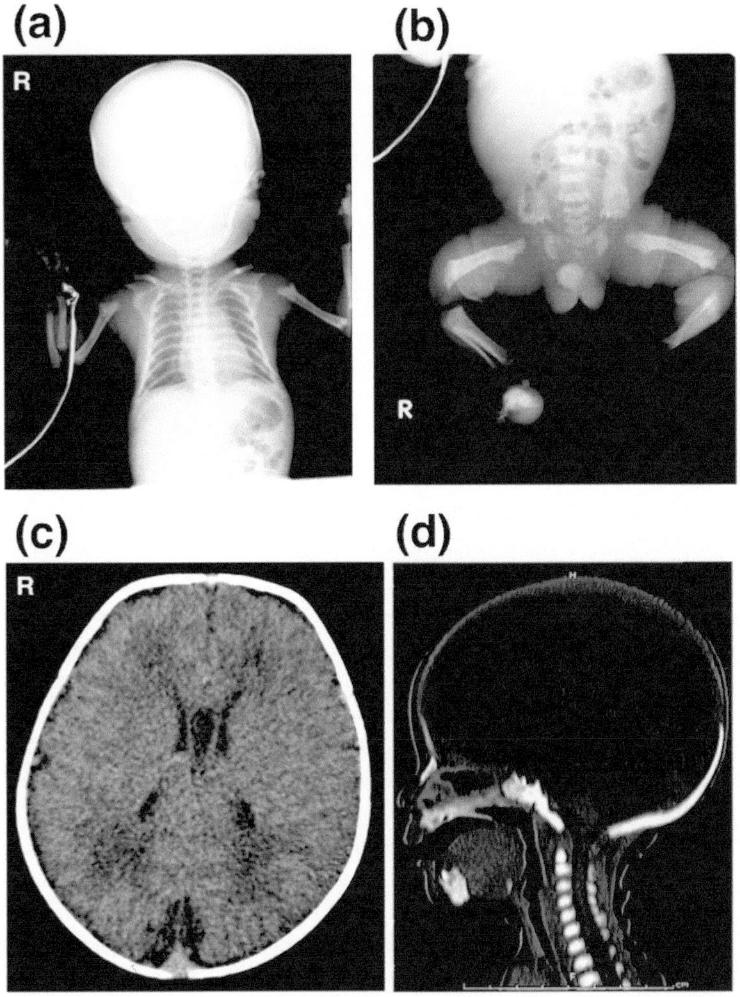

Fig. 14.1 Roentgenogram at birth (**a**, **b**) and head CT analysis at 1 week after birth (**c**, **d**). (**a**) Chest is narrow; metaphysis of the ulna and radius is flared. (**b**) Iliac wing and greater sciatic notches are hypoplastic. The femur is short and radiolucency of the femoral neck is apparent. Metaphyses in long bones of lower limbs are flared, as they are in upper limbs. (**c**) Hydrocephalus is not observed. (**d**) Foramen magnum was narrow. Cervicomedullary compression is not observed. The cranial base is hypoplastic

cal examinations described above led to his diagnosis as an achondroplasia patient.

Dyspnea improved gradually after 4 h. It was not observed again. Sucking of breast milk was not weak. He was discharged on hospital day 8. Achondroplasia results from genetic mutation of fibroblast growth factor receptor type 3 (*FGFR3*). Therefore, genetic analysis of *FGFR3* gene was conducted at 1 year of age; heterozygous nucleotide substitution of guanine at position 1138 to adenine was identified (Fig. 14.2). This substitution results in amino acid change from glycine at position 380 to arginine (p.Gly380Arg), a common genetic mutation in achondroplasia (Rousseau et al. 1994, Shiang et al. 1994, Bellus et al. 1995). His height growth after birth was stunted markedly (Fig. 14.3a), in contrast to his head growth: it gradually became larger for healthy infants. At 2 years of age, his head was large (52.5 cm: 2.0 SD) for his stature. Head MRI showed cervicomedullary compression (Fig. 14.4a). Decompression by laminectomy was necessary (Fig. 14.4b) at 2 years of age.

Because snoring and sleep apnea became severe from 2 years and 6 months old, polysomnography was conducted, which indicated obstructive sleep apnea (his apnea–hypopnea index was 34.2/h). An otolaryngologist resected adenoids and tonsils. At the time of his tonsillectomy, aeration tubes were inserted into both ears because of repeated otitis media.

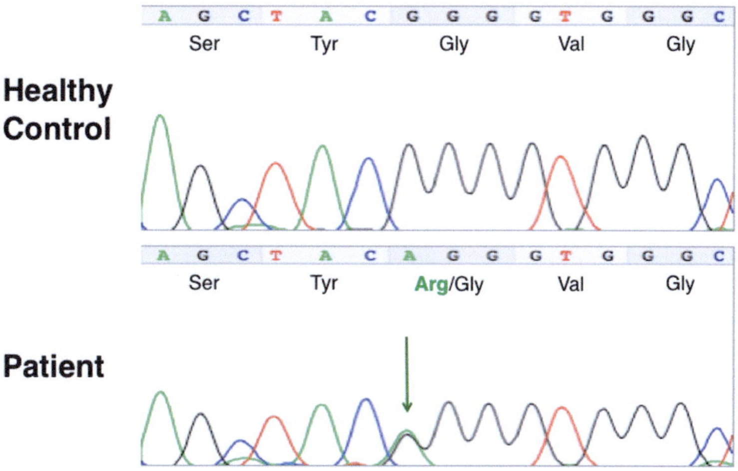

Fig. 14.2 Result of genetic analysis of *FGFR3* gene. In patient, heterozygous substitution of guanine to adenine is observed at nucleotide 1138 (arrow). This substitution results in amino acid change from glycine at 380 to arginine

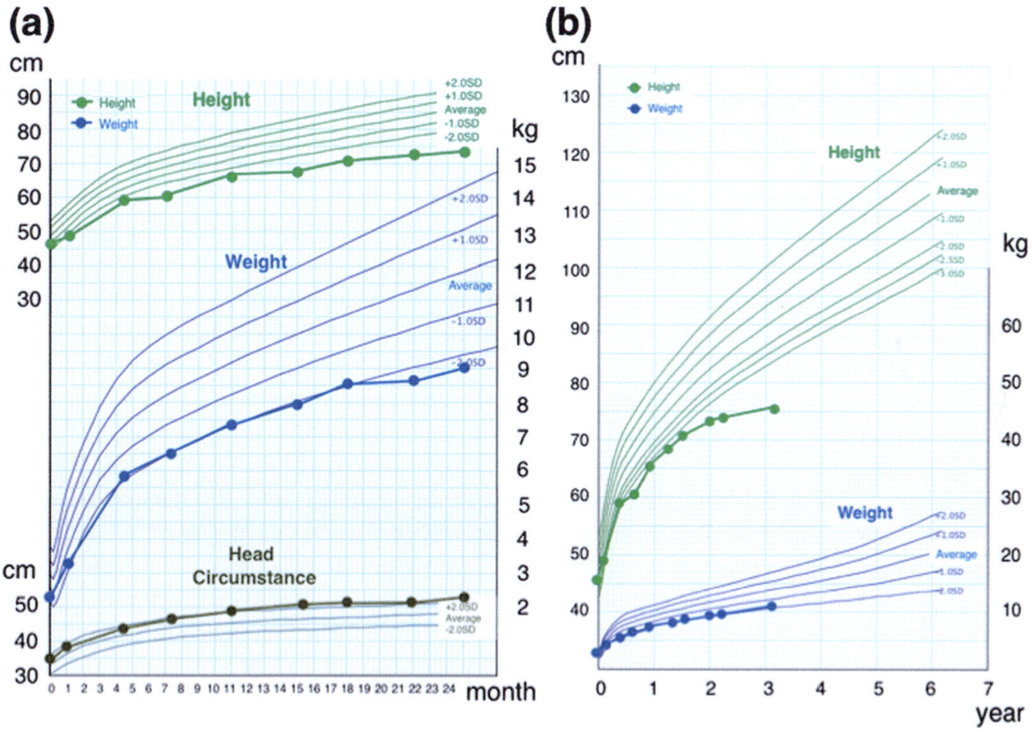

Fig. 14.3 Growth chart of patient. (**a**) Growth chart from birth to 2 years old. Height and weight growth are severely stunted while head circumstances are consistently large compared with the height and weight SD score. (**b**) Growth chart until 3 years old. Height growth is severely impaired after 3 years

His motor development before 3 years old was delayed: head control at 7 months, rolling over at 7 months, sitting alone at 12 months, standing alone at 20 months, and walking unaided at 24 months. In contrast to motor development, his neurological development was not delayed: he said one word at 11 months; he said two-word phrases at 24 months. At 3 years and 1 month old, his height was 75.8 cm (−5.2 SD). His arm span was 69.0 cm (Fig. 14.3b).

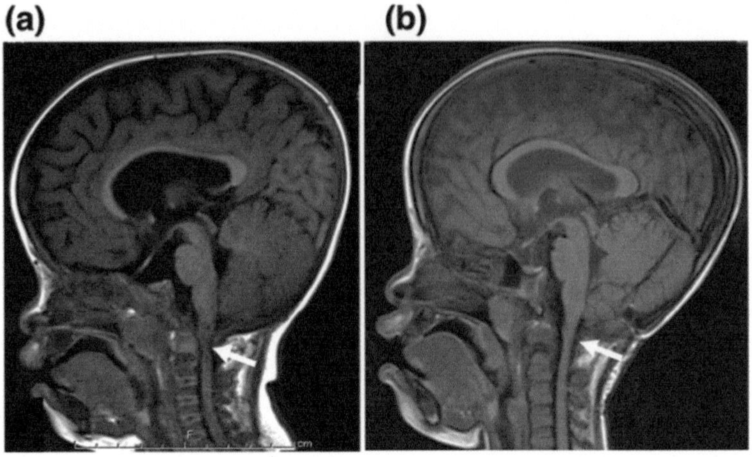

Fig. 14.4 Head MRI findings at 2 years old (T1-weighted images). (**a**) Before laminectomy. Cervicomedullary compression is observed at the foramen magnum (arrow). (**b**) After laminectomy. Compression is released (arrow).

14.2 Diagnosis

Achondroplasia, the most common congenital skeletal disorder, occurs with frequency of about 1 in 20,000. Most cases are sporadic, although some cases inherit an autosomal dominant trait. Its penetrance is 100%.

Diagnosis of achondroplasia is made mainly based on physical findings such as rhizomelic short stature, midface hypoplasia, and frontal bossing. The results of radiological examination are presented in Fig. 14.5. In some cases, diagnosis is made prenatally by ultrasound findings and 3D-CT analysis because shortening of long bones and macrocephaly are observed from the prenatal period, as in this patient.

FGFR3 gene analysis is helpful for the confirmation of clinical and radiological diagnosis because strong genotype–phenotype correlation is observed in FGFR3-related skeletal disorders described below. This patient was diagnosed as having achondroplasia from clinical findings and the heterozygous p.Gly380Arg mutation.

14.3 Biochemical and Molecular Perspectives

Bone is formed through two mechanisms: endochondral ossification and intramembranous ossification. Development of long bones such as the femur and humerus occurs through endochondral ossification. Endochondral ossification is a multistep process consisting of condensation of mesenchymal cells and proliferation of chondrocyte, differentiation to hypertrophic chondrocyte, apoptosis of hypertrophic chondrocyte, vascular invasion, and recruitment of osteoblasts. Chondrocytes in the growth plate are related to longitudinal bone growth (Fig. 14.6). These processes are regulated spatiotemporally by many hormones, growth factors, and transcription factors; expressions of chondrocyte-specific proteins also change according to the differentiation of chondrocytes. Abnormalities of these regulatory factors and chondrocyte-specific proteins produce various congenital skeletal disorders. During endochondral ossification, FGFR3 acts as "bone growth suppressor" by inhibiting chondrocyte proliferation and differentiation in the growth plate as explained below.

Fibroblast growth factor (FGF) has 22 isoforms, each of which exerts a different role through receptors of four types including FGFR3 (Brewer et al. 2016). The *FGFR3* gene is located on chromosome 4. Aside from bone, FGFR3 is expressed in various tissues such as the brain, skin, and testes. The FGFR3 molecule consists of an extracellular immunoglobulin-like domain and an intracellular tyrosine kinase domain. These two domains are connected by a transmembrane domain (Fig. 14.7). Ligand binding to FGFR3 at the extracellular domain causes dimerization of two FGFR3s and autophosphorylation at the intracellular tyrosine kinase domain, which results in subsequent signal transduction

Prenatal findings in ultrasound

- Shortened femur length
- Large head

Physical examination

- Rhizomelic short stature
- Frontal bossing
- Depressed nasal bridge
- Relatively macrocephaly
- Trident hand
- Kyphosis
- Lumbar lordosis
- Narrow chest

Radiological examination

- Cranial base hypoplasia
- Narrowing of foramen magnum
- Shortened long bones with metaphyseal flaring
- Radiolucency of femoral neck from neonate to infant
- Shortened femoral neck
- Squared iliac wing
- Hypoplasia of greater sciatic notches
- Narrowing of interpedicular distance of lumbar spine

FGFR3 gene analysis

Fig. 14.5 Diagnostic procedure of achondroplasia. Diagnosis of achondroplasia is made based on the clinical and radiological information. *FGFR3* gene analysis is helpful to define the clinical diagnosis

(Horton et al. 2007). Among the FGFR3 downstream pathway, activation of signal transducer and activator of the transcription 1 (STAT1) pathway decreases chondrocyte proliferation. Mitogen-activated protein kinase (MAPK) pathways inhibit proliferation and terminal differentiation and matrix synthesis. By these functions, FGFR3 suppresses growth plate chondrocyte proliferation and differentiation and acts as a "bone growth suppressor" during the endochondral bone formation process. In fact, target deletion of FGFR3 in mouse shows overgrowth and expanded growth plate (Deng et al. 1996). Furthermore, in humans, loss-of-function mutation of FGFR3, p.Arg621His, results in CATSHL syndrome, of which the major feature includes **ca**mptodactyly, **t**all **s**tature, and **h**earing **l**oss in heterozygous state (Toydemir et al. 2006). Another mutation, p.Thr546Lys, reportedly causes tall stature, severe lateral tibial deviation, scoliosis, hearing impairment, camptodactyly, and arachnodactyly in a homozygous state (Makrythanasis et al. 2014). These two human disorders also support the fact that the function of FGFR3 is as a "bone growth suppressor."

In "nosology and classification of genetic skeletal disorders" which classifies 436 genetic skeletal disorders into 42 groups (Bonafe et al. 2015), FGFR3-related disorders group appears first of all. In this group, thanatophoric dysplasia (TD) type I and type II, **s**evere **a**chondroplasia with **d**evelopmental **d**elay and **a**canthosis **n**igricans (SADDAN), achondroplasia, hypochondroplasia, and CATSHL syndrome are involved. In other groups of this classification, some FGFR3-related disorders such as Crouzon-like craniosynostosis with acanthosis nigricans, craniosynostosis, Muenke type (craniosynostosis syndrome group), and lacrimo-auriculo-dento-digital syndrome (polydactyly-syndactyly-triphalangism group)

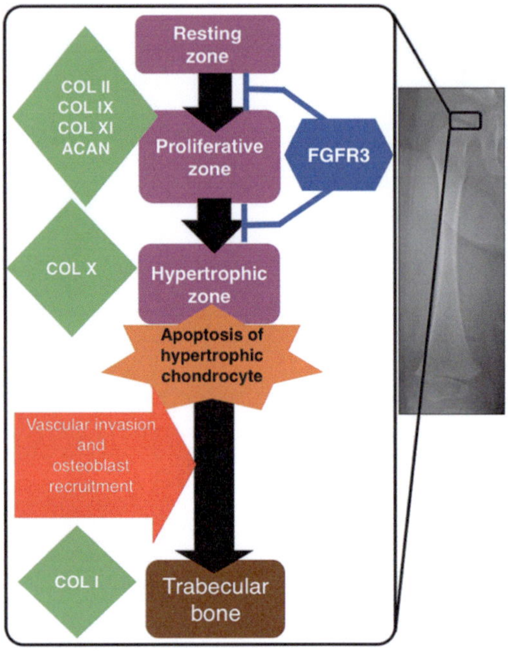

Fig. 14.6 Schematic representation of endochondral ossification in growth plate of femoral neck and action of FGFR3: *COL II* type II collagen, *COL IX* type IX collagen, *COL XI* type XI collagen, *ACAN* aggrecan, *COL X* type X collagen, *COL I* type I collagen

are involved. Each disorder shows some FGFR3 genotype–phenotype correlation (Fig. 14.7).

Actually, the p.Gly380Arg mutation in the transmembrane domain of *FGFR3* gene is found in more than 95% of achondroplasia patients. Other rare mutations such as p.Ser217Cys (Zhang et al. 2007), p.Ser279Cys (Heuertz et al. 2006), p.Ser344Cys (Takagi et al. 2015), p.Ser348Cys (Hasegawa et al. 2016), and p.Gly375Cys (Ikegawa et al. 1995) are also identified in achondroplasia patients. In p.Gly380Arg mutant FGFR3, constitutive activation of receptor occurs, producing excessive signal transduction that results in the inhibition of chondrocyte proliferation and differentiation through activation of STAT1. By activation of the ERK pathway, constitutive activation of FGFR3 also decreases chondrocyte matrix accumulation (Harada et al. 2009). This excessive signaling inhibits chondrocyte proliferation and differentiation severely and results in shortening of long bones and marked short stature. This mechanism is reproduced by a mouse model of achondroplasia (Naski et al. 1998). This mouse shows severely stunted growth and a narrow growth plate because of decreased proliferation and slowing of chondrocyte differentiation.

14.4 Clinical Features and Current Therapy of Achondroplasia

No fundamental treatment for patients with achondroplasia exists. Therapy is now designed to address the respective clinical symptoms such as short stature, neurosurgical complications, orthopedic complications, and otolaryngological complications.

Achondroplasia patients present extreme short stature because impairment of endochondral ossification causes growth disturbance of long bones. During childhood, height is about −5.0 SD. The absence of a growth spurt during puberty increases the difference from average height of healthy children. Adult height of Japanese achondroplasia patient with no treatment is about 130 cm in men (−7.0 SD) and 123 cm in women (−6.6 SD) (Tachibana et al. 1997). Not only legs but also arms become short, especially in the upper arms and femur. For that reason, achondroplasia is designated as “rhizomelic short stature” (“rhizomelic” means disproportion of the length of the proximal limbs). To treat short stature of achondroplasia, daily subcutaneous injection of recombinant human growth hormone (rhGH) was started in Japan. Short-term improvement of growth was observed (Tanaka et al. 1998), although it remains unknown how tall rhGH-treated achondroplasia patients will be as adults. Orthopedically, leg lengthening using external skeletal fixation devices such as an Ilizarov device has been conducted for the short stature treatment of achondroplasia. Infection, pain, and the long treatment duration are difficulties related to this procedure.

Impairment of endochondral ossification causes not only growth disturbance of long bones but also causes cranial base hypoplasia, which engenders hypoplasia of the foramen magnum in the occipital bone. One complication of achon-

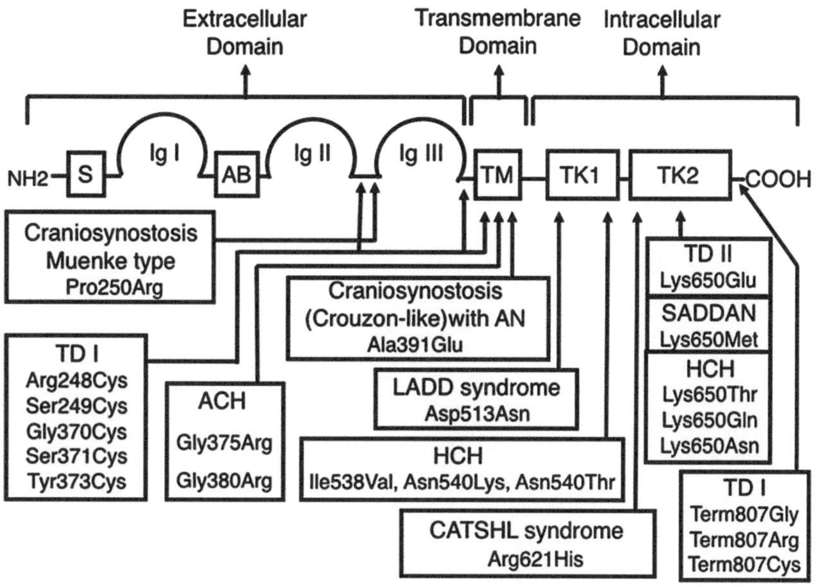

Fig. 14.7 Structural presentation of FGFR3 and representative FGFR3 mutations in FGFR3-related skeletal disorders: *NH2* amino terminal, *COOH* carboxy terminal, *S* signal sequence, *AB* acid box, *TM* transmembrane domain, *TK* tyrosine kinase domain, *Ig* immunoglobulin-like domain, *ACH* achondroplasia, *HCH* hypochondroplasia, *TD I* thanatophoric dysplasia type I, *TD II* thanatophoric dysplasia type II, *SADDAN* severe achondroplasia with developmental delay and acanthosis nigricans, *LADD* lacrimo-auriculo-dento-digital, *CATSHL* camptodactyly, tall stature, and hearing loss, *AN* acanthosis nigricans

droplasia is compression of the cervicomedullary junction caused by the foramen magnum stenosis. Cervicomedullary compression can cause sudden death and spastic paralysis. Therefore, cervicomedullary decompression by laminectomy for severe foramen magnum stenosis is necessary to release compression. Hydrocephalus is also observed occasionally in achondroplasia because narrowing of the craniocervical junction can engender increased intracranial venous pressure, which alters cerebrospinal fluid dynamics. A ventriculoperitoneal shunt is occasionally necessary for continuously worsening hydrocephalus.

Midface hypoplasia is a physical characteristic of achondroplasia. It frequently causes otolaryngological complications such as obstructive sleep apnea. Midface hypoplasia and physiological lymphoid tissue hyperplasia such as those of the tonsils and adenoids during childhood narrow the airway further. Severe airway obstruction engenders obstructive sleep apnea and subsequent life-threatening pulmonary hypertension. Narrowing of the airway also causes repetitive otitis media, which engenders persistent conductive hearing loss in adulthood if treated inappropriately.

Spinal canal stenosis can occur in achondroplasia patients from adolescence to adult because pedicles are short. The interpedicular distance is narrow compared with those of healthy subjects (Fig. 14.8a, b) and because kyphosis gradually becomes worse at the thoracolumbar junction (Fig. 14.8c, d). These orthopedic characteristics of achondroplasia engender spinal nerve compression (Fig. 14.8e), which causes bladder and rectal disturbance and intermittent claudication because of tenderness or numbness. Metaphyseal change in the femur, tibial bowing, change in bony alignment, and tendency of obesity in achondroplasia patients engender osteoarthritis of knee joints after adolescence.

Because of these various clinical features, achondroplasia patients should be managed by a multidisciplinary team from the fetal period to adulthood.

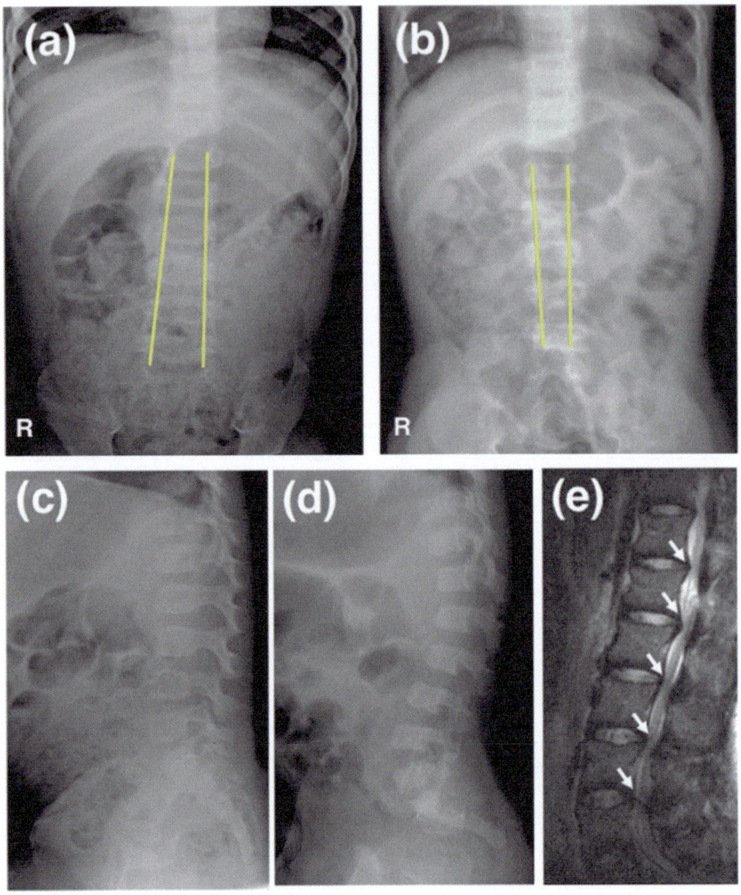

Fig. 14.8 Radiological finding of lumbar spine in a 3-year-old healthy subject (**a, c**) and 3-year-old girl with achondroplasia (**b**, **d**). Interpedicular distance (ID) of achondroplasia patient decreases from the upper to lower lumbar spine (**b**, yellow lines) in contrast to a healthy subject whose ID increases from upper to lower (**a**, yellow lines). In achondroplasia patients, pedicles are short (**d**) compared to those of healthy subject (**c**). Lordosis and gibbus might be readily apparent at the thoracolumbar and lumbosacral junction in achondroplasia patients (**d**) in contrast to healthy subject whose thoracolumbar and lumbosacral junctions are straight (**c**). (**e**) MR imaging of lumbar spine of adult patient with achondroplasia (T2-weighted image). Multiple lumbar hernias (arrow) are observed

14.5 Future Therapy Candidate of Achondroplasia

As explained above, patients with achondroplasia must cope with many complications, but no fundamental treatment exists. To find drug candidates to treat achondroplasia fundamentally, many trials are now underway.

C-type natriuretic peptide, CNP (coded by NPPC gene), belongs to the natriuretic peptide family to which atrial natriuretic peptide (ANP) and brain natriuretic peptide (BNP) also belong. CNP acts through its receptor NPR2 (coded by NPR2 gene and also designated as NPRB), its second messenger cyclic GMP, and subsequent intracellular signaling. Because targeted ablation of CNP in mouse results in impaired longitudinal bone growth and severe short stature, CNP-NPR2 axis is found to be involved in endochondral ossification (Chusho et al. 2001). In humans, disorders by genetic mutation of NPR2 gene results not only in short stature but also in tall stature. Biallelic loss-of-function mutation of *NPR2* gene results in acromesomelic dysplasia type Maroteaux (AMDM) (Bartels et al. 2004), which is characterized by extreme short stature, acromelia, and mesomelia. Mono-allelic loss-of-function mutation of *NPR2* gene results in idiopathic short stature with a family history (Vasques et al. 2013). In contrast to loss-of-function mutations of *NPR2* gene, gain-of-function mutation of *NPR2* gene results in epiphyseal chondrodysplasia, Miura type, which is characterized by extreme tall stature, scoliosis, and hallux valgus (Miura et al. 2014). These results show that CNP–NPR2 axis acts as a stimulator of endochondral ossification. Overexpression of CNP and CNP analog (Vosoritide: BMN 111) corrected the short stature phenotype of achondroplasia model mice (Yasoda et al. 2004, Lorget et al. 2012). A CNP analog is undergoing clinical trials for human patients with achondroplasia.

Statins, which inhibit the mevalonic acid pathway and which are used as lipid-lowering agents, recover the abnormal cartilage capacity of iPS cell from TD type I cell which could not generate normal cartilage and recover the skeletal phenotype of achondroplasia model mice (Yamashita et al. 2014).

Meclizine, an over-the-counter H1 receptor inhibitor, is used as an anti-motion sickness medication. By a drug repositioning strategy, this drug was found to facilitate chondrocyte proliferation and to recover extracellular matrix synthesis (Matsushita et al. 2013). As CNP and statins, meclizine also increases the longitudinal skeletal growth in achondroplasia model mice (Matsushita et al. 2015).

Through basic research, other drugs such as parathyroid hormone (Ueda et al. 2007, Xie et al. 2012), tyrosine kinase inhibitor (Jonquoy et al. 2012, Komla-Ebri et al. 2016), FGFR3-binding peptide (Jin et al. 2012), and soluble FGFR3(Garcia et al. 2013) were identified as candidate drugs for achondroplasia treatment.

End of Chapter Questions

1. What are the roles of fibroblast growth factors and their receptors in early human development?
2. What diseases, respectively, result from genetic mutations of *FGFR1*, *FGFR2*, and *FGFR4*?
3. What differences in clinical and radiological features are observed among achondroplasia, hypochondroplasia, and TD? Is there any correlation between clinical symptoms and the degree of receptor phosphorylation?
4. Except FGFR3, what factors are involved in the process of chondrocyte proliferation and differentiation in growth plate?
5. Is there any relation between parent age and achondroplasia?
6. In achondroplasia patients, are there any characteristics of motor and psychosocial development?
7. How do new drug candidates such as CNP, statin, and meclizine ameliorate the phenotype of achondroplasia model mice?

References

Bartels CF, Bukulmez H, Padayatti P et al (2004) Mutations in the transmembrane natriuretic peptide receptor NPR-B impair skeletal growth and cause acromesomelic dysplasia, type Maroteaux. Am J Hum Genet 75:27–34

Bellus GA, Hefferon TW, Ortiz De Luna RI et al (1995) Achondroplasia is defined by recurrent G380R mutations of FGFR3. Am J Hum Genet 56:368–373

Bonafe L, Cormier-Daire V, Hall C et al (2015) Nosology and classification of genetic skeletal disorders: 2015 revision. Am J Med Genet A 167A:2869–2892

Brewer JR, Mazot P, Soriano P (2016) Genetic insights into the mechanisms of Fgf signaling. Genes Dev 30:751–771

Chusho H, Tamura N, Ogawa Y et al (2001) Dwarfism and early death in mice lacking C-type natriuretic peptide. Proc Natl Acad Sci U S A 98:4016–4021

Deng C, Wynshaw-Boris A, Zhou F et al (1996) Fibroblast growth factor receptor 3 is a negative regulator of bone growth. Cell 84:911–921

Garcia S, Dirat B, Tognacci T et al (2013) Postnatal soluble FGFR3 therapy rescues achondroplasia symptoms and restores bone growth in mice. Sci Transl Med 5:203ra124

Harada D, Yamanaka Y, Ueda K et al (2009) FGFR3-related dwarfism and cell signaling. J Bone Miner Metab 27:9–15

Hasegawa K, Fukuhara R, Moriwake T et al (2016) A novel mutation p.Ser348Cys in FGFR3 causes achondroplasia. Am J Med Genet A 170A:1370–1372

Heuertz S, Le Merrer M, Zabel B et al (2006) Novel FGFR3 mutations creating cysteine residues in the extracellular domain of the receptor cause achondroplasia or severe forms of hypochondroplasia. Eur J Hum Genet 14:1240–1247

Horton WA, Hall JG, Hecht JT (2007) Achondroplasia. Lancet 370:162–172

Ikegawa S, Fukushima Y, Isomura M et al (1995) Mutations of the fibroblast growth factor receptor-3 gene in one familial and six sporadic cases of achondroplasia in Japanese patients. Hum Genet 96:309–311

Jin M, Yu Y, Qi H et al (2012) A novel FGFR3-binding peptide inhibits FGFR3 signaling and reverses the lethal phenotype of mice mimicking human thanatophoric dysplasia. Hum Mol Genet 21:5443–5455

Jonquoy A, Mugniery E, Benoist-Lasselin C et al (2012) A novel tyrosine kinase inhibitor restores chondrocyte differentiation and promotes bone growth in a gain-of-function Fgfr3 mouse model. Hum Mol Genet 21:841–851

Komla-Ebri D, Dambroise E, Kramer I et al (2016) Tyrosine kinase inhibitor NVP-BGJ398 functionally improves FGFR3-related dwarfism in mouse model. J Clin Invest 126:1871–1884

Lorget F, Kaci N, Peng J et al (2012) Evaluation of the therapeutic potential of a CNP analog in a Fgfr3 mouse model recapitulating achondroplasia. Am J Hum Genet 91:1108–1114

Makrythanasis P, Temtamy S, Aglan MS et al (2014) A novel homozygous mutation in FGFR3 causes tall stature, severe lateral tibial deviation, scoliosis, hearing impairment, camptodactyly, and arachnodactyly. Hum Mutat 35:959–963

Matsushita M, Kitoh H, Ohkawara B et al (2013) Meclozine facilitates proliferation and differentiation of chondrocytes by attenuating abnormally activated FGFR3 signaling in achondroplasia. PLoS One 8:e81569

Matsushita M, Hasegawa S, Kitoh H et al (2015) Meclozine promotes longitudinal skeletal growth in transgenic mice with achondroplasia carrying a gain-of-function mutation in the FGFR3 gene. Endocrinology 156:548–554

Miura K, Kim OH, Lee HR et al (2014) Overgrowth syndrome associated with a gain-of-function mutation of the natriuretic peptide receptor 2 (NPR2) gene. Am J Med Genet A 164A:156–163

Naski MC, Colvin JS, Coffin JD et al (1998) Repression of hedgehog signaling and BMP4 expression in growth plate cartilage by fibroblast growth factor receptor 3. Development 125:4977–4988

Rousseau F, Bonaventure J, Legeai-Mallet L et al (1994) Mutations in the gene encoding fibroblast growth factor receptor-3 in achondroplasia. Nature 371:252–254

Shiang R, Thompson LM, Zhu YZ et al (1994) Mutations in the transmembrane domain of FGFR3 cause the most common genetic form of dwarfism, achondroplasia. Cell 78:335–342

Tachibana K, Suwa S, Nishiyama S, Matsuda I (1997) Height of Japanese achondroplasia patients based on a nationwide investigation. J Pediatr Pract 60:1363–1369 in Japanese

Takagi M, Kouwaki M, Kawase K et al (2015) A novel mutation Ser344Cys in FGFR3 causes achondroplasia with severe platyspondyly. Am J Med Genet A 167A:2851–2854

Tanaka H, Kubo T, Yamate T et al (1998) Effect of growth hormone therapy in children with achondroplasia: growth pattern, hypothalamic–pituitary function, and genotype. Eur J Endocrinol 138:275–280

Toydemir RM, Brassington AE, Bayrak-Toydemir P et al (2006) A novel mutation in FGFR3 causes camptodactyly, tall stature, and hearing loss (CATSHL) syndrome. Am J Hum Genet 79:935–941

Ueda K, Yamanaka Y, Harada D et al (2007) PTH has the potential to rescue disturbed bone growth in achondroplasia. Bone 41:13–18

Vasques GA, Amano N, Docko AJ et al (2013) Heterozygous mutations in natriuretic peptide receptor-B (NPR2) gene as a cause of short stature in patients initially classified as idiopathic short stature. J Clin Endocrinol Metab 98:E1636–E1644

Xie Y, Su N, Jin M et al (2012) Intermittent PTH (1-34) injection rescues the retarded skeletal development and postnatal lethality of mice mimicking human achondroplasia and thanatophoric dysplasia. Hum Mol Genet 21:3941–3955

Yamashita A, Morioka M, Kishi H et al (2014) Statin treatment rescues FGFR3 skeletal dysplasia phenotypes. Nature 513:507–511

Yasoda A, Komatsu Y, Chusho H et al (2004) Overexpression of CNP in chondrocytes rescues achondroplasia through a MAPK-dependent pathway. Nat Med 10:80–86

Zhang SR, Zhou XQ, Ren X et al (2007) Ser217Cys mutation in the Ig II domain of FGFR3 in a Chinese family with autosomal dominant achondroplasia. Chin Med J 120:1017–1019

15 Acute Myeloid Leukemia: Mutations Blocking Differentiation Lead to Distinct Leukemic Subtypes

Amy L. Cummings, Darren Pan, and Gary J. Schiller

Keywords
Acute myeloid leukemia (AML) · Acute promyelocytic leukemia (APL) · Promyelocytic leukemia protein-retinoic acid receptor alpha (PML-RARa) · Isocitrate dehydrogenase (IDH) · Differentiation

15.1 Case Reports

15.1.1 Acute Promyelocytic Leukemia (APL)

A 70-year-old African-American male presented to the emergency room for 1 month of fatigue, progressive dyspnea on exertion, and easy bruising. He endorsed a past medical history significant for insulin-dependent diabetes, chronic kidney disease, and nonischemic cardiomyopathy. He denied a family history of malignancy and noted he was a retired army mechanic with prior exposure to Agent Orange. At triage, his vital signs were within normal limits. His physical exam revealed pale mucous membranes, a normal cardiopulmonary exam, no hepatosplenomegaly, and trace peripheral edema. A complete blood count showed leukocytosis of 26,450 cells/μL comprised of 2.1% neutrophils, 11.8% lymphocytes, and 86.1% blasts. He had a normocytic anemia with hemoglobin 9.3 g/dL and thrombocytopenia of 18,000 platelets/μL. Coagulation studies identified an elevated international normalized ratio (INR) of 1.6 (normal ≤1.1) and prothrombin time of 17.1 s (normal 8.9–11.4 s) and a decreased fibrinogen of 106 mg/dL (normal 217–420 mg/dL). A peripheral blood smear presented numerous circulating immature myeloid cells intermediate in size with scant cytoplasm, azurophilic granules, and folded nuclei (Fig. 15.1).

The patient was started empirically on hydroxyurea and all-trans-retinoic acid (ATRA). A bone marrow biopsy showed a hypercellular marrow involving 95% of the sample with clusters and sheets of immature cells consistent with acute myeloid leukemia (AML). Flow cytometric results showed a large population of immature myeloid cells expressing cell surface markers CD13, CD33, CD34, and CD117. Karyotyping revealed a translocation of chromosomes 15 and 17 (t15;17) in 100% of cells examined (Fig. 15.2a); fluorescence in situ hybridization (FISH) molecular studies revealed a promyelocytic leukemia protein-retinoic acid receptor alpha (PML-RARα) fusion transcript (Fig. 15.2b), consistent with acute promyelocytic leukemia (APL). The patient started arsenic trioxide (ATO) in addition to ATRA given his age

A. L. Cummings (✉) · D. Pan · G. J. Schiller
Aramont Program for Clinical/Translational Research in Human Malignancies, David Geffen School of Medicine at UCLA, Los Angeles, CA, USA
e-mail: alcummings@mednet.ucla.edu; dpan@mednet.ucla.edu; gschiller@mednet.ucla.edu

T. Oohashi et al. (eds.), *Human Pathobiochemistry*, https://doi.org/10.1007/978-981-13-2977-7_15

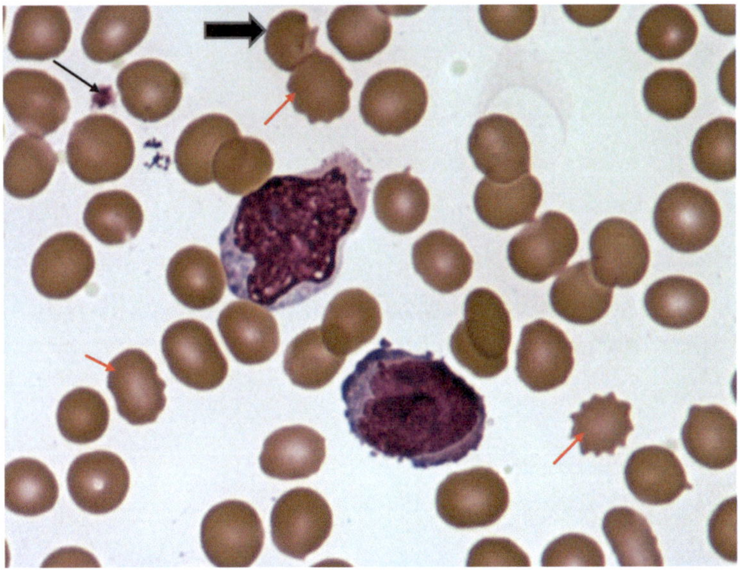

Fig. 15.1 Peripheral smear, 4000×. Myeloid blasts are highlighted midfield with irregularly folded and bi-lobed nuclei, open chromatin, and multiple nucleoli. Microgranules stain the cytoplasm azure. A unique feature of promyelocytes is that they may include numerous granules and Auer rods, which contain proteolytic enzymes. The release of these enzymes induces severe coagulopathy and fibrinolysis, predisposing patients to both hemorrhage and thrombosis, reiterating the importance of screening for coagulopathy at suspected diagnosis. There is a paucity of platelets (small black arrow), one fragmented cell (large black arrow), and scattered echinocytes (red arrows), which are manifestations of red blood cell membrane irregularity

and history of cardiomyopathy. Two weeks into his hospital stay, his course was complicated by cardiac arrest, renal failure, diffuse alveolar hemorrhage, multi-organism bacteremia, and critical illness polyneuropathy. ATRA and ATO were held. Although additional treatment was recommended after stabilization, the patient declined and was sent home on hospice.

15.1.2 Isocitrate Dehydrogenase (IDH)-Mutated Acute Myeloid Leukemia

A 79-year-old Iranian man in his normal state of health presented to his primary care doctor for a routine checkup. He noted mild fatigue and chronic back and leg pain but otherwise had no symptoms. His past medical history was significant only for hyperlipidemia and lumbar disc herniation, managed with low-dose simvastatin, low-dose aspirin, and gabapentin for pain. He denied a family history of hematologic malignancies. His physical exam was unremarkable with a normal cardiopulmonary exam, no hepatosplenomegaly, no lymphadenopathy, and skin exam without petechiae or ecchymoses. A routine complete blood count was significant for a white blood cell count of 2500 cells/μL, hemoglobin 11.7 g/dL, and platelet count 115,000 cells/μL. A subsequent bone marrow biopsy revealed a hypocellular marrow with decreased multilineage maturation with 30–35% myeloblasts in the aspirate supported by flow cytometry with a blast-enriched gate positive for CD34 and CD117 (Fig. 15.3), diagnostic of AML. Cytogenetic studies were significant for an abnormal duplication of chromosome 11q (+11q) on FISH.

The patient was started on a clinical trial with an oral Hedgehog signaling pathway inhibitor and subcutaneous cytarabine that offered several months of disease control, but then he developed refractory pancytopenia with complaints of severe fatigue, fevers, hyperhidrosis, and chills. A bone marrow biopsy revealed hypercellularity with 60% blasts (Fig. 15.4), confirmed by flow

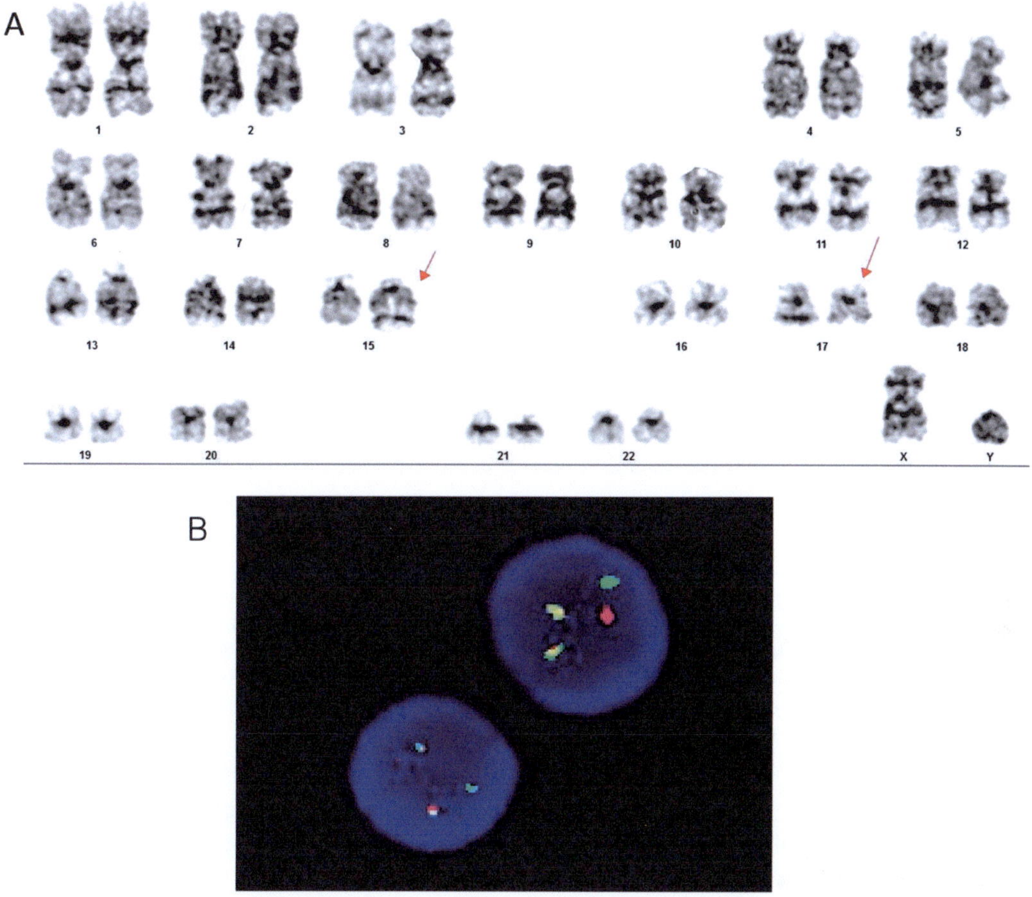

Fig. 15.2 (**a**) Karyotype of acute promyelocytic leukemia. Forty-six chromosomes are depicted as homologous pairs with 22 somatic and 1 XY diploids. Red arrows depict the patient's abnormal fusion of chromosomes 15 and 17. Note the right chromosome 15 is larger than its homologous pair, while the right chromosome 17 is smaller than its homologous pair. (**b**) Fluorescent in situ hybridization of PML-RARα. The rhodamine (red) signal represents PML on chromosome 15; the fluorescein (green) signal represents RARα signal from chromosome 17; yellow indicates a merged signal and PML-RARα fusion. A cell with normal karyotype would have two red signals and two green signals; this patient's FISH has one red, one green, and two yellow indicating a t(15;17)

cytometry. His karyotype again showed +11q, and molecular studies demonstrated mutations in deoxyribonucleic acid-methyltransferase-3-beta (DNMT3), a gene that encodes an enzyme that catalyzes the transfer of a methyl group to DNA; STAG2, a gene that encodes a protein cohesion subunit SA-2 involved in sister chromatid cohesion; and IDH2, which will be explained later in this chapter. The patient then received 5 days of decitabine, a hypomethylating agent, and was enrolled in another clinical trial for IDH2-mutated AML comparing a novel inhibitor of mutant IDH2 to conventional regimens.

15.2 Diagnosis

AML comprises a group of heterogeneous hematopoietic malignancies in which precursor cells committed to the myeloid lineage are arrested in an undifferentiated stage of development (Table 15.1) (Vardiman et al. 2009). These cells, known as leukemic blasts, accumulate in the peripheral blood, bone marrow, and other tissues, causing inhibition of normal hematopoiesis characterized by anemia, neutropenia, and thrombocytopenia. Confirmation of a diagnosis of AML requires observation of 20% blasts or more on

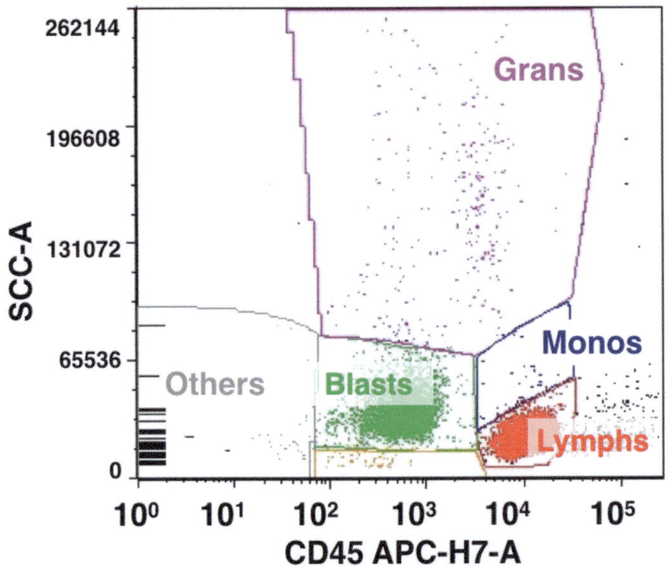

Fig. 15.3 Flow cytometry indicating an elevated blast population. Cells are gated by cluster of differentiation 45 (CD45), which is a leukocyte common antigen that is a type I transmembrane protein present on all hemopoietic cells except erythrocytes. These are then analyzed based on side-scattered light (SSC-A), which is proportional to cell granularity or internal complexity. Blasts which have scant cytoplasm and open chromatin tend to scatter less light and are depicted by the green color and are out of proportion to the pink granulocytes (grans), blue monocytes (monos), and red lymphocytes (lymphs) in this patient's sample

bone marrow biopsy or peripheral blood smear with evidence that the blasts are myeloid in origin or cytogenetics associated with the differentiation of hematopoietic stem cells, such as RUNX1-RUNX1T1 (Runt-related transcription factors), CBFB-MYH11 (core-binding factor subunit beta-myosin 11), or PML-RARα fusion (Table 15.2) (Vardiman et al. 2009). Cell lineage is most commonly determined by flow cytometry of peripheral blood or bone marrow cells, a method of immunophenotyping, which determines expression of important biomarkers using antibodies directed against surface antigens and/or intracellular proteins. Most cases of AML express antigens characteristic of neutrophilic or monocytic differentiation, including CD13, CD15, CD33, CD64, CD117, myeloperoxidase, and CD34. These antigens may be found on normal hematopoietic cells but are usually present in abnormal combinations or amounts, allowing for characterization as myeloid blasts.

Further characterization of AML involves cytogenetic analysis, FISH, and molecular sequencing. Karyotyping involves culturing dividing cells from a patient's biopsy and arresting the cells in metaphase with a mitotic inhibitor like colchicine for visual review. FISH incubates the patient's biopsy with fluorescent-labeled probes of interest, which determine the presence and abundance of an identical DNA species in the sample by fluorescence microscopy. While processes for detection of molecular mutations are varied and often determined institutionally, most employ directed polymerase chain reaction (PCR)-based analyses or genome-wide high-throughput DNA analysis, although capillary gel electrophoresis, restriction endonuclease digestion, and high-resolution melting analysis remain alternatives (Patel et al. 2011). Molecular sequencing is an increasingly important tool in acute leukemia diagnosis, prognosis, and therapeutics (Vardiman et al. 2009; Dohner et al. 2015). There have been over 5000 driver mutations identified in AML with two or more drivers identified in over 80% of patients (Papaemmanuil et al. 2016). Key mutations involve proliferative signaling, evasion of apoptosis, epigenetic regulation of clonal hema-

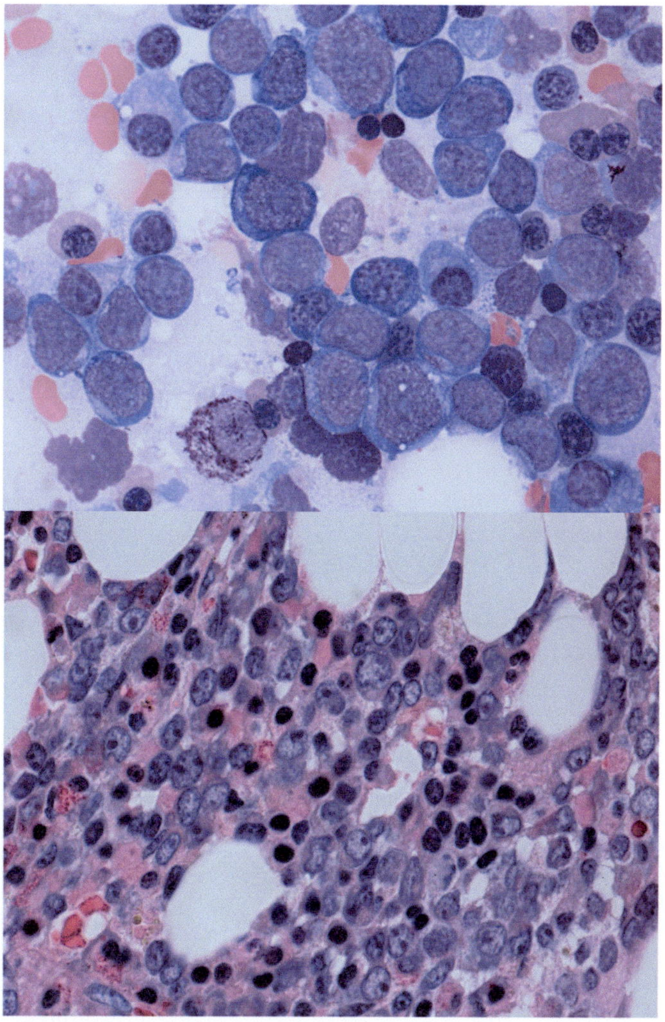

Fig. 15.4 Bone marrow aspirate 1000× (top), core biopsy 400× (bottom). Blasts exhibit scant cytoplasm with large, slightly irregular nuclei and open chromatin with prominent single to multiple nucleoli. Bone marrow cortex exhibits a hypercellular marrow with excess blasts in large aggregates comprising 50–60% of the marrow cellularity, indicative of acute myeloid leukemia. Adipose cells are large ovoid areas of white in the sample

topoiesis, transcription deregulation, dysregulated transcriptional splicing, and aberrant chromatin modification (Dohner et al. 2015; Viny and Levine 2016).

15.3 Biochemical and Molecular Perspectives

15.3.1 APL

APL is a clinically, cytogenetically, and prognostically distinct subtype of AML that accounts for ~5–15% of all adult AML cases (Dohner et al. 2015). There are five morphological distinct stages of myeloid precursors, including the myeloblast, promyelocyte, myelocyte, metamyelocyte, and band form; and APL represents a leukemia with clonal arrest at the promyelocyte stage. More than 95% of patients with APL have a balanced reciprocal translocation between chromosomes 15q22 and 17q21, which results in the fusion of promyelocytic leukemia (PML) and retinoic acid receptor-α (RARα) genes (Melnick and Licht 1999). The resulting chimeric protein prevents activation of key target genes required for normal myeloid differentiation by inhibiting normal RARα function. Other variant translocations that involve the RARα gene but produce alternate fusion products with other tumor suppressors include t(11;17)(q23;q21) that fuses RARα with the promyelocytic leukemia zinc fin-

Table 15.1 Types of acute leukemia

Causes	Known modifications/diseases
Chromosome abnormalities	t(1;22)(p13;q13)
	t(6;9)(p23;q34)
	t(8;21)(q22;q22)
	t(9;11)(p22;q23)
	t(15;17)(q22;q12)
	inv(3)(q21q26.2) or t(3;3)(q21;q26.2)
	inv(16)(p13.1q22) or t(16;16)(p13.1;q22)
Gene fusions	CBFB-MYH11
	DEK-NUP214AML
	MLLT3-MLL
	RBM15-MKL1
	RPN1-EVI1
	RUNX1-RUNX11
Gene mutations	CEBPα
Histology associations	AML with myelodysplasia-related changes
	AML with minimal differentiation
	AML without maturation
	AML with maturation
	Acute myelomonocytic leukemia
	Acute monoblastic/monocytic leukemia
	Acute erythroid leukemia
	Acute megakaryoblastic leukemia
	Acute basophilic leukemia
	Acute panmyelosis with myelofibrosis
	Blast plasmacytoid dendritic cell neoplasm
	Myeloid sarcoma
Other	Myeloid leukemia associated with Down syndrome
	Therapy-related myeloid neoplasms
	Transient abnormal myelopoiesis

ger (PLZF), t(5;17)(q35;q12-21) that binds nucleophosmin of the NPM1 gene, and the t(11;17)(q13;q21) that fuses with the nuclear matrix-mitotic apparatus (NuMA) protein (Melnick and Licht 1999).

Retinoids are crucial for normal myeloid differentiation and act as a ligand via RARs and retinoid X receptors (RXRs) to induce transcription factors. By forming a heterodimer with RXR, the DNA-binding domain of RARα can efficiently bind to retinoic acid response elements

Table 15.2 Findings diagnostic of acute myeloid leukemia

Finding	Details
Auer rods	Present in blasts on microscopy, suggestive of APL
Bone marrow infiltration	Bone marrow biopsy demonstrating at least 20% leukemic blasts
Blasts of myeloid origin	Flow cytometry including CD13, CD33, and CD117
Cytogenetic abnormalities with myelodysplasia-related changes	Complex karyotype[a]
Or	
Cytogenetic abnormalities diagnostic regardless of blast count	t(8;21)(q22;q22); RUNX1-RUNX1T1
	inv(16)(p13.1q22) or t(16;16)(p13.1;q22); CBFB-MYH11
	t(15;17)(q24.1;q21.1); PML-RARα

[a]Three or more unrelated abnormalities

(RAREs), which control the expression of target genes (Zhou et al. 2005). Without retinoic acid, RARα recruits corepressors, downregulating target gene expression. When retinoic acid is present at physiologic levels, the protein undergoes a conformational change that causes the corepressors to dissociate, enabling replacement with coactivators that promote gene expression and driving differentiation of promyelocytes into neutrophils (Fig. 15.5a, b) (Mistry et al. 2003). In APL, there are four chromosomal translocations that result in fusion proteins, in which the B through F domains of RARα, including the DNA-binding and ligand-binding domains of the protein, are linked through the C-terminal to four different nuclear proteins containing self-association domains. These fusion proteins retain their functional components but have an enhanced affinity for corepressors compared to the wild-type (non-mutated) RARα. Thus, the fusion protein acts as a "dominant-negative" mutant that antagonizes RARα function by sequestering its cofactors, binding RAREs in place and recruiting corepressors such as N-CoR-mSin3-histone deacetylase complex, which decreases histone acetylation and produces repressive chromatin organization to further enhance transcriptional

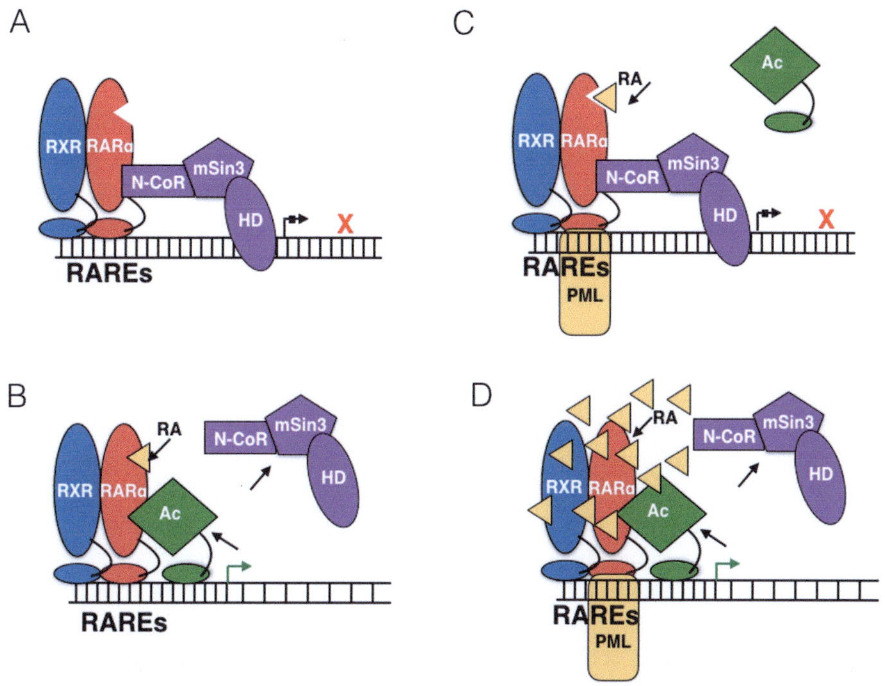

Fig. 15.5 RARα function. (**a**) Without retinoic acid (RA), heterodimers of RXR and RARα interact with corepressors including N-CoR-mSin3-histone deacetylase (HD) complex, decreasing histone acetylation and causing transcriptional repression. (**b**) When RA is present at physiologic levels, RARα undergoes a conformational change that causes the corepressors to dissociate, enabling replacement with coactivators (Ac) that work to unwind chromatin and enable transcription. (**c**) PML-RARα sequesters cofactors, binds RAREs in place, and interacts with HD complex, enhancing transcriptional repression in the presence of physiologic levels of RA. (**d**) Supraphysiologic levels of RA (ATRA) cause PML-RARα to release corepressors and recruit activators for transcription

repression, freezing the cell as a promyelocytic blast (Fig. 15.5c) (Melnick and Licht 1999; Grignani et al. 1993).

The PML protein is a tumor suppressor protein required for the assembly of PML nuclear bodies that act at the chromatin level and affect regulatory cellular functions, including programmed cell death, genome stability, and control of cell division (Bernardi and Pandolfi 2007). Wild-type PML protein localizes to discrete nuclear domains in a speckled pattern, but in t(15;17) APL cells, the PML-RARα binds to the PLZF protein, which causes disappearance of nuclear bodies from the nucleus (Zhou et al. 2005). Studies by Melnick and Licht and Bernardi and Pandolfi have suggested this leads to a loss of transcriptional and RNA splicing control, resulting in a failure to halt oncogenic transformation (Melnick and Licht 1999; Bernardi and Pandolfi 2007). Alternate fusion products, including nucleophosmin and NuMA, are closely associated with the nuclear matrix and may disrupt its function, while PLZF is closely associated with PML function (Melnick and Licht 1999). Interestingly, transgenic mice expressing the PML-RARα mutation manifest defects in myeloid development and are prone to neutropenia but do not immediately develop APL, suggesting that the RARα fusions may potentiate oncogenesis but ultimately require a “second hit” or additional leukemogenic mutation to engender APL (Melnick and Licht 1999).

15.3.2 IDH-Mutated AML

IDH is a metabolic enzyme involved in the citric acid cycle. There are three isoforms: IDH1 located in the cytoplasm, IDH2 located in the mitochondria, and IDH3, which functions as part of the tricarboxylic acid cycle. These enzymes catalyze the conversion of isocitrate to alpha-ketoglutarate (αKG) while utilizing nicotinamide adenine dinucleotide phosphate (NADP+) as a cofactor and generating cellular NADP-oxidase (NADPH) in most tissues, which defends against oxidative damage through reduction of glutathione and thioredoxins (Dang et al. 2016). IDH plays an important role in exchanging key metabolites and shuttling electrons between the mitochondria and the cytosol, and lost or reduced function results in impaired detoxification, which promotes DNA damage and genomic instability (Dang et al. 2016).

Mutations in IDH1 and IDH2 are reported in 5–20% of AML cases with clinically significant mutations found within the active site, in which three arginines are critical for isocitrate binding: R132 of IDH1 and R140 and R172 of IDH2 (Patel et al. 2011). $IDH1^{R132}$ and $IDH2^{R172}$ mutations generally result in amino acid substitution for arginine, while $IDH2^{R140}$ typically switches arginine to glutamine (Patel et al. 2011). The loss of the positive polarity of arginine changes protein binding, and using cellular metabolic profiling and structural biochemistry, it was demonstrated that these mutant proteins result in a gain of function, leading to excess conversion of αKG to D-2-hydroxyglutarate (2-HG) using NADPH as a cofactor (Dang et al. 2016; Falini et al. 2015). The compound 2-HG has no known metabolic function in mammals and is normally produced at a low level by errors of catalysis and rapidly cleared; however, intracellular 2-HG production by mutant IDH overwhelms the normal clearance mechanism and competitively inhibits αKG-dependent dioxygenases. These dioxygenases include ten-eleven translocation (TET) enzymes that mediate DNA demethylation (Fig. 15.6) and Jumonji C domain histone lysine demethylases that play an important role in chromatin regulation. Other affected proteins include prolyl and lysil hydroxylases that are required for proper collagen folding, hypoxia-inducible factors (HIF) that improve cell survival, and cytochrome C oxidase (COX), which induces antiapoptotic B-cell CLL-lymphoma 2 protein (Bcl-2) dependence (Dang et al. 2016; Falini et al. 2015; DiNardo et al. 2013; Stein 2016; Sasine and Schiller 2015).

The resultant hypermethylation affects CpG islands in promotor regions of tumor suppressor genes, methylation of DNA and histones in chromatin, and recruitment of coactivators and corepressors (Im et al. 2014). The addition of methyl groups to cytosine residues on DNA blocks regulatory protein access, while the methylation of histone lysine residues affects histone-histone and histone-DNA interactions affecting the structure of chromatin. The inability to demethylate these structures leads to closed chromatin, which prevents the binding of regulatory proteins and inactivation of genes, resulting in dysregulated epigenetic programming that blocks terminal cellular differentiation (Dang et al. 2016). As methylated cytosine has a greater propensity to undergo spontaneous deamination forming thymine, hypermethylation may lead alternatively to oncogenesis via point mutations in tumor suppressor genes, such as p53 (Im et al. 2014). The inhibition of prolyl and lysil hydroxylases further perpetuates dysmorphia, while Bcl-2 dependence leads to a strong antiapoptotic signal, decreasing the likelihood of cell death (Falini et al. 2015). However, this is not sufficient in and of itself to cause proliferation of the leukemic cell. Using mouse transplantation assays, it was elucidated further that mutant IDH1 alone led to a long latency and incomplete tumorigenesis, requiring secondary mutations to fully drive tumor progression, such as increased expression of homeobox protein HoxA9, a transcription factor that increases cell cycling, decreases cyclin-dependent kinase gene expression affecting cell cycle regulation, and increases mitogen-activated protein kinase activity boosting growth factor signaling (Chaturvedi et al. 2013). Thus, while the deregulation of cellular energetics through IDH1 and IDH2 mutations halts terminal differentiation, its promotion of genomic instability likely inspires clonal evolution in which additional mutations optimize the production of myeloid blasts that manifest AML. Alteration of methylation is a common pattern of oncogenesis in AML, and mutations in DNMT3A also have

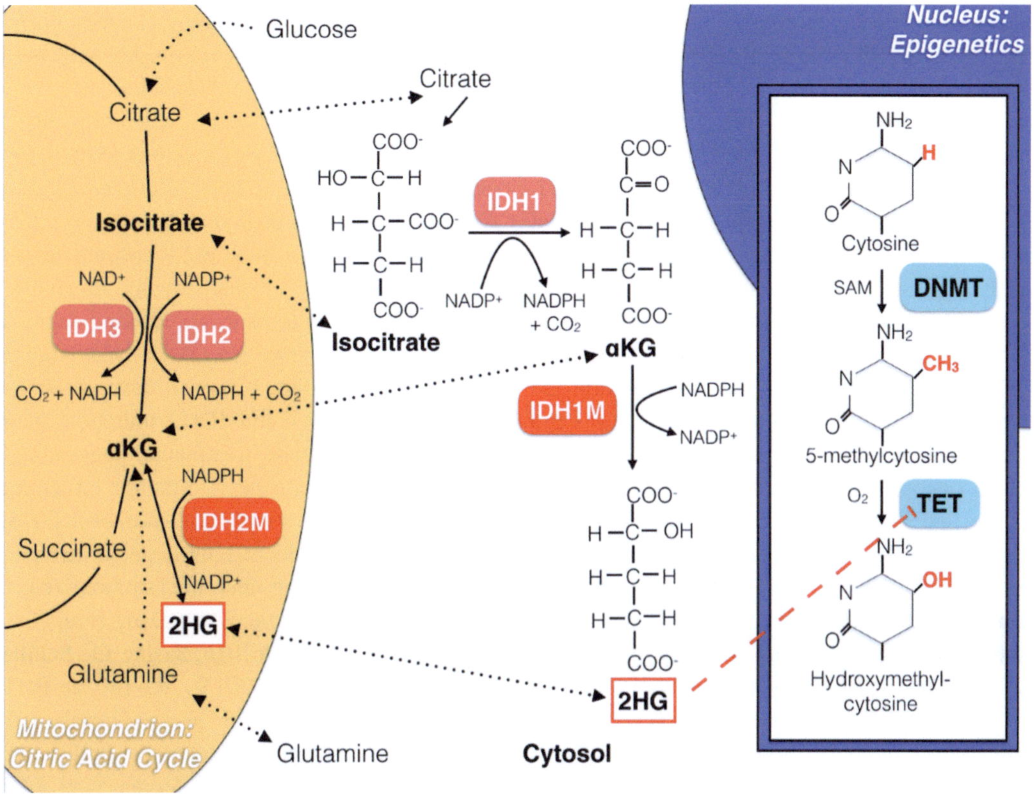

Fig. 15.6 IDH biochemistry. Mutated IDH1 (cytosol) and IDH2 (mitochondrial) lead to increased formation of 2-hydroxyglutarate (2-HG), which antagonizes TET enzymes that demethylate cytosine residues, leading to DNA hypermethylation by DNMT enzymes. This results in a perturbation of epigenetics, closed chromatin, and decreased transcription of genes required for white blood cell differentiation (Dang et al. 2016; Falini et al. 2015)

been proven as a mechanism of leukemogenesis (Im et al. 2014), as exemplified by the introductory case presentation.

15.4 Therapy

15.4.1 APL

APL is one of the rare types of leukemia that can be routinely and regularly cured (Tallman and Altman 2009). Although retinoic acid is a natural product of vitamin A metabolism, physiologic levels of retinoic acid are insufficient to induce myeloid differentiation in the presence of PML-RARα. However, the pharmacologic dose of all-trans-retinoic acid (ATRA), 45 mg/m^2 per day, causes the fusion protein to release corepressors of RARα target genes (Fig. 15. 5d) (Melnick and Licht 1999; Lo-Coco et al. 2013). Studies have shown that in addition to corepressor release, ATRA also promotes synthesis of proteins which selectively degrade PML-RARα, hampering its ability to exert a dominant-negative effect and shifting the balance toward wild-type RARα (Raelson et al. 1996). Through these effects, ATRA treatment reverses the differentiation block and allows the promyelocytic blasts to proceed down their differentiation pathway. Given the early mortality related to hemorrhage and coagulopathy, it is crucial that ATRA be started as soon as the diagnosis of APL is suspected, even before pathologic confirmation.

For induction or initial therapy, ATRA is combined with either arsenic trioxide (ATO) or anthracycline chemotherapy with the goal to achieve complete remission or undetectable PML-RARα with restoration of normal hematopoiesis (Tallman and Altman 2009). Typically, in low- to intermediate-risk APL (WBC less than 10,000/μL), ATRA is used for 1 to 3 days prior to ATO or chemotherapy; but in high-risk APL (WBC greater than 10,000/μL), ATRA and anthracycline chemotherapy are started concurrently to prevent the replication of rapidly growing cells (Tallman and Altman 2009; Sanz et al. 2010). Anthracyclines intercalate base pairs of DNA and RNA and prevent DNA replication and transcription, while at low doses, ATO binds and degrades the PML moiety, cleaving the PML-RARα oncoprotein and inducing partial differentiation and apoptosis of the APL blasts (Zhou et al. 2005; Tallman and Altman 2009). Currently, the standard of care for induction therapy is ATRA and ATO or ATRA with limited doses of anthracycline based on the clinician's discretion and patient's risk, ability to tolerate respective toxicities, and the availability of ATO, which is limited in certain regions worldwide (Tallman and Altman 2009). A bone marrow biopsy is completed 30 days after the start of treatment, and those who have undetectable PML-RARα by FISH are considered in complete remission. If patients survive induction, 80–90% are cured (Tallman and Altman 2009). Consolidation strategies, or treatment after complete remission to prevent relapse, depend on risk classification and age. Allogeneic transplant remains an option for those with refractory or recurrent disease (Tallman and Altman 2009).

15.4.2 IDH-Mutated AML

As targeted therapies for IDH-mutated AML are currently under investigation, IDH-mutated leukemia is treated with standard AML therapy, including induction chemotherapy with cytarabine and an anthracycline followed by consolidation and then maintenance chemotherapy. The presence of the $IDH2^{R172}$ mutation has limited effect on survival, although some retrospective studies have suggested poorer outcomes following conventional chemotherapy (Papaemmanuil et al. 2016; Dang et al. 2016). As epigenetic modification through hypermethylation has been proven to be a significant mechanism by which IDH-mutated AML arrests differentiation, hypomethylating agents have been of clinical interest. At low doses, hypomethylating agents such as decitabine and azacitidine are incorporated into DNA and trap DNA methyltransferases (DNMTs), leading to their proteosomal degradation and a decrease in DNA methylation. Retrospective studies, however, have shown no link between IDH1 and IDH2 mutation status and response to these treatments, suggesting a more complex pathogenesis, although larger prospective studies are warranted (Falini et al. 2015).

Preliminary results of IDH-targeted therapeutics are encouraging and are summarized in Table 15.3 (Dohner et al. 2015; Dang et al. 2016; Falini et al. 2015; Stein 2016; Sasine and Schiller 2015; Chaturvedi et al. 2013). AG-120, an IDH1 inhibitor, is currently in phase I trials, and AG-221, an IDH2 inhibitor, has preliminary phase I results with dramatic decreases in 2-HG levels in both R140- and R172-mutated IDH2 AML (Dang et al. 2016; Stein 2016). Of the 158 patients who received at least 1 dose of AG-221, 16.5% achieved a complete remission, 23.5% had a partial response, and 45% had stable disease (Stein 2016). In vitro, a disruptor of telomeric silencing 1-like (DOT1L) inhibitor EPZ004777 was proven to interfere with the influence of 2-HG on histone methylation (Falini et al. 2015); an IDH1 inhibitor HMS-101 was shown to block colony formation of IDH1-mutated AML cells and decrease ERK signaling (Chaturvedi et al. 2013); and a selective $IDH2^{R140}$ inhibitor AGI-6780 induced changes in DNA methylation (Falini et al. 2015).

IDH1- and IDH2-altered metabolic pathways are potential targets for therapy as well. Production of the oncometabolite 2-HG is dependent on glutaminolysis as a source of αKG, and glutaminase inhibitors, such as CPTEX, zaprinast, and CB-839, could be efficacious. The significant reduction of NADPH production also could make the cells potentially vulnerable to

Table 15.3 Selected agents available or in development for IDH-mutated AML

Drug Class	Agent	Mechanism
Epigenetic modifiers	Decitabine (Dacogen)	Hypomethylation
	Azacitidine (Vidaza)	Hypomethylation
	Oral azacitidine (CC-486)	Hypomethylation
	Guadecitabine (SGI-110)	Hypomethylation
IDH inhibitors	AG-120	IDH1 inhibitor
	HMS-101	IDH1 inhibitor
	AG-221	Reversible IDH2 inhibitor
	AGI-6780	$IDH2^{R140Q}$ inhibitor
IDH-mutated metabolic pathway modifiers	EZP004777	DOT1L inhibitor, interrupts 2-HG histone methylation interference
	BPTES	Glutaminase inhibitor
	Zaprinast	Glutaminase inhibitor
	CB-839	Glutaminase inhibitor
	EGCG	NADPH production inhibitor
Miscellaneous	Hydroxyurea	Inhibits ribonucleotide reductase, decreasing production of deoxyribonucleotides
	Venetoclax	BCL-2 inhibitor inhibiting anti-apoptosis

further reduction of this metabolic activity by using inhibitors of NADPH production, such as epigallocatechin-3-gallate (EGCG). Additionally, as IDH1- and IDH2-mutated AML has been proven to be dependent on Bcl-2 expression, these cells may also be more sensitive to the Bcl-2 inhibitor venetoclax that may be used in combination therapy (Dohner et al. 2015).

15.5 Conclusion

Acute myeloid leukemia represents a spectrum of disease united by the arrested development of myeloid precursor cells. The subtypes of AML characterized by either PML-RARα or by IDH mutation are characterized by distinct blocks in maturation of the neoplastic myeloid clone. Relieving these blocks through pharmacologic means can induce differentiation of the neoplastic clone into mature neutrophils and is associated with restoring the cell program of apoptosis. Discovery of drugs that target the product of the mutant genes in these distinct forms of AML represents a novel form of managing malignancy by inducing normal differentiation of the neoplastic clone. This form of differentiation therapy remains to be exploited in the management of other cancers but offers an opportunity to direct therapy to the molecular basis of oncogenesis.

End-of-Chapter Questions

1. How would one make an efficient and cost-effective diagnosis of APL? Of IDH-mutated AML?
2. Describe the differences and similarities of the driving mechanisms of APL and IDH-mutated AML.
3. Is there a chance these subtypes of leukemia could be familial or inherited genetically?
4. What types of side effects could be expected with targeted therapy for APL? For IDH-mutated AML?
5. There is a differentiation syndrome associated with treatment of APL. Would you expect a similar syndrome with use of IDH inhibitors?

Answers

1. *Diagnosis of AML requires either peripheral blood or bone marrow flow cytometry to confirm at least 20% of myeloid blasts prior to additional testing. APL diagnosis relies on karyotyping/FISH to identify t(15;17) or PML-RARα transcript. Karyotyping is typically less costly than FISH, but FISH is more accurate. Identification of an IDH mutation, however, requires molecular sequencing. PCR-based sequencing is more costly than karyotyping/FISH, and often, exome-only sequencing is used to decrease cost. While theoretically one could do exome-only sequencing of the specific locations of IDH1 and IDH2 on chromosomes 2 and 15 to identify a mutation, it is more common to send peripheral blood or bone marrow samples for sequencing of a few hundred cancer-associated genes of interest that encode known or likely targets of leukemia therapies or drivers of oncogenesis.*
2. *APL and IDH-mutated AML are similar in that each is defined by genetic changes that result in a blockade of transcription that prevents myeloid differentiation. They are also similar in that they require a "second hit" or additional mutation to enable clone propagation and clinical leukemia. APL is unique in that a gene translocation causes a fusion protein that acts as a "dominant negative" that directly inhibits transcription, freezing the cell specifically as a promyelocyte, while IDH mutations are single base pair missense mutations that lead to a gain of function and overproduction of a competitive inhibitor of enzymes involved in epigenetic modification, leading to closed chromatin and general inhibition of transcription, leaving myeloid blasts unable to differentiate.*
3. *Familial forms of AML are rare but have been described in germline mutations of transcription factors RUNX1 and CCAAT/enhancer-binding protein alpha (CEBPα) (Owen et. al. Br J Haematol 2008 Jan;140 (2):123–32). RUNX1 and CEBPα function as transcription factors in myeloid differentiation, but because of transcription factor redundancy, these individuals initially could have normal myeloid maturation, enabling procreation and propagation of the genetic defect. Theoretically, a germline t(15;17) would lead to an accumulation of fetal promyelocytes, while an IDH1/IDH2 mutation could result in a blockade of other myeloid cell lines, both of which likely would be incompatible with normal fetal development and trigger demise. As such, APL and IDH-mutated AML have been identified only as acquired mutations.*
4. *Treatment side effects are varied, and this does not represent an exhaustive list. Upon initiation of treatment, patients with both APL and IDH-mutated AML may develop differentiation syndrome, which will be more fully addressed in the next question. As a retinoid, ATRA may produce symptoms of retinoid toxicity including headache, fever, dry mucous membranes, nausea and vomiting, rashes, and changes in vision. ATO has similar side effects and has been associated with nausea and vomiting, cough, electrolyte abnormalities, rashes, arthralgias, and diarrhea. Arsenic toxicity and death occur only when given in doses larger than 600 mg/kg/day (APL treatment doses are 0.15 mg/kg/day). An important complication for which special monitoring is performed is QT prolongation, as ATO is a potent blocker of cardiac rapidly activating potassium channels and can*

(continued)

trigger tachyarrhythmias and sudden cardiac death. Note a corrected JT interval (QTc-QRS) is used when baseline QT interval is greater than 120 milliseconds. IDH1/IDH2-targeted agents are under current investigation with less known about their toxicities. Preliminary data from phase I studies of AG-210 and AG-221 indicate that the drugs are generally well-tolerated; rare side effects include atrial flutter with AG-210 and QT prolongation with AG-120 treatment.

5. *Differentiation syndrome is a potentially fatal consequence of inducing rapid differentiation of promyelocytes into neutrophils. Although rare, it is well described in APL treatment, and a similar syndrome has been noted with the use of IDH1 inhibitors. Although the pathogenesis has not been elucidated completely, it is suspected that the sudden increase of neutrophils causes a cytokine cascade which induces endothelial damage with capillary leak, occlusion of microcirculation, and tissue infiltration. The syndrome often presents with nonspecific symptoms like cough, shortness of breath, fever, and weight gain but can lead to a severe systemic inflammatory response syndrome with multi-organ failure. Responding appropriately requires that physician maintain a high index of suspicion for the condition, especially in the early stages of treatment. Patients suspected of suffering from the syndrome should be promptly treated with corticosteroids, which demarginalize neutrophils and reduce their adhesion to vessels in the lung and reduce release of inflammatory cytokines. In cases of severe leukocytosis, physicians may choose to start corticosteroid therapy prophylactically.*

References

Bernardi R, Pandolfi PP (2007) Structure, dynamics and functions of promyelocytic leukaemia nuclear bodies. Nat Rev Mol Cell Biol 8:1006–1016

Chaturvedi A, Araujo Cruz MM, Jyotsana N et al (2013) Mutant IDH1 promotes leukemogenesis in vivo and can be specifically targeted in human AML. Blood 122:2877–2887

Dang L, Yen K, Attar EC (2016) IDH mutations in cancer and progress toward development of targeted therapeutics. Ann Oncol Off J Eur Soc Med Oncol/ESMO 27:599–608

DiNardo CD, Propert KJ, Loren AW et al (2013) Serum 2-hydroxyglutarate levels predict isocitrate dehydrogenase mutations and clinical outcome in acute myeloid leukemia. Blood 121:4917–4924

Dohner H, Weisdorf DJ, Bloomfield CD (2015) Acute Myeloid Leukemia. N Engl J Med 373:1136–1152

Falini B, Sportoletti P, Brunetti L, Martelli MP (2015) Perspectives for therapeutic targeting of gene mutations in acute myeloid leukaemia with normal cytogenetics. Br J Haematol 170:305–322

Grignani F, Fagioli M, Ferrucci PF, Alcalay M, Pelicci PG (1993) The molecular genetics of acute promyelocytic leukemia. Blood Rev 7:87–93

Im AP, Sehgal AR, Carroll MP et al (2014) DNMT3A and IDH mutations in acute myeloid leukemia and other myeloid malignancies: associations with prognosis and potential treatment strategies. Leukemia 28:1774–1783

Lo-Coco F, Avvisati G, Vignetti M et al (2013) Retinoic acid and arsenic trioxide for acute promyelocytic leukemia. N Engl J Med 369:111–121

Melnick A, Licht JD (1999) Deconstructing a disease: RARalpha, its fusion partners, and their roles in the pathogenesis of acute promyelocytic leukemia. Blood 93:3167–3215

Mistry AR, Pedersen EW, Solomon E, Grimwade D (2003) The molecular pathogenesis of acute promyelocytic leukaemia: implications for the clinical management of the disease. Blood Rev 17:71–97

Papaemmanuil E, Gerstung M, Bullinger L et al (2016) Genomic classification and prognosis in acute myeloid leukemia. N Engl J Med 374:2209–2221

Patel KP, Barkoh BA, Chen Z et al (2011) Diagnostic testing for IDH1 and IDH2 variants in acute myeloid leukemia an algorithmic approach using high-resolution melting curve analysis. J Mol Diagn 13:678–686

Raelson JV, Nervi C, Rosenauer A et al (1996) The PML/RAR alpha oncoprotein is a direct molecular target of retinoic acid in acute promyelocytic leukemia cells. Blood 88:2826–2832

Sanz MA, Montesinos P, Rayón C et al (2010) Risk-adapted treatment of acute promyelocytic leukemia based on all-trans retinoic acid and anthracycline with addition of cytarabine in consolidation therapy for high-risk patients: further improvements in treatment outcome. Blood 115:5137–5146
Sasine JP, Schiller GJ (2015) Emerging strategies for high-risk and relapsed/refractory acute myeloid leukemia: novel agents and approaches currently in clinical trials. Blood Rev 29:1–9
Stein EM (2016) Molecular pathways: IDH2 mutations-co-opting cellular metabolism for malignant transformation. Clin Cancer Res 22:16–19
Tallman MS, Altman JK (2009) How I treat acute promyelocytic leukemia. Blood 114:5126–5135
Vardiman JW, Thiele J, Arber DA et al (2009) The 2008 revision of the World Health Organization (WHO) classification of myeloid neoplasms and acute leukemia: rationale and important changes. Blood 114:937–951
Viny AD, Levine RL (2016) Roads diverge--a Robert frost view of leukemia development. N Engl J Med 374:2282–2284
Zhou GB, Zhao WL, Wang ZY, Chen SJ, Chen Z (2005) Retinoic acid and arsenic for treating acute promyelocytic leukemia. PLoS Med 2:e12

α1-Antitrypsin Deficiency

16

Nobuaki Miyahara, Kuniaki Seyama,
and Erwin W. Gelfand

Keywords
Emphysema · COPD · Elastase · Liver disease · Augmentation therapy

16.1 Case Report

The patient was a 37-year-old Japanese male with α1-antitrypsin (AAT) deficiency. He had a 3-year history of progressive dyspnea on exertion. He smoked 10 cigarettes daily from ages 15 to 20 and then 40 cigarettes daily until the age of 34 years when he quit because of dyspnea. He worked as an automobile parts salesman and denied any history of exposure to noxious gases. Family history was not contributory concerning emphysema, asthma, or other respiratory disorders. Physical examination on admission disclosed no cyanosis, tachypnea, or tachycardia. No clubbing of the fingers was present. Breath sounds were diffusely diminished in both lungs. Laboratory data such as blood cell count and results of routine biochemical screening tests were within the normal range except for mild erythrocytosis with an elevated hemoglobin value (17.8 g/dl) and red blood cell count (637×10^4/μl). Liver function tests results were normal. Serum protein electrophoresis failed to detect a peak corresponding to α1-globulin; subsequent determination of serum levels of AAT disclosed severely decreased concentrations (20 mg/dl; normal range, 170–274; measured by nephelometry, SRL, Tokyo, Japan). Arterial blood gas levels sampled with the patient breathing room air showed moderate hypoxemia (PaO_2, 60 Torr; $PaCO_2$, 37.5 Torr). A chest roentgenogram disclosed hyperinflated lungs and bilaterally flattened hemidiaphragms (Fig. 16.1). Computed tomography of the chest demonstrated panlobular emphysema with lower lobe preponderance (Fig. 16.2). Pulmonary function tests indicated severe obstructive ventilatory impairment with forced expiratory volume in 1 sec (FEV_1) (1.13 L, 30.3% of the predicted value) and FEV_1/forced vital capacity (33.6%). Residual volume was increased (3.27 L; 228% of the predicted value), and the diffusing capacity of carbon monoxide was impaired (6.45 ml/min/mm Hg; 26.5% of the predicted value). The patient was diagnosed with severe pulmonary emphysema resulting from

N. Miyahara (✉)
Department of Allergy and Respiratory Medicine, Okayama University Hospital, Okayama, Japan

Department of Medical Technology, Okayama University Graduate School of Health Sciences, Okayama, Japan
e-mail: miyahara@okayama-u.ac.jp

K. Seyama
Division of Respiratory Medicine, Juntendo University Faculty of Medicine and Graduate School of Medicine, Tokyo, Japan

E. W. Gelfand
Division of Cell Biology, Department of Pediatrics, National Jewish Health, Denver, CO, USA

T. Oohashi et al. (eds.), *Human Pathobiochemistry*, https://doi.org/10.1007/978-981-13-2977-7_16

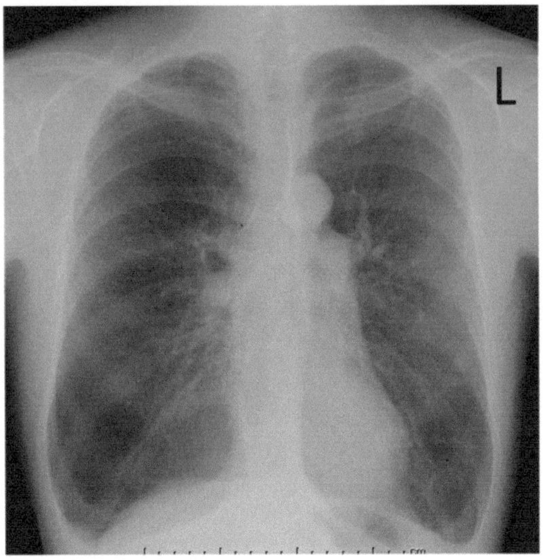

Fig. 16.1 A chest roentgenogram of the proband shows hyperinflated lungs and bilaterally flattened hemidiaphragms (Miyahara et al. (2001) Intern Med. 40:336–40)

AAT deficiency, and home oxygen therapy was initiated to ameliorate hypoxemia.

The patient gave informed consent for blood sampling and genetic analysis. To identify a mutation involving the *SERPINA1* gene, total RNA was isolated from a sample of his peripheral blood mononuclear cells, and cDNA then was synthesized. The entire coding region of the *SERPINA1* gene was amplified by a standard reverse transcription-polymerase chain reaction (RT-PCR) method, and the nucleotide sequence was determined using a ^{32}P-labeled primer for Sanger sequencing. Two different mutations were identified. The proband was found to carry both the S_{iiyama} and $QO_{clayton}$ mutations; S_{iiyama} is a deficient variant resulting from a missense mutation of Ser^{53} (TCC) -> Phe^{53} (TTC), while QO_{clayto} is a null variant arising from a C insertion after Pro^{362} (CCC). Cloning of RT-PCR products revealed that the S_{iiyama} and $QO_{clayton}$ mutations each were located on different cDNAs, indicating that the patient was a compound heterozygote. The backbone structure of AAT showed no difference among cDNAs with respect to the combination of Arg^{101}-Val^{213}-Glu^{376}. Based on thorough interviews, the family pedigree included no consanguineous marriage. Serum concentrations of AAT determined in all family members available were approximately half of the normal values, and each individual was considered a carrier of either the S_{iiyama} or $QO_{clayton}$allele. After informed consent was obtained from family members, mutation analysis of their *SERPINA1* gene was performed, confirming that the patient's uncles, father, and elder son inherited the S_{iiyama} allele, while the patient's mother, brother, and younger son carried the $QO_{clayton}$ allele (Fig. 16.3).

Although the patient ceased smoking and had best supportive care for chronic obstructive pulmonary disease (COPD) including treatment with inhaled cholinergic receptor antagonist and home oxygen therapy, his lung function deteriorated considerably with an FEV_1 value of only 0.58 L, and respiratory symptoms including dyspnea on effort progressively worsened. Ultimately, he was accepted as a candidate for a lung transplant. After a 2.5-year wait for an immunologically compatible donor, the lung transplant was successfully performed at age 44 years at the Okayama University Hospital.

16.2 Diagnosis

AAT is a serine protease inhibitor, and its deficiency can lead to pulmonary disease and hepatotoxicity (Carrell amd Lomas 2002). It is inherited as an autosomal-codominant condition for which more than 500 single nucleotide polymorphisms have been identified, although not all of these have been validated. The most common alleles include M (normal allele), Z (leading to low AAT levels associated with severe AAT deficiency), S (leading to a mild decrease in circulating AAT), and a null allele (no detectable AAT synthesis).

Those homozygous for the M allele (i.e., normal) are labeled *PI*MM* (PI stands for "protease inhibitor"), and those who are homozygous for the Z allele are noted as *PI*ZZ*. More than 90% of patients with severe AAT deficiency are *PI*ZZ*.

AAT deficiency is associated with two major clinical manifestations: emphysema, resulting from the loss of proteolytic protection of the lung by AAT and liver diseases such as cirrhosis and

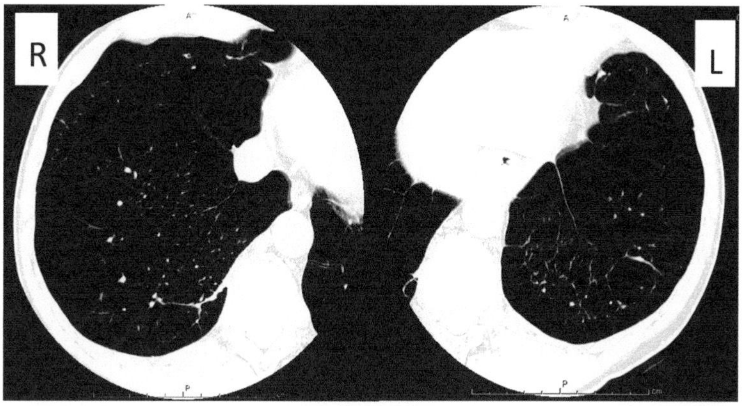

Fig. 16.2 Computed tomography of the chest demonstrates panlobular emphysema with lower lobe preponderance (Miyahara et al. (2001) Intern Med. 40:336–40)

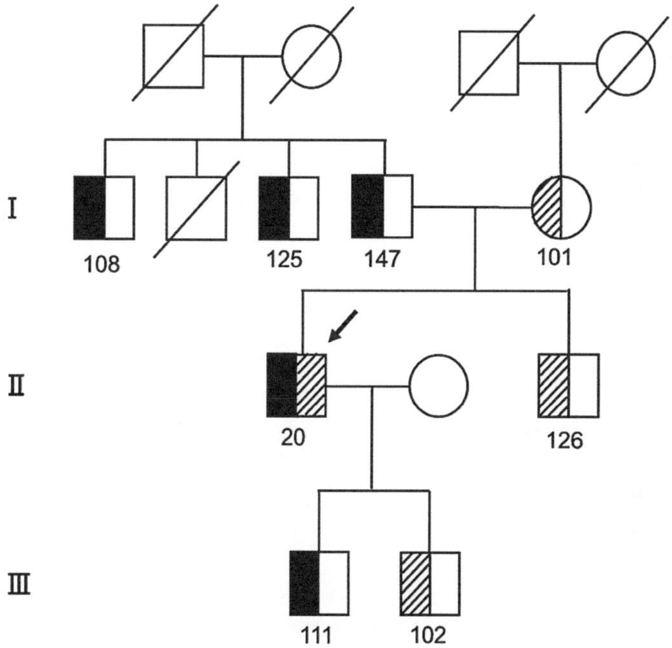

Fig. 16.3 **The family pedigree of a case of alpha-1-antitrypsin (AAT) deficiency carrying S_{iiyama} and $QO_{clayton}$ alleles.** The proband is indicated by an arrow. Inheritance of the S_{iiyama} allele is indicated by the black boxes and the $QO_{clayton}$ allele by the oblique lines. The level of individual serum AAT concentration is shown in milligrams per deciliters below each symbol. Note the absence of consanguineous marriage in this family (Miyahara et al. (2001) Intern Med. 40:336–40)

chronic hepatitis, which result from abnormal accumulation of AAT within hepatocytes, and hepatoma. Other clinical manifestations of AAT deficiency include bronchiectasis, panniculitis, and cytoplasmic anti-neutrophil cytoplasmic antibody-positive vasculitis.

In Europe, the highest prevalence of the *PI*Z* mutation has been reported in North-Western European countries with a gene frequency between 0.026 and 0.049% (Dahl et al. 2002). The prevalence in North America seems to be similar. In contrast, the frequency of *PI*Z* allele in eastern Asian countries is low and reported to be 0.006%. Although AAT deficiency was estimated to affect 100,000 Americans, fewer than 15,000 have received a clinical diagnosis. Because it is under-recognized, the average diagnostic delay was reported to be 7.2 years; as a result, many patients experience significant delays between first symptoms and diagnosis. The European Respiratory Society and American Thoracic Society Guidelines recommend testing of all symptomatic adults with persistent airway obstruction (Am J Respir Crit Care Med 2003).

COPD due to AAT deficiency is difficult to distinguish from "usual" COPD on a clinical basis. Compared to the usual forms of COPD, in AAT deficiency the lower zones of the lungs are more commonly involved resulting in basal emphysema. Computed tomography of the lungs is usually sensitive in detecting early stages of the lung destruction. Pulmonary function tests may also be able to identify patients with early emphysematous changes. Because COPD due to AAT deficiency is difficult to distinguish from usual COPD, all individuals with COPD regardless of age or ethnicity should be tested for AAT deficiency.

Neutrophil elastase has been implicated in the mucus hypersecretion associated with chronic bronchitis. Therefore, there is a high prevalence of anatomic bronchiectasis in AAT deficiency, with some individuals manifesting signs and symptoms of clinical bronchiectasis without emphysema. Thus, individuals with undefined chronic bronchitis should also be tested for AAT deficiency.

Quantitative measurements of serum level of AAT are used as the initial screening test, most often done by nephelometry. Radial immunodiffusion and latex-enhanced immunoturbidimetry are also available. The use of dried blood samples (DBSs) for AAT has also become widely available, with moderate to good correlations with serum levels of AAT (Costa et al. 2000). Normal serum levels generally range from 100 to 220 mg/dl by nephelometry.

Knowledge of the patient's specific phenotype is important for appropriate counseling and disease management. Phenotyping is usually performed by isoelectric focusing which can identify different band patterns associated with different alleles. The nomenclature of the *PI* locus is based on the electrophoretic mobility of the AAT at pH 4.9. Letters alphabetically before M designate anodal variants and letters after M designate cathodal variants. Genotyping involves determining which AAT alleles are present, most often using polymerase chain reaction testing targeting the S and Z alleles and occasionally set up to detect less common alleles such as F and I. Gene sequencing is occasionally necessary to achieve a definitive diagnosis. Recently, analytical genotyping kits have become available that measure 25 AAT alleles (Alpha-1-Antitrypsin Quantitation and Mutation Analysis (Quest Diagnostics, Madison, NJ) and Alpha 1 Antitrypsin Mutation Analysis (Geisinger Medical Laboratories, Danville, PA)). Next-generation diagnostic kits will soon be available that can measure most, if not all, mutations of the *SERPINA1* gene (McElvaney NG 2015).

In addition to the proband, parents, siblings, and children, as well as extended family members identified with an abnormal gene for AAT, should be provided genetic counseling and offered testing for AAT deficiency. For family testing after a proband is identified, simply determining AAT levels is not recommended because it does not fully characterize disease risk from AAT deficiency. For diagnostic testing of symptomatic individuals, genotyping for at least the S and Z alleles is recommended. Advanced or confirmatory testing should include *PI*-typing, AAT level testing, and/or expanded genotyping.

Once the diagnosis of AAT deficiency is made, complete lung function testing including diffusing capacity is recommended. Annual follow-up of adults with at least a spirometry test is recommended. In general, emphysema develops in the third or fourth decade of life in majority of young symptomatic *PI*ZZ* individuals who have a history of cigarette smoking. On the other hand, it has been reported that the onset of emphysema in nonsmokers homozygous for AAT deficiency living in areas free of air pollution was later when compared to smokers and may appear in the fifth or sixth decade of life (Tanash et al. 2008). It is controversial whether heterozygote (*PI*MZ*) individuals have a higher risk for lung disease. Some studies have shown that *PI*MZ* individuals who smoke are at increased risk of emphysema, whereas other studies have shown no increased risk (Dahl et al. 2002; Molloy et al. 2014). Further studies including prospective cohorts may be required to clarify this important issue.

The role of heterozygous *PI*SZ* in disease pathogenesis is also uncertain. In the Danish AAT deficiency registry, only a small proportion of *PI*SZ* individuals appeared to be at increased risk for developing pulmonary emphysema and at an older age than in *PI*ZZ* patients (Seersholm and Kok-Jensen 1998). Additionally, *PI*SZ* patients with active disease have reduced survival compared with the general population. In smokers, the *PI*SZ* phenotype may confer a significant risk to the development of COPD; in nonsmokers there may be no such added risk.

The clinical course of liver involvement is variable. Neonatal hepatic syndrome represents prolonged jaundice after birth with conjugated hyperbilirubinemia and abnormal liver enzymes. Approximately 20% of *PI*ZZ* patients develop it in the first weeks of life. Around 5% of *PI*ZZ* patients carry a risk for developing more severe forms of the disorder such as liver failure and cirrhosis in childhood/adolescence. Monitoring for liver disease at annual intervals with physical examination including focused examination for signs of liver disease, liver ultrasound, and laboratory monitoring of aspartate transaminase, alanine aminotransferase, γ-glutamyl transpeptidase, albumin, bilirubin, international normalized ratio, and platelets is recommended. Liver transplantation is often required for these patients. These patients are also more likely to develop hepatocellular carcinoma. There is no specific treatment or prevention for this condition, only liver transplantation.

In summary, the diagnosis of AAT deficiency and testing for AAT should be considered in all COPD patients regardless of age or ethnicity, in unexplained chronic liver disease, as well as necrotizing panniculitis, for granulomatosis with polyangiitis, and for unexplained bronchiectasis. Family members should also be offered testing for AAT deficiency including genotyping.

16.3 Biochemical and Molecular Perspectives

AAT deficiency was first described in 1963 (Laurell and Eriksson 1963). The gene coding for AAT is located on chromosome 14q32.1 as part of a gene cluster called serine protease inhibitor (SERPIN). The gene comprises four coding exons and three additional exons, and more than 150 alleles of this gene have been identified. The normal allele is denoted M. The most common and clinically most important alleles leading to AAT deficiency are the Z and S allele. The frequency of the Z allele is highest in Scandinavian countries, and the frequency of the S allele is highest in Spain and Portugal.

The AAT molecule includes 394 amino acids and 3 glycosylated side chains coupled to an asparagine residue. The molecule has a globular tertiary structure, and its active site to inhibit enzymes is on a surface protrusion where the most important amino acid is present, a methionine in position 358, an amino acid susceptible to conversion to methionine sulfoxide by oxidants from cigarette smoke, rendering it much less potent as an inhibitor of neutrophil elastase.

The Z allele has a single amino acid substitution (glutamic acid to lysine at position 342), which results in abnormal folding and formation of polymers of the Z molecule within hepatocytes, and it accumulates within the liver (Fig. 16.4). The hepatic retention and degradation of AAT both as monomers and polymers increase, and these polymers are recognized on liver biopsy as periodic acid-Schiff diastase-resistant eosinophilic inclusion bodies on histologic staining. Because Z molecule polymers are retained in the liver, concentrations of AAT in the bloodstream fall, which lead to proteolytic burdens of neutrophil elastase, especially for people who smoke. The mutant protein can also accumulate dramatically as inclusion bodies within hepatocytes, but it develops clinically significant liver disease in a relatively few subjects (Lomas and Mahadeva 2002).

It has been reported that some of the polymerized Z protein can escape the liver and circulate in the blood and that alveolar macrophages may also produce Z polymers. These Z polymers are chemotactic for neutrophils and recruit more neutrophils to the lung enhancing the inflammatory cascade, aggravating the proteolytic burden in the lung, and, consequently, the risks of developing emphysema. Z monomers that circulate can bind neutrophil elastase, but their binding avidity to neutrophil elastase is substantially lower than that of M-type AAT (Mahadeva et al. 2005).

The S allele is a single amino acid mutation in position 264 in the protein where a glutamine is

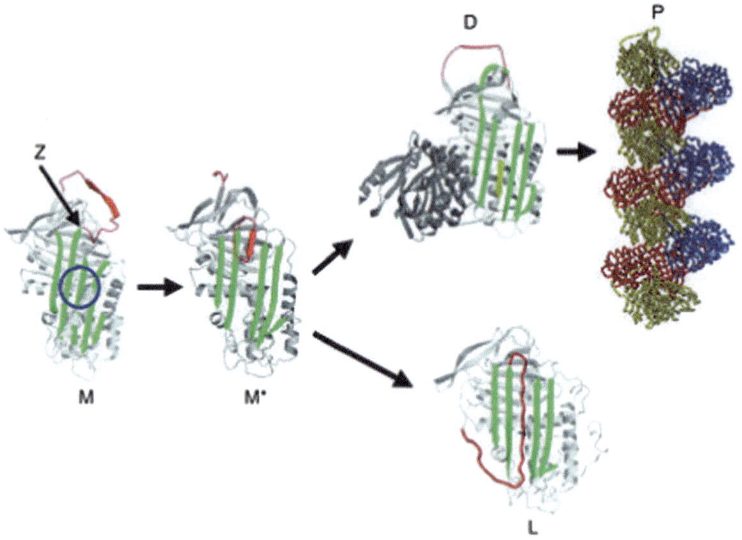

Fig. 16.4 **The structure of α1-antitrypsin is centered on β-sheet A (green) and the mobile reactive center loop (red).** Polymer formation results from the Z variant of α1-antitrypsin (Glu342Lys at 17 residues proximal to the P1 reactive center; arrow) or mutations in the shutter domain (blue circle) that open β-sheet A to favor partial loop insertion and the formation of an unstable intermediate (M*). The patent β-sheet A can accept either the loop of another molecule, to form a dimer (D), which then extends into polymers (P), or else its own loop, to form a latent conformation (L). The individual molecules of α1-antitrypsin within the polymer are colored red, yellow, and blue (Lomas DA and Mahadeva 2002). Copyright © 2002, with permission from American Society for Clinical Investigation

replaced by valine. The homozygous S variant of AAT results in an instable protein that is easily degraded outside the hepatocyte and affects the circulating half-life. The null alleles (*PI*Q*) have deletions of large portions of the AAT gene or nonsense mutations and are associated with no detectable AAT in the serum. There are several such *QO* allelic variants reported including $QO_{clayton}$ as described in this case report (Miyahara et al. 2001, Seyama et al. 1991).

It has been shown that the negative predictive value of the AAT serum level is 113 mg/dl. If the serum level exceeds this threshold, the probability to carry a homozygous mutation for AAT is extremely low. Therefore, if the primary goal is to detect all homozygous individuals, this can be done effectively without conducting PCR as a first step. However, in order to detect heterozygous carriers, it is difficult to determine a threshold value. Recent laboratory analyses of high numbers of patient samples demonstrated 97.5% percentiles of *PI*MZ* of approximately 150 mg/dl (220 mg/dl for *PI*MS*). Thus, the serum level has only limited value if the goal is to exclude heterozygous carriers. The threshold that is used to exclude severe deficiency is different from the so-called protective threshold. The protective threshold is reported at 50 mg/dl measured by nephelometry or immunoturbidimetry (Stoller and Aboussouan 2012).

The effects of the intracellular accumulation of the mutant Z protein in the liver include the formation of protein polymers, activation of autophagy, mitochondrial injury, and caspase activation, which progress in a cascade causing hepatocellular injury (Lomas and Mahadeva 2002). Liver disease can occur at any age, although the majority of children are free of significant liver dysfunction. The variable clinical presentations suggest an important contribution of genetic and environmental disease modifiers. The heterozygous carrier state for the *PI*Z* gene, present in 1.5%–3% of the population, is not itself a common cause of liver injury but may be a modifier gene for other liver diseases.

16.4 Therapy

The treatment of patients with severe deficiency of AAT associated with emphysema is similar to that of patients with usual COPD (Table 16.1). Specifically, smoking cessation, bronchodilators, occasionally inhaled steroids, supplemental oxygen, and pulmonary rehabilitation are included. Early and aggressive use of antibiotics when patients present with respiratory infections may reduce the potential for lung damage. Annual influenza vaccination and pneumococcal vaccination every 5 years should be considered to reduce the risk and severity of these infections.

Liver transplantation, the definitive treatment for advanced liver disease, appears to restore serum AAT levels to normal because a liver from a donor with a *PI*MM* genotype produces qualitatively and quantitatively normal AAT. However, there is a study showing that most of patients nevertheless experienced a decline in FEV_1 following liver transplantation, and significant benefits on lung function were not seen (Carey et al. 2013). The risk for childhood-onset liver disease in infants with the *PI*ZZ* genotype who are breast-fed during the first months of life was reported to be reduced; however, breast-feeding does not confer absolute protection against the development of severe liver disease (Pferdmenges et al. 2013).

Table 16.1 Current available therapeutic approaches for alpha-1-antitrypsin deficiency

Lifestyle modifications
Smoking cessation
Environmental risks avoidance
Pulmonary rehabilitation
Pharmacological treatment for COPD
Inhaled bronchodilators
Inhaled steroids
Oral corticosteroids
Antibiotics
Oxygen therapy
Influenza vaccine
Pneumococcal vaccination
Augmentation therapy
Intravenous injection of purified alpha-1-antitrypsin
Organ transplant
Lung for severe COPD
Liver for severe liver cirrhosis

COPD chronic obstructive pulmonary disease

Specific therapy for AAT deficiency currently involves weekly intravenous infusions of purified, pooled human plasma-derived AAT, augmentation therapy. The current US Food and Drug Administration-approved intravenous augmentation therapy dose for chronic administration is 60 mg/kg body weight, administered weekly. Four drugs have been approved for use in the United States: Prolastin-C (Grifols, Barcelona, Spain), Aralast NP (Baxalta, Bonneckborn, IL), Zemaira (CSL Behring, King of Prussia, PA), and Glassia (Baxalta, Bonneckborn, IL, and Kamada, Ness Ziona, Israel).

All of these were approved for use in the United States on the basis of biochemical efficacy. Specifically, infusion of these drugs has been shown to raise serum AAT levels above a protective threshold value (generally considered 57 mg/dL, the value below which the risk of developing emphysema increases). Randomized controlled trials have addressed the efficacy of intravenous augmentation therapy, and although there have been no definitive trials, the weight of evidence shows that augmentation therapy can slow the progression of emphysema. For example, augmentation therapy was associated with a slower progression of emphysema as assessed by the rate of loss of lung density on computed tomography (Dirksen A et al. 2009).

On the basis of the available evidence, the American Thoracic Society, the European Respiratory Society, and the Medical and Scientific Advisory Committee of the Alpha-1 Foundation have recommended augmentation therapy in individuals with established airflow obstruction from AAT deficiency. Their guidelines describe that the evidence that augmentation therapy is beneficial is stronger for individuals with moderate airflow obstruction (e.g., FEV_1 35%–65% of predicted) than for those with severe airflow obstruction. For those with an FEV1 greater than 65%, discussion with each individual is needed regarding the potential benefits of reducing lung function decline while at the same time considering the cost of therapy and lack of evidence for benefit (Gildea et al. 2003). Augmentation therapy is also recommended for individuals with necrotizing panniculitis.

Dosing schedules other than what is currently approved (e.g., 120 mg/kg every 2 weeks) appear to provide less complete biochemical protection against lung destruction in some individuals. Higher doses (i.e., 120 mg/kg/week) appear to be safe and increase AAT levels toward more physiologic values, but clinical efficacy has not been proven (Sandhaus et al. 2016).

On the other hand, intravenous augmentation therapy is not currently recommended for individuals with the *MZ* genotype of AAT deficiency, individuals with lung disease due to AAT deficiency who continue to smoke, and individuals with AAT deficiency and emphysema or bronchiectasis who do not have airflow obstruction. It is also not currently recommended for the treatment of liver disease due to AAT deficiency and for individuals who have undergone liver transplantation. The guidelines recognize that although augmentation therapy does not satisfy the usual criteria for cost-effectiveness due to its high cost, it is recommended for appropriate candidates because it is the only available specific therapy for severe deficiency of AAT.

Other approaches, including gene and stem cell therapies, are currently being investigated (Wozniak et al. 2015). In vitro and in vivo experimental AAT gene transfer with gamma-retroviral, lentiviral, adenoviral, and adeno-associated viral vectors has resulted in enhanced AAT serum levels and a promising safety profile. Human clinical trials using intramuscular viral transfer with adeno-associated viral vectors of the AAT gene demonstrated its safety, but did not achieve a protective level of AAT in serum. Further studies are needed to optimize the procedures, such as vector dose or site of administration, and the applicability of capsids other than adeno-associated viral, to avoid the neutralizing effects of the human immune system.

Questions

1. What specific populations should be tested for AAT deficiency?
2. If you found the serum level of AAT to be 75% of normal, how would you explain it?
3. How do you manage COPD in AAT deficiency?
4. What are the recommendations for intravenous augmentation therapy in AAT deficiency?

References

American Thoracic Society, European Respiratory Society. American Thoracic Society/European Respiratory Society statement: standards for the diagnosis and management of individuals with alpha-1 antitrypsin deficiency (2003) Am J Respir Crit Care Med 168:818–900

Carey EJ, Iyer VN, Nelson DR et al (2013) Outcomes for recipients of liver transplantation for alpha-1-antitrypsin deficiency–related cirrhosis. Liver Transpl 19(12):1370–1376

Carrell RW, Lomas DA (2002) Alpha1-antitrypsin deficiency--a model for conformational diseases. N Engl J Med 346(1):45–53

Costa X, Jardi R, Rodriguez F et al (2000) Simple method for alpha1-antitrypsin deficiency screening by use of dried blood spot specimens. Eur Respir J 15:1111–1116

Dahl M, Tybjaerg-Hansen A, Lange P et al (2002) Change in lung function and morbidity from chronic obstructive pulmonary disease in alpha1-antitrypsin MZ heterozygotes: a longitudinal study of the general population. Ann Intern Med 136:270–279

Dirksen A, Piitulainen E, Parr DG et al (2009) Exploring the role of CT densitometry: a randomised study of augmentation therapy in alpha1-antitrypsin deficiency. Eur Respir J 33:1345–1353

Gildea TR, Shermock KM, Singer ME et al (2003) Cost-effectiveness analysis of augmentation therapy for severe alpha1-antitrypsin deficiency. Am J Respir Crit Care Med 167:1387–1392

Laurell CB, Eriksson S (1963) The electrophoretic alpha-1-globulin pattern of serum alpha-1-antitrypsin deficiency. Scan J Clin Lab Invest 15:132

Lomas DA, Mahadeva R (2002) Alpha-1-antitrypsin polymerization and the serpinopathies: pathobiology and prospects for therapy. J Clin Invest 110:1585–1590

Mahadeva R, Atkinson C, Li Z et al (2005) Polymers of Z alpha1-antitrypsin co-localize with neutrophils in emphysematous alveoli and are chemotactic in vivo. Am J Pathol 166:377–386

McElvaney NG (2015) Diagnosing α1-antitrypsin deficiency: how to improve the current algorithm. Eur Respir Rev 24:52–57

Miyahara N, Seyama K, Sato T et al (2001) Compound heterozygosity for ala-1-antitrypsin (S(iiyama) and QO(clayton)) in an Oriental patient. Intern Med 40:336–340

Molloy K, Hersh CP, Morris VB et al (2014) Clarification of the risk of chronic obstructive pulmonary disease in alpha1-antitrypsin deficiency PiMZ heterozygotes. Am J Respir Crit Care Med 189:419–427

Pferdmenges DC, Baumann U, Müller-Heine A et al (2013) Prognostic marker for liver disease due to alpha1-antitrypsin deficiency. Klin Padiatr 225(5):257–262

Sandhaus RA, Turino G, Brantly ML et al (2016) The diagnosis and management of alpha-1 antitrypsin deficiency in the adult. COPD: J COPD Found 3:668–682

Seersholm N, Kok-Jensen A (1998) Intermediate alpha 1-antitrypsin deficiency PiSZ: a risk factor for pulmonary emphysema? Respir Med 92(2):241–245

Stoller JK, Aboussouan LS (2012) A review of alpha1-antitrypsin deficiency. Am J Respir Crit Care Med 185:246–259

Seyama K, Nukiwa T, Takabe K et al (1991) (serine 53 (TCC) to phenylalanine 53 (TTC)). A new alpha 1-antitrypsin-deficient variant with mutation on a predicted conserved residue of the serpin backbone. J Biol Chem 266:12627–12632

Tanash HA, Nilsson PM, Nilsson JA et al (2008) Clinical course and prognosis of never-smokers with severe alpha-1-antitrypsin deficiency (PiZZ). Thorax 63:1091–1095

Wozniak J, Wandtke T, Kopinski P et al (2015) Challenges and prospects for alpha-1 antitrypsin deficiency gene therapy. Hum Gene Ther 26(11):709–718

Hereditary Anticoagulant Deficiencies

17

Masataka Ishimura and Shouichi Ohga

Keywords
Antithrombin deficiency · Deep venous thrombosis · Protein C deficiency · Protein S deficiency · Purpura fulminans

Abbreviations

APC	Activated protein C
APTT	Activated partial thromboplastin time
AT	Antithrombin
CT	Computed tomography
DIC	Disseminated intravascular coagulation
DVT	Deep venous thrombosis
OR	Odds ratio
PC	Protein C
PIVKAII	Vitamin K absence or antagonists II
PS	Protein S
PT	Prothrombin time
rr	Reference range
VTE	Venous thromboembolism

17.1 Case Report

A full-term Japanese pregnant woman was transferred to a general hospital because of non-reassuring fetal status. She had two healthy siblings and had experienced twice spontaneous abortions. A fetal echogram revealed enlargement of the bilateral cerebral ventricles with periventricular hyper-echoic lesion. An emergency caesarean delivery was performed under the suspicion of fetal intracranial hemorrhage. A male infant was born weighing 2912 g, with APGAR scores 8 and 10 after 1 and 5 min, respectively. No skin lesion was observed at birth. A computed tomography (CT) showed high-density areas in bilateral periventricles. Multiple purpuric lesions appeared a few hours after birth with progressive disseminated intravascular coagulation (DIC). Because of the suspected diagnosis of purpura fulminans, he was transferred to our neonatal intensive care unit 1 day after birth.

On admission, multiple ecchymoses emerged on his extremities despite the plasma transfusion and anticoagulation therapy (Fig. 17.1). Blood tests revealed white blood cells 11.51 × 10^9/L, hemoglobin 11.0 g/dL, platelets 101 × 10^9/L, fibrinogen 80 mg/dL (reference range (rr), 200–400), prothrombin time (PT) 14.5 sec (rr, 11.6–14.4; PT-% 70%), activated partial thromboplastin time (APTT) 43.3 second (rr, 37.1–48.7), fibrinogen degradation products 142.0 μg/mL (rr, <10), D-dimer 79.7 μg/mL (rr, <1), thrombin–anti-

M. Ishimura (✉) · S. Ohga
Department of Pediatrics, Graduate School of Medical Sciences, Kyushu University, Fukuoka City, Japan
e-mail: ischii@pediatr.med.kyushu-u.ac.jp

T. Oohashi et al. (eds.), *Human Pathobiochemistry*, https://doi.org/10.1007/978-981-13-2977-7_17

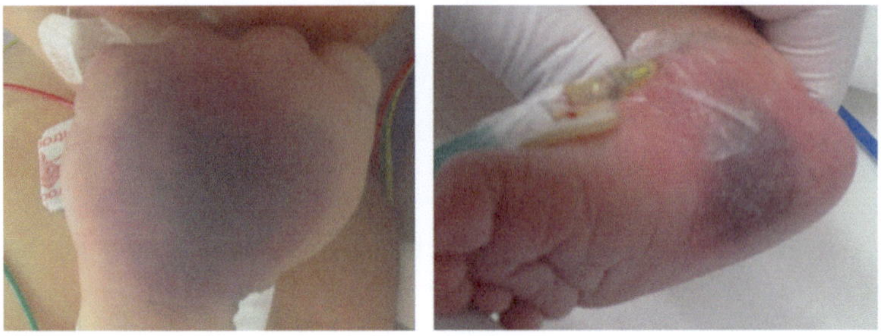

Fig. 17.1 The multiple ecchymoses in the dorsum of the left hand (Fig. 17.1) and in the right sole (Fig. 17.1) (on admission)

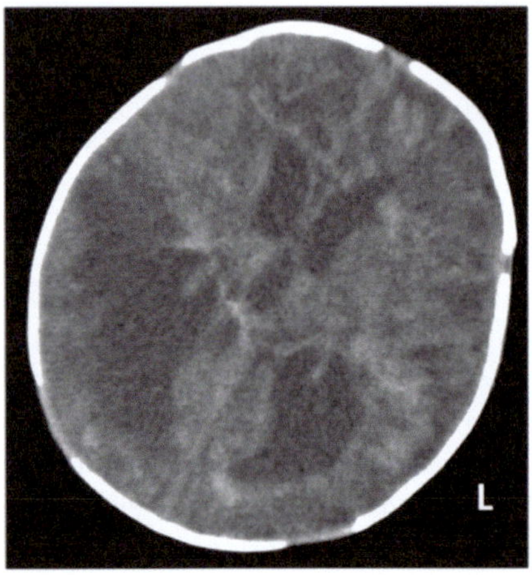

Fig. 17.2 The brain CT scan on admission
Diffuse intracranial hemorrhage indicates cerebral venous sinus thromboses

thrombin complex 7.2 ng/mL (rr, <4), and plasmin α2–antiplasmin complex 4.7 μg/mL (rr, <0.8). Protein C (PC) and protein S (PS) activity before the first plasma transfusion were <5% and 86%, respectively. Brain CT scan showed the diffuse intracranial hemorrhage, indicating cerebral venous sinus thromboses (Fig. 17.2).

He was diagnosed as having purpura fulminans caused by PC deficiency. Plasma-derived activated PC (AnactC®, Teijin Pharma, Japan) therapy led to the drastic improvement of his skin lesions and coagulopathy. A gene analysis revealed that he had compound heterozygous PC gene (*PROC)* mutations (c.236G > A, p.Glu49Lys and c.296G > A, p.Glu68Lys). His parents had either heterozygous *PROC* mutation. This patient was discharged from the hospital after receiving ventriculoperitoneal shunt. Unfortunately, he died suddenly at home on 9 months of warfarin therapy.

17.2 Diagnosis

Thromboembolisms occur as a consequence of genetic predispositions, underlying disorders, and direct triggers such as dehydration, infection, or injury. The formation of thrombosis depends on the Virchow's triad: the vessel wall, blood flow, and blood components. Heritable defects in the coagulant and anticoagulant pathways result in the development of venous thromboembolism (VTE) with or without apparent triggers. The genetic risks of VTE include the deficiency of anticoagulant factors of PC, PS, and antithrombin (AT), as well as the variants of coagulation factors of factor V G1691A (factor V Leiden) and prothrombin (factor II) G20210A (Caspers et al. 2012). Hereditary anticoagulant deficiencies are suspected in young adult VTE, neonatal purpura fulminans, patients having recurrent VTE, and/or family history of thrombosis or anticoagulant deficiencies. Hereditary anticoagulant deficiencies are also associated with a tripling risk for the late fetal loss, but no risk for the first-trimester loss. High risk of thrombophilias was reported in association with obstetric complications other than preeclampsia and intrauterine growth retardation.

The hereditary anticoagulant deficiencies are difficult to diagnose in infants and young children without the genetic tests, because these activities are physiologically lower in childhood than in adulthood (Fig. 17.3) (Ichiyama et al. 2016; Takahashi and Yoshioka 1994). Both PC antigen and activity levels increase after birth and reach the lower limit of adult references (~50 IU/dL) during 6 months–1 year of age. Prior to the genetic screening, anticoagulant activity levels

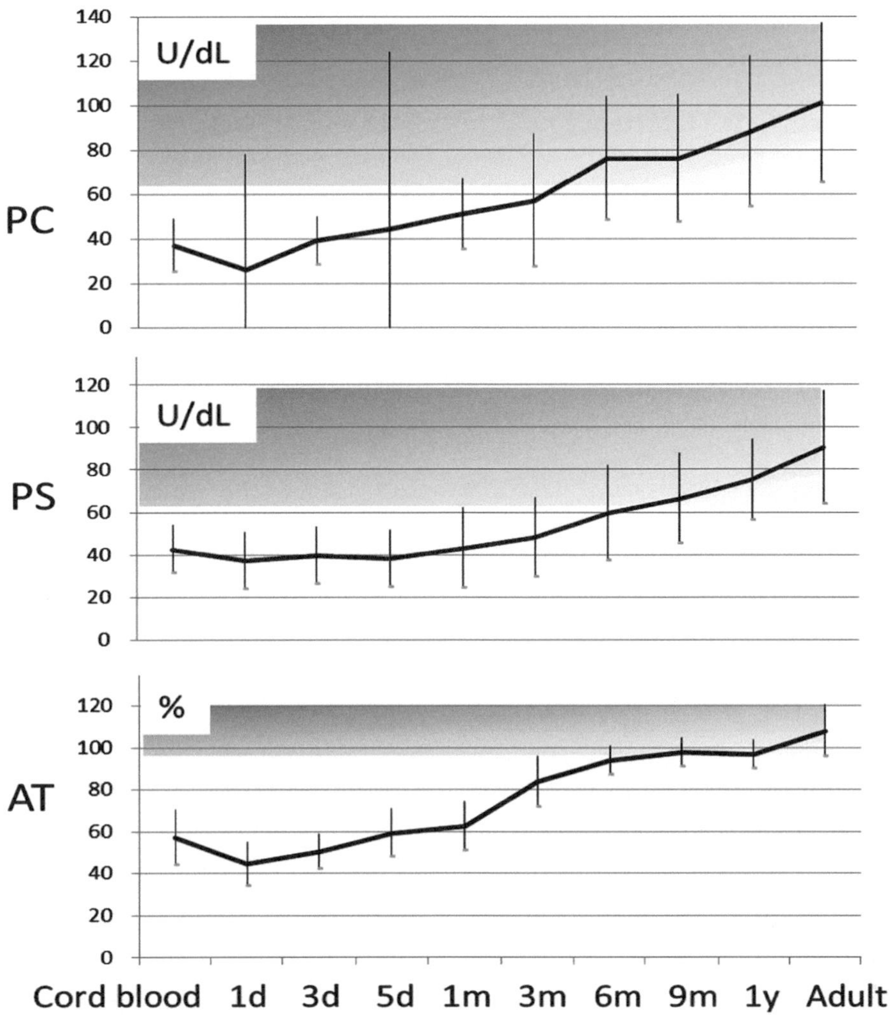

Fig. 17.3 Physiological changes in the plasma activity of protein C (PC), protein S (PS), and antithrombin (AT) in the first year of life
The vertical lines and gray backgrounds show the range of standard deviations of the plasma activity of PC, PS, and AT and the normal range in adults, respectively. Each measurement was not completed by using the same sample concurrently obtained from the subjects (Ref. Ichiyama et al. 2016; Takahashi and Yoshioka 1994)

are assessed with protein induced by vitamin K absence or antagonists II (PIVKAII), D-dimer, and antiphospholipid antibodies. PC and PS activities are recommended to follow with the concurrent measurements of factor VII, a vitamin K-dependent zymogen with a similar short plasma half-life, to exclude the effects of consumption coagulopathy. Unexplained dissociation among PC, PS, and factor VII activity levels may portend a diagnosis of heritable PC or PS deficiency (Ichiyama et al. 2016). The coagulation studies by using the age-dependent standards have then high sensitivity in screening for anticoagulant deficiencies. Nevertheless, repeated sampling, family studies, and genetic analyses along with the detailed information of reported cases are essential for the diagnosis for anticoagulant deficiency.

17.3 Biochemical and Molecular Perspectives

17.3.1 Protein C (PC) Deficiency

PC is a vitamin K-dependent anticoagulant factor, which is synthesized by the liver and circulates as a zymogen. It exerts the proper anticoagulant function after the cleavage of the molecule to the serine protease, activated PC (APC). This activation is mediated by thrombin alone, but occurs more efficiently when thrombin is bound to thrombomodulin on the endothelial cell receptor. APC inactivates factor V and factor VIII by cleaving the critical sites of these activated molecules. This reaction is enhanced by PS, factor V, and lipid cofactors of lipoproteins and phospholipids. Reduced concentration of circulating PC fails to control the propagation of thrombin generation by activated factor V and activated factor VIII, even if PC deficiency originates from the decreased production and/or increased consumption. APC has also anti-inflammatory as well as cytoprotective functions (Griffin et al. 2015). These are directly driven by the endothelial and immunocompetent cells via the specific receptor on the given cells. The protective effect of APC in animal models of sepsis depends on its capacity to activate protease-activated receptor-1 and not on its anticoagulant properties. The pleiotropic effects are expected to be beneficial for the treatment of sepsis. Clinical trials of recombinant APC were, however, withdrawn for the risk of bleeding (Ranieri et al. 2012). Refined molecules of APC are being explored for the effective treatment of sepsis and DIC (Quinn et al. 2015).

Plasma levels of PC activity are measured by chromogenic (amidolytic) or coagulometric (clotting) assays. The standard value in healthy adults ranges from 0.65 to 1.35 IU/mL (65–135% of normal). "Mild," "moderately severe," and "severe" PC deficiencies are defined as the range of >20% (>0.2 IU/mL), 1–20% (0.01–0.2 IU/mL), and <1% (<0.01 IU/mL), respectively (Ohga et al. 2013a).

PC gene (*PROC*) is located on chromosome 2q13-q14. Heterozygous *PROC* mutation inherits "mild" or at most "moderately severe" PC deficiency in an autosomal-dominant manner. The first presentation of PC deficiency includes the newborn-onset and the teen-onset modes. Neonatal purpura fulminans and stroke are the distinct manifestations of PC deficiency (Caspers et al. 2012; Ohga et al. 2013b). Thereafter, deep venous thrombosis (DVT) in the legs and pulmonary thromboembolism are commonly affected sites. Approximately 70% of patients first present in teens spontaneously and the remaining 30% with the risk factors. Nonhemorrhagic arterial stroke is also associated with pediatric PC deficiency. Severe PC deficiency is inherited in an autosomal-recessive fashion. The complete defects exclusively arise from biallelic *PROC* mutation, presenting neonatal purpura fulminans and/or intracranial hemorrhage.

The prevalence of heterozygosity for PC deficiency is considered to be 1/200–500 (Tait et al. 1995). Heterozygous PC-deficient adults have a sevenfold increased risk for an initial episode of DVT compared with those having normal PC activity. On the other hand, asymptomatic carriers have low annual incidence of thrombosis at less than 1.0%. The odds ratio (OR) of pediatric thrombosis with PC deficiency is estimated at 7.7 (Table 17.1) (Young G et al. 2008).

Heterozygous PC deficiency is classified into two types. The common form of type I represents an equal reduction in both immunologic and biologic activities. More than 200 mutations and rare deletions were reported. Type II-deficient subjects show normal antigen and decreased activity levels. However, several cases show normal levels of antigen and amidolytic activity, but with reduced levels of clotting activity. It is ascribed to a reduced ability of APC to interact with the platelet membrane or its substrates of activated factor V and activated factor VIII. Recently, K193del has been recognized as the most common variant in Chinese thromboembolisms (Yin and Miyata 2014). On the other hand, it was considered as a polymorphism at first in Japan, because the heterozygotes often show normal activity for PC in resting conditions. It was then reported as PC-Tottori, in the homozygote of a young adult patient with DVT.

Table 17.1 Reported odds ratio and 95% confidence interval [in brackets] for genetic traits associated with a first onset of venous thromboembolism in children and in pediatric patients with recurrent deep venous thrombosis

		Deep venous thrombosis	
Genetic predisposition	Cerebrovascular occlusion (first onset)	First onset	Recurrence
Protein C deficiency	9.3 [4.8–18.0]	7.7 [4.4–13.4]	2.4 [1.2–4.4]
Protein S deficiency	3.2 [1.2–8.4]	5.8 [3.0–11.0]	3.1 [1.5–6.5]
Antithrombin deficiency	7.1 [2.4–22.4]	9.4 [3.3–26.7]	3.0 [1.4–6.3]
Factor V Leiden	3.3 [2.6–4.1]	3.8 [3.0–4.8]	0.6 [0.4–1.2]
Factor II G20210A	2.4 [1.7–3.5]	2.6 [1.6–4.4]	2.1 [1.01–3.5]

Ref. Kenet G et al. 2010; Young G et al. 2008

We experienced that the heterozygote infants for PC-Tottori showed slowly age-dependent increase of PC activity. There is increasing evidence in Asian countries on the double mutants who escaped neonatal thromboses. In Caucasians, factor V Leiden and factor II variant might mask the effects of *PROC* variants on the thrombotic risk of patients (Yin and Miyata 2014).

17.3.2 Protein S (PS) Deficiency

PS is a vitamin K-dependent protein that enhances the anticoagulant effect of APC. This coenzyme is primarily synthesized by hepatocytes and also by endothelial cells, megakaryocytes, and brain cells. It serves as a cofactor for APC in the setting of activated factor V and activated factor VIII inactivation (Fig. 17.4). The inactivation of activated factor V occurs, at first, in the rapid cleavage at Arg506 (factor V Leiden: Gln506) of the molecules, followed by the slower cleavage at Arg306 (second binding site of PS with APC; factor V Cambridge: Thr306) and then Arg679 (Fig. 17.5). PS with APC increases the affinity for phospholipids to enhance activated factor V inactivation. Approximately 40% of circulating PS molecules is in the free form, and the remainder 60% is bound to C4b-binding protein not to interact with APC. Free PS levels are responsible for the direct anticoagulant effects.

Heterozygous mutations in the gene encoding protein S (*PROS1*) on chromosome 3q11 cause autosomal-dominant thrombophilia. Homozygotes or compound heterozygotes for the mutation are rarely found as an autosomal-dominant thrombophilia. However, most cases of neonatal purpura fulminans had PC but not PS deficiency (Ichiyama et al. 2016; Ohga et al. 2013a). It might account for the relative increase of free PS concentrations (by physiologically low levels of C4b-binding protein), the narrower ranged activity, and/or shorter half-life of PS during the early neonatal period. PS deficiency conveys a risk of thrombosis similar to PC deficiency. The clinical findings in the heterozygotes for PS deficiency are similar to those for AT or PC deficiency. Pulmonary emboli are the common affected lesions. PS-deficient patients experience an initial thrombotic event at approximately 25–30 years of age. Over half of episodes occurred spontaneously, and the remainder had certain risk factors for thrombosis. In pediatric cases, the OR for venous thrombosis due to PS deficiency is estimated to be 5.8 (Table 17.1) (Young G et al. 2008). Several reports suggest an association between PS deficiency and arterial thromboses including ischemic stroke in infants and children (Kenet et al. 2010).

On the other hand, adults with low PS levels but no family history of venous thrombosis have minimal risk of VTE. Low free PS or total PS level (both <0.10th percentile) was not associated with an increased risk of VTE. Young patients with recurrent VTE are associated with double mutations of *PROS1*. Their parents were reportedly asymptomatic although they carried type I PS deficiency. PS deficiency is classified into three types, according to the levels of total and free antigens, along with functional activity. Type I deficiency shows about half levels of normal PS antigen, although free PS antigen and functional activity levels are greatly reduced. Most patients have the missense mutations and base pair

Fig. 17.4 Blood clotting cascade and anticoagulation factors Each Roman numeral represents coagulation factor. The "a" means activated form of coagulation factors. *APC* activated protein C, *AT* antithrombin, *PC* protein C, *PS* protein S, *TF* tissue factor, *TM* thrombomodulin

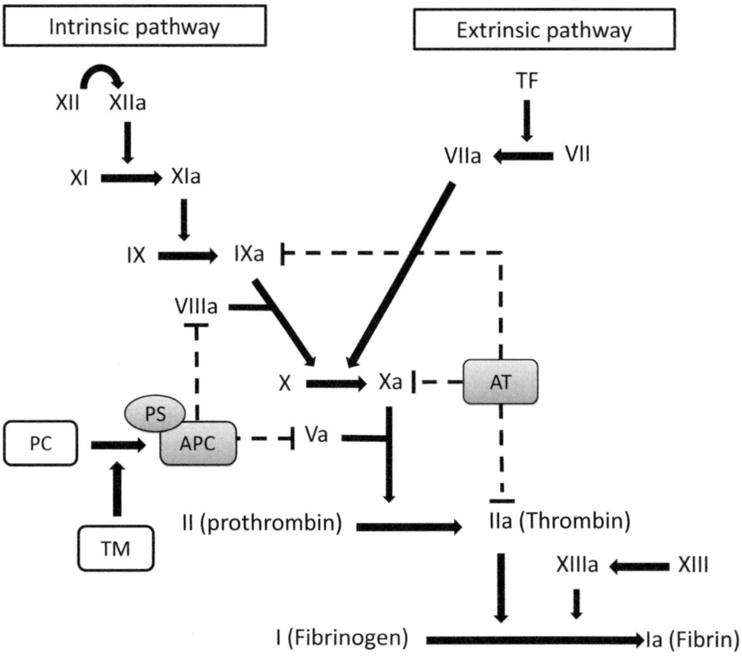

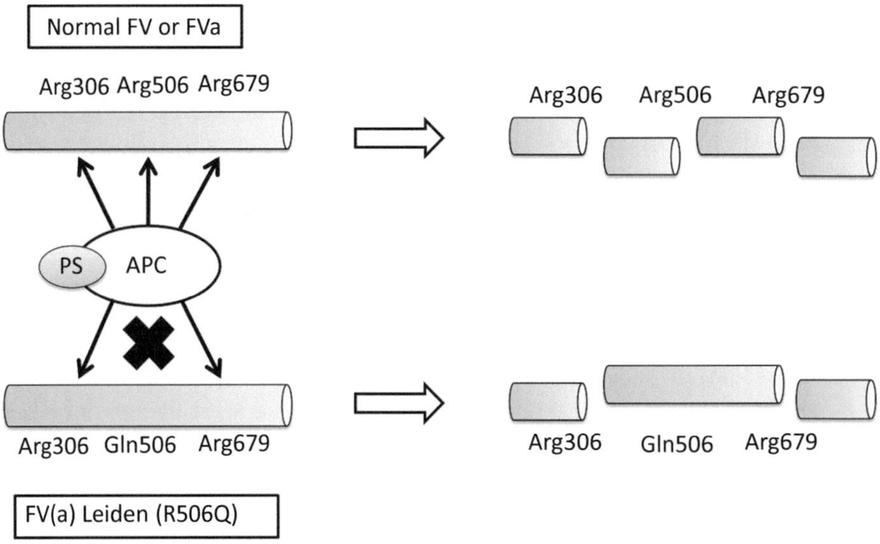

Fig. 17.5 APC resistance of factor V Leiden Factor V Leiden (Arg506Gln) abolishes the Arg506 cleavage site by APC in factor V and factor Va. *APC* activated protein C, *PS* protein S

insertions. Type II is a qualitative deficiency with normal levels of total antigen and free PS, but impaired function. The rarity of type II-deficient patients indicates the insufficient screening power of functional assays. Type III is characterized by normal levels of total PS antigen, but reduced levels of free PS and functional activity. Plasma activity levels rather than antigen levels are then preferable for screening PS deficiency. Although the biologic basis of type III PS deficiency remains unclear, *PROS1* mutations were responsible for the type I deficiency, but not the type III phenotype of age-dependent free PS deficiency (Alhenc-Gelas et al. 2016). The prevalence of PS deficiency depends on the ethnicity. In Asian countries, PC and PS deficiencies are dominantly reported in adults and children with thrombosis. PS-Tokushima (Lys196Glu) is found in 1 per 55 healthy Japanese adults as a common low-risk variant for thrombosis in Japan (Ikejiri et al. 2010).

The lower limit of plasma levels of total and free PS (~65%) in heterozygous deficiency considerably overlaps the ranges of healthy controls (Ichiyama et al. 2016). PS levels are lower in female than male and increase with advanced age. PS, PC, and AT levels are reduced in inflammation. Because C4b-binding protein is an acute phase reactant, it shifts PS to the complexed inactive form leading to the reduced PS activity. On the other hand, PS but not PC levels decrease in pregnancy.

17.3.3 Antithrombin (AT) Deficiency

AT, previously called ATIII, is a plasma protein synthesized by the liver. The anticoagulant factor binds and neutralizes the serine proteases of thrombin, activated factor X, and activated factor IX, which are generated by the coagulation cascade. The AT-mediated effects on the inhibition of these proteases are enhanced by binding to heparin (Fig. 17.4). AT has two major active sites: the reactive center toward the carboxyl terminus and the heparin-binding site at the amino terminus. Thrombin cleaves the reactive site, and the inactive complex molecules are then cleared from circulation. Heterozygous AT deficiencies are found in about 1% of patients having the first episode of DVT. The thrombotic risk of inherited AT-, PS-, or PC-deficient patients is higher than that of factor V Leiden or factor II variant even in infants and children. They have also an eight- to tenfold higher risk for thrombosis than noncarriers (Young G et al. 2008). More than 50% of the first-degree relatives of each deficient patients experience VTE until the first 25 years of life (Holzhauer et al. 2012). The first thrombotic episode of AT deficiency occurs in more than 40% of patients spontaneously. The remaining patients have some triggers including pregnancy, parturition, contraceptive use, surgery, or trauma. The OR of thrombosis with AT deficiency is estimated to be 9.4, although the risk of thromboembolic events depends on the population selected (Table 17.1) (Young G et al. 2008). Heritable AT deficiency is an autosomal-dominant disease with equal sex distribution. Common affected sites of thrombosis include deep veins in the legs, mesenteric veins, and pulmonary embolism. Complete AT defects due to null mutation lead to the fetal loss. Therefore, neonatal purpura fulminans is not the regular presentation of AT deficiency.

Heritable AT deficiency is classified into two types. Type I-deficient individuals show paralleled reduction of both antigen and activity levels of AT, representing a reduced synthesis of biologically normal protease inhibitor molecules. AT deficiency arises from mostly point mutations of the AT gene (*SERPINC1*) located on chromosome 1q23-25 and rarely a deletion encompassing the gene. Type II-deficient subjects show reduced levels of plasma activity but normal antigen levels of AT, representing a discrete molecular defect. Type II deficiency is subcategorized into three groups assessed by progressive AT assays and heparin-binding assays: the molecular abnormality in reactive site, heparin-binding site, and pleiotropic effect.

17.4 Factor V Leiden

The resistance to APC in an APTT-based clotting assay accounts for familial thrombophilia (Dahlback et al. 1993). The major genotype of APC resistance is factor V Leiden (G1691A, p. Arg506Gln) (Fig. 17.5) (Bertina et al. 1994), being found in more than 80% of APC-resistant patients. In Western countries, factor V Leiden was found in 20–50% of adult patients with VTE. Factor V Leiden is a risk factor for venous and arterial thrombosis. Retrospectively, heterozygosity for factor V Leiden was identified in 12% of patients who had a first episode of DVT or pulmonary embolism and in 6% of controls. In the elderly (>60 years of age) who suffered from the first VTE without triggers, 26% were heterozygotes for factor V Leiden. Pediatric studies found that factor V Leiden heterozygosity conferred an OR of 3.8 for a first episode of VTE and 0.6 for recurrent VTE in children (Table 17.1) (Young G et al. 2008). The OR for pediatric cerebrovascular occlusion in Factor V Leiden carriers was 3.3 (Kenet G et al. 2010). On the other hand, the annual incidence of VTE in asymptomatic carriers of the factor V Leiden is low (0.58%), which raises questions about the screening asymptomatic family members (Middeldorp et al. 2001). Factor V Leiden was not associated with the risk of myocardial infarction or stroke, but potentially perinatal and pediatric arterial ischemic stroke. Factor V Leiden homozygotes may have the similar high risk of VTE to the heterozygotes for AT, PC, or PS deficiency. As factor V and AT gene are located on chromosome 1p, the coinheritance of factor V Leiden and AT mutations leads to a more severe thrombotic diathesis.

The prevalence of heterozygosity for factor V Leiden ranges 1–9% in Caucasians, but quite rare (<1%) in African blacks, Chinese, Japanese, or Native American ancestries. Factor V Leiden is more prevalent in northern Europe than the southern countries. A single founder allele was suggested among whites of differing ethnic backgrounds. Severe infections might shape the distribution of factor V Leiden, because of the survival advantage inferred by the mouse models. Co-segregation of heterozygous APC resistance due to factor V Leiden and type I factor V deficiency results in severe APC resistance in APTT assays, as found in homozygous factor V Leiden patients (pseudohomozygous). Several polymorphisms in factor V gene include a HR2 haplotype containing the R2 polymorphism (His1299Arg) with mild APC resistance. Some rare factor V mutations other than factor V Leiden showed also APC resistance (factor V Cambridge, Arg306Thr; factor V Liverpool, Ile359Thr; factor V Bonn, Ala512Val; and factor V Nara, Trp1920Arg).

17.4.1 Factor II (Prothrombin) Variant

Factor II (prothrombin)-related thrombophilias are encountered in adults with VTE. The factor II G20210A mutation located at the 3′ untranslated region is considered a gain-of-function mutation, which leads to the increase of plasma prothrombin activity level (Soria et al. 2000). For the diagnosis of factor II G20210A, it should be kept in mind that the range of plasma activity level of prothrombin in heterozygotes or homozygotes overlaps the ranges of normal controls. Many individuals heterozygous and homozygous for factor II G20210A develop no thrombosis. Most patients of the heterozygotes also remain asymptomatic until adulthood. On the other hand, the relative risk of thrombosis in the heterozygotes is increased to two- to fivefold in adults and three- to fourfold in children. The recurrent risk for VTE is modest in factor II G20210A carriers. Pregnant loss and obstetric complications in the carriers are similar to those in factor V Leiden carriers.

The diagnosis depends on the genetic study. The factor II variant is not found in Asian ancestries, similarly in factor V Leiden, although factor II Yukuhashi (Arg596Leu) with AT resistance has been recently identified in a Japanese family. Because of the relatively low risk of thrombosis in factor V Leiden and factor II G20210A carriers, the routine screening is not recommended for healthy individuals as long as they have no positive family history of recurrent or young thromboses.

17.5 Prophylaxis and Treatment of Thromboses

Standard management for the first acute thrombosis in adults consists of a course of low-molecular-weight heparin or fondaparinux and concurrent warfarin therapy for at least 5 days on the monitoring for international normalized ratio.

Rivaroxaban, a factor Xa inhibitor, is approved for treatment of acute VTE and prevention of recurrence in adult cases. Factor Xa inhibitors raise no chance of developing heparin-induced thrombocytopenia. However, the pediatric use of factor Xa inhibitors is challenging, because the practical monitoring is on controversy. The standard therapy using tissue plasminogen activator for arterial thrombosis has not been established in the newborn infants.

The concept of anticoagulation therapy depends on the potential risks for (1) VTE recurrence and (2) therapy-related bleeding. Individuals with a spontaneous thrombosis with no identifiable provoking factors and those with persistent risk factors are candidates for the long-term anticoagulation therapy. Three-month treatment is recommended for individuals with transient (reversible) risk factors such as surgery. Graduated compression stockings should be worn for at least 2 years following an acute DVT in adult cases.

No consensus for the use of factor Xa inhibitors alternate for conventional heparin, and warfarin therapy exists on the optimal management for pediatric patients with PC, PS, and AT deficiencies. Fresh frozen plasma should be sufficiently administered to keep at least the lower limit of standard ranges based on the age. PC levels may be required to exceed 20% levels for the effective use of recombinant thrombomodulin, although the effective application is limited to patients with sepsis-induced DIC (Umemura et al. 2016). Thrombomodulin therapy has not been then established in patients with defective PC–PS pathway. Replacement therapy by using plasma-derived or recombinant PC and antithrombin concentrates is useful for the treatment of acute thrombosis as well as the prophylaxis for recurrence.

End-of-Chapter Questions

1. What are required to make the correct diagnosis for hereditary anticoagulant deficiencies?
2. Describe the difference of anticoagulant activities between infants and adults.
3. Describe the difference of genetic background between Caucasian and Asian population in hereditary anticoagulant deficiencies.

References

Alhenc-Gelas M, Plu-Bureau G, Horellou MH et al (2016) GEHT genetic thrombophilia group. PROS1 genotype phenotype relationships in a large cohort of adults with suspicion of inherited quantitative protein S deficiency. Thromb Haemost 115:570–579

Bertina RM, Koeleman BP, Koster T et al (1994) Mutation in blood coagulation factor V associated with resistance to activated protein C. Nature 369(6475):64–67

Caspers M, Pavlova A, Driesen J et al (2012) Deficiencies of antithrombin, protein C and protein S – practical experience in genetic analysis of a large patient cohort. Thromb Haemost 108:247–257

Dahlback B, Carlsson M, Svensson PJ (1993) Familial thrombophilia due to previously unrecognized mechanism characterized by poor anti-coagulant response to activated protein C: prediction of a cofactor to activated protein C. Proc Natl Acad Sci U S A 90:1004–1008

Griffin JH, Zlokovic BV, Mosnier LO (2015) Activated protein C: biased for translation. Blood 125:2898–2907

Holzhauer S, Goldenberg NA, Junker R et al (2012) Inherited thrombophilia in children with venous thromboembolism and the familial risk of thromboembolism: an observational study. Blood 120:1510–1515

Ichiyama M, Ohga S, Ochiai M et al (2016) Age-specific onset and distribution of the natural anticoagulant deficiency in pediatric thromboembolism. Pediatr Res 79:81–86

Ikejiri M, Wada H, Sakamoto Y et al (2010) The association of protein S Tokushima-K196E with a risk of deep vein thrombosis. Int J Hematol 92:302–305

Kenet G, Liitkhoff LK, Albisetti M et al (2010) Impact of thrombophilia on risk of arterial ischemic stroke or cerebral sinovenous thrombosis in neonates and children: a systematic review and meta-analysis of observational studies. Circulation 121:1838–1847

Middeldorp S, Meinardi JR, Koopman MM et al (2001) A prospective study of asymptomatic carriers of the factor V Leiden mutation to determine the incidence of venous thromboembolism. Ann Intern Med 135:322–327

Ohga S, Ishiguro A, Takahashi Y et al (2013a) Protein C deficiency as the major cause of thrombophilias in childhood. Pediatr Int 55:267–271

Ohga S, Kang D, Kinjo T et al (2013b) Paediatric presentation and outcome of congenital protein C deficiency in Japan. Haemophilia 19:378–384

Quinn LM, Drakeford C, O'Donnell JS et al (2015) Engineering activated protein C to maximize therapeutic efficacy. Biochem Soc Trans 43:691–695

Ranieri VM, Thompson BT, Barie PS et al (2012) Drotrecogin alfa (activated) in adults with septic shock. N Engl J Med 366:2055–2064

Soria JM, Almasy L, Souto JC et al (2000) Linkage analysis demonstrates that the prothrombin G20210A mutation jointly influences plasma prothrombin levels and risk of thrombosis. Blood 95:2780–2795

Tait RC, Walker ID, Reitsma PH et al (1995) Prevalence of protein C deficiency in the healthy population. Thromb Haemost 73:87–93

Takahashi Y, Yoshioka A (1994) Hemostasis and its regulation system in childhood. Jpn J Pediatr Hematol 8:389–397 in Japanese

Umemura Y, Yamakawa K, Ogura H et al (2016) Efficacy and safety of anticoagulant therapy in three specific populations with sepsis: a meta-analysis of randomized controlled trials. J Thromb Haemost 14:518–530

Yin T, Miyata T (2014) Dysfunction of protein C anticoagulant system, main genetic risk factor for venous thromboembolism in northeast Asians. J Thromb Thrombolysis 37:56–65

Young G, Albisetti M, Bonduel M et al (2008) Impact of inherited thrombophilia on venous thromboembolism in children: a systematic review and meta-analysis of observational studies. Circulation 118:1373–1382

Cherubism

18

Noriaki Shoji, Ernst J. Reichenberger, and Yasuyoshi Ueki

Keywords
SH3BP2 · Autoinflammation · Macrophages · Osteoclasts · Bone destruction

18.1 Case Report

A 21-year-old female visited our hospital with cosmetic concerns regarding her expanded mandible (Fig. 18.1A). Although she had noticed some facial deformity since she was 6 years old, she did not seek medical attention. When she was around 20 years old, she became more concerned about the noticeable gradual expansion of the mandible. There was no family history of enlarged mandibles. Physical examination revealed a bilateral hard bony swelling of the face and a right exophthalmos. There were no signs of pain or paralysis in the orofacial area nor abnormalities in visual field and acuity. Dental examination revealed (i) three missing teeth other than wisdom teeth, (ii) tooth displacement with an impacted right upper canine, and (iii) microdontia with malformed tooth crowns (Fig. 18.1B, C).

A panoramic radiograph (Fig. 18.1D) showed bilateral mandibular expansion with many multilocular radiolucent lesions spreading from the mandibular body to rami on both sides. Obvious radiolucent lesions were also seen in the maxilla. Root resorption of mandibular incisors was noticeable (Fig. 18.1D). There was an impacted tooth in the lower left mental region that appeared to be equivalent to the first premolar. Right first (or second) lower premolar, left lower second molar, and four wisdom teeth were missing. In agreement with these findings, axial computed tomography (CT) images revealed bilateral multilocular and expansile lesions with heterogeneous density destructing the mandible and maxilla. Notably, the bilateral mandibular body and ramus were substantially affected by these lesions (Fig. 18.2A). The right maxillary lesion was significantly larger than the left and progressed into maxillary sinus, orbit, and pterygopalatine fossa (Fig. 18.2B, C). The three-dimensional CT image showed remarkable expansion and disruption of mandibular and maxillary cortical bone as well as an elevated orbital floor on the right side (Fig. 18.2D). Magnetic resonance (MR) imaging was performed to clarify the extent of lesions and to particularly evaluate the orbital involvement

N. Shoji
Division of Oral Diagnosis, Department of Oral Medicine and Surgery, Tohoku University Graduate School of Dentistry, Sendai, Japan

E. J. Reichenberger
Department of Reconstructive Sciences, School of Dental Medicine, UConn Health, Farmington, CT, USA

Y. Ueki (✉)
Indiana Center for Musculoskeletal Health, Department of Biomedical and Applied Sciences, School of Dentistry, Indiana University, Indianapolis, IN, USA
e-mail: uekiy@iu.edu

T. Oohashi et al. (eds.), *Human Pathobiochemistry*, https://doi.org/10.1007/978-981-13-2977-7_18

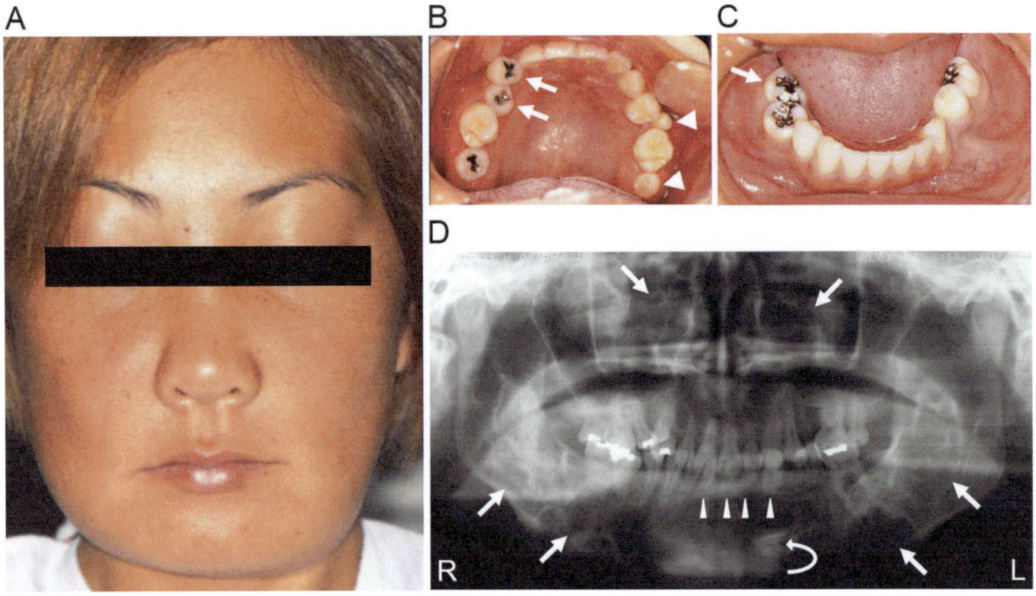

Fig. 18.1 Facial appearance of patient with cherubism (**a**), photograph of oral cavity (**b**, **c**), and panoramic radiograph showing bilateral multilocular radiolucent lesions in the mandible and the maxilla (**d**, arrows). Right upper canine is invisible due to infralabioversion (**b**). Right lower premolar and left lower second molar are missing (**c**). Arrows and arrowheads show displacement of teeth and microdontism with crown malformation, respectively (**b**, **c**). Root resorption of mandibular incisors is noted (**d**, arrowheads). Left lower premolar is impacted (**d**, curved arrow)

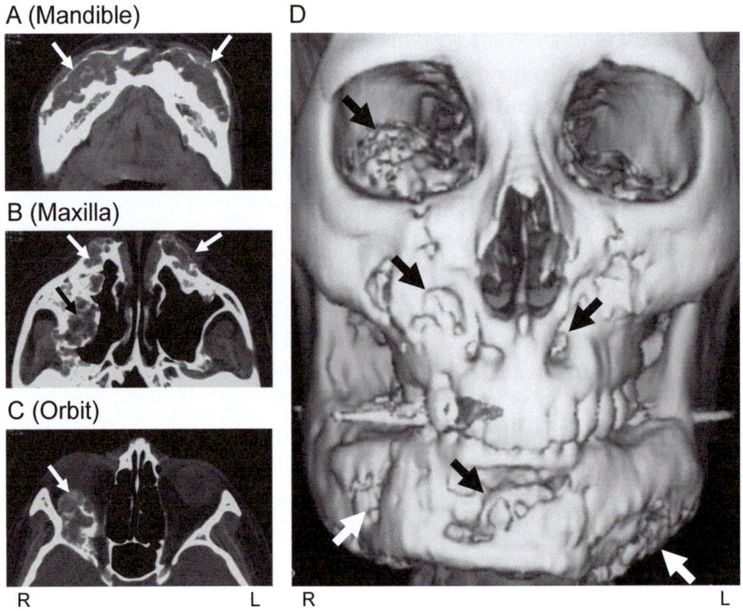

Fig. 18.2 Axial CT images showing multilocular and expansile lesions of the mandible and the maxilla with extension into the right orbit (**a**, **b**, **c**, arrows). Three-dimensional computed tomography image of facial bones of a cherubism patient showing destruction of the mandible and maxilla and elevation of the orbital floor (**d**, arrows). The expansion is more severe in the buccal cortical plates than in the lingual cortical plates. Disruption of the cortex is more severe in the labial side than in the lingual side of the mandible. Coarse bony septa are seen in the mandible, while there is relatively fine trabeculation in the maxilla. D: Courtesy of Dr. Imai (Imai et al. 2003)

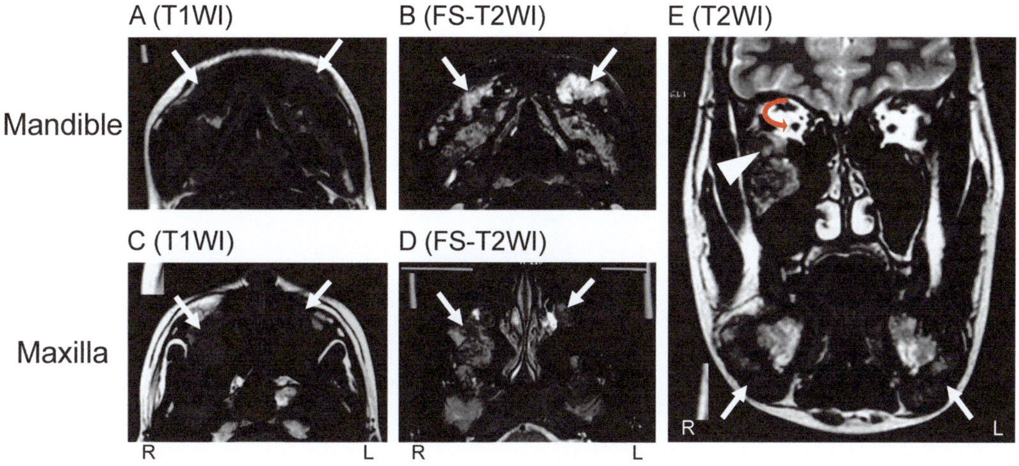

Fig. 18.3 Magnetic resonance imaging detecting lesions (arrows) in both the mandible and maxilla with low-signal intensity on axial T1-weighted images (**a**, **c**: T1WI) and intermediate to high heterogeneous signal intensity on both axial fat-suppressed T2-weighted images (**b**, **d**: FS-T2WI) and a coronal T2-weighted image (**e**: T2WI). The lesion extended into the right orbit (E, arrowhead); however optic nerve is intact on coronal T2WI (curved arrow)

that is likely responsible for the exophthalmos. The mandibular and maxillary lesions showed low-signal intensity on axial T1-weighted MR images (Fig. 18.3A, C) and a heterogeneous intermediate- to high-signal intensity on axial fat-suppressed T2-weighted images (Fig. 18.3B, D). The lesion extended into the right orbit; however, the optic nerve was not damaged on the coronal T2-weighted image (Fig. 18.3E).

A bone scintigram with technetium-99m-labeled methylene diphosphonates (^{99m}Tc-MDP) was performed to investigate skeletal abnormalities beyond the orofacial region. This imaging technique allows for efficient screening of many pathologic conditions due to its high sensitivity. Uptake levels of this radiotracer reflect metabolic and remodeling activity of bone. Results showed increased accumulation of ^{99m}Tc-MDP not only in the mandible and maxilla but also in the right orbital cavity (Fig. 18.4A), while no obvious increased uptake was seen in other regions of the body (Fig. 18.4B). Peripheral blood was sampled for examination of some serum markers for bone turnover. While the level of alkaline phosphatase (ALP) exceeded the reference range (374 IU/L, normal, 112 to 330 IU/L), levels of parathyroid hormone and calcium were within normal limits.

After 2 months of her initial visit, the patient underwent plastic surgery to correct cosmetic deformities by removing bilateral fibrous lesions of the mandible and maxilla. Hematoxylin and eosin (H&E) staining of surgical specimen revealed fibrous stromal tissue that contained a large number of multinucleated giant cells (Fig. 18.5A). The same histological features were observed in specimens of the mandible and maxilla. Malignant characteristics were not seen. These histopathological features were suggestive of giant cell granuloma. Immunohistochemical staining with an antibody against tartrate-resistant acid phosphatase (TRAP), a marker for osteoclasts, showed that the giant cells are strongly positive for TRAP (Fig. 18.5B). Sequencing of genomic DNA from blood cells detected a heterozygous missense mutation in the gene coding for SH3 domain-binding protein 2 (SH3BP2) that causes an amino acid substitution from proline to arginine at amino acid position 418 (Fig. 18.6), confirming cherubism. This non-familial Pro418Arg mutation is most likely a de novo mutation although germline mosaicism cannot be excluded. This patient had no further progression or recurrence of lesions after surgery.

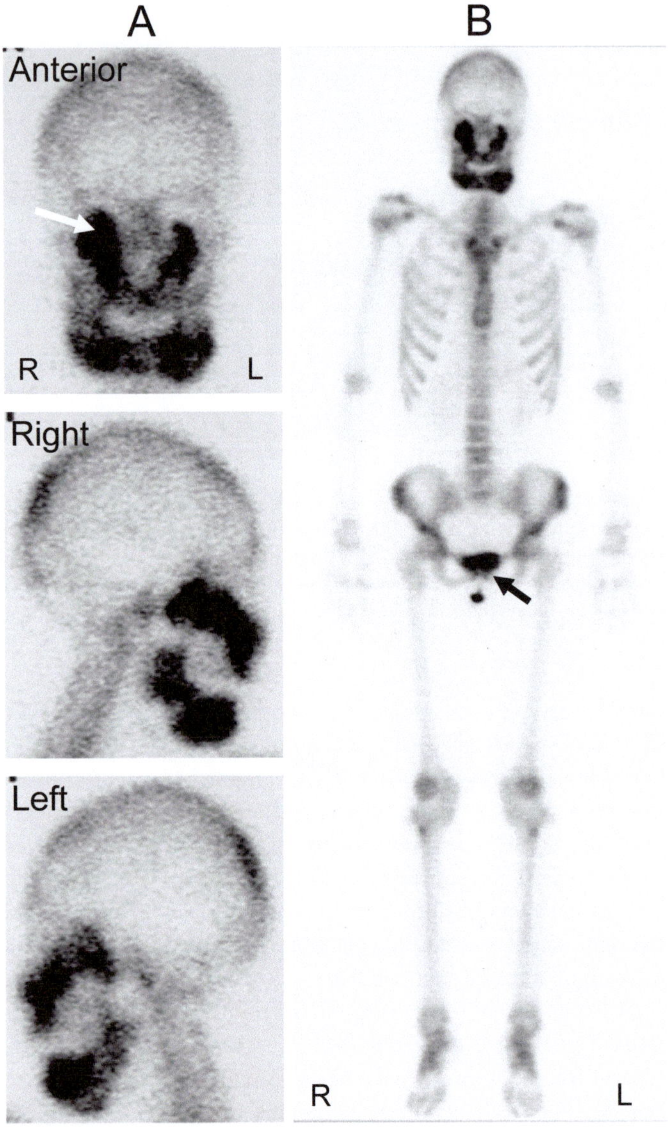

Fig. 18.4 Bone scintigram showing strong radiotracer accumulation in multiple bilateral regions of the mandible and maxilla as well as in the right orbital cavity (**a**, arrow). Anterior whole-body scintigram shows symmetric distribution of radioactivity throughout the skeletal system except the head (**b**). Increased symmetric uptake at the scapulae, clavicles, sternum (angle of Louis), and iliac bones is normal. Increased accumulation in urinary bladder by urine that contains radiotracer is also normal (arrow)

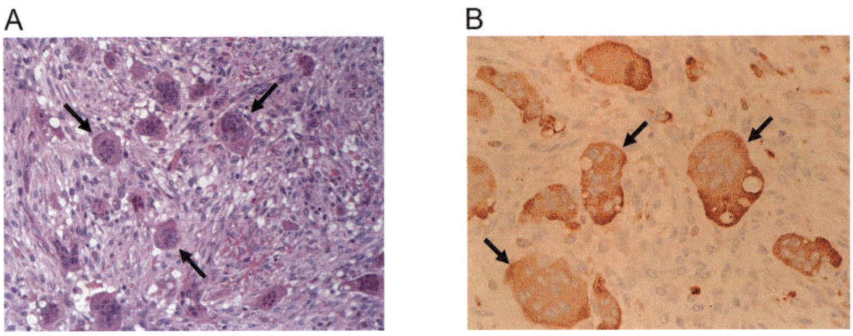

Fig. 18.5 Histological image of surgical specimen demonstrating osteoclast-like multinucleated giant cells within cellular and fibrous tissue (**a**, arrows: H&E staining X100). Immunohistochemical analysis with an anti-TRAP antibody showing TRAP-positive multinucleated giant cells (**b**, arrows: X200). Courtesy of Dr. Imai (Imai et al. 2003)

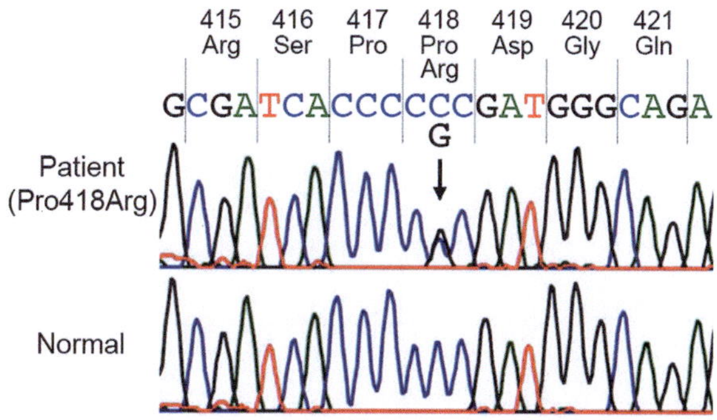

Fig. 18.6 Sequence analysis of the *SH3BP2* gene showing heterozygous Pro418Arg mutation compared to normal sequence

18.2 Diagnosis

The hallmark of cherubism (OMIM#118400) is bilateral symmetrical osteolytic expansion of mandibular and maxillary bone in children that begins around 2–6 years of age. The swelling of the lower face progresses until puberty and often regresses spontaneously thereafter. Typically, X-ray images show multilocular symmetrical cystic changes in the mandible and maxilla. Cherubism usually begins with a swelling of submandibular lymph nodes that may precede and/or exaggerate the enlargement of the lower face, although sometimes this sign goes unnoticed. Cysts most often first develop at the gonial angle of the jaws. The cysts fill with a fibro-osseous tissue mass that can expand significantly, and the bones are further resorbed as the tissue mass expands. This leads to the characteristic facial appearance of full cheeks often accompanied by upward turning of the patient's eyes that are reminiscent of cherubic angels in Renaissance paintings. The upward-turning eyes are caused by invasion of fibrous tissues into the orbital floor. The fibrotic lesions are nonneoplastic. Cherubism is often first diagnosed in children after taking dental radiographs. Progression of lesions varies between cases, even within families, and can be mild. A large number of osteoclastic multinucleated giant cells within fibro-osseous lesions (not adjacent to bone) that have TRAP enzyme activity support the diagnosis of cherubism (Fig. 18.7A). No serological marker for the diagnosis of cherubism has been established.

18.3 Molecular Genetics

Linkage and haplotype analysis for 12 families affected with autosomal dominant cherubism refined the location of the causative gene to a 1.5-megabase-pair (Mbp) interval between genetic markers D4S127 and D4S115 on the short arm of human chromosome 4p16.3 (Ueki et al. 2001). Exon sequencing of candidate genes within the interval detected point mutations in the *SH3BP2* gene of affected individuals in the families. These mutations were not detected in unaffected family members or in 200 healthy controls and cosegregated with cherubism symptoms in all families. All mutations were heterozygous, and the affected amino acids were exclusively located within a six-amino acid sequence (RSPPDG) in exon nine of the *SH3BP2* gene between the SH3-binding domain and SH2 domain of the SH3BP2 protein (Fig. 18.7B). Therefore, mutation analysis of exon nine of the *SH3BP2* gene can be an important tool for confirming the diagnosis of cherubism (Ueki et al. 2001). *SH3BP2* lies within a region that is frequently deleted in individuals with Wolf-Hirschhorn syndrome (WHS; OMIM#194190), but haploinsufficiency of SH3BP2 in individuals with WHS does not result in cherubism or cherubism-like characteristics. This finding and the clustering of amino acid missense mutations in SH3BP2 suggest that the mutations lead to a gain of function (Ueki et al. 2001).

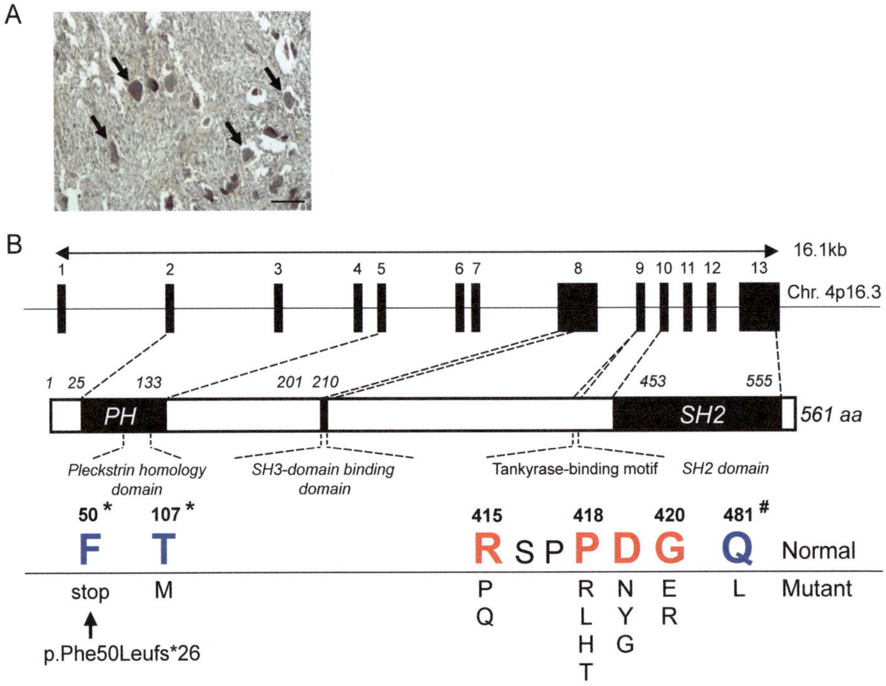

Fig. 18.7 TRAP enzyme activity assay showing TRAP-positive multinucleated osteoclasts in cherubism lesion (**a**, arrows). Gene structure of human SH3BP2 indicating cherubism mutations in SH3BP2 in a six-amino acid interval (415–420) (**b**). Mutations for unusual cherubism cases (*) and for one case of central giant cell granuloma (#) have been reported outside of this region. Modified from Ueki et al. (Ueki et al. 2001). Bar = 100 μm

18.4 Differential Diagnosis

Cherubism cannot be diagnosed by histology alone because histological features of fibrous stromal cells mixed with osteoclastic giant cells are indistinguishable from other giant cell lesion syndromes of bone. Details of histological findings at the various stages of cherubism are rarely described. Differential diagnosis of unusual cases of giant cell lesions, brown tumor of hyperparathyroidism (OMIM#145001), Noonan/multiple giant cell lesion syndrome (OMIM#163950), fibrous dysplasia (OMIM#174800), ossifying fibroma, aneurysmal bone cyst (OMIM#606179), neurofibromatosis type 1 (OMIM#162200), and the hyperparathyroidism-jaw tumor syndrome (HPT-JT) (OMIM#145001) can be challenging. No mutation in *SH3BP2* has been reported in other giant cell lesion syndromes of bone (Reichenberger et al. 2012; Papadaki et al. 2012).

18.5 Molecular Perspectives

SH3BP2 is an adapter protein that plays key roles in the regulation of intracellular signal transduction by coordinating protein–protein or protein–lipid interaction. SH3BP2 was first identified as a protein binding to the SH3 domain of the protein tyrosine kinase c-ABL (Bell et al. 1997). Since then, many binding partners and cell type-specific functions of SH3BP2 have been identified (Reichenberger et al. 2012). Cherubism mutations are located in the RSPPDG hexapeptide region of SH3BP2 between SH3-binding and SH2 domains. The RSPPDG sequence is a consensus motif important for interaction with tankyrase, a member of the poly(ADP-ribose) polymerase (PARP) family that catalyzes post-translational protein modification (Guettler et al. 2011). Tankyrase ADP ribosylates SH3BP2 and regulates SH3BP2 stability through subsequent

SH3BP2 ubiquitylation by ubiquitin ligase RNF146. The ubiquitylated SH3BP2 protein undergoes proteasomal degradation in myeloid cells. Since cherubism mutations uncouple SH3BP2 from tankyrase, the tankyrase-mediated SH3BP2 degradation is inhibited, and mutant SH3BP2 protein is stabilized. This leads to the accumulation of mutant SH3BP2 protein, resulting in elevation of downstream signaling activity in a gain-of-function effect (Levaot et al. 2011).

18.6 Cellular and Biochemical Perspectives

The cherubism knock-in mouse model that harbors a proline to arginine substitution at amino acid position 416 (P416R) of SH3BP2, which is equivalent to the most common P418R mutation in human cherubism patients, replicates many features of cherubism (Ueki et al. 2007). The mouse model shows swelling of the face, mandibular bone erosion, and cervical lymphadenopathy due to postnatal development of inflammatory lesions. This inflammatory lesion develops not only in the face but also in many organs such as the lung, liver, and stomach of the mice. Joint inflammation causes severe erosive bone destruction with increased osteoclast numbers on the bone surface, which is similar to human rheumatoid arthritis. Serum levels of tumor necrosis factor-alpha (TNF-α) are elevated in cherubism mice compared to wild-type control mice. This inflammation is suppressed in cherubism mice lacking TNF-α (Ueki et al. 2007). These phenotypes can be recapitulated in wild-type mice transplanted with fetal liver cells from cherubism mice. Therefore, animal studies suggest that cherubism is an inflammatory bone disease of the face resulting from increased activation of macrophages and osteoclasts and that cherubism is a hematopoietic disorder of myeloid lineage cells. However, while human cherubism mutations are heterozygous (autosomal dominant inheritance), homozygosity is required in mice to achieve a severe cherubism-like phenotype, suggesting that the damaging threshold for SH3BP2 levels in mice may be different from humans. Therefore, most pathological and translational studies in cherubism have been conducted in homozygous mutant mice where the phenotype is strongly expressed.

18.6.1 Pathological Role of Mutant SH3BP2 in Macrophages

Macrophages are cells that phagocytose foreign bodies including bacteria and play an important role in inflammatory regulation and fibrosis through functional changes (Wynn and Vannella 2016). Recognition of microbial components mainly occurs via toll-like receptors (TLRs) on cell membranes of immune cells. Thus, TLRs play a key role in host defense against pathogenic microorganisms that produce pathogen-associated molecular patterns (PAMPs) such as lipopeptides of Gram-positive bacteria and LPS of Gram-negative bacteria. They also recognize cellular debris released from injured or necrotic tissues that is called damage-associated molecular patterns (DAMPs). Bone marrow-derived M-CSF-dependent macrophages (BMMs) from cherubism mice are highly responsive to the ligands that stimulate TLRs by producing increased amounts of TNF-α (Yoshitaka et al. 2014b).

Myeloid differentiation primary response 88 (MYD88) is an essential signal transducer of TLR2 and TLR4, the receptor for lipopeptides and LPS, respectively. It was found that cherubism mice deficient in MYD88 do not develop inflammation responsible for swelling in the face (Yoshitaka et al. 2014b) (Fig. 18.8). Consistent with these findings, cherubism mice crossed with TLR2 and TLR4 knockout mice were rescued from development of inflammation in the face (Fig. 18.8). It was also shown that cherubism inflammation developed even in a germ-free environment (Fig. 18.8), suggesting that DAMPs are sufficient to cause cherubism. Furthermore, elevated levels of mutant SH3BP2 protein enhance the downstream NF-κB signaling pathway through spleen tyrosine kinase (SYK) in macrophages stimulated by TLR ligands, leading to increased TNF-α levels in

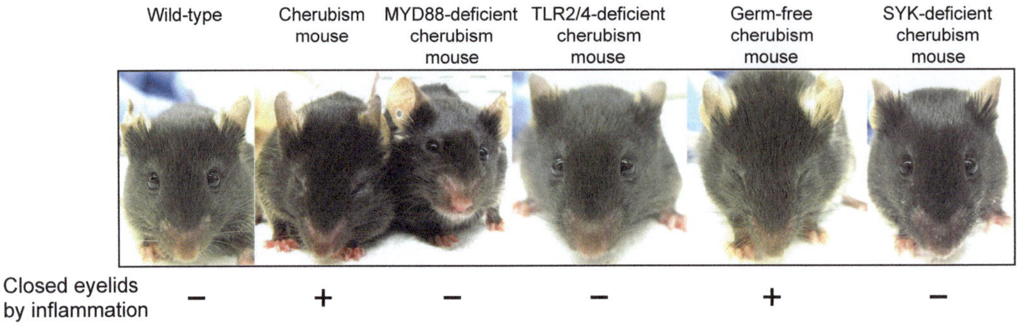

Fig. 18.8 Facial appearance of wild-type and homozygous cherubism mice that develop closed eyelids due to facial swelling from inflammation. Note that eyelid closure in cherubism mice disappears in MYD88-, TLR2/4-, and myeloid cell-selective SYK-deficient cherubism mice, but not in germ-free cherubism mice. Modified from Yoshitaka et al. (Yoshitaka et al. 2014b)

serum responsible for systemic inflammation in cherubism mice (Yoshitaka et al. 2014b). In addition, inflammation in cherubism mice had been shown to occur even in the absence of T- and B-lymphocytes that are important for the adaptive immune system (Ueki et al. 2007). Based on these results, it was concluded that hyperactivation of the TLR-mediated signaling pathway in macrophages by mutant SH3BP2 is responsible for the inflammation in cherubism mice. Thus, cherubism belongs to the spectrum of autoinflammatory disorders and should also be considered a homeostatic inflammatory disease due to the chronic inflammatory reaction against DAMPs (Yoshitaka et al. 2014b). The fibrotic phenotype of human cherubism lesions may be due to impaired regulation of extracellular matrix synthesis mediated by inflammatory macrophages carrying the SH3BP2 mutation.

18.6.2 Pathological Role of Mutant SH3BP2 in Osteoclasts

Osteoclasts are multinucleated giant cells that resorb organic and inorganic mineral components of bone. Like macrophages, osteoclasts originate from the myeloid cell lineage of hematopoietic cells. Receptor activator of nuclear factor kappa-B ligand (RANKL) is a key cytokine that is needed to induce osteoclast differentiation. BMMs from cherubism mutant mice show increased responsiveness to the stimulation with RANKL, resulting in the formation of hyperactive osteoclasts using a mechanism that enhances the induction of nuclear factor of activated T-cells, cytoplasmic, calcineurin-dependent 1 (NFATc1), a master transcription factor of osteoclastogenesis (Ueki et al. 2007; Aliprantis et al. 2008). BMMs expressing mutant SH3BP2 also show an increased responsiveness to TNF-α and can differentiate into osteoclasts with TNF-α alone, without the need for RANKL stimulation. This increased sensitivity to stimulatory cytokines is attributed to increased amounts of SH3BP2 in the cells. Therefore, it was concluded that SH2BP2 is an important regulator of osteoclast differentiation and that mutant SH3BP2 causes inflammatory jawbone destruction via elevated bone-resorbing capacity of osteoclasts in cherubism.

From animal studies, it is concluded that the increased inflammatory response of macrophages is likely responsible for the development of fibrotic lesions in cherubism patients and the cause for lower facial swelling and enlargement of submandibular lymph nodes. Jawbone destruction in cherubism patients is therefore likely attributable to the increased sensitivity of osteoclast precursors to low amounts of cytokines that leads to the formation of osteoclasts with high capacity of bone resorption (Fig. 18.9). Studies of cherubism mice further suggest that the presence of large amounts of TLR ligands in the oral cavity, presumably oral bacteria and DAMPs released during the process of jawbone

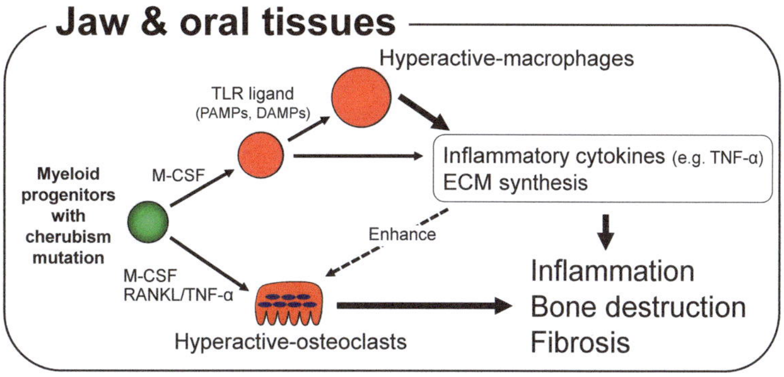

Fig. 18.9 Proposed mechanism for inflammation, bone destruction, and development of fibrous lesion of the jaw in cherubism. ECM, extracellular matrix. Modified from Ueki et al. (Ueki et al. 2007)

remodeling, may contribute to the jaw-specific development of human cherubism lesions. The age-dependent regression of cherubism may be explained with reduced DAMP production in orofacial regions after the cessation of tooth eruption and after the slowdown of jaw remodeling after puberty.

18.7 Therapy and Prevention

Because cherubism is usually self-limiting after puberty, surgical treatment may not be necessary. In most cases, longitudinal observation and follow-up is the initial management. Surgical intervention with curettage, contouring, or resection may be performed for functional, aesthetic, or cosmetic reasons. Surgical procedures are usually performed after the disease becomes quiescent. Aggressive lesions that cause severe functional problems such as airway obstruction justify early surgical intervention (Papadaki et al. 2012). Nonsurgical treatments have been tried in a few instances, but outcomes are not consistent. It is expected that translational studies with mouse models for cherubism may lead to therapeutic options that could either prevent or ameliorate cherubism symptoms.

18.7.1 Calcitonin Administration

Calcitonin is a peptide hormone produced in the thyroid. This hormone is known to inhibit osteoclast function by binding to its cellular receptor. Successful treatment of central giant cell granuloma with calcitonin (Harris 1993) and in vitro inhibition of bone resorption of cells harvested from cherubism lesions with calcitonin (Southgate et al. 1998) suggested effectiveness of calcitonin administration for cherubism treatment. Three cases of cherubism patients treated with calcitonin have been reported. In an 11-year-old boy with rapidly growing lesions in the mandible and maxilla, regression of lesions and restoration of normal contour of the mandible were seen after daily administration of salmon calcitonin by nasal spray for 15 months (De Lange et al. 2007). In a 14-year-old boy who had intraosseous cherubism lesions in the mandible, the lesions regressed after 30 months of salmon calcitonin therapy with nose spray every other day (Etoz et al. 2011). However, a 7-year-old boy with significant mandibular enlargement and facial deformity due to presumed cherubism failed to show clinical or radiological improvement after calcitonin treatment for 6 months (Lannon and Earley 2001). It could be argued that the regression of cherubism lesions after calcitonin treatment may in part be due to the natural course of cherubism. Therefore, further studies are required to assess the efficacy of calcitonin in the treatment of cherubism before it can be used as a general therapy.

18.7.2 Neutralization of TNF-α

Cherubism mice lacking TNF-α do not exhibit inflammation and bone loss (Ueki et al. 2007), suggesting that TNF-α is a key inflammatory mediator in the cherubism mouse model. Administration of the TNF-α blocker, etanercept, to neonatal cherubism mice prevented the cherubism phenotypes of facial swelling and jawbone destruction (Yoshitaka et al. 2014a). In contrast, administration of the same TNF-α blocker to mice that already had developed cherubism phenotypes failed to improve the inflammation and bone destruction. The results suggest that anti-TNF-α therapy may be effective in young cherubism patients, if treated before the inflammatory phase or bone resorption occurs. Consistent with the results, administration of a TNF-α antagonist adalimumab to two 4-year-old cherubism patients for approximately 2.5 years did not result in lesion regression or prevent lesion expansion in active cherubism. However, histologically the treatment resulted in a significant reduction in the numbers of multinucleated giant cells and TNF-α-positive cells in both cases (Hero et al. 2013). Therefore, early genetic mutation screening in *SH3BP2* followed by early treatment with TNF-α antagonists might prevent or ameliorate the development of cherubism lesions.

18.7.3 Blocking of NFAT Transcription Factors

Tacrolimus (also known as FK-506) is an immunosuppressive drug that is commonly used to lower the risk for organ rejection after transplantation and treatment of various immune diseases. This drug is a calcineurin inhibitor and suppresses immune cell activation and osteoclast formation by inhibiting NFAT-mediated transcriptional activation. Recently, it was reported that tacrolimus enhanced bone formation in a 4-year-old boy with aggressive cherubism by inhibiting osteoclastogenesis and stimulating osteogenesis (Kadlub et al. 2015). After tacrolimus therapy for 1 year, the patient showed significant clinical improvement, including stabilization of jaw size and intraosseous osteogenesis. Immunohistological analysis of granulomatous lesion showed that tacrolimus caused a significant reduction in the number of TRAP-positive osteoclasts and NFATc1 nuclear staining in the cells. Clinical and functional improvement associated with these biological results suggests that blocking of NFAT-mediated transcriptional activation by tacrolimus administration might prevent the need for surgical removal of growing lesions and recurring maxillofacial surgeries in severe cherubism cases, but might not be effective in all patients. Further clinical trials will be necessary to confirm the efficacy and safety of tacrolimus for mild and severe cherubism.

18.7.4 Bone Marrow Transplantation

Because cherubism phenotypes can be transferred through fetal liver cells in mice (Ueki et al. 2007), effectiveness of transplantation of wild-type bone marrow cells to cherubism mice was examined. Transplantation of bone marrow cells from wild-type mice to 6-week-old cherubism mice with developing inflammation as well as to 10-week-old cherubism mice with established inflammation improved facial swelling and jawbone destruction. Elevation of serum TNF-α levels was not detected after wild-type bone marrow transplantation in cherubism mice (Yoshitaka et al. 2015). These studies suggest that bone marrow transplantation might potentially be effective for cherubism patients who are developing cherubism or are already in the active phase.

18.7.5 Inhibition of SYK

Cherubism mice deficient in SYK in myeloid cells do not exhibit inflammation (Fig. 18.8). This result suggests that SYK is a key mediator for macrophage inflammation in cherubism (Yoshitaka et al. 2014b). Several SYK inhibitors have been developed and are currently in clinical trial. Indeed, administration of a novel SYK inhibitor entospletinib improves inflammatory bone destruction in the mouse model of cherubism (Yoshimoto et al. 2018). Effects of SYK inhibitors need to be confirmed by a controlled clinical trial to determine the efficacy and safety in cherubism patients.

In summary, early genetic diagnosis before disease onset will be an important factor for individuals born with cherubism to receive preventive care. It is anticipated that postnatal genetic testing as part of personalized and precision medicine approaches will help to identify SH3BP2 mutations early and that further clinical studies based on translational studies with animal models described here will lead to novel pharmacological therapies that can ameliorate cherubism symptoms or even prevent the development of lesions.

End-of-Chapter Questions

1. When you see an individual who has bilateral jaw swelling that is reminiscent of cherubism, how do you diagnose cherubism histologically and genetically?
2. What will be the difference in treatment strategy between sporadic cherubism patients who already have active lesions and affected newborns with a SH3BP2 mutation born to a cherubism family?
3. Which features in the orofacial region are likely responsible for the jaw-dominant manifestation of cherubism symptoms?
4. How could meticulous oral hygiene potentially impact cherubism?
5. What age-related changes in the orofacial region presumably explain the typical regression of cherubism lesions after puberty?

References

Aliprantis AO, Ueki Y, Sulyanto R, Park A, Sigrist KS, Sharma SM, Ostrowski MC, Olsen BR, Glimcher LH (2008) NFATc1 in mice represses osteoprotegerin during osteoclastogenesis and dissociates systemic osteopenia from inflammation in cherubism. J Clin Invest 118:3775–3789

Bell SM, Shaw M, Jou YS, Myers RM, Knowles MA (1997) Identification and characterization of the human homologue of SH3BP2, an SH3 binding domain protein within a common region of deletion at 4p16.3 involved in bladder cancer. Genomics 44:163–170

De Lange J, Van Den Akker HP, Scholtemeijer M (2007) Cherubism treated with calcitonin: report of a case. J Oral Maxillofac Surg 65:1665–1667

Etoz OA, Dolanmaz D, Gunhan O (2011) Treatment of cherubism with salmon calcitonin: a case report. Eur J Dent 5:486–491

Guettler S, Larose J, Petsalaki E, Gish G, Scotter A, Pawson T, Rottapel R, Sicheri F (2011) Structural basis and sequence rules for substrate recognition by Tankyrase explain the basis for cherubism disease. Cell 147:1340–1354

Harris M (1993) Central giant cell granulomas of the jaws regress with calcitonin therapy. Br J Oral Maxillofac Surg 31:89–94

Hero M, Suomalainen A, Hagstrom J, Stoor P, Kontio R, Alapulli H, Arte S, Toiviainen-Salo S, Lahdenne P, Makitie O (2013) Anti-tumor necrosis factor treatment in cherubism--clinical, radiological and histological findings in two children. Bone 52:347–353

Imai Y, Kanno K, Moriya T, Kayano S, Seino H, Matsubara Y, Yamada A (2003) A missense mutation in the SH3BP2 gene on chromosome 4p16.3 found in a case of nonfamilial cherubism. Cleft Palate Craniofac J 40:632–638

Kadlub N, Vazquez MP, Galmiche L, L'hermine AC, Dainese L, Ulinski T, Fauroux B, Pavlov I, Badoual C, Marlin S, Deckert M, Leboulanger N, Berdal A, Descroix V, Picard A, Coudert AE (2015) The calcineurin inhibitor tacrolimus as a new therapy in severe cherubism. J Bone Miner Res 30:878–885

Lannon DA, Earley MJ (2001) Cherubism and its charlatans. Br J Plast Surg 54:708–711

Levaot N, Voytyuk O, Dimitriou I, Sircoulomb F, Chandrakumar A, Deckert M, Krzyzanowski PM, Scotter A, Gu S, Janmohamed S, Cong F, Simoncic PD, Ueki Y, La Rose J, Rottapel R (2011) Loss of Tankyrase-mediated destruction of 3BP2 is the underlying pathogenic mechanism of cherubism. Cell 147:1324–1339

Papadaki ME, Lietman SA, Levine MA, Olsen BR, Kaban LB, Reichenberger EJ (2012) Cherubism: best clinical practice. Orphanet J Rare Dis 7(Suppl 1):S6

Reichenberger EJ, Levine MA, Olsen BR, Papadaki ME, Lietman SA (2012) The role of SH3BP2 in the pathophysiology of cherubism. Orphanet J Rare Dis 7(Suppl 1):S5

Southgate J, Sarma U, Townend JV, Barron J, Flanagan AM (1998) Study of the cell biology and biochemistry of cherubism. J Clin Pathol 51:831–837

Ueki Y, Tiziani V, Santanna C, Fukai N, Maulik C, Garfinkle J, Ninomiya C, Doamaral C, Peters H, Habal M, Rhee-Morris L, Doss JB, Kreiborg S, Olsen BR, Reichenberger E (2001) Mutations in the gene encoding c-Abl-binding protein SH3BP2 cause cherubism. Nat Genet 28:125–126

Ueki Y, Lin CY, Senoo M, Ebihara T, Agata N, Onji M, Saheki Y, Kawai T, Mukherjee PM, Reichenberger E, Olsen BR (2007) Increased myeloid cell responses to M-CSF and RANKL cause bone loss and inflammation in SH3BP2 "cherubism" mice. Cell 128:71–83

Wynn TA, Vannella KM (2016) Macrophages in tissue repair, regeneration, and fibrosis. Immunity 44:450–462
Yoshimoto T, Hayashi T, Kondo T, Kittaka M, Reichenberger EJ, Ueki Y (2018) Second-generation SYK inhibitor entospletinib ameliorates fully established inflammation and bone destruction in the cherubism mouse model. J Bone Miner Res 33(8):1513–1519
Yoshitaka T, Ishida S, Mukai T, Kittaka M, Reichenberger EJ, Ueki Y (2014a) Etanercept administration to neonatal SH3BP2 knock-in cherubism mice prevents TNF-alpha-induced inflammation and bone loss. J Bone Miner Res 29:1170–1182
Yoshitaka T, Mukai T, Kittaka M, Alford LM, Masrani S, Ishida S, Yamaguchi K, Yamada M, Mizuno N, Olsen BR, Reichenberger EJ, Ueki Y (2014b) Enhanced TLR-MYD88 signaling stimulates autoinflammation in SH3BP2 cherubism mice and defines the etiology of cherubism. Cell Rep 8:1752–1766
Yoshitaka T, Kittaka M, Ishida S, Mizuno N, Mukai T, Ueki Y (2015) Bone marrow transplantation improves autoinflammation and inflammatory bone loss in SH3BP2 knock-in cherubism mice. Bone 71:201–209

19 Cancer and Excess Iron

Shinya Toyokuni

Keywords
Iron · Asbestos · Mesothelioma · Oxygen · Reactive oxygen species (ROS)

19.1 Case Report

A 62-year-old male presented transient loss of consciousness during dialysis and was transferred to the emergency room. Chest X-ray and computed tomography revealed prominent right pleural effusion with large tumor mass surrounding the contour of the right lung. The patient had been involved in demolition work of old buildings and water pipes for more than 25 years, where he was often exposed to asbestos fibers in the air. The patient suffered from diabetes mellitus for 18 years and started dialysis three times a week since 7 years ago due to diabetic nephropathy.

The patient was diagnosed as probable malignant mesothelioma at an advanced stage, based on needle biopsy of the lung tumor, and thereafter received chemotherapy. However, the chemotherapy was not very effective, and the patient died 3 months later due to pulmonary and renal failure. An autopsy was performed.

S. Toyokuni (✉)
Department of Pathology and Biological Responses, Nagoya University Graduate School of Medicine, Nagoya, Japan
e-mail: toyokuni@med.nagoya-u.ac.jp

19.2 Diagnosis

In general, diagnosis of cancer is based on histological examination of formalin-fixed paraffin-embedded sections under a microscope. Cancer is a word, representing malignant neoplasm, and is classified according to the differentiation of the cancer cells. If the cancer cells are differentiated toward glands, the cancer is called "adenocarcinoma," and if the cancer cells are differentiated toward stratified squamous cells, such as epidermis, the cancer is called "squamous cell carcinoma." Here the cancer cells were differentiated into mesothelial cells, so this cancer is called as "malignant mesothelioma." Mesothelial cells are flat one-layer lining cells of somatic cavities. The humans have three somatic cavities in the body, namely, pleural cavity, peritoneal cavity, and pericardial cavity, in which the lung, abdominal organs (stomach, small and large intestine, spleen, liver, etc.), and heart are accommodated, respectively. One of the major functions of mesothelial cells is to lubricate the somatic cavities for the movement of organs inside by secreting hyaluronic acid.

Autopsy examination revealed that the tumor was located at the periphery of the lung, thus consistent with malignant mesothelioma (Fig. 19.1). The histology of the tumor showed proliferation of atypical polygonal cells with a sheet or glandular structure (Fig. 19.2). Immunohistochemical analyses by monoclonal antibodies of these

T. Oohashi et al. (eds.), *Human Pathobiochemistry*, https://doi.org/10.1007/978-981-13-2977-7_19

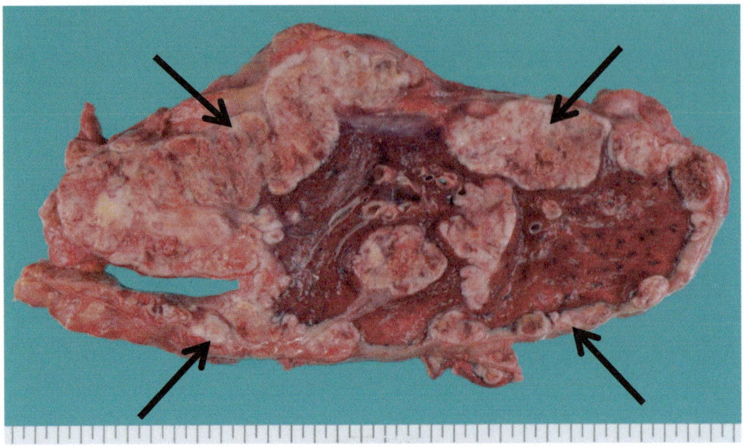

Fig. 19.1 Macroscopic appearance of pleural malignant mesothelioma in a human autopsied case. Note that whitish mass (indicated by arrows) surrounding the lung tissue is the invading malignant mesothelioma

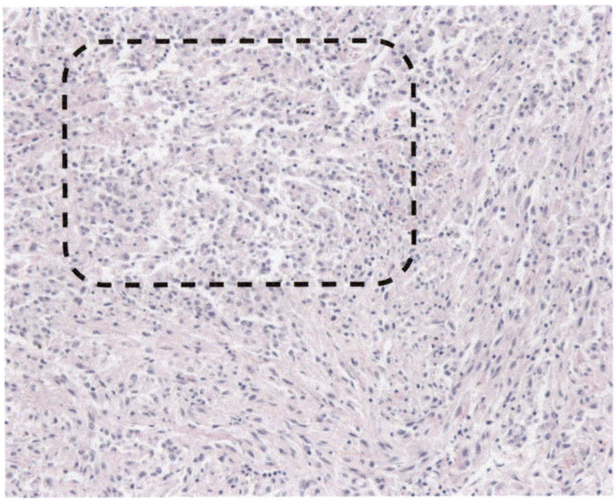

Fig. 19.2 Microscopic appearance of pleural malignant mesothelioma. Atypical mesothelial cells are proliferating with a sheet or glandular (areas surrounded by dotted line) structure (hematoxylin and eosin staining)

Table 19.1 Classification of carcinogenic agents

Habits	Smoking, alcohol
Living environments	UV, radiation, arsenic, PM2.5
Work environments	Asbestos, benzene, dichloromethane
Diet	High fat, low fiber
Infection	HTLV-1, hepatitis C virus, papilloma virus, helicobacter pylori
Chronic inflammation	Autoimmune diseases (Hashimoto thyroiditis, Crohn's disease, ulcerative colitis)
Hereditary (genetic)	Allelic loss/inactivation of tumor suppressor genes (*p53, VHL, MUTHY, BRCA*)

The items and carcinogenic agents are representative. *UV* ultraviolet, *PM* particulate matter, *HTLV* human T-cell leukemia virus

tumor cells were positive for cytokeratin (clones, AE1/3 and CAM5.2), calretinin, and podoplanin (clone D2–40) and negative for TTF-1 (a molecular marker of adenocarcinoma of lung origin). These results were consistent with mesothelial differentiation, and the diagnosis of malignant mesothelioma was established.

19.3 Biochemical and Molecular Perspectives

Cancer is one of the top causes of death in virtually all of the developed countries (Siegel et al. 2015) after the conquest of major infectious diseases, such as tuberculosis (Heesterbeek et al. 2015). Cancer patients are increasing now partially due to prolonged average lifetime (80.5 years for males and 86.8 years for females in Japan in 2016). Table 19.1 summarizes the classification of carcinogenic agents, most of which are associated with oxidative stress.

Oxidative stress is an imbalance between oxidants and antioxidants in favor of the oxidants, leading to a disruption of redox signaling and control and/or molecular damage. We cannot live without oxygen even for 5 min, but we do not obtain energy from oxygen. We obtain energy through diet. Oxygen works as a medium for electron flow. We use the characteristics of oxygen to be reduced four times independently to

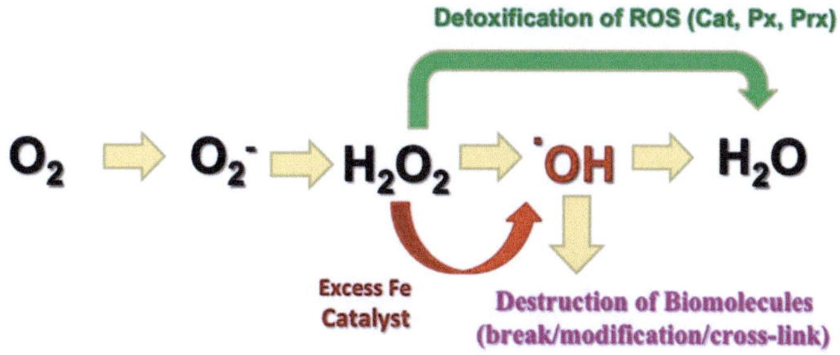

Fig. 19.3 Role of oxygen in the biological system. *Cat* catalase, *Px* peroxidase, *Prx* peroxiredoxin, *ROS* reactive oxygen species

Table 19.2 Association of excess iron with carcinogenesis in humans

Hereditary hemochromatosis	Liver	Hepatocellular carcinoma
Viral hepatitis B/C	Liver	Hepatocellular carcinoma
Ovarian endometriosis	Ovary	Adenocarcinoma
Exposure to asbestos	Somatic cavity (mesothelial cell), lung	Malignant mesothelioma, lung cancer

water (Fig. 19.3). During these reactions, reactive oxygen species (ROS) are generated, which account for a few % of oxygen consumed (Toyokuni 1999). Here superoxide (O_2^-) and hydrogen peroxide (H_2O_2) themselves are not so reactive but are used as signaling molecules in the cell. However, hydroxyl radical (·OH) is the most reactive chemical species in the biological system and damages all sorts of biomolecules nearby. If ·OH attacks DNA, several different kinds of DNA damages, strand breaks, base modifications, etc. are induced, which may lead to mutations (alteration of genome information). Generally, five or six cumulative mutations at oncogene or tumor suppressor genes in a single cell are necessary for carcinogenesis (Weinberg 2013). Importantly, a generation of ·OH is a chemical reaction catalyzed by Fe(II) and is called Fenton reaction: $Fe(II) + H_2O_2 \rightarrow Fe(III) + {\cdot}OH + OH^-$ (Toyokuni 1996).

Iron is the most abundant transition metal in our body, and adult male humans contain approximately 4 grams of iron. Sixty percentage of them are present in red blood cells as heme in hemoglobin. Usually catalytic or free iron scarcely exists in or outside of the cell except for lysosomes, because of efficient clearing systems such as serum transferrin (iron transport) and cytosolic ferritin (iron storage) proteins. However, excess iron produces catalytic iron in vivo and is significantly associated with carcinogenesis (Toyokuni 2009b). Table 19.2 summarizes excess iron-associated carcinogenesis in humans, supported by epidemiological data (Toyokuni 2016).

Asbestos is a natural fibrous mineral, found in asbestos mines (Fig. 19.4). Because it is a stone, it is resistant to heat, acid, and friction, so it has been used in a huge amount worldwide in the last century. It is a fiber of micrometer-order length and several hundred nanometer-order diameter, so it can easily fly in the air and miners and workers are exposed at the respiratory tract (IARC 2012; Oury et al. 2014). Asbestos was used even for the wall surface of houses in the 1970s to be preventive against fire, and the general population was also exposed. Epidemiologists noticed that asbestos is responsible for malignant mesothelioma in the 1970s, but it took for a while to ban asbestos legally. For example, Japan banned all kinds of asbestos in 2006 and Canada in 2012. There are two different types of asbestos: serpentine (chrysotile, white asbestos) and amphibole (crocidolite, blue asbestos; amosite, brown asbestos). Amphiboles show more rigidity, and ~30% of its content is iron (IARC 2012; Oury et al. 2014). However, it has become clear that white

Fig. 19.4 Standard asbestos fibers distributed for experiments by UICC (Union for International Cancer Control)

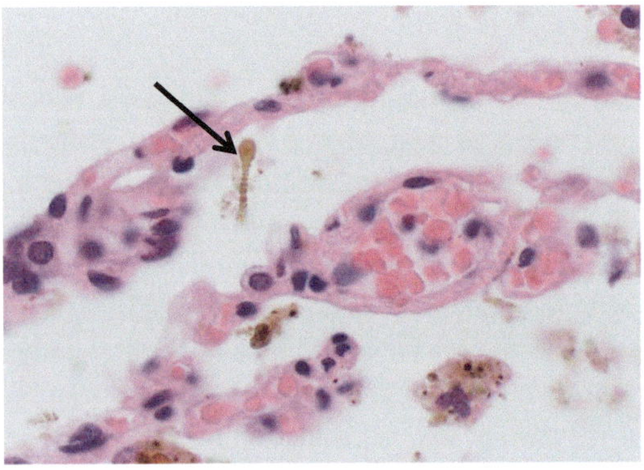

Fig. 19.5 Asbestos body (arrow) formed in a human lung (hematoxylin and eosin staining). Note the dumbbell-like appearance and golden brown color of asbestos body, which accumulated iron as hemosiderin

asbestos (chrysotile) which contains no iron in its content is also potently carcinogenic (Jiang et al. 2012). It is well known that there is a long incubation period of 30–40 years for the induction of malignant mesothelioma after asbestos exposure (IARC 2012; Oury et al. 2014).

Now we know what is happening, based on the data of animal experiments. Very long incubation period is the time required for the fibers to pass through the lung tissue into the pleural cavity. Because asbestos is foreign to our body, alveolar macrophages engulf asbestos fibers as soon as they find the fibers and try to digest them or to carry them to regional lymph nodes. However, neither of them is possible for the macrophages, and they die after engulfing asbestos fibers because they are sharp and long enough to break the plasma membrane of macrophages. Due to the negative pressure of the pleural cavity, the asbestos fibers go peripherally over years as they kill the macrophages and other cells (Toyokuni 2009a; Chew and Toyokuni 2015).

It is now known that all the asbestos fibers have affinity to hemoglobin and histones (Nagai et al. 2011a). During the journey of asbestos fiber in the lung to the pleural cavity, they collect iron from hemoglobin and other proteins on the surface. Those are famous hallmark of asbestos exposure called asbestos body, which is asbestos fiber coated with massive iron (Fig. 19.5). Mesothelial cells have phagocytic activity in continuation with lymphatic vessels. Finally, those iron-coated asbestos fibers are taken up by the parietal mesothelial cells, and fibers tangle with chromosomal structure of mesothelial cells, which are thought to induce mutations in the mesothelial cells (Toyokuni 2009a).

There are characteristic mutational patterns in malignant mesothelioma: homozygous deletion of $p16^{INK4A}$ tumor suppressor gene, inactivation of Hippo pathway including *NF2* tumor suppressor gene, and inactivation of *BAP1* tumor suppressor gene (Chew and Toyokuni 2015). Malignant mesothelioma is divided into epithelioid, sarcomatoid, and biphasic (both epithelioid and sarcomatoid) subtypes, where sarcomatoid subtype reveals the worst prognosis and ~100% incidence of homozygous deletion of $p16^{INK4A}$ tumor suppressor gene (Oury et al. 2014). This fact is used

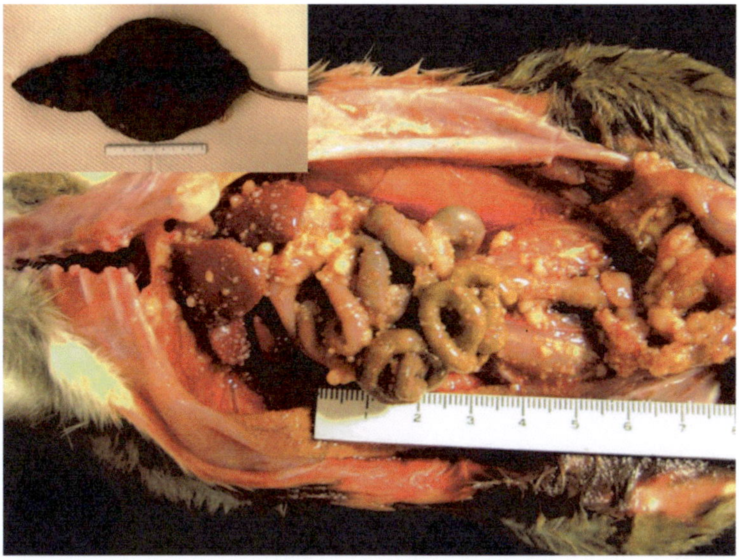

Fig. 19.6 Peritoneal mesothelioma induced by intraperitoneal administration of chrysotile (white asbestos; Jiang et al. 2012)

for the diagnosis of malignant mesothelioma with fluorescent in situ hybridization (FISH) analysis (Hwang et al. 2016), and the same mutation occurs in a rat model by intraperitoneal administration of asbestos fibers (Fig. 19.6) (Jiang et al. 2012). Of note, if mesothelial cells are directly exposed to asbestos fibers, only 1.5 years is necessary to induce malignant mesothelioma in virtually all the rats. There was some warning from the scientists that some carbon nanotubes (synthetic nanomaterial) but not all of them have the same capacity for mesothelial carcinogenesis as asbestos (Nagai et al. 2011b). Therefore, ample care has to be taken for the workers dealing with those nanotubes.

19.4 Prevention

What happens if Fenton reaction is repeated many times in an organ *in vivo*? Dr. Shigeru Okada and Prof. Osamu Midorikawa found the answer in 1982 with serendipity (Ebina et al. 1986; Toyokuni 2016). With repeated intraperitoneal injections of an iron chelate, ferric nitrilotriacetate, Fenton reaction was unexpectedly induced in the renal proximal tubules, which eventually led to a high incidence of adenocarcinoma (Fig. 19.7), originating from renal proximal tubules (Toyokuni 2016). This is distinct in that (1) most of the tumors induced by this way were highly aggressive despite the fact that only wild-type animals were used and (2) genomic alterations induced here were very similar to those in human cancers and showed a high incidence of homozygous deletion of $p16^{INK4A}$ tumor suppressor gene, suggesting that this is a footmark of excess iron-induced carcinogenesis (Akatsuka et al. 2012).

Iron has been a very precious metal for all the lives on earth. Independent life on earth should have been born with the reaction with iron. Therefore, we do not have any active pathway to excrete iron once it is absorbed through duodenal villous glandular cells into the blood (Toyokuni et al. 2017). Therefore, the only way to remove iron from our body would be phlebotomy or blood donation. Phlebotomy has been used as an official therapy in Japan for chronic active hepatitis by hepatitis C virus when other therapies did not work and showed effects to decrease activity of hepatitis and an incidence of hepatocellular carcinoma (Kato et al. 2007).

Because major infectious diseases, such as tuberculosis, have been conquered during the latter half of the twentieth century, now the tear and wear by the use of iron and oxygen is an unavoidable process in our body, which eventually causes human mortality. Most probably, the major factor to cause mutations in our genome is iron and

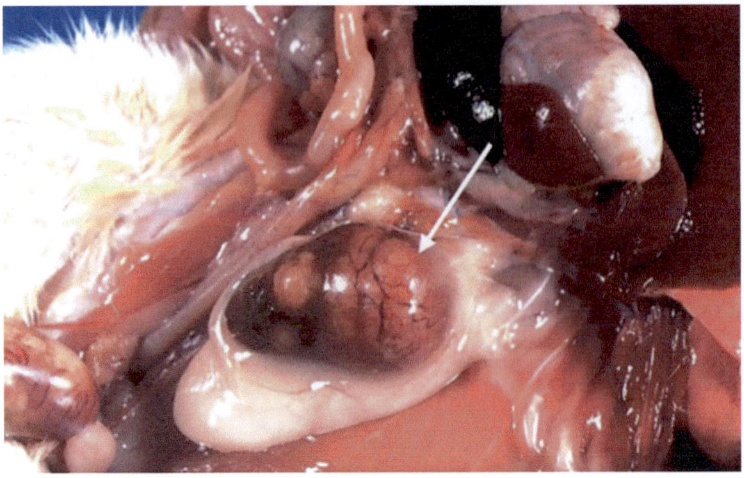

Fig. 19.7 Ferric nitrilotriacetate-induced renal cell carcinoma in a rat. Whitish mass (arrow) is the renal cell carcinoma in a rat kidney

oxygen, which is the major pathology of carcinogenesis with aging (Toyokuni 2011; Toyokuni et al. 2017). Recently, we successfully prolonged the survival and reduced the tumor mass of animals with phlebotomy in crocidolite-induced mesothelial carcinogenesis as a preclinical experiment (Ohara 2018; Toyokuni 2018).

Questions

1. Why is iron excess associated with carcinogenesis?
2. Why is asbestos fiber carcinogenic to humans?
3. Give another example of iron-induced carcinogenesis in humans, and explain the mechanism.

References

Akatsuka S, Yamashita Y, Ohara H, Liu YT, Izumiya M, Abe K, Ochiai M, Jiang L, Nagai H, Okazaki Y, Murakami H, Sekido Y, Arai E, Kanai Y, Hino O, Takahashi T, Nakagama H, Toyokuni S (2012) Fenton reaction induced cancer in wild type rats recapitulates genomic alterations observed in human cancer. PLoS One 7:e43403

Chew SH, Toyokuni S (2015) Malignant mesothelioma as an oxidative stress-induced cancer: an update. Free Radic Biol Med 86:166–178

Ebina Y, Okada S, Hamazaki S, Ogino F, Li JL, Midorikawa O (1986) Nephrotoxicity and renal cell carcinoma after use of iron- and aluminum- nitrilotriacetate complexes in rats. J Natl Cancer Inst 76:107–113

Heesterbeek H, Anderson RM, Andreasen V, Bansal S, De Angelis D, Dye C, Eames KT, Edmunds WJ, Frost SD, Funk S (2015) Modeling infectious disease dynamics in the complex landscape of global health. Science 347:aaa4339

Hwang HC, Sheffield BS, Rodriguez S, Thompson K, Christopher HT, Gown AM, Churg A (2016) Utility of BAP1 immunohistochemistry and p16 (CDKN2A) FISH in the diagnosis of malignant mesothelioma in effusion cytology specimens. Am J Surg Pathol 40:120–126

IARC, WHO 2012. Asbestos (chrysotile, amosite, crocidolite, tremolite, actinolite, and anthophyllite). IARC monographs on the evaluation of carcinogenic risks to humans a review of human carcinogens; part C: arsenic, metals, fibres, and dusts. Lyon, France

Jiang L, Akatsuka S, Nagai H, Chew SH, Ohara H, Okazaki Y, Yamashita Y, Yoshikawa Y, Yasui H, Ikuta K, Sasaki K, Kohgo Y, Hirano S, Shinohara Y, Kohyama N, Takahashi T, Toyokuni S (2012) Iron overload signature in chrysotile-induced malignant mesothelioma. J Pathol 228:366–377

Kato J, Miyanishi K, Kobune M, Nakamura T, Takada K, Takimoto R, Kawano Y, Takahashi S, Takahashi M, Sato Y, Takayama T, Niitsu Y (2007) Long-term phlebotomy with low-iron diet therapy lowers risk of development of hepatocellular carcinoma from chronic hepatitis C. J Gastroenterol 42:830–836

Nagai H, Ishihara T, Lee WH, Ohara H, Okazaki Y, Okawa K, Toyokuni S (2011a) Asbestos surface provides a niche for oxidative modification. Cancer Sci 102:2118–2125

Nagai H, Okazaki Y, Chew S, Misawa N, Yamashita Y, Akatsuka S, Yamashita K, Ishihara T, Yoshikawa Y, Jiang L, Ohara H, Takahashi T, Ichihara G, Kostarelos

K, Miyata Y, Shinohara H, Toyokuni S (2011b) Diameter of multi-walled carbon nanotubes is a critical factor in mesothelial injury and subsequent carcinogenesis. Proc Natl Acad Sci U S A 108:E1330–E1338

Ohara Y, Chew S-H, Shibata T, Okazaki Y, Yamashita K, Toyokuni S (2018) Phlebotomy as a preventive measure for crocidolite-induced mesothelioma in male rats. Cancer Sci 109(2):330–339

Oury TD, Sporn TA, Roggli VL (2014) Pathology of asbestos-associated diseases, 3rd edn. Springer-Verlag Berlin Heiderberg, New York

Siegel RL, Miller KD, Jemal A (2015) Cancer statistics, 2015. CA Cancer J Clin 65:5–29

Toyokuni S (1996) Iron-induced carcinogenesis: the role of redox regulation. Free Radic Biol Med 20:553–566

Toyokuni S (1999) Reactive oxygen species-induced molecular damage and its applicaton in pathology. Pathol Int 49:91–102

Toyokuni S (2009a) Mechanisms of asbestos-induced carcinogenesis. Nagoya J Med Sci 71:1–10

Toyokuni S (2009b) Role of iron in carcinogenesis: cancer as a ferrotoxic disease. Cancer Sci 100:9–16

Toyokuni S (2011) Iron as a target of chemoprevention for longevity in humans. Free Radic Res 45:906–917

Toyokuni S (2016) The origin and future of oxidative stress pathology: From the recognition of carcinogenesis as an iron addiction with ferroptosis-resistance to non-thermal plasma therapy. Pathol Int 66:245–259

Toyokuni S (2018) Iron addiction with ferroptosis-resistance in asbestos-induced mesothelial carcinogenesis: toward the era of mesothelioma prevention. Free Radic Biol Med. https://doi.org/10.1016/j.freeradbiomed.2018.10.401

Toyokuni S, Ito F, Yamashita K, Okazaki Y, Akatsuka S (2017) Iron and thiol redox signaling in cancer: an exquisite balance to escape ferroptosis. Free Radic Biol Med 108:610–626

Weinberg R (2013) The biology of cancer. Garland Science, New York

Fukuyama Congenital Muscular Dystrophy and Related Diseases

20

Motoi Kanagawa, Hideki Tokuoka, and Tatsushi Toda

Keywords
Muscular dystrophy · Glycosylation · Dystroglycanopathy · Fukutin · Ribitol phosphate

20.1 Case Report

A 1-year-old Japanese female patient with motor retardation was brought to our outpatient clinic by her mother. The child was found to have high serum creatine kinase (CK) level according to the 1-year-old medical checkup and was recommended further evaluation. The girl, the second child in her family, was born at full term in a local hospital and, after birth, was transferred to a general hospital because of muscle hypotonus, poor eye opening and sucking, and a high serum CK level (5461 IU/L: normal ranges are between 40 and 310 IU/L). She was followed up because her doctors considered these symptoms to arise from neonatal asphyxia. The infant developed quickly after 2 weeks, and her serum CK level decreased to normal. She grew up slightly slowly and acquired head control at 5 months of age (commonly acquired at 3–4 months after birth); she could not roll over at 7 months and rehabilitation was started. At 1 year of age, her muscle tonus remained decreased, and her serum CK level was elevated again (7532 IU/L). She could not stand up without support.

Physical examination revealed that the child could not fully raise her arms or sit for >10 s. She presented general hypotonia and depressed deep tendon reflex and joint contracture of both ankles, although her hip and knee joints still retained full range of movement. Laboratory studies revealed high levels of serum CK, lactate dehydrogenase, aspartate aminotransferase, and alanine aminotransferase. However, normal results were obtained in other laboratory tests, such as complete blood count and analyses of renal function and electrolytes. Results of Giemsa banding analysis, used for detecting chromosomal abnormality, were also normal. Brain MRI showed symmetrical unmyelinated changes in frontal white matter in T2-weighted images, and pachygyria was observed predominantly in frontotemporal lobes (Fig. 20.1). To identify the underlying genetic cause, commercially available genetic testing was performed.

The child was brought to our hospital every 6 months after definitive diagnosis. At age 18 months, she began uttering a few words without meaning, such as "Dah." She could also roll

M. Kanagawa (✉) · H. Tokuoka
Division of Neurology/Molecular Brain Science, Kobe University Graduate School of Medicine, Kobe, Hyogo, Japan
e-mail: kanagawa@med.kobe-u.ac.jp

T. Toda
Department of Neurology, Division of Neuroscience, Graduate School of Medicine, The University of Tokyo, Tokyo, Japan

T. Oohashi et al. (eds.), *Human Pathobiochemistry*, https://doi.org/10.1007/978-981-13-2977-7_20

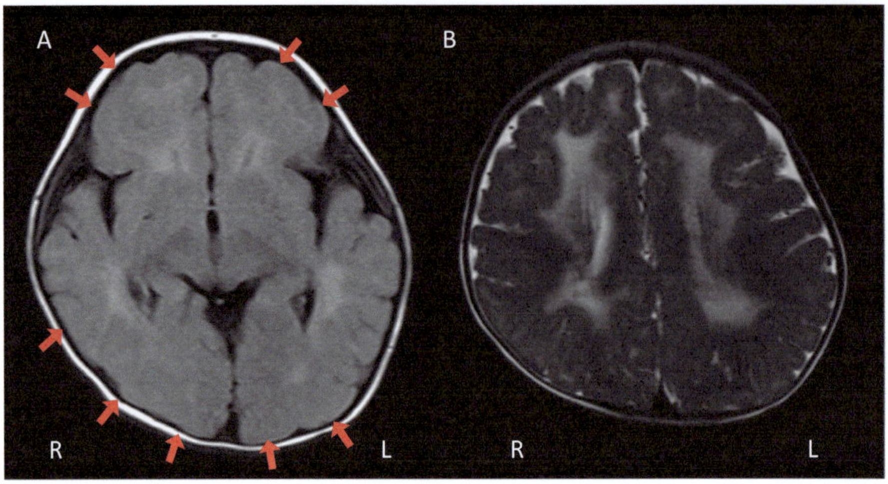

Fig. 20.1 (**a**) FLAIR image showing pachygyria (red arrows) in frontotemporal and occipital cortices. (**b**) T2-weighted image showing symmetrical high intensity in the white matter

over by herself and raise her arms fully. At age 2 years, she exhibited febrile seizures lasting several minutes when she was infected with roseola infantum. However, electroencephalography results were normal, and her seizures were considered typical febrile seizures that were unrelated to the disease. Echocardiography results were also normal in her 2-year-old medical checkup. At age 3 years, the child could speak two-word sentences such as "Where mom?;" such two-word sentences are commonly spoken starting at 2 years of age. She could move on her knees, but could not stand or walk unassisted. Overall, she grew up slower than normal infants in terms of motor function and intelligence and continuously presented elevated levels of serum CK.

20.2 Diagnosis

20.2.1 Approach to Diagnosis

Congenital muscle weakness is caused by diverse neuromuscular and connective tissue diseases (Table 20.1). The most critical consideration for diagnosis is the determination of the neuroanatomical location of the deficit, based on various medical histories and examinations. Motor deficit

Table 20.1 Cause of congenital hypotonia (floppy infant)

Origin (localization)	Disease
Brain	Hypoxic-ischemic encephalopathy
	Intracranial hemorrhage
	Chromosomal abnormalities
	Down syndrome
	Prader-Willi syndrome
	Metabolic disorders
	Krabbe disease, Menkes disease
	Mitochondrial disease
	Benign congenital hypotonia
Spinal cord	Werdnig-Hoffmann disease (spinal muscular atrophy type I)
	Traumatic myelopathy
	Hypoxic-ischemic myelopathy
Peripheral nerve	Charcot-Marie-Tooth disease
	Congenital neuropathy
	Guillain-Barré syndrome
Neuromuscular junction	Congenital myasthenic syndromes
	Myasthenia gravis
Muscle	Congenital myopathy
	Nemaline myopathy
	Mitochondrial myopathy
	Metabolic myopathy
	Pompe disease
	Congenital muscular dystrophy
	Fukuyama-type congenital muscular dystrophy
	Congenital myotonic dystrophy
Connective tissue	Ehlers-Danlos syndrome

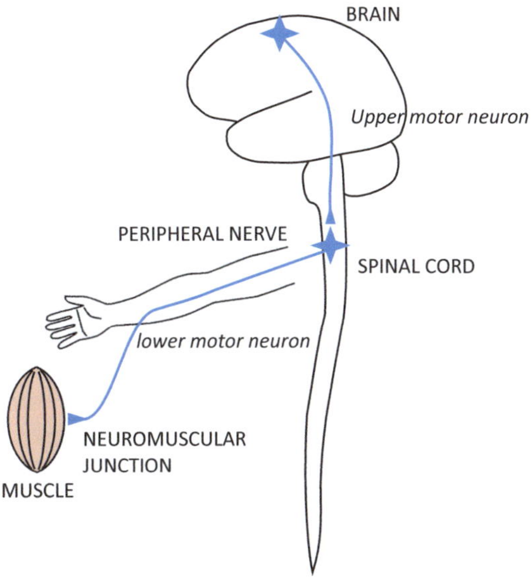

Fig. 20.2 Anatomical schema of the motor system and common cause of hypotonia

can be caused by diseases of the brain, spinal cord, peripheral nerve, neuromuscular junction, and muscle (Fig. 20.2). The most important examination here is for deep tendon reflexes: If the deep tendon reflex is depressed, the responsible lesion can be narrowed down to being below the spinal cord (peripheral nerve, neuromuscular junction, or muscle). For differentiating between the diseases in adults, the crucial analyses are electromyography (EMG) and nerve conduction study. However, EMG studies cannot be readily used for neonates and infants because they cannot move in response to the doctor's instructions. In such cases, marked elevation of CK levels suggests muscle diseases (myopathies) such as muscular dystrophies.

Myopathies can be divided into two major categories: inherited and acquired. Acquired myopathies are further classified by etiology into, for example, inflammatory myopathies, infectious myopathies, toxic myopathies, and myopathies associated with systemic diseases. Conversely, inherited myopathies are classified into muscular dystrophies, congenital myopathies, metabolic myopathies, and mitochondrial myopathies. Most cases of infant myopathies are inherited, particularly congenital myotonic dystrophy and congenital myopathies, and these myopathies cannot be differentiated through EMG studies and routine laboratory testing. Certain myopathies present characteristic symptoms, such as a high-arched palate in congenital myopathies, but these are not definitive symptoms. Genetic testing and/or muscle biopsies are frequently necessary for definitive diagnosis, and electroencephalography and neuroimaging are also useful for diagnosis. The recent, unprecedented progress made in molecular-genetic, ultrastructural, and biochemical studies on myopathies, especially muscular dystrophies, has revealed the mechanisms underlying these diseases and provided diagnostic tools. In the case reported here, a combination of motor deficits resulting from muscle involvement and brain malformation strongly suggested Fukuyama congenital muscular dystrophy (FCMD).

20.2.2 Epidemiology of FCMD

Muscular dystrophy represents a group of genetic disorders that cause progressive weakness and loss of muscle volume. FCMD is an autosomal recessive muscular dystrophy and is the most common childhood muscular dystrophy after Duchenne muscular dystrophy (DMD) in Japan. FCMD, which was first characterized by Dr. Yukio Fukuyama, is caused by mutation in the gene *fukutin* (*FKTN*). The disease incidence is ~3/100,000, and one person in approximately 90 is expected to be a heterozygous carrier. A few cases of muscular dystrophy patients harboring *fukutin* mutations have also been reported from countries other than Japan.

20.2.3 Physical Symptoms and Signs

FCMD is characterized by severe congenital muscular dystrophy (CMD), abnormal neuronal migration associated with mental retardation and epilepsy, and frequent eye abnormalities (Fukuyama et al. 1981). Patients with FCMD manifest muscle weakness and hypotonia by early infancy (floppy infant). In approximately half of the cases, poor sucking and weak crying are documented during the neonatal period. The infants also present motor development delay, which is why parents bring them to the clinic. Most FCMD patients can remain sitting by the age of 2 years, but they rarely show the ability to remain standing or walking unassisted. They also present pseudohypertrophy in the calves by early childhood. Joint contractures are not detected at birth, but most patients reveal hip, knee, and ankle contracture before they are 1 year old. Myopathic face resulting from facial muscle involvement gradually strengthens, and a tendency for the mouth to remain open persists from infancy onward. FCMD patients who live more than 10 years tend to develop slowly progressive myocardial involvements, including fibrosis of the myocardium, which finally result in cardiac failure.

Brain structural defects—such as brain malformations characterized by micropolygyria of the cerebrum and cerebellum, type II lissencephaly (cobblestone lissencephaly), and cerebellar cysts—are associated with FCMD. Mental retardation is also a characteristic of FCMD patients, whose IQ scores range between 30 and 50. Few patients speak only two-word sentences (such as "Mom come") throughout life. Seizures occur in nearly 70% of the cases, and eye abnormalities also occur, including myopia, cataract, abnormal eye movement, pale optic disc, and retinal detachment. Thus, FCMD is a genetic disorder that includes the involvement of the muscle, eye, and brain. Most patients do not develop the ability to work, and they die by the age of 20 years.

20.2.4 Laboratory Findings

Serum CK levels of FCMD patients are elevated >tenfold above normal, which means that muscle tissue is being destroyed in these patients. The levels of aspartate aminotransferase, alanine aminotransferase, and lactate dehydrogenase also increase together with marked CK elevation, because these enzymes are present in not only hepatocytes but also skeletal muscle tissue. The results of muscle imaging, such as MRI, ultrasound imaging, and CT scanning, frequently show signal abnormalities due to fatty infiltration, which follows the muscle destruction. These imaging studies also help in identifying the muscle that is optimal for biopsy, and muscle biopsies reveal myopathic changes suggesting muscular dystrophy, including necrotic and regenerating processes, fibrosis, central nuclei, and variation in fiber size. However, given the development of molecular-genetic testing, muscle biopsy is no longer necessary for FCMD diagnosis. Furthermore, brain MRI reveals brain malformations, which include polymicrogyria, pachygyria, and agyria of the cerebrum and cerebellum. T2-weighted high intensity in the white matter, absence of the corpus callosum, and hypoplasia of the pons are also observed. The major mutation in FCMD is a 3-kb SINE-VNTR-Alu (SVA) retrotransposal insertion into the 3′-UTR of *fukutin* in Japanese patients (described in detail in the following section) (Fig. 20.3). PCR-based molecular-genetic testing for the presence of this mutation is frequently performed for the molecular diagnosis of FCMD (Watanabe et al. 2005). The results of this genetic test revealed that our patient carried a homozygous retrotransposon insertion.

20.3 Biochemical and Molecular Perspectives

20.3.1 Identification of Fukutin, the Gene Responsible for FCMD

The causative gene for FCMD was identified through positional cloning by Toda and colleagues and named *fukutin* (Kobayashi et al. 1998). Fukutin, a 461-amino-acid protein, contains a hydrophobic transmembrane region followed by a stem region and a putative catalytic domain. Fukutin also contains a conserved DxD motif that is present in several glycosyltransferases. Bioinformatic analysis has revealed that fukutin shows sequence similarities to bacterial proteins involved in polysaccharide/phosphorylcholine modification and to a yeast protein

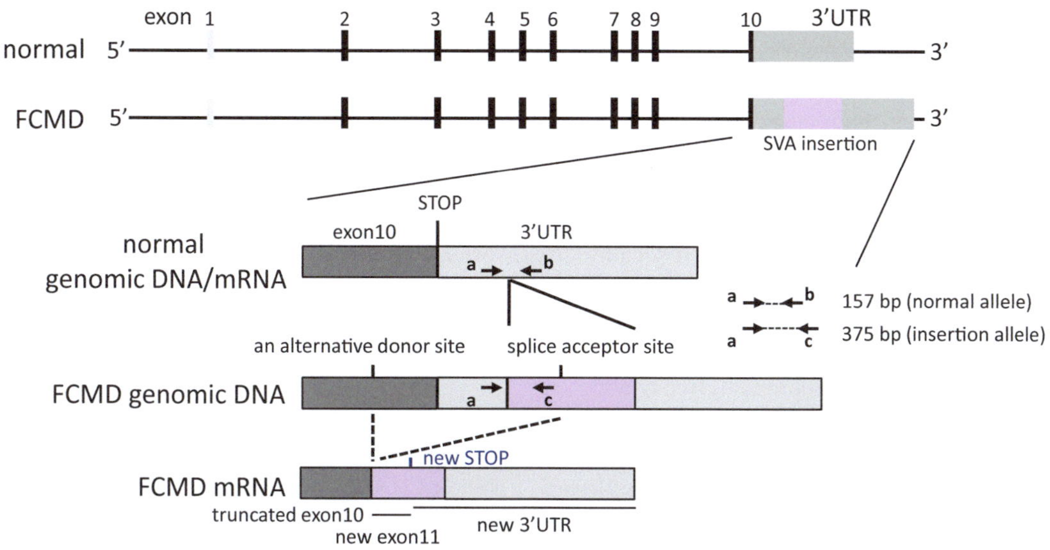

Fig. 20.3 Schematic representation of the gene *fukutin* and rapid PCR-based diagnosis of FCMD. Representation of genomic DNA and mRNA in FCMD. In FCMD diagnosis, PCR performed using three primers (a, b, and c) concurrently detects both the normal (157 bp) and insert-harboring (375 bp) alleles

involved in mannosyl phosphorylation of oligosaccharides; however, the activity of fukutin long remained unidentified.

The major mutation in FCMD was identified as a 3-kb SVA retrotransposon insertion in the 3′ noncoding region of *fukutin*, which accounts for ~87% of the FCMD chromosome (Kobayashi et al. 1998). FCMD is the first human disease identified to be caused by an ancient retrotransposal integration. The retrotransposal insertion leads to abnormal splicing through exon trapping, which results in aberrant mRNA splicing (Taniguchi-Ikeda et al. 2011). A rare alternative donor site in exon 10 is activated and trapped by an alternative acceptor site in the SVA retrotransposon, creating an additional and aberrant exonic sequence (exon 11); the abnormal splicing excises the authentic stop codon and produces another stop codon located in exon 11 (Fig. 20.3). The resulting product truncates the fukutin C-terminus and adds 129 amino acids encoded by the SVA retrotransposon.

According to genetic calculation, the FCMD founder retrotransposal mutation was introduced into the Japanese population 100 generations ago, i.e., the founder lived approximately 2000–2500 years ago. This was the period during which the Yayoi people migrated from Korea and China and intermingled with the indigenous Jomon population, generating the modern Japanese people. Other mutations in *fukutin*, such as frameshift, nonsense, and missense mutations, have been also reported. Patients harboring compound heterozygous SVA insertion and other pathological mutations frequently develop more severe or milder phenotypes than other FCMD patients depending on the deleterious effects of the mutations. Non-Japanese FCMD patients carry either homozygous or heterozygous mutations other than the founder SVA insertion.

20.3.2 Abnormal Glycosylation in FCMD

In 2001, Hayashi and colleagues reported that staining by the monoclonal antibody IIH6, which recognizes glycosylated α-dystroglycan (α-DG), is decreased in FCMD biopsies (Hayashi et al. 2001). Subsequently, the Campbell group reported similar abnormal glycosylation in patients with muscle-eye-brain disease (MEB) and Walker-Warburg syndrome (WWS) and showed that the abnormal glycosylation leads to a loss of the ligand-binding activity of α-DG (Michele et al. 2002). The abnormal glycosylation of α-DG was also later reported in patients with CMD and limb-girdle muscular dystrophy (LGMD). These discoveries established a newly recognized disease entity, muscular dystrophy caused by abnormal glycosylation of α-DG, referred to as dystroglycanopathy (DGpathy). As described in the following sections, 18 genes are currently recognized as DGpathy causative genes. Although DGpathies are biochemically characterized by abnormal glycosylation of α-DG and are clinically associated with severe CMD and brain abnormalities, a considerable number of patients exhibit abnormal glycosylation of α-DG but show only mild muscular dystrophy without brain malformations. Numerous studies on the genotype-phenotype correlation within large cohorts of DGpathy patients have confirmed that a wide clinical spectrum is associated with mutations in specific causative genes. The most severe end of the spectrum is characterized by CMD with extensive structural abnormality in the brain and eye, which typically results in early infantile death. Conversely, patients at the mildest end of the spectrum might present in adult life with LGMD without brain or eye involvement.

In 2007, Muntoni and colleagues proposed a clinical classification system comprising seven broad phenotypic categories (Godfrey et al. 2007). DGpathy was first categorized into four groups: (a) WWS/WWS-like, (b) MEB/FCMD-like, (c) CMD, and (d) LGMD. Categories (c) and (d) were further classified into these subgroups: (c-1) CMD with mental retardation and cerebellar involvement as the only structural brain abnormality, (c-2) CMD with mental retardation and a structurally normal brain, (c-3) CMD with no abnormal cognitive development, (d-1) LGMD with mental retardation, and (d-2) LGMD without mental retardation. Later, new Online Mendelian Inheritance in Man (OMIM) entries created a simplified classification scheme for DGpathy (MDDG, muscular dystrophy-dystroglycanopathy) by combining three broad phenotypic groups and gene defects (Godfrey et al. 2011). First, DGpathy was divided into three groups: (A) CMD with brain/eye abnormalities, (B) CMD with milder brain structural abnormalities, and (C) LGMD. Second, the causative gene was indicated numerically; for example, *POMT1* is "1" and *fukutin* is "4." According to this classification, typical FCMD is named "MDDG type A4."

20.3.3 DG

DG was originally identified from skeletal muscle as a component of the dystrophin-glycoprotein complex (DGC). Mutations in dystrophin cause Duchenne and Becker muscular dystrophies, the most frequently occurring muscular dystrophies worldwide. The DGC forms a structural linkage between the basement membrane and the actin cytoskeleton, and this provides physical strength to the skeletal muscle cell membrane. Within the DGC, DG functions as a central axis by anchoring dystrophin beneath the cell membrane and binding to basement membrane proteins (such as laminin) on the cell exterior (Fig. 20.4). DG is composed of α and β subunits, both of which are expressed from a single mRNA and cleaved into α-DG and β-DG during posttranslational modification. α-DG is a highly glycosylated extracellular protein, and it functions as a ligand-binding subunit that interacts with several ligand proteins, which commonly contain laminin G-like (LG) domains. The ligands known to bind to α-DG include laminins, perlecan, agrin, neurexin, and pikachurin. β-DG is a transmembrane subunit that anchors α-DG on the cell surface and binds to dystrophin intracellularly. α-DG consists of three domains: two globular domains at the N- and C-termini and a mucin-like domain between them. The mucin-like domain contains >40 Ser/Thr residues that form *O*-glycan clusters. The N-terminal globular domain is essential for producing *O*-glycosylation in the mucin-like domain; however, the N-terminal domain is cleaved and shed from the α-DG core portion after the completion of glycosylation and therefore is not present in mature α-DG.

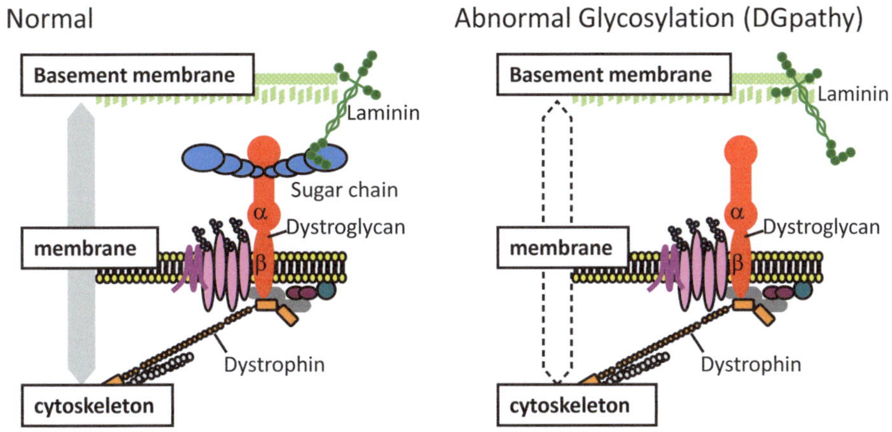

Fig. 20.4 Schematic representation of the dystrophin-glycoprotein complex (DGC). DGC links the basement membrane to the cytoskeleton (left). DG functions as the cellular receptor for matrix proteins, and the sugar chains on α-DG are essential for ligand binding. Abnormally glycosylated α-DG loses the ligand-binding activities and is associated with DGpathy (right)

20.3.4 Sugar-Chain Structure of α-DG and Its Biosynthetic Enzymes

In the early 2000s, in addition to *fukutin*, several genes were identified whose mutations cause abnormal glycosylation of α-DG: *POMT1*, *POMT2*, *POMGNT1*, *fukutin-related protein* (*FKRP*), and *LARGE*. The genes have been identified as causative genes for WWS, MEB, CMD, or LGMD (MDC1C and LGMD2I). After 2012, additional genes responsible for DGpathy were identified, and as of 2017, the list includes 18 genes (Table 20.2). Currently, DGpathies can be classified as primary DGpathies (affecting the DG gene directly), secondary DGpathies (affecting genes that directly modify DG), and tertiary DGpathies (caused by mutations in genes that indirectly affect DG function). As the DGpathy genes were identified, the sugar-chain structure of α-DG was revealed in a stepwise manner. In this chapter, we introduce the sugar-chain structure of α-DG and the enzymes involved in its biosynthesis.

1. **CoreM1**

In 1997, Endo and colleagues identified a unique *O*-mannose (*O*-Man)-type glycan from bovine peripheral nerve α-DG (Gal-GlcNAc-Man-*O*). This structure was later named CoreM1 (Fig. 20.5). CoreM1 is the first *O*-Man-type glycan identified in mammals. The initial Man transfer to the Thr/Ser residues of α-DG is catalyzed by a POMT1/POMT2 (protein *O*-mannosyl transferase 1/2) heterocomplex, which uses dolichol-phosphate-mannose (Dol-P-Man) as the Man donor substrate (Manya et al. 2004). The GlcNAcβ1-2Man linkage is synthesized by POMGnT1, which transfers GlcNAc from a UDP-GlcNAc to *O*-Man (Yoshida et al. 2001). CoreM1 was originally considered a ligand-binding moiety; however, enzymatic removal of Gal and GlcNAc from CoreM1 does not affect the ligand-binding activity of the protein. Conversely, disruption of CoreM1 biosynthesis in either human (POMGnT1-mutant-carrying patients) or mouse (POMGnT1-deficient mice) results in loss of ligand-binding activity of α-DG, which indicates that CoreM1 plays a critical role in the functional maturation of α-DG.

2. **CoreM3**

In 2010, Campbell and colleagues identified another type of *O*-Man glycan in recombinant α-DG expressed in HEK293 cells

Table 20.2 Dystroglycanopathy genes and their functions

Dystroglycanopathy genes	Gene functions	MDDG number
POMT1	*O*-Mannosyl transferase (POMT1/2 complex)	1
POMT2	*O*-Mannosyl transferase (POMT1/2 complex)	2
POMGNT1	Protein *O*-mannose β1,2-GlcNAc transferase	3
FKTN	Ribitol phosphate transferase	4
FKRP	Ribitol phosphate transferase	5
LARGE	α3-Xylosyl- and β3-glucuronyl-transferase (synthase of Xyl-GlcA repeating units)	6
ISPD	CDP-ribitol synthase	7
GTDC2/POMGNT2	Protein *O*-mannose β1,4-GlcNAc transferase	8
DAG1	Dystroglycan	9
TMEM5/RXYLT1	β1,4-Xyl transferase	10
B3GALNT2	β1,3-GalNAc transferase	11
SGK196/POMK	*O*-Man kinase	12
B3GNT1/B4GAT1	β1,4-GlcA transferase	13
GMPPB	GDP-mannose pyrophosphorylase B	14
DPM1	Dol-P-Man synthesis	–
DPM2	Dol-P-Man synthesis	–
DPM3	Dol-P-Man synthesis	–
DOLK	Dolichol phosphate synthesis	–

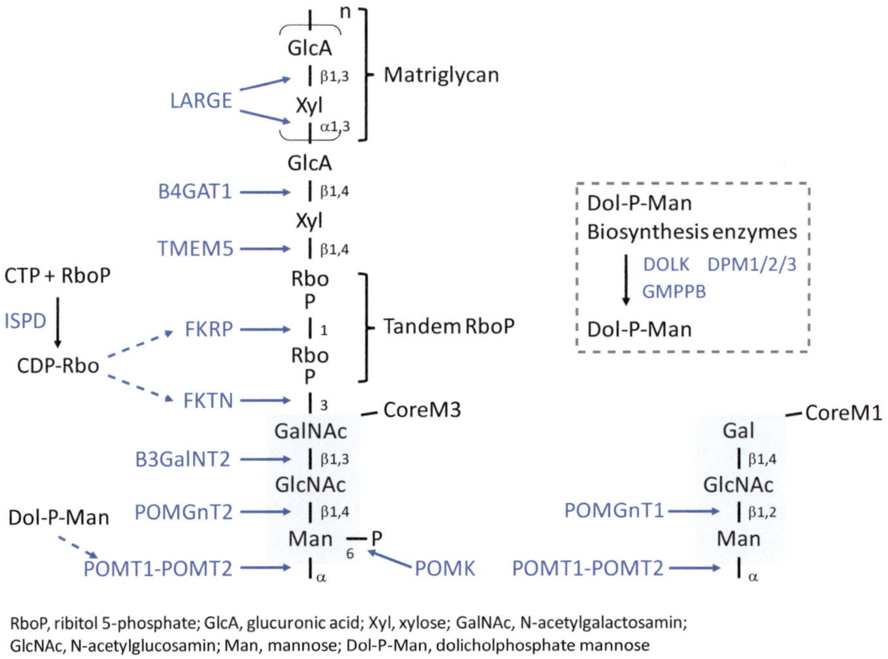

Fig. 20.5 Schematic representation of the sugar-chain structure of α-DG and DGpathy gene functions. *Rbop* ribitol 5-phosphate, *GlcA* glucuronic acid, *Xyl* xylose, *GalNAc* N-acetylgalactosamin, *GlcNAc* N-acetylglucosamin, *Man* mannose, *Dol-P-Man* dolicholphosphate mannose

(GalNAc-GlcNAc-Man-*O*), which was later named CoreM3 (Fig. 20.5). CoreM3 serves as a scaffold for the building of the ligand-binding moiety on the core protein of α-DG (as described below) (Yoshida-Moriguchi et al. 2013; Yoshida-Moriguchi and Campbell 2015). The initial Man transfer is catalyzed by the enzymatic action of the POMT1/POMT2 complex as in the case of CoreM1. The GlcNAcβ1-4Man linkage is synthesized by POMGnT2, which transfers GlcNAc from a UDP-GlcNAc. The GalNAcβ1-3GlcNAc linkage is synthesized by B3GalNT2, which transfers GalNAc from a UDP-GalNAc. The *O*-Man residue in CoreM3 is phosphorylated at the 6 position, and the phosphate group is added from ATP by protein *O*-mannose kinase (POMK). The physiological role of this phosphorylation is unknown, but it is likely required for subsequent modification of CoreM3 with the ribitol phosphate group (see below).

3. **Matriglycan (LARGE-glycan)**

The overexpression of LARGE, whose mutations were identified in MDC1D patients and in the spontaneous mutant *Large*^{myd} mouse, increases the glycosylation and ligand-binding activities of α-DG. This unique feature suggested that LARGE is involved in the biosynthesis of the ligand-binding moiety of α-DG. In 2012, LARGE was identified to possess two glycosyltransferase activities, α3-xylosyltransferase and β3-glucuronyltransferase activities, which generate the [−3GlcAβ1-3Xylα1] (GlcA-Xyl) repeat from UDP-Xyl and UDP-GlcA (Inamori et al. 2012). Enzymatically synthesized [−3GlcAβ1-3Xylα1] repeats interact with laminin and are recognized by IIH6 antibody, and therefore the repeat structure serves as the ligand-binding motif of α-DG and is named "matriglycan" (Fig. 20.5). A recent crystal structure analysis revealed the atomic details of DG-ligand binding. A single GlcA-Xyl disaccharide straddles a Ca^{2+} ion in the LG4 domain of laminin α2 (Briggs et al. 2016). This chelating binding mode explains the high affinity of this protein-carbohydrate interaction. Moreover, multiple GlcA-Xyl repeats are predicted to increase the apparent affinity for the LG domains present in ligand proteins by favoring rapid rebinding after dissociation. Accordingly, long repeats (>13) effectively bind to laminins, although the number of repeats present in a single matriglycan remains unknown.

For the initiation of matriglycan formation, an acceptor primer structure, GlcAβ1-4Xyl, is necessary. This structure is synthesized by B4GAT1, previously known as β3GnT1, which exhibits β4-glucuronyltransferase activity by using a donor UDP-GlcA to form the GlcAβ1-4Xyl linkage that can be elongated by LARGE with matriglycan (Yoshida-Moriguchi and Campbell 2015). For the matriglycan modification, in addition to the LARGE enzyme activities and the acceptor primer structure, physical interaction between LARGE and the α-DG N-terminal domain is required (Kanagawa et al. 2004). Thus, the N-terminal domain serves as a recognition motif for LARGE, thereby providing substrate specificity for this unique modification. A mutation in the N-terminal domain of α-DG was identified in a patient with LGMD with cognitive impairment, and the mutation reduced LARGE recognition and therefore resulted in diminished LARGE-dependent matriglycan modification (Hara et al. 2011). This was the first case in which a mutation in the DG gene (*DAG1*) itself was shown to cause muscular dystrophy ("primary" DGpathy).

4. **Tandem ribitol phosphate**

In 2016, Toda and colleagues showed the presence of novel moiety between matriglycan and CoreM3 by mass spectrometry analysis (Kanagawa et al. 2016). The moiety contains a GlcA-Xyl disaccharide and a tandem ribitol 5-phosphate (RboP) (Fig. 20.5). Whereas the GlcA-Xyl disaccharide serves the acceptor primer for LARGE, the tandem RboP links the primer GlcA-Xyl to the GalNAc residue in CoreM3. Ribitol is a sugar alcohol that was not known to be used in mammals before this finding, whereas RboP was initially recognized as a component of the teichoic acids present in bacterial cell walls. In bacteria, the RboP polymer is synthesized by enzymes that use CDP-Rbo as a donor substrate, and CDP-Rbo is synthesized by the enzyme TarI from cytidine triphosphate (CTP) and RboP. TarI shares homology with a DGpathy protein, isoprenoid synthase domain containing (ISPD), and, notably, ISPD was shown to possess CDP-Rbo synthase activity. Subsequently, the tandem RboP structure was shown to be synthesized through the sequential enzymatic actions of fukutin and FKRP: Fukutin transfers RboP from a CDP-Rbo to GalNAc in CoreM3, and then FKRP transfers RboP from a CDP-Rbo to the first RboP. Missense mutations found in DGpathy patients carrying mutant *fukutin* or *FKRP* reduced the RboP transferase activities. NMR analysis revealed the anomeric conformation to be Rbo5P-1Rbo5P-3GalNAc. Lastly, TMEM5 was shown to exhibit Xyl transferase activity by using UDP-Xyl as a donor substrate and to form a Xylβ1-4Rbo5P linkage that connects matriglycan and the tandem Rbo5P (Manya et al. 2016). This Xyl residue is a part of the acceptor primer for matriglycan formation. TMEM5 was recently renamed RXYLT1 on the basis of its function (Rbo5P β1,4-Xyl transferase).

5. **Dol-P-Man synthesis pathway**

As mentioned above while describing CoreM1, the initial Man transfer to the Thr/Ser residues in α-DG requires the POMT1/POMT2 enzyme complex and its donor substrate Dol-P-Man. Thus, defects in the Dol-P-Man biosynthesis pathway can result in abnormal *O*-mannosylation of α-DG, and, accordingly, mutations in certain proteins involved in this pathway cause DGpathies (Table 20.2). The synthesis of Dol-P-Man from GDP-Man and dolichol phosphate is catalyzed by the DPM synthase complex, which consists of the catalytic component DPM1 and the ER-localized transmembrane proteins DPM2 and DPM3. Dolichol kinase (DOLK) is responsible for the formation of dolichol phosphate. GDP-mannose pyrophosphorylase B (GMPPB) is the β-subunit of the essential enzyme GDP-mannose pyrophosphorylase, which catalyzes the formation of GDP-mannose from mannose-1-phosphate and GTP. Consequently, mutations in DOLK or GMPPB could result in a decrease in Dol-P-Man levels.

Dol-P-Man also acts as a Man donor in *N*-glycosylation and in glycophosphatidylinositol-anchor biosynthesis. Therefore, defects in the Dol-P-Man pathway affect these modifications and are associated with human diseases. Congenital disorder of glycosylation (CDG) is a heterogeneous group of rare genetic disorders that were originally defined as diseases caused by defects in the *N*-glycosylation process, and CDG has now been reclassified to include *O*-linked and lipid glycosylation defects. Notably, mutations in *DPM1*, *DPM2*, or *DPM3* have been identified in patients presenting DGpathy with type I CDG, and mutations in *DOLK* have been discovered in patients with CDG and dilated cardiomyopathy, whose heart tissue showed abnormal *O*-mannosylation of α-DG. Overall, mutations in the genes introduced in this section and in *ISPD* indirectly affect α-DG function; thus, the diseases caused by these gene mutations are categorized as tertiary DGpathies.

20.4 Therapy

20.4.1 Cellular Pathogenesis of FCMD

Studies conducted using genetically modified disease-model animals have contributed substantially to our understanding of the molecular and cellular pathogenesis of single-gene disorders. Here, we introduce the DGpathy cellular pathogenesis revealed by studies performed using *fukutin*-deficient FCMD model animals. Because embryonic lethality was exhibited by fukutin or DG gene-knockout mice that were generated through targeted disruption of *fktn* or *dag1*, respectively, several lines of tissue-selective conditional knockout (cKO) mice have been generated for disrupting the expression of *fktn* or *dag1*. In DGpathy, the loss of ligand-binding activity of α-DG due to abnormal glycosylation leads to a disruption of the linkage between the basement membrane and the cytoskeleton. Accordingly, ultrastructural abnormality, in which the basement membrane was detached from the muscle cell membrane, was detected in the skeletal muscle of both myofiber-selective DG-deficient mice and mice that were spontaneously *Large*-defective, the *Large*^{myd} DGpathy mice. These structural abnormalities render muscles prone to contraction-induced injuries, eventually leading to the necrosis of muscle cells. Membrane fragility that triggers disease manifestation is also observed in myofiber-selective *fukutin* cKO mice; however, mice with myofiber-selective loss of DG or fukutin exhibit only mild muscular dystrophy, which indicates that membrane fragility is not sufficient for explaining the severe muscle pathology of DGpathy.

Myofiber damage leads to the activation of satellite cells, which are muscle-specific stem cells; activated satellite cells differentiate into muscle precursor cells (MPCs) and then into myoblasts, which eventually fuse into myotubes. Thus, through this process, skeletal muscle can undergo regeneration. MPC-selective *fukutin*-cKO mice show severe muscular dystrophy with reductions in the number of satellite cells, MPC proliferation/differentiation activities, and muscle regeneration activity (Kanagawa et al. 2013). The severity of these abnormalities increases as the disease progresses. Therefore, α-DG plays critical roles in the maintenance of satellite cell viability and MPC activities, and defects in the functions of these cells correlate with disease severity. Moreover, the absence of α-DG glycosylation during postnatal/juvenile muscle growth and development has been proposed to potentially affect muscle degeneration and/or dystrophic pathology in later stages. In FCMD and *Large*^{myd} mice, neuromuscular junction formation is aberrant and numerous immature muscle fibers are present, and this suggests that impaired differentiation signals from the aberrant neuromuscular junctions and the maturational delay in muscle fibers underlie the etiology of DGpathy. Thus, in addition to muscle membrane fragility, muscle maturation and regeneration processes appear to be involved in the pathogenesis of DGpathy.

Abnormal glycosylation of α-DG is also regarded as one of the main causes of brain malformation, characterized by polymicrogyria. The surface of the cerebral cortex is covered by the glia-limitans/basement-membrane complex, which prevents the overmigration of neurons. In FCMD and other DGpathy models, breakdown of the basement membrane leads to a protrusion of neurons into the subarachnoid space where the basement membrane breaches are observed. These abnormalities are possibly the underlying causes of cortical dysplasia and type II lissencephaly (Michele et al. 2002). α-DG is expressed in radial glia and presumably plays a role in physically connecting the glia limitans to the basement membrane. During the developmental growth of the embryonic brain, abnormally glycosylated α-DG might be incapable of maintaining sufficient physical strength and/or plasticity of the glia-limitans/basement-membrane complex against an expanding cortical surface area. Intriguingly, neuron-selective DG cKO and *Large*^{myd} mice show impairment of long-term potentiation at CA3–CA1 synapses. DG is known to be expressed at the postsynaptic apparatus, but the precise mechanism that mediates the impairment of synaptic plasticity remains to be elucidated.

20.4.2 Therapeutic Strategies

Effective therapies for FCMD or DGpathies have not yet been established. However, because the molecular basis of FCMD and related DGpathies has been clarified, effective clinical treatments should be developed in the future. Conversely, at the laboratory level, several therapeutic strategies have been proposed. Gene replacement is one of the most rational strategies for DGpathies because these are single-gene disorders. Supporting this view, adeno-associated virus vector-mediated *fukutin* delivery results in improvement of the disease pathology in fukutin-deficient model mice (Kanagawa et al. 2013). Here, we introduce innovative treatment strategies based on the molecular pathogenesis of FCMD and DGpathy with RboP deficiency.

Most FCMD patients carry the founder SVA retrotransposal insertion in the 3′-UTR of *fukutin*.

This insertion contains a strong splice acceptor site, which induces a rare alternative donor site in the final exon, thus causing abnormal mRNA splicing (exon trapping) (Taniguchi-Ikeda et al. 2011). Notably, introduction of antisense oligonucleotides that target the splice acceptor, the predicted exonic splicing enhancer, and the intronic splicing enhancer prevented the pathogenic exon trapping by SVA in the cells of FCMD patients and model mice. This resulted in the rescue of normal *fukutin* mRNA expression and protein production and, consequently, the restoration of α-DG glycosylation and laminin-binding activity. This treatment strategy can potentially be applied to almost all FCMD patients in Japan and can therefore be the first radical clinical treatment for DGpathies.

As described in the previous section, CDP-Rbo is an essential donor substrate for fukutin and FKRP to build the tandem RboP structure. Thus, reduction or loss of cellular CDP-Rbo due to ISPD mutations impairs RboP modifications, and this results in a failure of functional glycan synthesis on α-DG. These abnormalities directly underlie the molecular pathology of ISPD-deficient-type DGpathy. In ISPD-deficient cells, CDP-Rbo supplementation was shown to rescue RboP modification of α-DG. Although the details of the biosynthetic pathway of Rbo5P, an essential metabolite required for ISPD-dependent CDP-Rbo synthesis, remain incompletely elucidated, potential pathways for Rbo5P synthesis have been proposed (Gerin et al. 2016). Addition of D-ribose or ribitol, which are candidate metabolites in these pathways, led to an increase in CDP-Rbo levels in cells overexpressing ISPD, and ribitol supplementation in drinking water led to an increase in muscle CDP-Rbo levels in wild-type mice. These data suggest that supplementation of D-ribose or ribitol might effectively increase cellular Rbo5P level and might also enhance CDP-Rbo synthase activity of disease-causing mutant ISPD proteins that exhibit reduced affinity for the substrate Rbo5P. Further studies involving the use of animal models will be necessary for developing CDP-ribitol or metabolite supplementation therapy.

End-of-Chapter Questions

1. Immunostaining performed on a patient's biopsy samples indicates abnormal glycosylation of α-DG. Clinical manifestation also suggests that the patient suffers from DGpathy. However, whole-exon sequencing of the 18 recognized DGpathy genes indicates no mutations in these genes. Why do you suspect that the patient has DGpathy?
2. Discuss the advantages of investigating the effects of disease-causing mutations in *ISPD*, *fukutin*, and *FKRP* on the activity of the encoded enzymes when considering therapy.
3. The major mutation in FCMD is a retrotransposon insertion in the gene *fukutin*, which accounts for ~87% of the FCMD chromosome. This is considered to be due to the founder effect. Discuss other examples of genetic disorders associated with the founder effect.

References

Briggs DC, Yoshida-Moriguchi T, Zheng T et al (2016) Structural basis of laminin binding to the LARGE glycans on dystroglycan. Nat Chem Biol 12:810–814

Fukuyama Y, Osawa M, Suzuki H (1981) Congenital progressive muscular dystrophy of the Fukuyama type – clinical, genetic and pathological considerations. Brain and Development 3:1–29

Gerin I, Ury B, Breloy I et al (2016) ISPD produces CDP-ribitol used by FKTN and FKRP to transfer ribitol phosphate onto α-dystroglycan. Nat Commun 7:11534

Godfrey C, Clement E, Mein R et al (2007) Refining genotype–phenotype correlations in muscular dystrophies with defective glycosylation of dystroglycan. Brain 130:2725–2735

Godfrey C, Foley AR, Clement E et al (2011) Dystroglycanopathies: coming into focus. Curr Opin Genet Dev 21:278–285

Hara Y, Balci-Hayta B, Yoshida-Moriguchi T et al (2011) A dystroglycan mutation associated with limb-girdle muscular dystrophy. N Engl J Med 364:939–946

Hayashi YK, Ogawa M, Tagawa K et al (2001) Selective deficiency of α-dystroglycan in Fukuyama-type congenital muscular dystrophy. Neurology 57:115–121

Inamori K, Yoshida-Moriguchi T, Hara Y et al (2012) Dystroglycan function requires xylosyl- and glucuronyltransferase activities of LARGE. Science 335:93–96

Kanagawa M, Saito F, Kunz S et al (2004) Molecular recognition by LARGE is essential for expression of functional dystroglycan. Cell 117:953–964

Kanagawa M, Yu CC, Ito C et al (2013) Impaired viability of muscle precursor cells in muscular dystrophy with glycosylation defects and amelioration of its severe phenotype by limited gene expression. Hum Mol Genet 22:3003–3015

Kanagawa M, Kobayashi K, Tajiri M et al (2016) Identification of a post-translational modification with ribitol-phosphate and its defect in muscular dystrophy. Cell Rep 14:2209–2223

Kobayashi K, Nakahori Y, Miyake M et al (1998) An ancient retrotransposal insertion causes Fukuyama-type congenital muscular dystrophy. Nature 394:388–392

Manya H, Chiba A, Yoshida A et al (2004) Demonstration of mammalian protein *O*-mannosyltransferase activity: coexpression of POMT1 and POMT2 required for enzymatic activity. Proc Natl Acad Sci U S A 101:500–505

Manya H, Yamaguchi Y, Kanagawa M et al (2016) The muscular dystrophy gene *TMEM5* encodes a ribitol β1,4-xylosyltransferase required for the functional glycosylation of dystroglycan. J Biol Chem 291:24618–24627

Michele DE, Barresi R, Kanagawa M et al (2002) Post-translational disruption of dystroglycan-ligand interactions in congenital muscular dystrophies. Nature 418:417–422

Taniguchi-Ikeda M, Kobayashi K, Kanagawa M et al (2011) Pathogenic exon-trapping by SVA retrotransposon and rescue in Fukuyama muscular dystrophy. Nature 478:127–131

Watanabe M, Kobayashi K, Jin F et al (2005) Founder SVA retrotransposal insertion in Fukuyama-type congenital muscular dystrophy and its origin in Japanese and Northeast Asian populations. Am J Med Genet 138A:344–348

Yoshida A, Kobayashi K, Manya H et al (2001) Muscular dystrophy and neuronal migration disorder caused by mutations in a glycosyltransferase, POMGnT1. Dev Cell 1:717–724

Yoshida-Moriguchi T, Campbell KP (2015) Matriglycan: a novel polysaccharide that links dystroglycan to the basement membrane. Glycobiology 25:702–713

Yoshida-Moriguchi T, Willer T, Anderson ME et al (2013) SGK196 is a glycosylation-specific *O*-mannose kinase required for dystroglycan function. Science 341:896–899

Hereditary Proteinuric Glomerular Disorders

21

Hiroyasu Tsukaguchi

Keywords
Glomerular disease · Proteinuria · Nephrtoic syndrome · Podocyte · Actin cytoskeleton

21.1 Case Report

A 26-year-old male first manifested asymptomatic proteinuria at age 16 years at a regular checkup. Because of persistent proteinuria, he was referred for a consultation with a nephrologist. Urinalysis by dipstick showed grade 3+ proteinuria (1.2 g/g creatinine) with no occult blood. Urinary sediment analysis revealed 1–5 red blood cells and no white blood cells per high-power field. His blood pressure was 107/56 mmHg. He had a sensorineural hearing disability on the left side. On physical examination, his height was 174 cm, weight 52.7 kg, and body mass index 17.4. He did not show any abnormalities in muscle or neurologic functions. There was no apparent peripheral edema or swelling in the lymph nodes or tonsils. The skin and joints appeared normal.

Blood chemistry revealed a total protein level of 7.5 g/dl (normal range, 6.7–8.3), albumin level of 4.8 g/dl (normal range, 3.8–5.2), blood urea nitrogen level of 10.7 mg/dl (normal range, 8–20), creatinine level of 0.99 mg/dl (normal range, 0.6–1.0), and estimated glomerular filtration rate (eGFR) of 77.0 ml/min/1.73 m^2 (normal range for his sex and age, 77–135). Serum levels of immunoglobulin (Ig)A (505 mg/dl; normal range 110–410 mg/dl) and IgM (284 mg/dl; normal range 33–190 mg/dl) were slightly elevated, while IgG levels were normal (979 mg/dl; normal range 870–1700 mg/dl). Levels of complement factors C3, C4, and CH50 were normal. Serologic tests for antibodies targeting double-stranded DNA, hepatitis B, and hepatitis C were negative.

A renal ultrasonographic examination demonstrated that the kidneys were normal in size and shape. No hydronephrosis or renal stone was found. Ultrasound cardiography results were normal. The patient had no history of diabetes or hypertension and had not been treated with any medications. The proband (III-2) has two other family members (mother and uncle) with kidney disease (Fig. 21.1). His mother (II-2) had not noticed any remarkable symptoms until age 36, when she developed end-stage renal disease (ESRD). His uncle (II-3) had also progressed to ESRD and underwent hemodialysis at age 26. He died from a cerebral hemorrhage at age 51. His father (II-1) and older brother (III-1) are healthy without any kidney disease.

H. Tsukaguchi (✉)
Second Department of Internal Medicine, Kansai Medical University, Osaka, Japan
e-mail: tsukaguh@hirakata.kmu.ac.jp

T. Oohashi et al. (eds.), *Human Pathobiochemistry*, https://doi.org/10.1007/978-981-13-2977-7_21

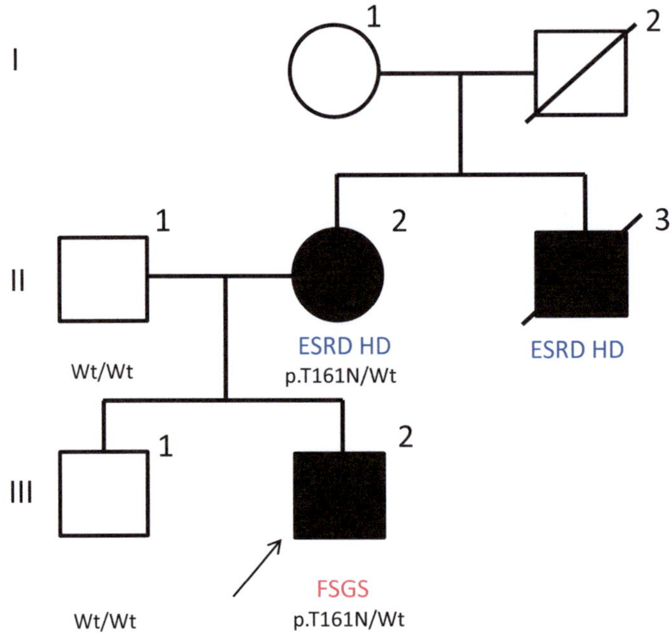

Fig. 21.1 Pedigree of familial focal segmental glomerulosclerosis (FSGS)
The proband, a 26-year-old-male who was histologically diagnosed with FSGS, is indicated by the arrow. His father (II-1) and brother (III-1) are healthy without any kidney disease. The proband (III-2) and his mother (II-2) were found to have a heterozygous missense *INF2* mutation, p.Thr161Asn (p.T161N/wild type [Wt]). His unaffected father (II-1) and brother (III-1) are homozygous for the wild-type allele (Wt/Wt). The uncle (II-3) developed end-stage renal disease (ESRD) at age 26 years but did not undergo renal biopsy. Males and females are shown by squares and circles, respectively. Opened and closed symbols indicate unaffected and affected individuals, respectively. HD, hemodialysis

In the proband (III-2) at age 26, light microscopy of a renal biopsy specimen demonstrated a histological pattern of focal segmental glomerulosclerosis (FSGS) (Fig. 21.2) (D'Agati et al. 2011). Among all 16 glomeruli, 2 showed segmental sclerosis with tuft adhesions and 5 global sclerosis, while the remaining 9 appeared normal. Mild-to-moderate tubulointerstitial changes, including fibrosis and tubular atrophy, occupied approximately 30–40% of the cortex. In an immunofluorescence study, IgM, IgA, and C3 were deposited in a coarse segmental pattern mainly around the tips of tufts (Fig. 21.3a). There were no significant depositions of IgG or the complement factor C1q. Immunostaining of the α5 chain of type IV collagen, a major component of the glomerular basement membrane (GBM), was homogenously positive. Electron microscopy revealed diffuse effacement of the podocyte foot processes (Fig. 21.3b), a hallmark of proteinuric disorders. Some electron-dense deposits were focally present along the mesangiocapillary interface (Fig. 21.3). Overall, the GBM appeared normal in thickness (310–400 nm, average 315 nm; normal range 239–453 nm), while some portions displayed slight irregularities with wrinkling. There were no morphologic abnormalities in the GBM including splitting and lamellation. The number and shape of mitochondria appeared normal in both podocytes and tubular epithelia.

For 2 years after biopsy, the patient was treated with an angiotensin type 2 receptor blocker (olmesartan 10 mg) alone. His proteinuria level remained in the subnephrotic range (1.4 g/g creatinine, serum albumin 3.9 g/dl). However, there was a slight decrease in renal function during this 2-year period (creatinine level, 1.05 mg/dl; eGFR, 70 ml/min/1.73 m^2).

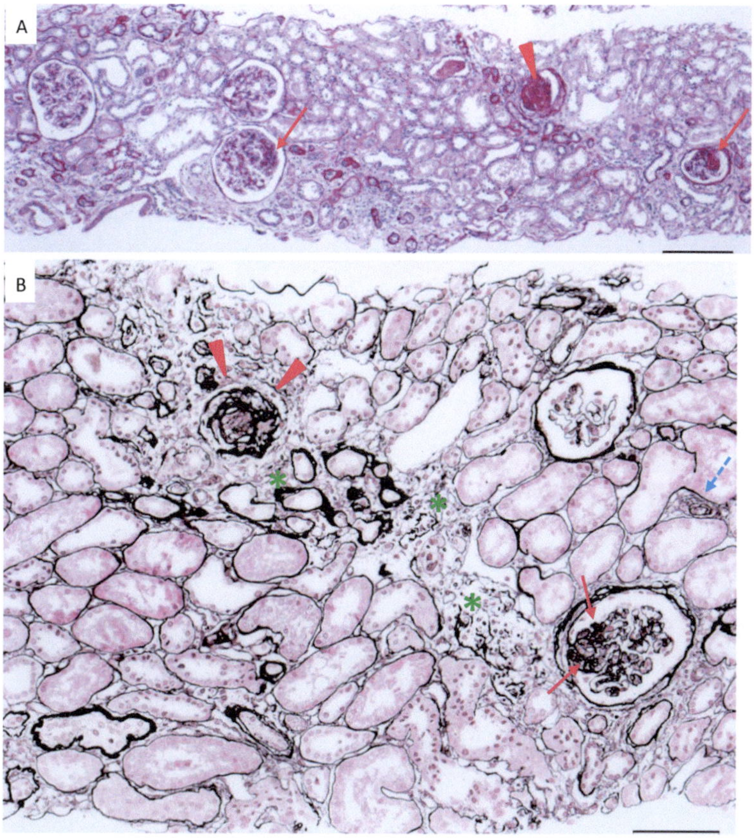

Fig. 21.2 Light microscopy of kidney biopsy specimens Renal histology images from the proband (III-2) at age 26 years. (**a**) Low-power magnification of periodic acid–Schiff staining of glomeruli showed varying degrees of sclerosis, ranging from segmental (*arrows*) to global patterns (*arrowhead*). Global sclerosis was found in ~ 30% of all glomeruli. Moderate tubular atrophy and interstitial fibrosis occupied 30–40% of the cortical area. (**b**). High-power magnification of periodic acid–methenamine silver staining revealed segmental sclerosis (*arrows*) coexisting with global obsolescence (*arrowheads*) in glomeruli. Tubulointerstitial scarring (*asterisks*) shows a striped pattern surrounding the sclerosing glomeruli. No sclerotic changes were seen in the arteriole (*dotted arrow*). Scale bars, 200 μm (**a**, 40×) and 100 μm (**b**, 100×)

21.2 Diagnosis

A 26-year-old male (III-2) presented with persistent moderate subnephrotic-range proteinuria with a unique histological pattern of glomerular injury, known as FSGS. He had no hematuria and maintained a normal eGFR. A history of end-stage renal disease in his mother (II-2) and uncle (II-3) suggests that the members of this family may share a common genetic etiology leading to progressive glomerulosclerosis.

21.3 Proteinuric Glomerular Diseases

Proteinuric disorders are classified based on the extent of proteinuria, underlying etiology, and histology. Urinary protein excretion >150 mg/day is regarded as pathological **proteinuria**. The most common cause of **proteinuria,** particularly in moderate to severe cases (urinary protein excretion >1 g/day), is glomerular injury that disrupts the filtration barrier (D'Agati et al. 2011; Kriz and Elger 2010). Plasma ultrafiltration in the kidney occurs via the capillary wall in the glomerulus. The filtration unit is composed of endo-

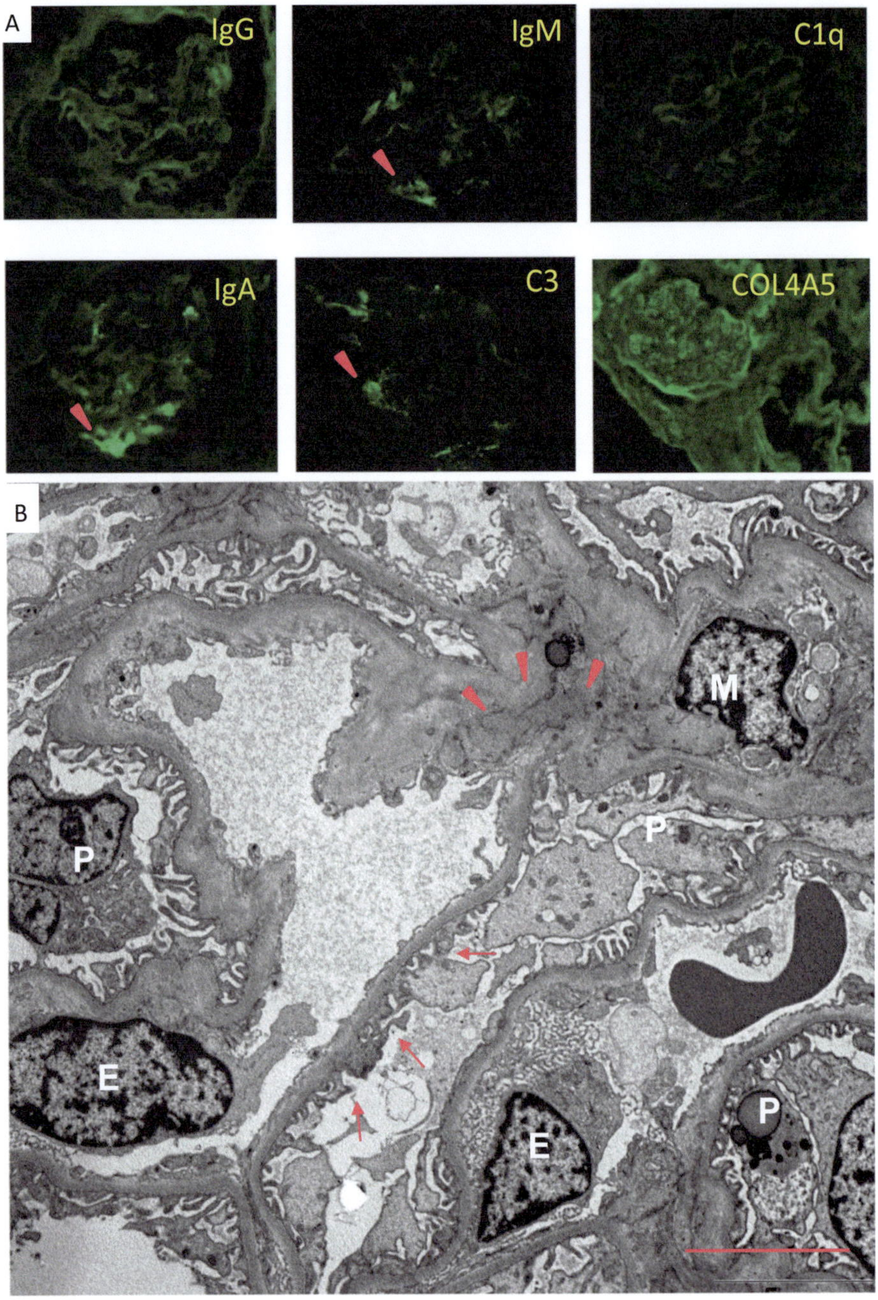

Fig. 21.3 Immunofluorescence and electron microscopy of kidney biopsy specimens
Renal biopsy samples from the proband (III-2) at age 26 years. (**a**) Immunofluorescence revealed IgM, IgA, and C3 deposition in a coarse segmental pattern mainly around the tip of tufts (*arrowheads*). Immunostaining of the α5 chain of type IV collagen (COL4A5) was homogeneously positive. (**b**) Transmission electron microscopy showed diffuse effacement of podocyte foot processes along the outer surface of the capillary wall. The inner contour and thickness of the GBM were normal overall, whereas some irregularities with wrinkling were seen in sclerosing tufts. The absence of pathologic GBM changes, i.e., splitting, lamellation, or fragmentation, excludes a primary GBM defect. No accumulation of storage materials or dysmorphic mitochondria was found in the podocyte cytoplasm. *Arrows* indicate effacement of podocyte foot processes, and *arrowheads* indicate focal electron-dense deposits at the mesangiocapillary interface. E, endothelial cell; M, mesangial cell; P, podocyte; GBM, glomerular basement membrane. Scale bar, 5 μm

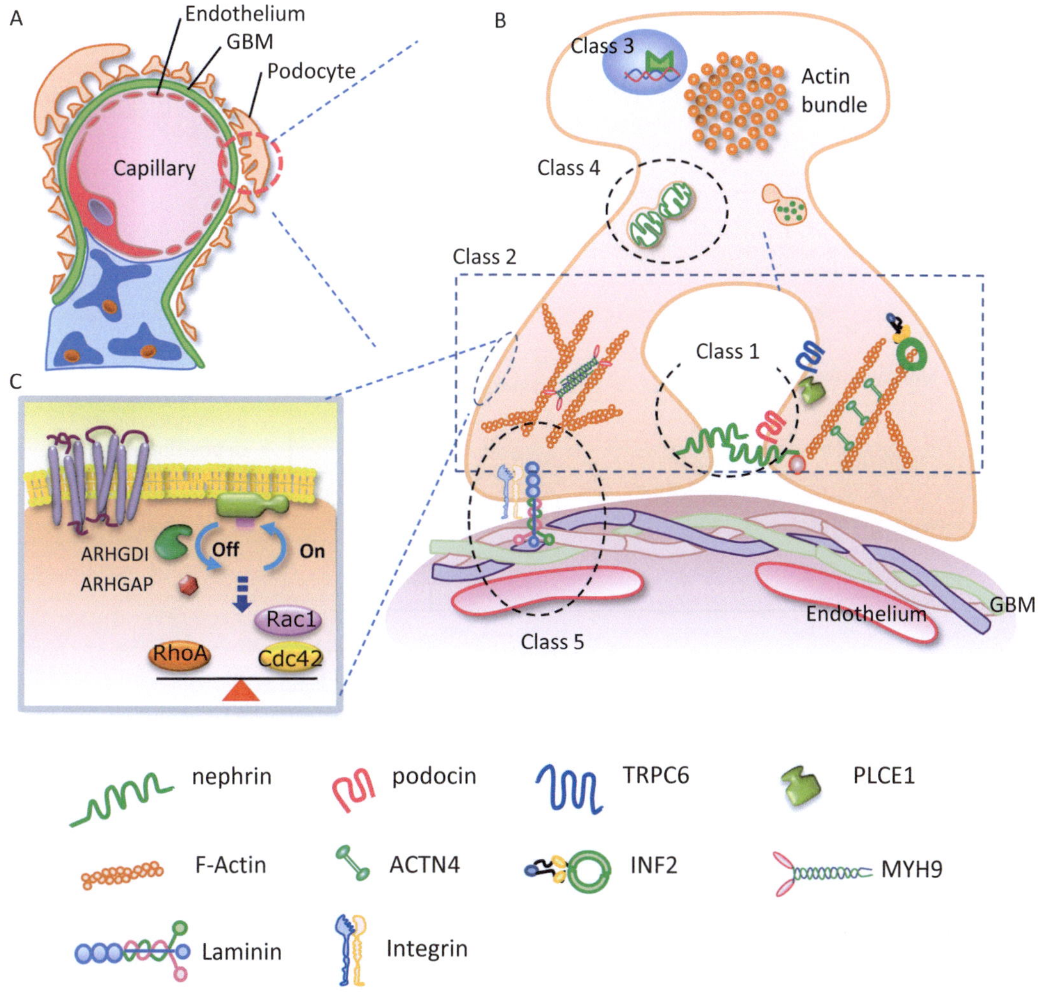

Fig. 21.4 Molecular causes and subclassification of FSGS
(**a**) The glomerular capillary wall, comprising three layers including the endothelium, GBM, and podocytes, functions as a filtration barrier. (**b**) Actin cytoskeletal dynamics in podocyte foot processes are regulated by the relative balance among Rho GTPase subfamily members (Rho, Rac, and Cdc42), as well as the interplay between Rho GTPases and effectors. Predominant Rac1 and Cdc42 activation induces the cytoskeleton to be "motile," whereas increased RhoA signaling leads to a "stationary" phenotype. (**c**) Most FSGS genes are expressed in the podocytes, indicating that the podocyte is the principal culprit for loss of slit diaphragm integrity. The genes can be categorized into five subclasses: (1) slit diaphragm complex, (2) actin cytoskeleton, (3) transcription factors and nuclear envelope, (4) mitochondria, (5) and GBM and basolateral anchoring of podocytes to the GBM (Lovric et al. 2016; Benoit et al. 2010; Sampson and Pollak 2015; Sadowski et al. 2015; Trautmann et al. 2015). The GBM is composed mainly of three type IV collagen chains (α3, α4, and α5), which assemble into a triple helical protomer. Podocyte basolateral membranes are anchored to the GBM and the surrounding extracellular matrix through the interaction of laminin-521 (α5, β2, and γ1) and integrin (α3 and β1)
Podocyte foot processes are always exposed to the capillary filtration pressure (40–50 mmHg) (Kriz and Elger 2010). Neighboring foot processes are aligned in an interdigitating fashion. The capillary walls are therefore susceptible to various injuries, particularly if genetic structural weaknesses are present. The intercellular space (20–50 nm) is bridged by the slit diaphragm complex (nephrin and podocin). *INF2* inverted formin 2, *TRPC6* transient potential cation channel, *PLCE1* phospholipase C epsilon 1, *MYH9* myosin heavy chain 9, *ACTN4* actinin-4

thelial cells, the GBM, and podocytes (Fig. 21.4). The **slit diaphragm,** a cell–cell junction between the apposing podocyte foot processes, plays a central role in size-selective macromolecule filtration (D'Agati et al. 2011; Kriz and Elger 2010). The filtration barrier is injured in both systemic and renal-specific disorders of various etiologies, including genetic defects, infections,

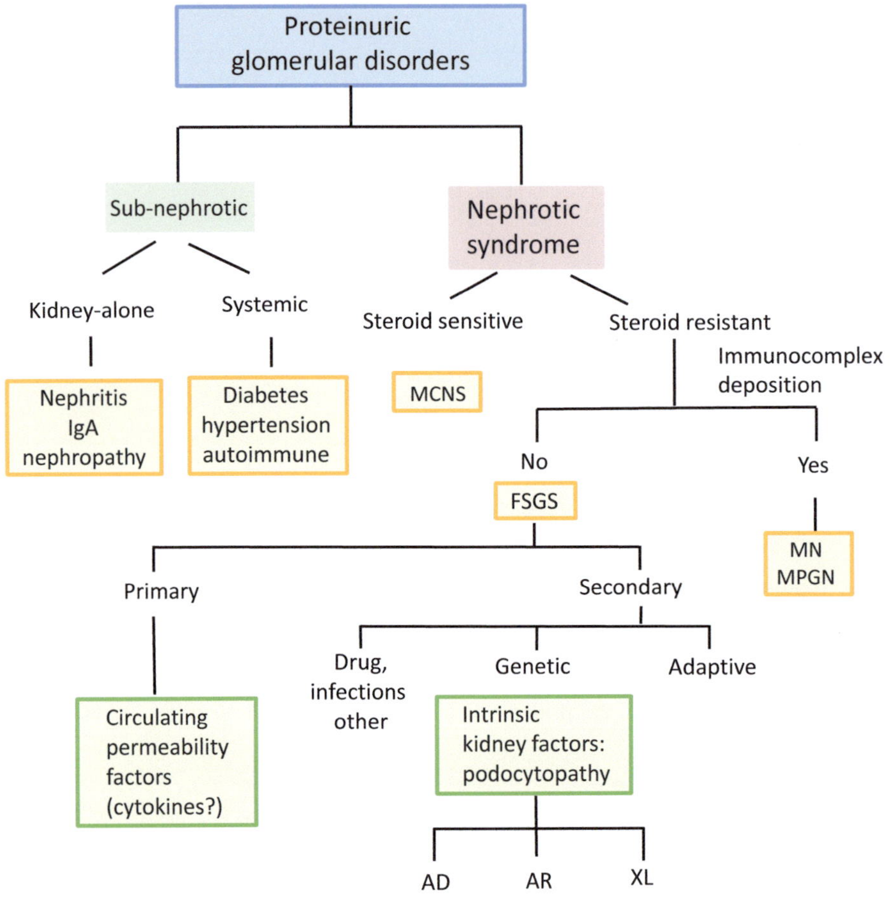

Fig. 21.5 Diagnostic flowchart of Proteinuric Glomerular Disorders
Massive proteinuria (urinary protein >40 mg/m^2/h in children or > 3.5 g/day in adults) is the hallmark of nephrotic syndrome (NS) (Radhakrishnan and Cattran 2012). Most children with NS are steroid sensitive and exhibit minimal histological alterations. However, 20% of children and 50% of adults with NS are steroid resistant and typically display FSGS. From an etiological point of view, FSGS is subdivided into primary and secondary forms (D'Agati et al. 2011). Primary FSGS is caused by a circulating factor that is associated with NS recurrence posttransplantation. Most of the genes associated with FSGS, more than 30 reported so far, are expressed in podocytes, indicating that genetic FSGS represents a primary podocytopathy. Three Mendelian modes of transmission of familial FSGS have been recognized (AD, AR, and XL). AD, autosomal dominant; AR, autosomal recessive; XL, X-linked; MCNS, minimal change nephrotic syndrome; MN, membranous nephropathy; MPGN, membranoproliferative glomerulonephritis

autoimmune diseases, diabetes, and hypertension (Fig. 21.5). **Proteinuric Glomerular Disorders** thus represent a clinically heterogeneous group of diseases with various histological patterns (Fig. 21.5). Barrier disruption subsequently leads to protein leakage into the urine (proteinuria) and progression to ESRD. Glomerular disorders are the most common primary cause of ESRD, and therefore it is critical to understand their basic molecular pathophysiology. **Proteinuric Glomerular Disorders** are typically sporadic; however, there are also rare familial forms involving monogenic diseases following a Mendelian pattern of inheritance.

Glomerular disorders may produce a wide range of proteinuria depending on the severity of barrier function damage (D'Agati et al. 2011; Kriz and Elger 2010). **Nephrotic syndrome (NS)** is defined as hypoalbuminemia, hypercholesterolemia, and peripheral edema. In children, NS is defined as proteinuria (>40 mg urine protein/m^2 body surface area/hour or a protein-to-creatinine ratio of 2000 mg/g) with hypoalbuminemia (serum albumin level <2.5 g/dl) (Radhakrishnan and Cattran 2012). In adults, NS is defined as a urine protein level of >3.5 g/day and an albumin level of <3.0 g/dl (Lovric et al. 2016). Most (~80%) children with NS are idiopathic and respond well to steroid therapy (steroid sensitive), while the remaining 20% are resistant to steroids (**steroid-resistant nephrotic syndrome, SRNS**) (Fig. 21.5) (Radhakrishnan and Cattran 2012). Renal histology shows varying patterns, ranging from minimal change nephrotic syndrome to diffuse mesangial sclerosis, or focal segmental glomerulosclerosis (**FSGS**) (Radhakrishnan and Cattran 2012; Lovric et al. 2016), which is the most common histological pattern of SRNS in both children and adults. FSGS accounts for 20% of NS cases in children and 40% in adults (D'Agati et al. 2011).

Clinically, there are two major entities of **SRNS**: (1) *early-onset SRNS*, which occurs earlier in glomerular development, manifesting either prenatally, directly after birth, or during early infancy, and (2) *late-onset SRNS*, in which affected individuals first manifest proteinuria from childhood to young adulthood and typically develop FSGS with NS during the second to third decade of life. Approximately one-third of SRNS cases, in whom the onset of proteinuria ranges from birth to 25 years of age, are caused by genetic defects (Lovric et al. 2016; Benoit et al. 2010; Sampson and Pollak 2015; Sadowski et al. 2015; Trautmann et al. 2015).

21.4 Focal Segmental Glomerulosclerosis (FSGS)

FSGS is not a single disease but rather a clinicopathological entity, characterized by proteinuria with progressive renal dysfunction and a unique histological pattern of glomerular injury. Glomerulosclerosis initially develops both focally, involving a few glomeruli, and segmentally, affecting a portion of the glomerular globe. With progression, FSGS develops into more diffuse and global glomerulosclerosis. These features of FSGS represent a final common pathway for glomerular injuries resulting from various causes and pathogenic mechanisms (Fig. 21.5) (D'Agati et al. 2011).

Approximately 80% of FSGS cases are the **primary (idiopathic) form,** in which a circulating factor likely disrupts the glomerular barrier. Primary FSGS is diagnosed only after **secondary causes**, including genetic factors, infections, drugs, and reduced nephron loss, are excluded (D'Agati et al. 2011). Secondary FSGS arises from structural and functional adaptations to glomerular hypertension, termed **adaptive FSGS**. Such maladaptive responses are caused by a reduction in the number of functioning nephrons (e.g., renal agenesis, reflux nephropathy, or low nephron formation due to very low birth weight) or hemodynamic stress on an initially normal nephron population (e.g., obesity, cyanotic congenital heart disease, or sickle cell anemia). **Primary FSGS** usually presents with classic NS, whereas **secondary FSGS** often shows milder proteinuria with normal serum albumin levels and slower disease progression. Approximately two-thirds of patients with FSGS are fully nephrotic at presentation, whereas the remaining one-third have subnephrotic-range proteinuria. Hypertension and hematuria are also found in 30–50% and 50% of FSGS patients, respectively. The GFR is lower than the standard age-specific rate (20–30%) at presentation and gradually decreases over the course of the disease. Approximately half of FSGS cases progress to ESRD. FSGS caused by genetic factors, particularly mutations affecting the actin cytoskeleton, tends to exhibit milder proteinuria but microscopic hematuria and hypertension, as observed for the present case (III-2) (Brown et al. 2010).

Most FSGS cases are sporadic, but a small subset (~1%) has a genetic cause. **Genetic FSGS** cases are clinically indistinguishable from common primary FSGS, suggesting the importance of genetic testing for a definitive diagnosis. Recent studies on the monogenic form of FSGS have significantly advanced our knowledge on the pathogenesis (Lovric et al. 2016; Benoit et al. 2010; Sampson and Pollak 2015; Sadowski et al.

2015; Trautmann et al. 2015). Emerging evidence indicates that the majority of the causative genes remarkably converge into the glomerular podocytes. Genetic FSGS occurs as a single renal disease but may occasionally accompany other extrarenal manifestations involving the eye, ear, and skeletal, central nervous, or peripheral nervous systems. Exploration of the molecular causes of such syndromic cases will increase understanding of the mechanisms common among various nonrenal tissues.

21.5 Differential Diagnosis

A clinical hallmark of FSGS is proteinuria. If hematuria is predominant over proteinuria, a diagnosis of primary GBM defects or glomerulonephritis (IgA nephropathy) is more likely rather than primary FSGS. Alport syndrome (AS) is a common GBM disorder caused by genetic alterations in any of the three α chains of type IV collagen located in the GBM (α3, α4, α5). The genetic alterations occur at a considerably high incidence in familial FSGS (10–40%) (Malone et al. 2014; Gast et al. 2016). About 80% of AS are caused by mutations in the *COL4A5* gene with the X-linked transmission. On the other hand, the remaining 20% cases result from the mutations in one or both copies of either *COL4A3* or *COL4A4* gene, thereby showing a dominant or recessive transmission pattern. In X-linked AS patients, ultrastructural GBM abnormalities are typically prominent, manifesting as a basket-weave appearance and/or lamellation. Moreover, α5 chain immunostaining is typically negative in the glomerular capillary. The GBM in the proband (III-2) appeared morphologically normal overall with positive α5 chain immunostaining (Fig. 21.3a), arguing against the diagnosis of X-linked AS. However, distinguishing pure genetic FSGS from primary GBM defects is sometimes difficult, because GBM abnormalities are weak or even absent during the earlier stages and develop non-specifically with progressing glomerular damage. Immunofluorescence revealed IgA deposition in a coarse segmental pattern in addition to IgM and C3 expression in the proband (III-2) (Fig. 21.3a). The macromolecules IgM and C3 are often non-specifically expressed in the sclerotic areas of FSGS (D'Agati et al. 2011). IgA deposits in IgA nephropathy are typically found more diffusely around the mesangial angle. Thus, the IgA deposits in this case appeared to be of no pathogenic significance as similar to those reported for other genetic FSGS cases (Xie et al. 2015), making the diagnosis of IgA nephropathy less likely.

Mitochondrial defects are relatively common causes of kidney disorders and are inherited maternally. Histologically, cellular respiration failure leads to mitochondrial dysmorphism or increasing numbers or fragmentations in glomerular podocytes and/or tubular cells (Koopman et al. 2012), cell types that demand more energy. The lack of mitochondrial alterations on renal biopsy (Fig. 21.3) suggested an absence of mitochondrial diseases. Taken together, the present case (III-2) manifested subnephrotic proteinuria with a histology suggesting FSGS and a mildly decreased GFR. The patient lacked any systemic disorders (e.g., diabetes and hypertension). The renal pathology (Figs. 21.2 and 21.3) excluded the possibility of immune-complex deposition disease, GBM defects, or mitochondrial cytopathies. Genetic testing was subsequently performed to clarify the molecular pathophysiology underlying the familial FSGS.

21.6 Molecular Diagnosis

The parent-to-child transmission (from II-2 to III-2) of FSGS in this family suggests the presence of autosomal dominant types of FSGS. Three genes, *TRPC6*, *INF2*, and *ACTN4*, account for approximately half of autosomal dominant FSGS cases (Fig. 21.4) (Brown et al. 2010; Barua et al. 2013). Sequence analyses of both the affected mother and proband revealed a heterozygous, single-nucleotide transversion of C (cytosine) to A (adenine) at cDNA position 482 (c.482 C > A), located in exon 3 of the inverted formin 2 (*INF2*) gene (RefSeq accession no. NM_022489) (Fig. 21.6a). This nucleotide change leads to a

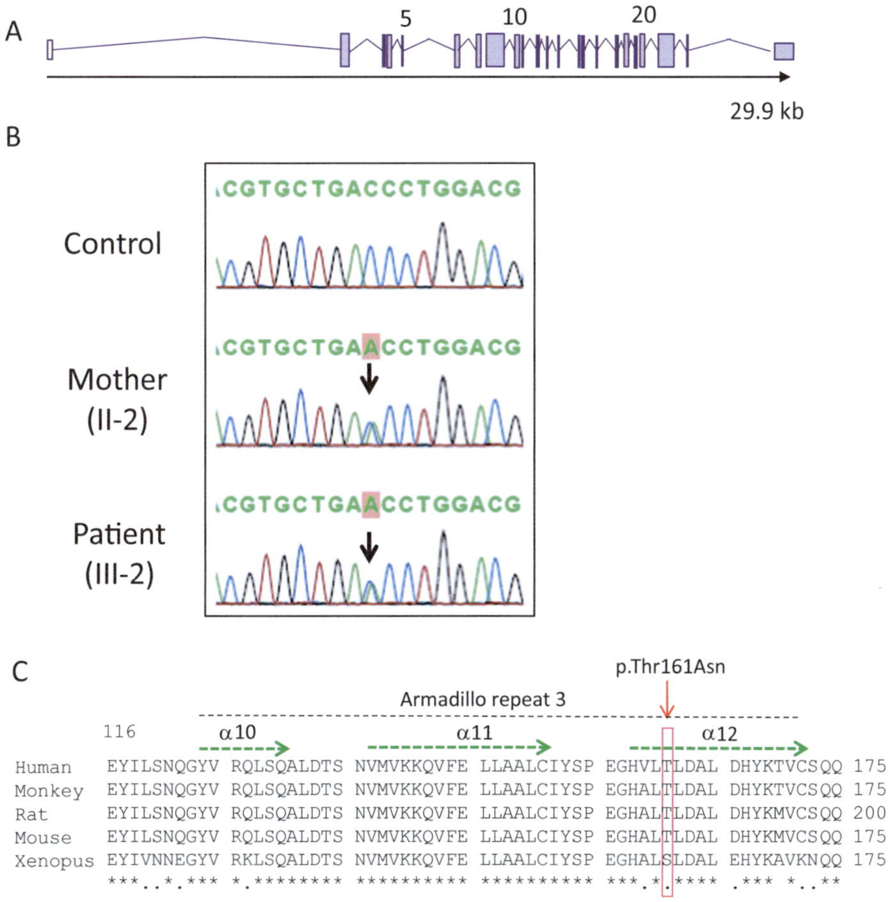

Fig. 21.6 Sequence analysis of *INF2* gene
(**a**) The exons (23 total) and introns of the *INF2* gene (29.9 kb) are shown by *boxes* and *lines*, respectively. The majority of mutations responsible for FSGS are clustered in exons 2–4. The *numbers* above the diagram indicate the exon number. Physical size is shown in kb. (**b**) Sequencing of *INF2* genomic DNA revealed a heterozygous, single-nucleotide, C-to-A transversion (c.482C>A, *arrow*), causing a threonine (ACC) to asparagine (AAC) substitution (p.Thr161Asn). cDNA numbering was in accordance with the reference sequence NM_022489.3. (**c**) Sequence alignment of *INF2* and its orthologues revealed that the threonine 161 residue (*box*) is highly conserved across various species. The *INF2* orthologous sequences from monkey, rat, mouse, and *Xenopus* were aligned using ClustalW. Portions of the different α helices are indicated by the *dotted arrows*. The number of α helices is depicted as the most N-terminal one is the first helix. The degree of conservation is classified as full (*), strong (:), or weak (.) conservation

threonine-to-asparagine missense substitution (p.Thr161Asn), located within the diaphanous inhibitory domain (DID) of INF2; this mutation is highly evolutionarily conserved among various species including chimpanzees, rats, and mice (Fig. 21.6b). The p.Thr161Asn mutation was found to be absent in over 1208 ethnically matched healthy control individuals and over 8600 controls in a public global database (ExAc). The p.Thr161Asn is a novel variant not found in mutation databases including ClinVar and the Human Gene Mutation Database.

In addition, in silico analyses using the PolyPhen-2 and SIFT algorithms predicted the p.Thr161Asn substitution to be "probably damaging" or "deleterious," with a score of 0.937 or 0.009, respectively. Homology modeling based on the mouse homolog of mammalian diaphanous-related formin 1 (mDia1) suggested that the p.Thr161Asn substitution alters the structure of the

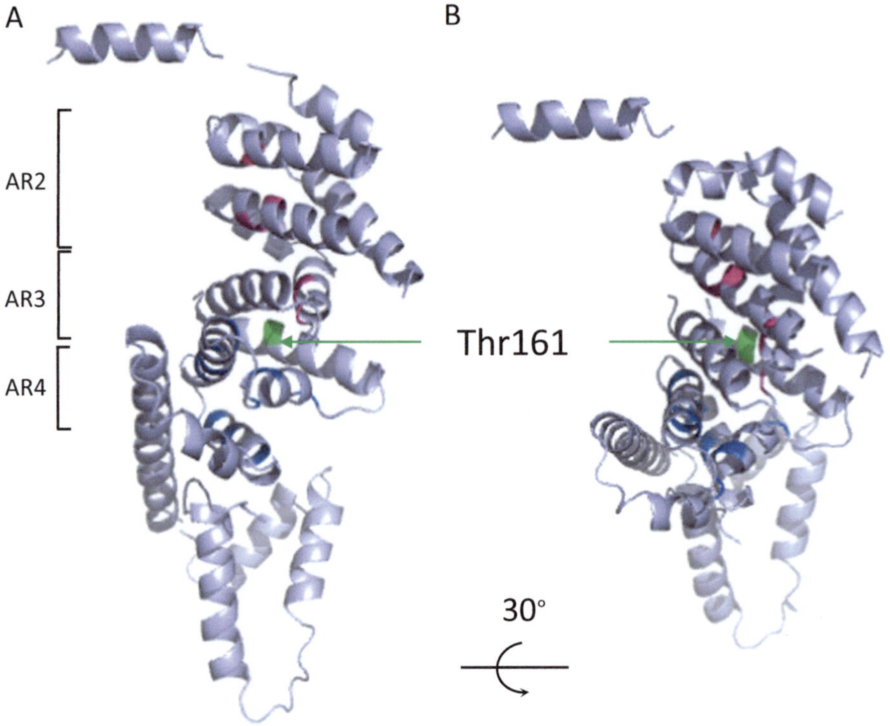

Fig. 21.7 Structural modeling of *INF2* mutations
Three-dimensional structure of the diaphanous inhibitory domain (DID) of human INF2 was modeled based on the reported crystal structure of the mouse homolog of mammalian diaphanous-related formin 1 (mDia1, PDB entry 2BNX). Amino acids 1–330 of INF2 (corresponding to amino acids 46–440 of mDia1) were analyzed using the Phyre protein threading program (Brown et al. 2010; Boyer et al. 2011). *Ribbon* presentation of the protein structure viewed from the front (**a**, *left*) and after a 30° vertical rotation (**b**, *right*), obtained using the PyMOL program. The DID is composed of four armadillo repeats (ARs; only ARs 2–4 are shown for clarity), each of which is formed by three adjacent α helices tertiarily folded into a superhelical coil (Boyer et al. 2011). The DAD of INF2 and mDia are thought to bind in the concave pocket of DID, which is formed by the central helices of ARs 2 and 3 (residues Gln 71–Cys172), thereby mediating self-auto-inhibitory and heteromeric regulatory interactions, respectively (Brown et al. 2010; Boyer et al. 2011). The positions of the altered residue in this family (Thr161, *green*, *arrow*), as well as those of reported pathogenic mutations causing the phenotypes of either isolated FSGS (*blue*, Arg177, Lys184, Ser186, Tyr193, Leu198, Asn202, Ala203, Arg214, Arg218, Lys 220) (Boyer et al. 2011) or FSGS with CMT disease [MIM614455] (*red*, Leu57, Cys104, Arg106, Leu128, Leu132) (Radhakrishnan and Cattran 2012; Lovric et al. 2016; Benoit et al. 2010; Sampson and Pollak 2015; Sadowski et al. 2015; Trautmann et al. 2015), are mapped to the structural model. CMT, Charcot–Marie–Tooth

12th α-helix (amino acids 158–172) of the DID of INF2 (Fig. 21.7). Structural studies have suggested that the α-helices of the DID form a concave pocket for the binding of interacting proteins (Brown et al. 2010). The p.Thr161Asn substitution may alter the protein interface for intramolecular (between DID and the diaphanous autoregulatory domain [DAD]) and/or intermolecular heteromeric (between DID and regulatory partners, e.g., mDia–DAD and RhoA) interactions (Sampson and Pollak 2015; Brown et al. 2010).

21.7 Molecular Perspectives

Understanding the pathomechanisms underlying genetic FSGS is one of the primary challenges in clarifying the etiology of progressive kidney disorders. Recent studies of monogenic FSGS revealed that the vast majority of FSGS genes are physiologically committed to control the development and/or structural function maintenance of podocytes. To date, over 30 genes have been associated with SRNS, thereby advancing our

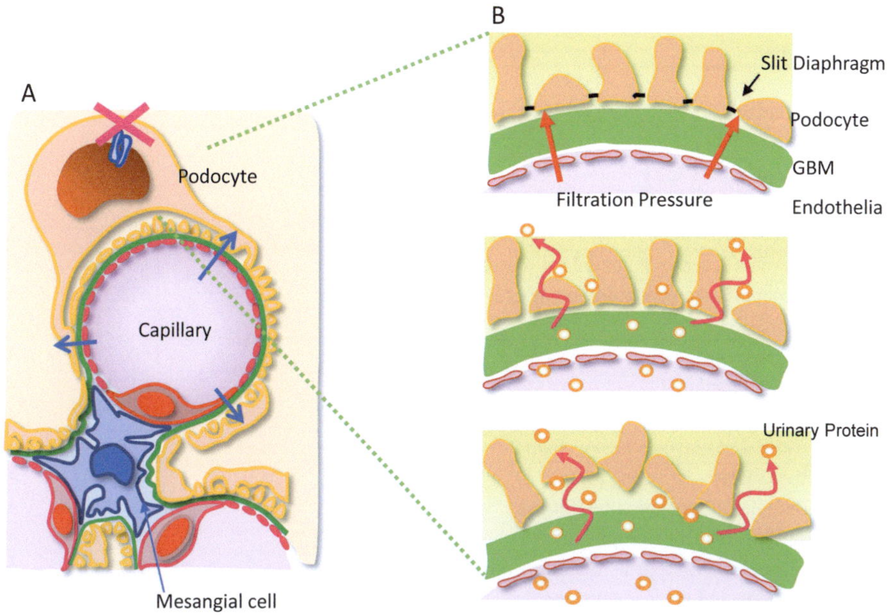

Fig. 21.8 Pathogenic mechanism of focal segmental glomerulosclerosis
(**a**) In the glomerular capillary, the filtration units are composed of the endothelium (*red*), glomerular basement membrane (GBM; *green*), and visceral epithelium (podocytes, *orange*). *Arrows* indicate the direction of filtration driven by the capillary pressure (reproduced with slight modification from Floege J et al., Comprehensive Clinical Nephrology, fourth edition, with permission from Elsevier). (**b**) A model of FSGS pathogenesis is presented schematically. *Upper panel:* under normal conditions, the slit diaphragm formed between apposing foot processes prevents urinary protein leakage. *Middle panel*: genetic defects or external insults affecting the podocytes disrupt the functional and structural integrity of the slit diaphragm, thereby allowing protein leakage into the urine. *Lower panel:* accumulating podocyte damage consequently leads to podocyte death and detachment from the GBM

knowledge of the pathogenic mechanisms involved in SRNS and podocyte development and function (Fig. 21.4, Table) (Lovric et al. 2016; Benoit et al. 2010; Sampson and Pollak 2015; Sadowski et al. 2015; Trautmann et al. 2015). These observations have led to the emerging concept that FSGS is a podocyte disease, or "podocytopathy," considering that the integrity of the glomerular slit diaphragm complex plays a key role in glomerular barrier function. Consistent with the paradigm that FSGS is a podocytopathy, the patient harbored a heterozygous missense mutation in the *INF2* gene. *INF2* mutations are major causes of familial autosomal dominant FSGS, with a prevalence of approximately 15% in all screened families (Barua et al. 2013).

Podocytes are terminally differentiated, high-energy-requiring cells that have no mitotic activity and do not proliferate after injury. Thus, once podocytes are injured and subsequently detach from the GBM, the denuded GBM is incapable of repair via the renewal of podocytes (Fig. 21.8). The unrepaired GBM will lead to adhesion between the capillary tuft and Bowman's capsule, a key event that triggers a series of events leading to progression of sclerosis. The initially focal and segmental pattern of sclerosis will gradually develop into a diffuse and global appearance. The histological progression is clinically correlated with a continual decrease in the GFR, ultimately leading to ESRD (Fig. 21.8).

From an etiological point of view, podocyte genes are classified into five subclasses (Fig. 21.4): (1) slit diaphragm, (2) actin cytoskeleton, (3) transcription factors and nuclear envelope, (4) mitochondria, and (5) GBM and basolateral anchoring of podocytes to the GBM (Radhakrishnan and Cattran 2012; Lovric et al. 2016; Benoit et al. 2010; Sampson and Pollak 2015; Sadowski et al. 2015; Trautmann et al. 2015) (Table 21.1). INF2

Table 21.1 Genes causing focal segmental glomerulosclerosis (FSGS)

Gene	Proteins	OMIM	Inheritance
I. Slit membrane			
NPHS1	Nephrin	256300	AR
NPHS2	Podocin	600995	AR
PLCE1	Phospholipase C ε1	610725	AR
CD2AP	CD2-associated protein	607832	AR (AD)
PTPRO	Protein tyrosine phosphatase receptor-O	600579	AR
TRPC6	Transient receptor potential cationic channel	603965	AD
CRB2	Drosophila crumbs family member 2	616220	AR
II. Actin cytoskeletons			
ACTN4	α-Actinin 4	603278	AD
INF2	Inverted formin 2	613237	AD
MYO1E	Non-muscle myosin 1E	614131	AR
MYH9	Non-muscle myosin heavy chain 9	155100	AD
ARHGAP24	Rho GTPase-activating protein 24	610586	AD
ARHGDIA	Rho GDP-dissociation inhibitor 1	615244	AR
ANLN	Anillin, actin-binding protein	616032	AD
KANK1,2,4	Kidney ankyrin repeat-containing protein	617783	AR
III. Nuclear transcription and nuclear envelope			
NUP107	Nucleoporin 107	616730	AR
NUP93	Nucleoporin 93	616892	AR
NUP205	Nucleoporin 205	616893	AR
EPO5	Exportin-5	607845	AR
WT1	WT tumor suppressor gene 1	194080	SD[a]
		136680	SD[a]
PAX2	Paired box gene 2	616002	AD
LMX1B	LIM homeobox transcriptional factor 1β	161200	AD
SMARCAL1	SWI/SNF-related, matrix-associated, actin-dependent regulator of chromatin, subfamily A-like 1	606622	AR
NXF5	Nuclear RNA export factor 5	300319	XL
IV. GBM and extracellular matrices			
LAMB2	Laminin β-2	609049	AR
ITGA3	Integrin α-3	614748	AR
ITGB4	Integrin β-4	147557	AR
V. Mitochondria			
COQ2	Coenzyme Q10 biosynthesis monooxygenase 2	607426	AR
COQ6	Coenzyme Q6 monooxygenase	614650	AR
PDSS2	Decaprenyl-diphosphate synthase subunit 2	607429	AR
MTTL1	Mitochondrially encoded tRNA leucine 1 (UUA/G)	590050	Maternal[b]
ADCK4	aarF domain-containing kinase 4	615573	AR
VI. Miscellaneous			
SCARB2	Scavenger receptor class B member 2	254900	AR
APOL1	Apolipoprotein L-1	612551	AR

AD autosomal dominant, *AR* autosomal recessive, *XR* X-linked, *SD* sporadic
[a] Autosomal dominant inheritance is reported on rare occasions
[b] maternal inheritance

is a regulator of actin filaments within the foot processes of podocytes. Since the glomerular filtration barrier is driven by an outward pressure gradient (40–50 mmHg) from the capillary, podocytes are usually exposed to this pressure (Kriz and Elger 2010). For these environmental reasons, podocytes are susceptible to injury, particularly if they have genetically determined intrinsic structural weaknesses. Mutations affecting the actin cytoskeleton, including those in *INF2*, are generally heterozygous missense substitutions likely to cause subtle changes in podocyte foot processes. Such mutations with mild consequences do not affect development of podocytes but rather impair repair and remodeling processes. Clinically overt proteinuria, due to aberrations in slit diaphragm integrity, develops during late childhood to adulthood once the accumulating stress exceeds the pathologic threshold, thereby triggering podocyte death and detachment. Although SRNS is the leading cause of ESRD in children worldwide, approximately 70% of those with childhood-onset SRNS are still genetically uncharacterized. Further efforts are therefore obviously needed to identify the genes responsible for SRNS.

21.8 Cellular and Biochemical Perspectives

21.8.1 Role of Actin Regulation in Proteinuric Glomerular Disorders

Podocytes are equipped with complex, actin-rich, interdigitating structures that enable their adaption to physical and chemical stresses. The foot processes can change their shape flexibly in a highly dynamic manner via fast reorganization of the actin-based cytoskeleton. The cellular processes are retracted (effacement) in the proteinuric state but quickly restore their normal interdigitating configuration in remission. Actin cytoskeletal dynamics are regulated mainly by the Rho family of small GTPases (Rho, Rac, and Cdc42), as well as their interacting partners (Breitsprecher and Goode 2013). Dysregulation of Rho GTPase activities has been shown to cause FSGS in mice and humans (Sadowski et al. 2015; Trautmann et al. 2015).

21.8.2 Molecular Basis of *INF2* Mutations Leading to FSGS

INF2 is a member of the formin family of actin-regulating proteins that mediates a variety of physiological process (i.e., maintenance of cell polarity, migration, and proliferation and microtubule assembly). INF2 serves to both elongate (polymerization) and shorten actin filaments (depolymerization). In addition to its direct role in mediating actin polymerization, INF2 acts as a downstream negative regulator of RhoA signaling by antagonizing the effects of mDia1. INF2 protein is ubiquitously expressed in tissues including podocytes and Schwann cells (Brown et al. 2010; Boyer et al. 2011). Podocytes bearing *INF2* mutations are expected to be highly susceptible to injuries because of an inability to restore the integrity of foot processes by fine-tuning actin remodeling.

Interestingly, the disease-causing *INF2* mutations reported to date are clustered in the DID, encoded by exons 2 and 3, within the N-terminus of INF2, suggesting specific roles for this domain in the pathogenesis of FSGS (Fig. 21.9) (Sampson and Pollak 2015; Brown et al. 2010; Barua et al. 2013). The N-terminal DID binds autonomously to the C-terminal DAD within the same INF2 molecule, thereby creating a closed loop configuration that autoinhibits depolymerization. It is therefore speculated that the mutations affecting DID function impair the autoinhibition ability of INF2, leading to impaired actin organization in the foot processes (Breitsprecher and Goode 2013). Recent studies using mice harboring the *Inf2* Arg218Gln mutation (Subramanian et al. 2016) showed neither renal phenotype nor histological abnormalities in either the heterozygous or homozygous state. However, homozygous Arg218Gln mice displayed delayed recovery of podocyte injury induced by secondary challenge with polycationic reagents. The observations suggest that the *INF2* Arg218Gln mutation does not alter slit diaphragm function under normal situations but rather impairs recovery processes in podocytes after injury. The phenotype of Arg218Gln mutant mice agrees with the late-onset nature of *INF2* mutation causing human FSGS, in that patients do not manifest overt proteinuria until adolescence, once the accumulation of injuries exceeds the pathogenic threshold.

Interestingly, *INF2* mutations have been shown to cause the peripheral neuropathy **Charcot-Marie-Tooth (CMT) disease**, present-

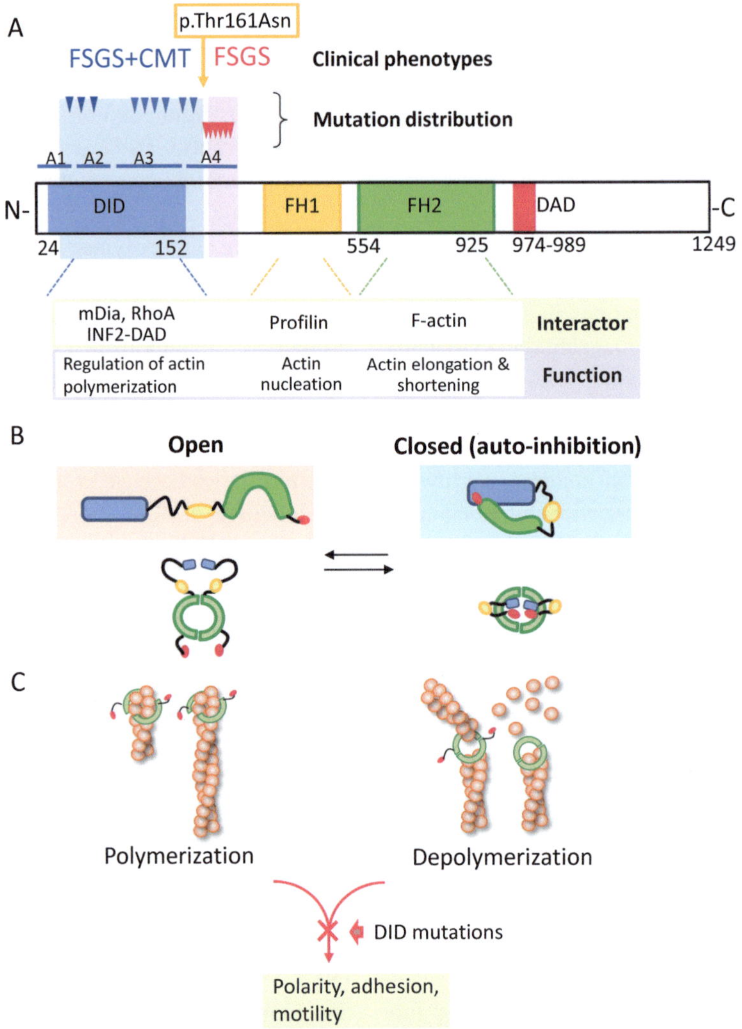

Fig. 21.9 Positions of *INF2* mutations and domain structure and functions
(**a**) ***INF2*** **mutations and domain structure**. The locations of p.Thr161Asn and reported *INF2* mutations are shown. *Arrowheads* indicate the distinct positions of mutations associated with the phenotypes of FSGS presenting with Charcot–Marie–Tooth disease (CMT; *blue arrowheads*) and isolated FSGS (*red*). *Bars* underneath the mutations indicate the positions of the four armadillo repeats (A1–4). The INF2 protein is a large multi-domain protein composed of three functional domains: N-terminal diaphanous inhibitory domain (DID, *blue*), formin homology (FH) domains (*orange and green*), and C-terminal diaphanous autoregulatory domain (DAD, *red*). The DID interacts with the DAD of the same molecule (intramolecular autoinhibition) as well as with other actin regulators (intermolecular interactions with RhoA and mDia) (Sampson and Pollak 2015; Brown et al. 2010). The FH domains mediate both assembly and disassembly of actin filaments. The functional domains were predicted by InterPro (Q27J81, EMBL). *Numbers* beneath the diagram of the protein indicate the amino acid positions of INF2. mDia, mammalian diaphanous-related formin 1. (**b**) **Conformational changes of INF2**. INF2 functions as a dimer, in which FH2 domains are aligned in a paired, head-to-tail configuration forming a doughnut structure (reproduced with slight modification from Breitsprecher and Goode (2013) with permission from The Company of Biologists). The INF2 molecule could form either *open* or *closed* configuration, the latter of which is mediated through the DAD–DID interaction. The intramolecular DID–DAD interaction normally silences actin depolymerization. Binding of the effector Rho A sterically regulates INF2 activity by changing an INF2 conformation from a close to open configuration (Otomo et al. 2005). (**c**) **Hypothetical model of actin regulation**. INF2 mediates both elongation (*left*, polymerization) and shortening (*right*, serving or depolymerization) of actin filaments. The dynamic actin organization and remodeling are regulated by the balance between the *open* and *closed* state of INF2. Mutations in the DID region of INF2 would disrupt INF2 functions including autoinhibition, homodimerization, and/or interaction with Rho GTPases. The resulting aberrant actin filament organization and remodeling impair polarity, adhesion, and proliferation of podocytes or Schwann cells

ing simultaneously with FSGS (Boyer et al. 2011). CMT is a monogenic neurodegenerative disorder affecting both motor and sensory neurons that ultimately leads to progressive distal muscle weakness and atrophy. Sensory deafness manifests in one-third of patients affected by both CMT and FSGS. The co-occurrence of CMT with FSGS, first described in 1967, is very rare, but more than 20 cases have been reported. The observations from these cases revealed an unexpected role of INF2 in maintaining Schwann cells in healthy individuals.

All the *INF2* mutations cluster in the DID domain but distinctly localize between the two groups of patients (Boyer et al. 2011) (Fig. 21.9). Mutations causing a dual CMT and FSGS disease localize predominantly in the second armadillo repeat of the DID (residue 71–119). Those in the third armadillo repeat (residues 124–172), including the present case (p.Thr161Asn), principally lead to an isolated FSGS. However, with lesser frequency, the third armadillo repeat could manifest a dual CMT/FSGS phenotype, suggesting that alterations in the second repeat give more susceptibility to CMT than those in the third repeat. In contrast, mutations in the fourth armadillo repeat (residues 179–224) invariably manifest FSGS only (Brown et al. 2010; Barua et al. 2013; Boyer et al. 2011). These positional effects of mutations may be explained by the distinctive structural roles of the second and third armadillo repeats, which form the central core of the DID to bind the DAD (Figs. 21.6 and 21.7). The aberrant DID–DAD interaction caused by mutations may impair the autoinhibition, thereby leading to more severe dual CMT/FSGS phenotypes. Alternatively, the N-terminal part DID comprising the second and third armadillo repeats may play a tissue-specific role in the maintenance of Schwann cells via interactions with myelin-specific Rho GTPases (e.g., Cdc42) and other proteins (e.g., the myelin and lymphocyte protein; MAL) (Boyer et al. 2011). These *INF2* DID mutations were not found in patients with isolated CMT disease, suggesting that podocytes are more sensitive to INF2 defects rather than Schwann cells.

The affected family members (i.e., the proband and mother), harboring the p.Thr161Asn mutation, lack any clinical or electrodiagnostic evidence of peripheral neuropathies so far when we follow up them on clinical visits. The absence of peripheral neuropathy reconciles with the view that the DID mutations involving the third armadillo repeat usually cause the FSGS phenotype only. However, the phenotypic expression and severity of peripheral neuropathy due to the DID third armadillo repeat may vary among patients, depending upon the locations of mutations, coexisting secondary modifiers, and environmental factors. For instance, the third armadillo repeat DID mutation p.Leu165Pro, which arises in four-residue upstream to Thr161, exhibited a CMT/FSGS double phenotype (Boyer et al. 2011). Careful neurological follow-up is therefore necessary to evaluate neuromuscular and auditory functions.

21.9 Treatment

The goal of therapy for FSGS is to reduce proteinuria and preserve renal function. Most untreated and unremitted patients will eventually develop ESRD within 5–20 years (D'Agati et al. 2011). The clinical responses depend on the age of onset as well as underlying etiologies. Distinguishing primary and secondary FSGS is crucial, because the treatment approach differs between the two types. For primary FSGS, some patients may respond to empirical steroid therapy. In contrast, for secondary FSGS, treatment of the underlying cause should be prioritized, and immunosuppression is not beneficial. Genetic testing to elucidate the molecular pathogenesis will help in deciding the treatment and assessing prognosis (Sadowski et al. 2015; Trautmann et al. 2015).

21.9.1 Corticosteroids

A corticosteroid is generally used as the initial therapy. Even partial remission can improve survival. The initial response to steroids is associated with the renal prognosis. Patients with genetic FSGS show a higher frequency of steroid resistance, compared with those with nongenetic FSGS, of whom 70–80% usually enter remission (Giglio et al. 2015; Buscher et al. 2016).

21.9.2 Immunosuppressive Therapy

Children and adults affected with FSGS generally show resistance to the standard steroid regimens according to the clinical guidelines of International Study of Kidney Disease in Children (ISKDC) or Kidney Disease: Improving Global Outcomes (KDIGO) (Giglio et al. 2015; Buscher et al. 2016). In most cases, calcineurin inhibitors are administered as second-line treatment (Radhakrishnan and Cattran 2012).

21.9.3 Other Medications

Angiotensin-converting enzyme inhibitors and/or angiotensin II receptor blockers are effective in reducing proteinuria and may be used initially for FSGS with mild proteinuria.

21.9.4 Transplantation

Renal transplantation is considered for children with FSGS who develop ESRD earlier in life. The genetic forms of FSGS usually exhibit steroid resistance and a progressive course but generally do not have recurrence of nephrotic syndrome after renal transplantation. This is because, in podocytopathies, the defects are intrinsic to the kidney and can be theoretically cured by the allograft transplantation. A genetic test is thus recommended to determine the underlying etiology and whether the FSGS arises from a podocytopathy.

The overall frequency of posttransplantation recurrence of primary FSGS is generally 30–40% and is often associated with a circulating permeability factor. In contrast, recurrence of NS posttransplantation in genetic FSGS is considerably low (~5%) (D'Agati et al. 2011; Radhakrishnan and Cattran 2012; Lovric et al. 2016; Benoit et al. 2010; Sadowski et al. 2015; Trautmann et al. 2015), because genetic FSGS is a podocyte-specific disease. However, it has been recognized that some genetic FSGS cases recur after transplantation. Posttransplantation recurrence was reported in FSGS patients with *INF2* mutations and coexisting risk-associated mutations in complement factor H and CD46, a complement regulatory protein that activates alternative pathways (Challis et al. 2017).

Questions

1. What are the basic mechanisms of proteinuria development?
2. What molecular pathways are involved in the pathogenesis of FSGS? Why are FSGS mutations clustered in podocytes?
3. How can podocyte mutations lead to an FSGS pattern of injury?
4. By what mechanisms do *INF2* mutations lead to two separate diseases: isolated FSGS or FSGS–CMT dual diseases?
5. How can genetic testing for SRNS clinically contribute to treatment?

Acknowledgments The author is grateful to Drs. Tatsuyo Takahashi and Masataka Miyoshi (Ajisu Kyoritsu Hospital, Japan) for referral of the patients and to Drs. Naoya Morisada and Kazutomo Iijima for the sequencing analysis. The author would also like to thank Dr. Satoshi Hisano (Department of Pathology, Faculty of Medicine, Fukuoka University) for the preparation of the renal pathology specimens and diagnosis.

References

Barua M, Brown EJ, Charoonratana VT et al (2013) Mutations in the INF2 gene account for a significant proportion of familial but not sporadic focal and segmental glomerulosclerosis. Kidney Int 83(2):316–322. https://doi.org/10.1038/ki.2012.349

Benoit G, Machuca E, Antignac C (2010) Hereditary nephrotic syndrome: a systematic approach for genetic testing and a review of associated podocyte gene mutations. Pediatr Nephrol 25(9):1621–1632. https://doi.org/10.1007/s00467-010-1495-0

Boyer O, Nevo F, Plaisier E et al (2011) INF2 mutations in Charcot-Marie-Tooth disease with glomerulopathy. N Engl J Med 365(25):2377–2388. https://doi.org/10.1056/NEJMoa1109122

Breitsprecher D, Goode BL (2013) Formins at a glance. J Cell Sci 126:1–7. https://doi.org/10.1242/jcs.107250

Brown EJ, Schlondorff JS, Becker DJ et al (2010) Mutations in the formin gene INF2 cause focal segmental glomerulosclerosis. Nat Genet 42(1):72–76. https://doi.org/10.1038/ng.505

Buscher AK, Beck BB, Melk A et al (2016) Rapid response to cyclosporin A and favorable renal outcome in nongenetic versus genetic steroid-resistant nephrotic syndrome. Clin J Am Soc Nephrol 11(2):245–253. https://doi.org/10.2215/CJN.07370715

Challis RC, Ring T, Xu Y et al (2017) Thrombotic microangiopathy in Inverted Formin 2-mediated renal disease. J Am Soc Nephrol 28(4):1084–1091. https://doi.org/10.1681/ASN.2015101189

D'Agati VD, Kaskel FJ, Falk RJ (2011) Focal segmental glomerulosclerosis. N Engl J Med 365(25):2398–2411. https://doi.org/10.1056/NEJMra1106556

Gast C, Pengelly RJ, Lyon M et al (2016) Collagen (*COL4A*) mutations are the most frequent mutations underlying adult focal segmental glomerulosclerosis. Nephrol Dial Transplant 31(6):961–970. https://doi.org/10.1093/ndt/gfv325

Giglio S, Provenzano A, Mazzinghi B et al (2015) Heterogeneous genetic alterations in sporadic nephrotic syndrome associate with resistance to immunosuppression. J Am Soc Nephrol 26(1):230–236. https://doi.org/10.1681/ASN.2013111155

Koopman WJ, Willems PH, Smeitink JA (2012) Monogenic mitochondrial disorders. N Engl J Med 366(12):1132–1141. https://doi.org/10.1056/NEJMra1012478

Kriz W, Elger M (2010) Renal anatomy. In: Floege J, Johnson RJ, Feehally J (eds) Comprehensive clinical nephrology, 4th edn. Elsevier Inc, St. Louis, pp 3–14

Lovric S, Ashraf S, Tan W et al (2016) Genetic testing in steroid-resistant nephrotic syndrome: when and how? Nephrol Dial Transplant 31(11):1802–1813. https://doi.org/10.1093/ndt/gfv355

Malone AF, Phelan PJ, Hall G et al (2014) Rare hereditary COL4A3/COL4A4 variants may be mistaken for familial focal segmental glomerulosclerosis. Kidney Int 86(6):1253–1259. https://doi.org/10.1038/ki.2014.305

Otomo T, Otomo C, Tomchick DR, et al (2005) Structural basis of Rho GTPase-mediated activation of the formin mDia1. Mol Cell 18(3):273–281. https://doi.org/10.1016/j.molcel.2005.04.002

Radhakrishnan J, Cattran DC (2012) The KDIGO practice guideline on glomerulonephritis: reading between the (guide)lines--application to the individual patient. Kidney Int 82(8):840–856. https://doi.org/10.1038/ki.2012.280

Sadowski CE, Lovric S, Ashraf S et al (2015) A single-gene cause in 29.5% of cases of steroid-resistant nephrotic syndrome. J Am Soc Nephrol 26(6):1279–1289. https://doi.org/10.1681/ASN.2014050489

Sampson MG, Pollak MR (2015) Opportunities and challenges of genotyping patients with nephrotic syndrome in the genomic era. Semin Nephrol 35(3):212–221. https://doi.org/10.1016/j.semnephrol.2015.04.002

Subramanian B, Sun H, Yan P et al (2016) Mice with mutant Inf2 show impaired podocyte and slit diaphragm integrity in response to protamine-induced kidney injury. Kidney Int 90(2):363–372. https://doi.org/10.1016/j.kint.2016.04.020

Trautmann A, Bodria M, Ozaltin F et al (2015) Spectrum of steroid-resistant and congenital nephrotic syndrome in children: the PodoNet registry cohort. Clin J Am Soc Nephrol 10(4):592–600. https://doi.org/10.2215/CJN.06260614

Xie J, Hao X, Azeloglu EU et al (2015) Novel mutations in the inverted formin 2 gene of Chinese families contribute to focal segmental glomerulosclerosis. Kidney Int 88(3):593–604. https://doi.org/10.1038/ki.2015.106

22 Marfan Syndrome

Francesco Ramirez and Julie De Backer

Keywords
Dissecting aneurysm · Fibrillin 1 · Marfan syndrome · Mechanosensing · TGFβ signaling

22.1 Case Report

22.1.1 Family 1

At 2.5 years of age, the proband (III.3 in Fig. 22.1) was referred to an orthopedic surgeon for evaluation of pronounced flat feet with medial displacement of the medial malleolus. In addition to flat feet and tall and slender stature (105 cm, >p97, 11 kg < p10), a mild pectus carinatum deformity and marked generalized joint laxity were noted (Fig. 22.2). The orthopedic surgeon considered Marfan syndrome (MFS) in his differential diagnosis, and the boy was referred for cardiac evaluation where mild aortic root dilatation at the level of the sinuses of Valsalva (20 mm, z-score 2.8) and mitral valve prolapse (MVP) were diagnosed on cardiac ultrasound. Lens luxation was excluded on a slit lamp study. Based on these findings, a diagnosis of MFS was "suspected." Regular cardiac and orthopedic surveillance was scheduled.

The proband had two older brothers (Fig. 22.1), both tall, and one of them had some stretch marks, while the other one had mild myopia. Cardiac ultrasound in both of them was normal. Their parents' medical history was uneventful, but evaluation in the father revealed the presence of a chronic and complex type B aortic dissection, deemed unsuitable for repair at that time. He had a history of severe flat feet, being a reason for excluding him from military service. He didn't have any clinically significant scoliosis and presented a much milder pectus excavatum than his son. He had severe myopia since childhood. Beta-blocker treatment was initiated. Seventeen months after diagnosing the dissection, the proband's father died during surgery from acute aortic rupture of the dissected vessel at 45 years of age.

Due to excessive growth, the proband received testosterone treatment during puberty. Beta-blockers had not been given up to that point for reasons of mild asthma but were initiated with the testosterone treatment. With growth his scoliosis worsened to the extent of requiring surgical correction at 14 years of age (Fig. 22.3).

F. Ramirez (✉)
Department of Pharmacological Sciences, Icahn School of Medicine at Mount Sinai, New York, NY, USA
e-mail: francesco.ramirez@mssm.edu

J. De Backer
Department of Cardiology and Center for Medical Genetics, Ghent University Hospital, Ghent, Belgium
e-mail: Julie.DeBacker@ugent.be

T. Oohashi et al. (eds.), *Human Pathobiochemistry*, https://doi.org/10.1007/978-981-13-2977-7_22

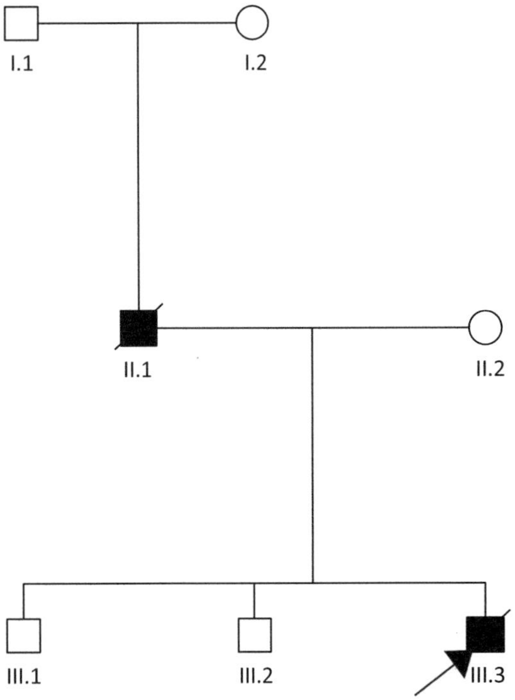

Fig. 22.1 Pedigree of Family 1 (F1). Proband (III.3) is indicated with an arrow; square symbols represent males, circles females; black symbols indicate affected individuals

Clinically, the boy developed additional characteristic skeletal manifestations, including marked pectus carinatum, arachnodactyly, decreased elbow extension, and distinctive facial features with downslanting palpebral fissures, malar hypoplasia, and a high-arched palate. His height at the age of 14 was 184 cm and weight 49 kg. Cardiac ultrasound revealed progressive dilatation of the aortic root at the level of the sinus of Valsalva (47 mm; z-score 6.8 at age 14–52 mm at age 15 and 56 mm at age 15.5) as well as mitral valve prolapse with mild mitral valve regurgitation.

Two years after scoliosis surgery, the proband underwent elective valve-sparing aortic root replacement. The surgical procedure was complicated by an episode of ventricular fibrillation requiring DC shock. After surgery, the proband developed left ventricular dilatation. A coronary angiography performed 10 months after surgery revealed normal coronary arteries. On follow-up at the age of 19, progressive dilatation with markedly depressed left ventricular systolic function was noted with a left ventricular end-diastolic diameter of 81 mm (normal 50 ± 4.1 mm) and an ejection fraction of 25% (normal 62%).

At the age of 24, the proband collapsed while driving his car and died. Autopsy revealed the absence of aortic dissection and no intracerebral abnormalities – sudden arrhythmic cardiac death was suspected as the cause of death. Genetic testing performed shortly before his death had revealed the presence of a mutation in the fibrillin-1(*FBN1*) gene (c.305_306del insAA; p.Cys102X in exon 3), which was not found in both of his brothers and his paternal grandparents.

22.1.2 Family 2

Cardiovascular evaluation for palpitations in the 62-year-old male proband in the second family (F2-II.1 in Fig. 22.4) revealed an aneurysm of the ascending aorta. The man had normal blood pressure, never smoked and had normal serum cholesterol levels. The diameter of the aortic root at the level of the sinuses of Valsalva was 57 mm (z-score 6.2). He underwent aortic root replacement using a composite graft (Bentall procedure with replacement of the proximal aorta by a Dacron graft and replacement of the aortic valve with a mechanical prosthesis). After surgery, the patient was referred to the medical genetics department for diagnostic work-up.

Medical history revealed that he had undergone ocular surgery at the age of 47 for bilateral cataract and 5 years later for glaucoma. Ocular examinations prior to these interventions had identified microspherophakia and very shallow anterior eye chambers, reminiscent for ocular features of Weill-Marchesani syndrome (WMS). On physical examination, the proband presented a stocky build with a height of 173 cm and weight of 110 kg (Fig. 22.5). His arm span-to-height ratio was greater than normal (armspan 183 cm, ratio 1.06, normal <1.05), but the upper/lower segment ratio was normal (0.92). He had a round face and a characteristically flat skull (brachycephaly). He presented marked stiffness of the

Fig. 22.2 Images of the proband in Family 1 as a child; his father is shown in the picture on the left

finger joints and brachydactylia. The chest wall, spine, elbows, and feet were all normal on inspection. He had no stretch marks.

The proband's family history revealed that his older brother and sister had a very similar phenotypic appearance. Reportedly, they both had ocular problems and had died suddenly at the age of 63 and 45, respectively. The proband's mother and maternal uncle also had been diagnosed with glaucoma. However, they were tall and had long, slender fingers. They both died suddenly at the age of 84 and 70, respectively, from an "unspecified cardiac cause."

Based on the presence of a thoracic aortic aneurysm with a family history of sudden death, the proband was tested for a probable *FBN1* mutation. A heterozygous 12 nucleotide in frame deletion was detected in exon 20 of the *FBN1* gene (c.2502-2513delT- GAAAGTACTTT; p.Glu 835-Leu838del). It was not until the identification of the genetic defect that the proband agreed to contact and examine his children. Physical examination of the eldest son (F2-III.1) revealed the presence of mild skeletal features of MFS (increased armspan, arachnodactylia), in addition to severe myopia, bilateral upward luxation of the lenses, and moderate dilatation of the proximal aorta at the level of the sinuses of Valsalva (43 mm, z-score 3.14). Examination of the proband's 7-year-old grandson (F2-IV.2) revealed several skeletal manifestations, including scoliosis, pectus deformity, arachnodactylia, and flat feet, as well as bilateral subluxation of the lenses and myopia. Echocardiographic evaluation was completely normal at that time. Molecular studies confirmed the presence of the *FBN1* deletion in the proband's son and grandson. Clinical evaluation of the proband's daughter (F2-III.4) showed no manifestations of MFS. Molecular analysis confirmed the absence of the *FBN1* mutation. The proband's granddaughter (F2-IV.1) was only evaluated molecularly and found not to carry the *FBN1* mutation. Over the years, the proband's grandson (F2-IV.2) developed progressive aortic root dilatation, which by the age of 16 had a diameter of 41 mm

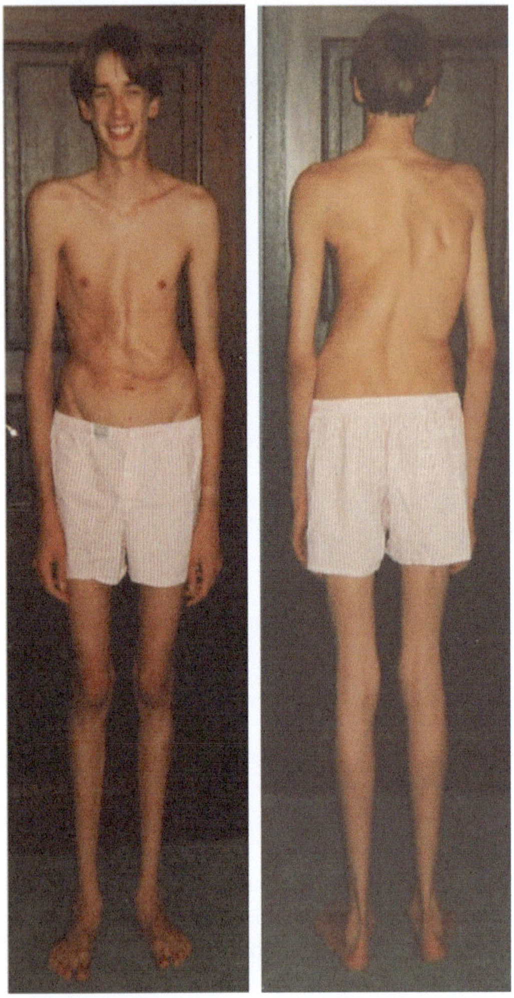

Fig. 22.3 Images of the proband in Family 1 at age 14 prior to scoliosis surgery

(z-score 4.8) at the level of the sinus of Valsalva – on the other hand, his father showed no additional growth of the aneurysm, remaining stable at 43 mm.

22.2 Diagnosis: Marfan Syndrome (MFS)

The first clinical description of MFS was at the end of the nineteenth century by the French pediatrician Antoine Bernard Marfan, who reported of a young girl presenting a combination of striking skeletal features. However, it is now well-established that the skeletal manifestations of MFS are not specific to this disease condition but show significant overlap with other connective tissue disorders and other genetic aortic disease entities (Table 22.1).

Many skeletal features of MFS can also occur in the general population, and conversely many MFS patients may not exhibit the full phenotype. As a result, MFS diagnosis relies on strict clinical criteria that have evolved over time. The current MFS nosology (aka, revised Ghent nosology; Loeys et al. 2010a) stipulates that diagnosis requires the presence of two cardinal clinical features (aortic root dilatation and ectopia lentis) and a systemic score based on manifestations in different organ systems. Depending on family history, seven diagnostic scenarios based on cardinal features are possible:

In the absence of family history:

1. Ao ($Z \geq 2$) + EL = MFS
2. Ao ($Z \geq 2$) + *FBN1* = MFS
3. Ao ($Z \geq 2$) + Syst (≥7 pts) = MFS
4. EL + *FBN1* with known Ao = MFS

In the presence of family history:

5. EL + FH of MFS (as defined above) = MFS
6. Syst (≥7 pts) + FH of MFS (as defined above) = MFS
7. Ao ($Z \geq 2$ in adults, $Z \geq 3$ in children) + FH of MFS (as defined above) = MFS

Ao aortic root dilatation defined as a z-score ≥2 for adults and ≥3 for children; *EL* ectopia lentis; *Syst* systemic score. “Adults” are defined when ≥20 years.

Systemic features used for MFS diagnosis are scored as follows:

Wrist and thumb sign – **3** (wrist or thumb sign – **1**)
Pectus carinatum deformity – **2** (pectus excavatum or chest asymmetry – **1**)
Hindfoot deformity – **2** (plain pes planus – **1**)
Protrusio acetabuli – **2**
Pneumothorax – **2**
Dural ectasia – **2**

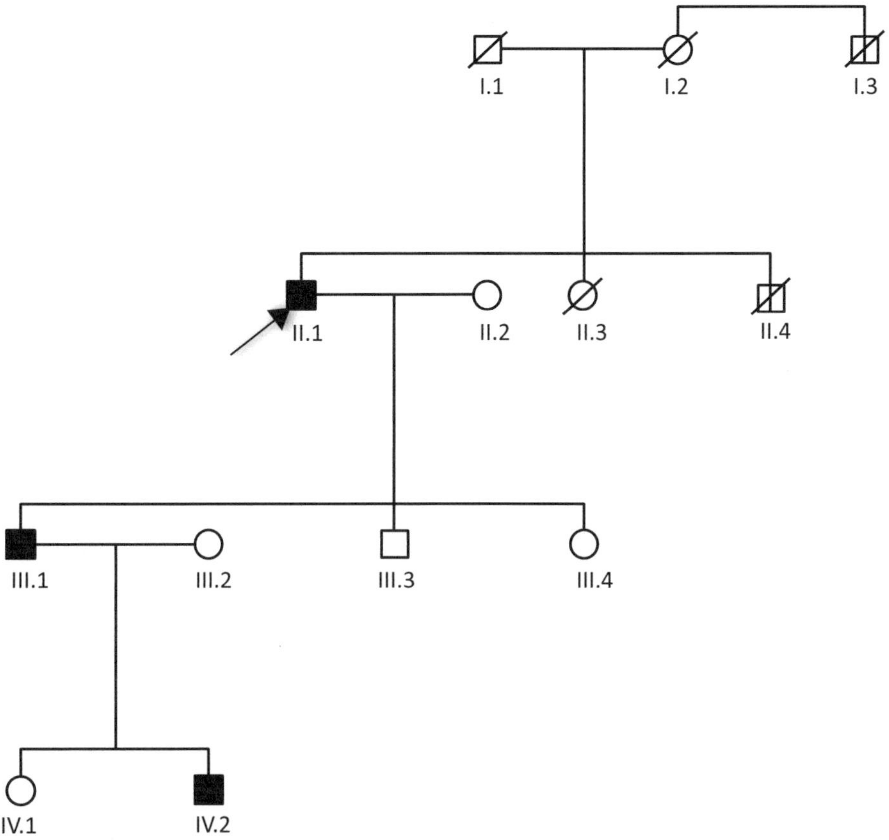

Fig. 22.4 Pedigree of Family 2; the proband (II.1) is indicated with an arrow

Reduced US/LS and increased arm/height and no severe scoliosis – **1**
Scoliosis or thoracolumbar kyphosis – **1**
Reduced elbow extension – **1**
Facial features (3/5) – **1**
Skin striae – **1**
Myopia >3 diopters – **1**
Mitral valve prolapse (all types) – **1**
TOTAL: 20 points

While improved, MFS diagnosis can still be a challenge particularly in young children who have not yet developed the full-blown phenotype. In these cases, intermediate diagnoses like "non-specific connective tissue disorder" and "potential MFS" can be used. Several issues related to the diagnostic difficulties in MFS are nicely illustrated by the cases presented above:

1. At the initial evaluation at 2.5 years of age, the boy of the first family did not fulfill the diagnostic criteria of MFS. With a systemic score of 5/20, an aortic root z-score of 2.8, and no ectopia lentis, he would have been classified as afflicted with a "non-specific connective tissue disorder."
2. As illustrated by the proband of Family 2 and the boy's father of F1, outward features of the disease may be absent or so subtle to go unnoticed throughout life – often until a fatal event occurs.
3. The variability of clinical features in patients with MFS may be striking, even within the same family. Case in point is the very pronounced skeletal abnormalities of the proband in Family 1 that were largely absent in his father. The variability in the second family is

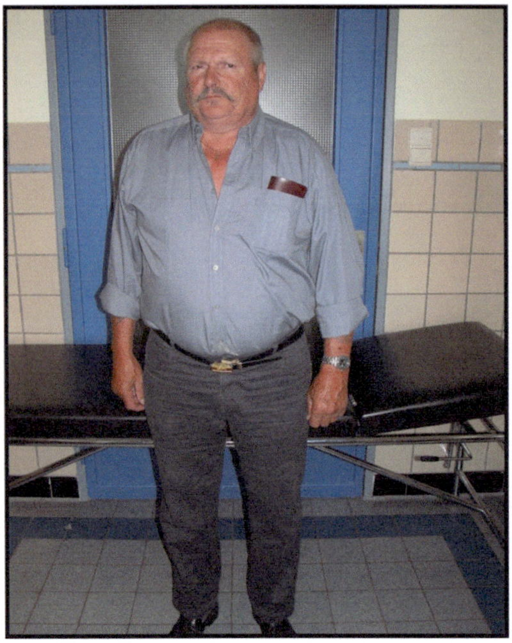

Fig. 22.5 Image of the proband in Family 2

even more striking and does not only apply to the skeletal system but also to the ocular and cardiovascular system.

4. Molecular testing is helpful for the diagnosis of affected family members *and* for exclusion of the diagnosis in unaffected family members. Both brothers of the proband in Family 1 had mild Marfanoid features – demonstrating the absence of the molecular defect was extremely important for reassuring them and their mother who had lost both her husband and son.

22.2.1 Thoracic Aortic Disease (TAD)

As already mentioned, aortic root dilatation in MFS is one of the cardinal features in the diagnostic setting and is the most important factor determining life expectancy. The diagnosis of aortic root dilatation requires careful measurement of the aortic diameter at the correct location and according to standardized methods. In MFS, the dilatation typically occurs at the level of the sinuses of Valsalva. Obtained values need to be corrected for the individual's age, gender, and body surface area (Devereux et al. 2012). Commonly, z-scores are calculated, taking into account the obtained and expected diameter for a specific individual: z-score = (obtained diameter – expected diameter)/standard deviation. A z-score is a measure of how many standard deviations below or above the population mean a raw score is – as such a z-score >2 indicates that the value obtained for that individual exceeds 2 standard deviations of what is expected.

While the technical details of aortic root z-score calculation are beyond the scope of this chapter, some points are nonetheless worthwhile to be mentioned: (1) the method used to calculate the z-score should be exactly the same as the one used to generate the reference values used in the calculation – methodological aspects such as imaging technique and location and timing of the measurement are critical; (2) the reference population should match and cover the age range of the patient; (3) assessment of aneurysm progression in the same patient can vary as the patient grows or gains or loses weight.

Similar to skeletal manifestations, aortic root dilatation alone is not diagnostic for MFS, and patients with a dilatation of the aortic root need to be carefully evaluated in order to define the underlying cause. In older patients, as in the proband of the second family, atherosclerosis commonly underlies aortic aneurysm formation, and assessment of conventional cardiovascular risk factors is important. Additionally, careful clinical evaluation of the proband, supplemented by a dedicated ocular exam, is mandatory. Last but not least, a detailed family history and clinical assessment of first-degree relatives in selected cases are necessary to confirm diagnosis.

The cases presented here nicely illustrate the importance of assessment of family members. Diagnosis in the young boy in Family 1 was uncertain, but assessment of his father further raised suspicion of MFS, even though he did not fulfill the diagnostic criteria. In the second case, the diagnosis in the proband could have been confirmed with assessment of his son, underscoring the increasing importance of molecular genetic testing. Based on clinical evaluation and family history in the proband, an underlying

Table 22.1 Skeletal features in MFS and related diseases

Skeletal features in MFS	Arachnodactylia, increased armspan, decreased upper to lower segment ratio, decreased elbow extension, pectus deformities, kyphoscoliosis, flat feet, protrusio acetabuli, facial characteristics	
Skeletal features in other connective tissue and genetic aortic disorders		
	Features in common with MFS	Discriminating features – not typical for MFS
Congenital contractural arachnodactyly	Arachnodactylia, pectus deformities, decreased elbow extension, scoliosis	Joint contractures of the hips and knees, crumpled ears, clinodactyly
Kyphoscoliotic Ehlers-Danlos syndrome	Kyphoscoliosis	Skin features
Homocystinuria	Tall stature, increased armspan	Developmental delay/ intellectual disability
Loeys-Dietz spectrum	Pectus deformities, scoliosis, arachnodactylia	Hypertelorism, split uvula, club feet

genetic entity may be suspected – in this respect both syndromic and non-syndromic forms are recognized. Identified genetic aortic entities, grouped under the denominator heritable thoracic aortic disease (H-TAD), are listed in Table 22.2 along with their respective genetic and clinical features (Pyeritz 2014).

MFS is the prototypical syndromic H-TAD. Other forms of syndromic H-TAD include Loeys-Dietz syndrome (caused by mutations in components of the TGFβ signaling pathway; Loeys et al. 2006), aneurysm-osteoarthritis syndrome (caused by mutations in *SMAD3*, an intracellular component of the TGFβ signaling pathway; van de Laar et al. 2011), and smooth muscle cell dysplasia syndrome (caused by the R179 mutation in *ACTA2*; Milewicz et al. 2010). Differentiating these syndromes is often difficult due to overlapping manifestations; similar considerations apply to non-syndromic forms of H-TAD. However, identifying the underlying mutation in H-TAD may be helpful in risk stratification and management of individual patients. Arterial disease in MFS vs. Loeys-Dietz syndrome best exemplifies this point. Whereas patients harboring mutations in *FBN1* generally have a lower risk for developing aortic dissection at diameters below 55 mm, those with mutations genes involved in the TGFβ pathway generally evolve more aggressively and require earlier surgery (Milewicz and Regalado 2015).

22.2.2 Other Cardiovascular Manifestations

With extended life expectancy thanks to improved treatment and management of aortic disease, an increasing amount of MFS patients develop aneurysms elsewhere in the arterial tree. Apart from increasing diameters, the aorta also elongates with age, which forces the anatomically fixed aorta to bend and become tortuous. By means of magnetic resonance imaging (MRI), the tortuosity index can be measured and used as a marker of aortic disease severity in MFS.

MVP is an established clinical feature of MFS that occurs in >60% of cases. MVP is often a reason for referral in young children and shows a less benign course in MFS patients than in the general population (Judge and Dietz 2005). MFS-related cardiomyopathy is subclinical in most cases although severe forms – even necessitating cardiac transplantation – have been reported. Indeed, heart failure resulting from cardiomyopathy has been reported as the main cause of death in both surgical and clinical follow-up data in MFS series. Underlying valvular heart disease may cause dilated cardiomyopathy (DCM) in MFS, but there is also experimental evidence for intrinsic myocardial dysfunction. Clinical evidence that left ventricular contractility and ventricular-vascular coupling are abnormal in MFS, independent of aortic stiffness, is

Table 22.2 Overview of clinical entities to be taken into account in the differential diagnosis of H-TAD (both syndromic and non-syndromic). Distinctive cardiovascular and other clinical features are indicated in bold

Disorder	Gene(s)	Main cardiovascular features	Additional clinical features
Syndromic H-TAD			
Marfan	*FBN1*, *TGFBR1&2* *SMAD3*, *TGFB2*	**Sinus of Valsalva aneurysm**, aortic dissection, mitral valve prolapse, main pulmonary artery dilatation, left ventricular dysfunction	**Lens luxation**, skeletal features (arachnodactylia, pectus deformity, scoliosis, flat feet, increased armspan, dolichocephalia), dural ectasia, striae
Loeys-Dietz	*TGFBR1&2* *SMAD3*, *TGFB2&3*	**Sinus of Valsalva aneurysm**, aortic dissection, arterial aneurysms and dissections, **arterial tortuosity**, patent ductus arteriosus, atrial septal defect, bicuspid aortic valve	**Bifid uvula/cleft palate**, **hypertelorism**, pectus abnormalities, scoliosis, club feet
Vascular Ehlers-Danlos syndrome	*COL3A1*	**Arterial rupture and dissection without preceding dilatation/aneurysm**	**Gastrointestinal rupture**, **thin and translucent skin**, dystrophic scars, facial characteristics (Madonna face, thin lips, deep set eyes), club feet, uterine rupture
Multi-systemic smooth muscle dysfunction syndrome	*ACTA2*	**Ascending aortic aneurysm**, aortic dissection, **patent ductus arteriosus**, aortic coarctation, aortopulmonary window, pulmonary arterial hypertension	**Congenital mydriasis**, malrotation of the gut, moyamoya disease, periventricular white matter hyperintensities
Shprintzen-Goldberg syndrome	*SKI*	**Mild aortic root dilatation**, mitral valve prolapse	**Craniosynostosis**, distinctive craniofacial features, skeletal changes, neurologic abnormalities, mild-to-moderate intellectual disability
Arterial tortuosity syndrome	*SLC2A10*	**Arterial tortuosity**, arterial stenoses and aneurysms, mild aortic root dilatation	Hyperlax skin and joints, beaked nose, elongated face, micrognatia
Cutis laxa syndromes (autosomal dominant and recessive)	*ELN*, *FBLN4*	Mild aortic dilatation and tortuosity	Skin hyperlaxity, emphysema, downslanting palpebral fissures, inguinal hernia
Non-syndromic H-TAD			
	ACTA2	Thoracic aortic aneurysm/dissection, cerebrovascular disease, coronary artery disease	Lack of Marfanoid skeletal features, **livedo reticularis**, **iris flocculi**, **coronary artery/cerebrovascular disease**
	TGFBR1/2	Thoracic aortic aneurysm/dissection	Lack of syndromal features
	FBN1	Sinus Valsalva aneurysm, mitral valve prolapse	**Lack of syndromal features**
	MYLK	Thoracic aortic aneurysm/dissections often at low aortic diameters	
	SMAD3	Intracranial and other arterial/visceral aneurysms	
	TGFB2	**Mitral valve prolapse**	
	MYH11	**Patent ductus arteriosus**	
	PRKG1	**Aortic dissection at young age**	
	MFAP5	**Lone atrial fibrillation**	

also consistent with intrinsic impairment of myocardial contractility. Probably related to both MVP and cardiomyopathy are cardiac arrhythmias and increased ventricular ectopia. As discussed in the next section, the emerging view is that structural abnormalities altering mechanosignaling underlie myocardial dysfunction in MFS.

While it is obvious that MFS and nonsyndromic H-TAD are part of one clinical spectrum, it is more difficult to reconcile how mutations in *FBN1* can also cause acromelic dysplasias, which manifest short stature, short hands and feet, stiff joints, and a hypermuscular build – in other words the opposite phenotype than MFS. Geleophysic dysplasia, acromicric dysplasia, and WMS are all part of the acromelic dysplasia group of diseases, and they can all be associated with rare mutations in *FBN1*.

22.3 Molecular Perspectives

Mutations that impair the structure or reduce the expression of the extracellular matrix (ECM) protein fibrillin-1 are responsible for the multisystem manifestations of MFS. Fibrillin-1 is an ubiquitous 350-kD glycoprotein made almost entirely of calcium-binding EGF (cb-EGF) repeats interspersed with a few 8-cysteine (TB/8-Cys) motifs (Fig. 22.6a; Sakai et al. 2016). Calcium binding stabilizes contiguous cb-EGF sequences into a rigid linear structure required for fibrillin-1 polymerization, interaction with other matrix proteins, and protection from proteolytic enzymes. Aside from fibrillin-1 and the structurally related fibrillin-2, TB/8-Cys modules are only found in latent TGFβ-binding proteins (LTBPs), key regulators of TGFβ bioavailability. LTBPs bind the dimeric pro-peptide of TGFβ non-covalently associated with the bioactive ligand (aka, small latent complex or SLC) so as to facilitate folding and secretion of the resulting tripartite complex (aka, large latent complex or LLC), which is then tethered to the ECM in part via LTBP association with fibrillin-1 assemblies (microfibrils and elastic fibers) (Fig. 22.6b; Robertson and Rifkin 2016).

Fibrillin-1 microfibrils are assembled through an undefined hierarchical process whereby individual molecules are organized into head-to-tail polymers that associate laterally with one another and incorporate other ECM proteins within them (Fig. 22.6a). Fibrillin-1 microfibrils can also serve as the scaffold guiding tropoelastin alignment and cross-linking in the elastic fibers (Fig. 22.6a). Fibrillin-1 assemblies therefore represent a unique example of a dual-function component of the architectural matrix. The first role is structural in that they endow tissues with tensile strength and elasticity, in addition to demarcating functionally distinct areas within them; the second role is instructive for they modulate the behavior of resident cells by interacting with integrin receptors and LLCs. TGFβ sequestration into the ECM promotes spatial distribution and proper concentration of bioactive ligands for either immediate presentation to cells or subsequent release during tissue remodeling/repair (Fig. 22.6b). All *FBN1* mutations translate into MFS tissues with less fibrillin-1 immunostaining as a result of impaired mRNA stability, reduced protein secretion, or/and increased protein degradation. While the mechanistic relationship between *FBN1* mutations and MFS severity remains largely undefined, studies of mice with genetically engineered *Fbn1* mutations have yielded insights into organ-specific disease processes. Examples relevant to cardiovascular manifestations in MFS are described below along with current speculations about how some domain-specific mutations in fibrillin-1 may cause non-MFS phenotypes.

22.3.1 TAD

Aortic aneurysms are characterized by progressive vessel dilation associated with smooth muscle cell (SMC) dysfunction, localized inflammatory infiltrates, and destructive (maladaptive) ECM remodeling that, together, predispose the arterial wall to tear (dissection) and rupture. Consistent with the degenerative nature of the disease, inherited forms of TAD are accounted for by mutations in molecules nor-

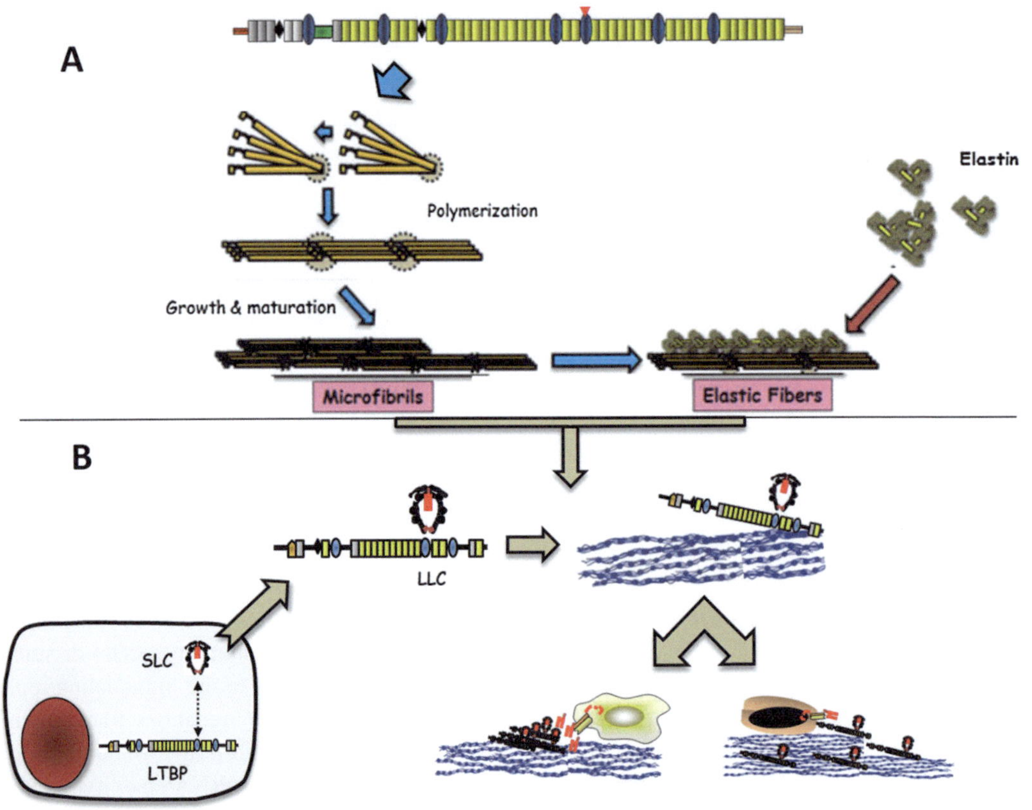

Fig. 22.6 Fibrillin microfibrils' biogenesis (**a**) and interaction with TGFβ complexes (**b**). At the top of Panel A is shown the primary structure of fibrillin-1 protein with yellow rectangles, blue ovals, and red triangle representing EGF-cb and TB/8-Cys motifs and RGD integrin-binding site, respectively. At the bottom right of Panel B are depicted how fibrillin-1 assemblies promote spatial distribution and proper concentration of bioactive ligands for either immediate presentation to cells or subsequent release during tissue remodeling/repair

mally implicated in supporting tissue integrity and homeostasis, such as components of the ECM, SMC cytoskeleton, and TGFβ signaling pathways (Milewicz et al. 2017). In this respect, MFS represents an informative model to characterize altered cell-matrix interactions leading to progressive degeneration and ultimately, structural collapse of the aortic wall. Fibrillin-1 assemblies are abundantly distributed throughout the vessel wall, as components of the elastic lamina underneath endothelial cells in the intima, as sheets of interconnected elastic fibers that together with concentric layers of SMCs constitute the contractile unit of the media, and as the loosely arranged meshwork intermixed with collagen fibrils that engulfs fibroblasts, nerve, and stem cells in the adventitia (Wagenseil and Mecham 2009).

In spite of significant research effort, the molecular mechanism of arterial disease in MFS remains controversial with negative implications for developing new evidence-based drug treatments. The major controversy centers on TGFβ's role in TAD pathogenesis and the mechanistic relationship between TGFβ over-activation and constitutively elevated signaling by the angiotensin II (AngII) type I receptor (AT1R) (Chen et al. 2013). Earlier studies with MFS mice manifesting TAD without dissection and rupture had concluded that AT1R-dependent TGFβ hyperactivity is the main driver of TAD development and, consequently, that AT1R antagonism may prevent

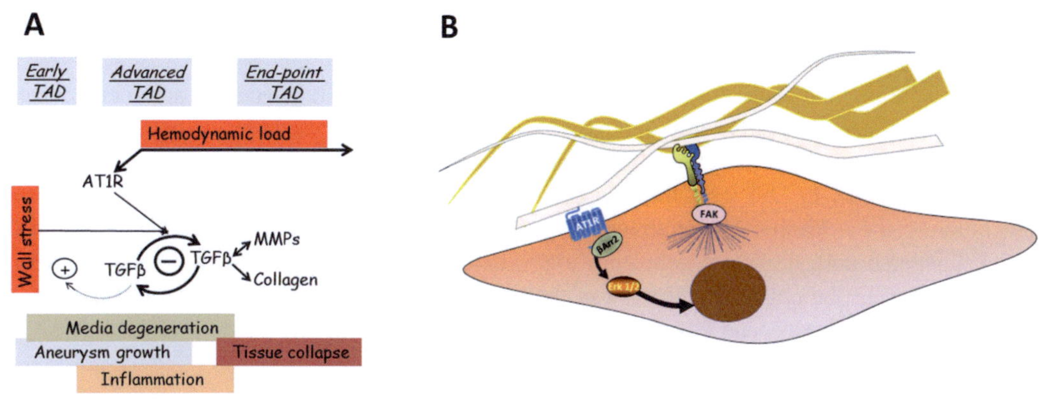

Fig. 22.7 Proposed models of aortic aneurysm (**a**) and cardiomyopathy (**b**) in MFS that are both based on findings from studies of mouse models of the disease (Cook et al. 2014, 2015)

TAD formation by blunting TGFβ signaling (Habashi et al. 2006). However, subsequent studies have shown partially overlapping roles of AT1R and TGFβ hyperactivity in driving TAD formation and progression, respectively, in addition to revealing a protective (AT1R-independent) role of baseline TGFβ signaling during postnatal vessel growth (Li et al. 2014; Cook et al. 2015) Accordingly, an alternative new model of TAD pathogenesis has emerged whereby a fibrillin-1-deficient aortic matrix communicates to resident cells signals of hemodynamic load and wall stress that are translated into biochemical responses aimed at maintaining physiological tone (Fig. 22.7a). Among others, the model predicts that the early response of the MFS aorta to physiological blood pressure increase includes AT1R over-activation and protective baseline TGFβ signaling whose effectiveness is impaired by fibrillin-1 deficiency. The model also predicts that dysregulated AT1R signaling, together with other wall stress-induced processes, such as inflammation and ECM remodeling, gradually establishes an irreversible TGFβ-centered loop that drives media degeneration ultimately leading to physical collapse of the vessel wall. Implicitly, this new disease model argues for a combinatorial drug treatment strategy that would target promiscuous AT1R and TGFβ signaling while sparing the early protective role of TGFβ activity.

22.3.2 DCM

Heart function largely depends on the ECM ability to transmit mechanical forces to cardiomyocytes where mechanosensors, including AT1R and ECM-binding integrins, convert stretch stimuli into biochemical signals that regulate muscle activity to maintain tissue homeostasis. Cardiomyocytes can also adapt to elevated workload by augmenting contractility and increasing their mass (hypertrophy) so as to normalize cardiac output. Genetic lesions or environmental insults that interfere with this physiological response lead to maladaptive tissue remodeling and cardiomyopathy. Cardiac valve disease and stiffening of the dilating aortic wall have long believed to cause heart dysfunction in MFS by imposing volume overload on the left ventricle. However, subsequent analyses of mice with tissue-specific inactivation of the *Fbn1* gene expression have refuted this notion by demonstrating that fibrillin-1 deficiency in the myocardium is necessary and sufficient to trigger DCM (Cook et al. 2014).

The myocardial matrix consists of highly specialized multi-protein structures that surround individual myocytes and interconnect them into a synchronously functioning unit. Fibrillin-1 microfibrils couple myocytes to the ECM, as associated components of basement membranes;

contribute to the mechanical properties of myocardial tissue, as obligatory constituents of interstitial elastic fibers; and weave together adjoining myocytes, as molecular bridges between macromolecular assemblies in the pericellular and interstitial matrices. In line with the structural integration of these fibrillin-1 assemblies into a single functional unit, fibrillin-1 deficiency significantly weakens the physical properties of the myocardium of MFS mice, and this in turn translates into impaired muscle contractility associated with abnormal myocyte mechanosignaling, as evidenced by AngII-independent AT1R hyperactivity and abated FAK-mediated integrin signaling (Fig. 22.7b). In contrast to TAD, TGFβ hyperactivity appears not to be involved in DCM formation. Consistent with these findings, systemic administration of losartan, an AT1R blocker, prevented DCM formation in MFS mice. Additionally, this drug treatment was also reported to mitigate ventricular dysfunction in MFS patients (Hartog et al. 2016).

22.3.3 MFS-Related Diseases

As already noted, the complex phenotype of the MFS/WMS family described in this chapter further underscores our limited understanding of genotype-phenotype relationships in MFS and related disorders, as well as of the role that genetic modifiers play in these different pathological correlates. This problem is particularly evident for the rare mutations in *FBN1* that have been associated with acromelic dysplasias, which manifest musculoskeletal abnormalities opposite of those normally seen in MFS (Sakai et al. 2016). The same consideration applies for mutations in *FBN1* that cause the tissue-specific phenotype of stiff skin syndrome (SSS; Loeys et al. 2010b). Unlike MFS mutations, which are spread throughout the entire fibrillin-1 molecule, those associated with acromelic dysplasias and SSS are clustered in domains believed to mediate interactions with cells and other matrix proteins. Indeed, experimental findings indicate that *FBN1* mutations causing acromelic dysplasias or SSS do not apparently affect the assembly or stability of fibrillin-1 microfibrils and that acromelic dysplasias can also be caused by mutations in molecules known to interact with fibrillin-1, such as LTBP3 and ADAMTSL2 (Sakai et al. 2016). However, this evidence does not explain how deletion of exon 20 would account for the variable phenotype of MFS/WMS family described here.

22.4 Treatment

We will limit our description to the treatment of cardiovascular abnormalities in MFS. It goes without saying that patients afflicted with this pleiotropic disorder require a multidisciplinary approach and appropriate treatment of major skeletal and ocular manifestations.

22.4.1 Medical Treatment

Slowing down aortic root growth rate and preventing dissection are the mainstays of treatment in patients with MFS. The most commonly prescribed drugs to achieve this are β-adrenergic blockers, which reduce the aortic dilation rate in patients with MFS due to their effects on aortic wall shear stress. Losartan, an AT1R blocker, might be an alternative or complementary therapy to β-blockers, since losartan reduces arterial pressure and potentially interferes with the pathophysiology of MFS by TGF-β antagonism. After evidence of losartan effectiveness in a mouse model of MFS (Habashi et al. 2006), several randomized clinical trials were launched to test losartan in various protocols. Despite the differences in study design and outcome, these studies consistently showed that losartan is not more effective in reducing the rate of aortic dilatation than a high dosage of β-blockers. Losartan may be regarded as a valid alternative in those patients that do not tolerate β-blockers, and hopefully, an ongoing large-scale meta-analysis will shed more light on possible subgroups that may benefit from specific treatments (Pitcher et al. 2015).

22.4.2 Surgical Treatment

The threshold aortic diameter for prophylactic surgery in MFS is 50 mm at any level of the vessel or 45 mm at the aortic root. Associated risk factors include family history of dissection, progressive dilation of more than 2 mm/year, severe aortic or mitral valve regurgitation, or pregnancy. Over the past 30 years, the composite replacement of the aortic valve and ascending aorta (aka, "Bentall procedure") has proved to be low risk and very durable operation for aortic root aneurysm in MFS. Valve-sparing operations with root replacement by a Dacron prosthesis and reimplantation of the coronary arteries into the prosthesis (aka, the "David procedure") have now become the preferred choice due to the inherent need for lifelong anticoagulation after a Bentall procedure.

Treatment of aneurysms/dissections in the distal aorta of MFS patients also consists of medical treatment in a first step and surgery with documented increased growth rate or organ ischemia. Due to an increased rate of stent-related complications, open surgery is the preferred method for treatment. Techniques have evolved significantly over the last decades, mainly aimed at avoiding organ damage during the procedure. The father of the proband in the first family described here was considered too high risk for surgery 25 years ago, but today he might have undergone surgery.

MVP with significant valvular regurgitation is another reason for primary surgery in approximately 15% of MFS patients. Forty-five percent of cases will also undergo aortic root surgery at that time. About 20% of MFS patients undergoing aortic root surgery will simultaneously undergo a mitral valve procedure. Technically, mitral valve repair is associated with better survival than mitral valve replacement. Heart failure in MFS patients should be treated according to conventional guidelines and should include heart transplantation when indicated.

End of Chapter Questions

1. Molecular genetic testing may be useful in some situations to confirm the diagnosis in MFS – can you give an example?
2. How would you link the various genes identified in H-TAD patients?
3. What are possible targets for treatment in MFS?
4. What are the major mechanisms responsible for TAD and DCM development in mouse models of MFS?
5. Which connective tissue diseases are associated with *FBN1* mutations?

References

Chen X et al (2013) Conundrum of angiotensin II and TGF-β interactions in aortic aneurysms. Curr Opin Pharmacol 13:180–185

Cook JR et al (2014) Abnormal muscle mechanosignaling triggers cardiomyopathy in mice with Marfan syndrome. J Clin Invest 124:1329–1339

Cook JR et al (2015) Dimorphic effects of TGFβ signaling during aortic aneurysm progression in mice suggest a combinatorial therapy for Marfan syndrome. Arterioscler Thromb Vasc Biol 35:911–917

den Hartog AW et al (2016) The effect of losartan therapy on ventricular function in Marfan patients with haploinsufficient or dominant negative FBN1 mutations. Neth Hear J 24:675–681

Devereux RB et al (2012) Normal limits in relation to age, body size and gender of two-dimensional echocardiographic aortic root dimensions in persons ≥15 years of age. Am J Cardiol 110:1189–1194

Habashi JP et al (2006) Losartan, an AT1 antagonist, prevents aortic aneurysm in a mouse model of Marfan syndrome. Science 312:117–121

Judge DP, Dietz HC (2005) Marfan's syndrome. Lancet 366:1965–1976

Li W et al (2014) Tgfbr2 disruption in postnatal smooth muscle impairs aortic wall homeostasis. J Clin Invest 124:755–767

Loeys BL et al (2006) Aneurysm syndromes caused by mutations in the TGF-b receptor. N Engl J Med 355:788–798

Loeys BL et al (2010a) The revised Ghent nosology for the Marfan syndrome. J Med Genet 47:476–485

Loeys BL et al (2010b) Mutations in fibrillin-1 cause congenital scleroderma: stiff skin syndrome. Sci Transl Med 2:23ra20

Milewicz DM, Regalado ES (2015) Use of genetics for personalized management of heritable thoracic aortic disease: how do we get there? J Thorac Cardiovasc Surg 149:S3–S5
Milewicz DM et al (2010) De novo ACTA2 mutation causes a novel syndrome of multisystemic smooth muscle dysfunction. Am J Med Genet A 152A:2437–2443
Milewicz DM et al (2017) Altered smooth muscle cell force generation as a driver of thoracic aortic aneurysm and dissections. Arterioscler Thromb Vasc Biol 37:26–34
Pitcher A et al (2015) Design and rationale of a prospective, collaborative meta-analysis of all randomized controlled trials of angiotensin receptor antagonists in Marfan syndrome, based on individual patient data: a report from the Marfan Treatment Trialists' Collaboration. Am Heart J 169:605–612
Pyeritz RE (2014) Heritable thoracic aortic disorders. Curr Opin Cardiol 29:97–102
Robertson IB, Rifkin DB (2016) Regulation of the bioavailability of TGF-b and TGF-b-related proteins. Cold Spring Harb Perspect Biol 8(6):a021907
Sakai LY et al (2016) FBN1: the disease-causing gene for Marfan syndrome and other genetic disorders. Gene 591:279–291
van de Laar IM et al (2011) Mutations in SMAD3 cause a syndromic form of aortic aneurysm and dissections with early-onset osteoarthritis. Nat Genet 43:121–126
Wagenseil JE, Mecham RP (2009) Vascular extracellular matrix and arterial mechanics. Physiol Rev 89:957–989

23 When Materials Are at Fault: The Skeletal Collagens, Osteogenesis Imperfecta and Chondrodysplasias

Andrea Superti-Furga

Keywords
Osteogenesis imperfecta · Chondrodysplasia · Type 1 collagen · Type 2 collagen

23.1 Case Report: Patient 1

Prenatal ultrasound done at week 34 in a 29-old primigravida and primipara reveals bowing of several long bones (tibiae, femurs, and humeri) as well as a suspected fracture of the left femur in the fetus. The woman and her partner are clinically healthy and non-consanguineous. At week 39, a boy is delivered vaginally. The newborn has bowed limbs, a triangular face, and blue sclerae. On palpation, his skull is soft. Tachypnea (rapid breathing) develops and he is given ventilator support for 24 h. He appears to be comfortable when left quiet but seems to have pain upon handling (Fig. 23.1). As his sucking is weak, a gastric tube is inserted for alimentation. Radiographs show generalized osteopenia; the long bones are expanded and show numerous fractures; the skull is very poorly mineralized (Fig. 23.1). A diagnosis of osteogenesis imperfecta (OI) is made; handling is reduced to a minimum; nonsteroidal anti-inflammatory drugs given for pain control when needed. The baby remains stable and his sucking gains strength, allowing for breastfeeding. He is discharged at age 3 weeks. Over the first months of life, he gains weight but remains intolerant to handling, and his spontaneous movements are very much reduced. His social development seems unaffected, and he follows well with is eyes and develops social smiling.

At age 3 months, a first intravenous infusion of pamidronate is given to promote bone mineralization; infusions will be repeated every 6 weeks up to age 6 months and then every 3 months. At age 2 years, the long bones remain fragile and legs are markedly deformed, making it impossible for the boy to stand; he can sit with support, can feed himself with a spoon (his teeth are small and discolored), and participates in the environment.

Note on the nomenclature of collagens: originally, collagen types have been distinguished using Roman numerals (I, II, III, IV, etc.). However, using Roman numerals for newer collagens, such as collagen 27 ("collagen XXVII"), is cumbersome; the collagen genes are numbered using Arabic numerals (e.g., COL27A1); and in the digital age, Roman numerals are easily misinterpreted. For these reasons, only Arabic numerals will be used here.

A. Superti-Furga (✉)
Division of Genetic Medicine, Lausanne University Hospital, University of Lausanne, Lausanne, Switzerland
e-mail: asuperti@unil.ch

T. Oohashi et al. (eds.), *Human Pathobiochemistry*, https://doi.org/10.1007/978-981-13-2977-7_23

23.2 Patient 1: Diagnosis

The combination of bowed limbs with fractured bones, a poorly mineralized skull, blue sclerae, and discolored teeth strongly suggests the diagnosis of *Osteogenesis Imperfecta* (also known as the "brittle bone disease" or "*Glasknochenkrankheit*" – the disease with the glass-like bones). The clinical severity of OI is highly variable; the most severe forms are incompatible with life, while mild forms may be diagnosed only in adult life or not at all. This boy has a severe form that was compatible with survival but resulted in severe deformities of the limbs with the ensuing disabilities. Of note, the more severe forms of OI can be suspected prenatally on the basis of some specific sonographic signs such as short and crumpled long bones, beaded ribs, poor mineralization of the skull with deformability of the cranial vault, and increased visualization of the brain on ultrasound because of reduced absorption of sound waves by the thin and non-mineralized skull bones. However, sonographic examinations during pregnancy are done according to a routine protocol, and these signs are often missed. In this boy, molecular analysis showed a heterozygous transversion in the *COL1A1* gene (c.1714C > G) predicting the substitution of glycine at position 572 with arginine (p.G572R). Neither parent carried the mutation, confirming its de novo origin. This type of mutation is typical for OI (see below).

The Clinical Variability of Osteogenesis Imperfecta The wide range of clinical expression of osteogenesis imperfecta (from mild cases to lethal ones) and its mode of inheritance (dominant families, sporadic cases, and some instances of sib recurrence) have long been a puzzle for clinicians. Attempts at a classification of OI in distinct "types" were published as early as 1959, 1967, and 1975. In 1979, Sillence and coworkers proposed a classification of OI in four types, and to each type, a specific inheritance mode was attributed (Sillence et al. 1979). Some instances of multiple sibs born to unaffected parents had been observed; and at that time, the paradigm of mild or moderate variants of disease often being "dominant", with severe variants often being "recessive", was widely accepted; also, severely affected individuals would hardly ever reproduce, hiding the dominant nature of their condition. Thus, the Sillence classification had the most common form, type 1, as mild and dominantly inherited; the most severe (at that time mostly "lethal") form, type 2, as recessive; and then a severe type 3 form as a likely recessive; and a moderate type 4 form as a dominant. The genetic implications of the Sillence classification are no longer valid today, but the classification is still useful to define the clinical severity of affected individuals (Table 23.1). Although named by its prominent osseous fragility, OI is not only a disease of bone; non-mineralized connective tissue is also affected, as revealed by joint laxity and propensity to dislocations, involvement of heart valves and of large vessels, and – last but not least – the conspicuous thinning of the ocular "connective tissue bag," the sclera, that allows for translucence of the underlying pigmented retinal epithelium, giving the impression of blue or gray sclerae.

23.3 Pathophysiology of Osteogenesis Imperfecta

"*Colla*" is the Latin word for glue, and "collagen" means literally "glue giving"; it has been known from ancient times that from boiling bones and skins from animals, one would obtain a sticky substance that, depending on the procedure and on the purification, would give gelatin or glue. Given that collagen is the most abundant protein in bone and connective tissue, it came natural to think of the brittle bone disease as a possible disease of collagen. Thus, in 1972, Victor McKusick, the father of modern medical genetics, wrote (McKusick 1972): "The collagen molecule has become more familiar in its physicochemical details, although it has far to go to match that paragon of protein molecules – hemoglobin. (…) Do the amino acid sequences of collagen, in OI, differ from those in the 'normal' by a single amino acid substitution, comparable to the differences demonstrated in many hemoglobin variants? If there is indeed a recessive form of OI, an enzyme defect is likely to be found, rather

Table 23.1 A modern version of the "Sillence classification" of osteogenesis imperfecta

Sillence type	Degree of severity	Main clinical features	Inheritance and molecular bases
OI type 1	Mild	Increased susceptibility to fractures; stature normal or mildly reduced, blueish sclerae, normal teeth; little or no restrictions in daily living	Dominant haploinsufficiency mutations in *COL1A1* (rarely *COL1A2* or other genes such as *WNT1*)
OI type 2	Formerly "lethal"	Easily recognized on antenatal scans; can result in fetal or neonatal demise; often respiratory insufficiency in the newborn period; flail chest, soft calvarium, short and deformed limbs, blue sclerae, inguinal herniae, sometimes avulsion of limbs or head during delivery; when survival, turns into severe type 3	Most frequently de novo dominant *COL1A1* or *COL1A2* mutations; in approx. 10% of cases, recessive mutations in other genes (see text)
OI type 3	Severe	Severe deformation of the limbs and of the cranium; dentinogenesis imperfecta; severely reduced stature; limited deambulation, severe restrictions in daily living	Similar as OI type 2
OI type 4	Moderate	Moderately reduced stature, limb deformities usually restricted to the legs, and dentinogenesis imperfecta. Deambulation usually possible, some restrictions in daily living	Similar as OI type 2 but more often familial mutations

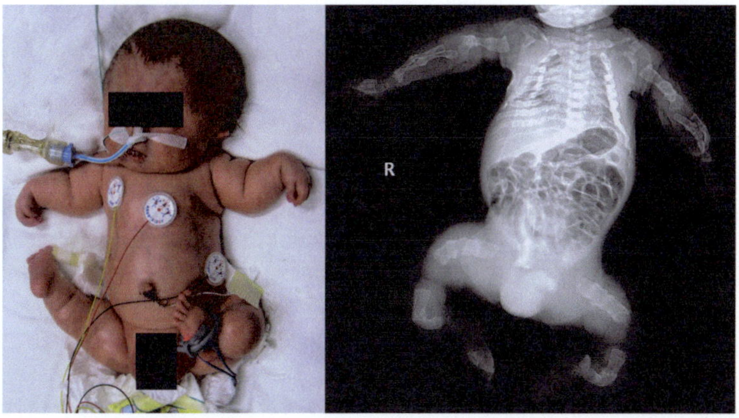

Fig. 23.1 The left panel shows patient 1 a few hours after birth. Note the abnormal positioning of the four limbs caused by underlying bone fractures. The newborn baby could be weaned from ventilatory support at day 5. The right panel shows the radiograph taken a few days later. Note the generalized osteopenia. The long bones in the limbs (humeri, radius and ulna, femurs, and tibiae) are expanded and have irregular contours because of several fractures occurring before birth. The vertebrae are flatter than normal, and the ribs are deformed

than a structural change in a connective tissue protein…". All these predictions have proven correct; let us see how (Figs. 23.1 and 23.2).

The fibroblast is the human cell type that is easiest to maintain in long-term culture, and fibroblast cultures have been instrumental in the study of metabolic and biosynthetic aspects of a large number of genetic disorders. Fibroblast cultures synthesize relatively large quantities of collagen type 1, as well as moderate amounts of collagen type 3 and type 5 and tiny amounts of other collagens and structural proteins such as fibronectin or proteoglycans. Incubation of fibroblasts with radioactively labeled glycine or proline, harvesting of the culture medium and cell layer, and digestion of the harvested materials with pepsin (to which the triple-helical collagens are resistant) is a simple method to obtain purified collagens; their structure can then be studied using denaturing gel electrophoresis (SDS-PAGE) before and after different procedures such as thermal denaturation or cyanogen bromide cleavage. Using these techniques, several structural abnormalities in collagen 1 have been identified in individuals

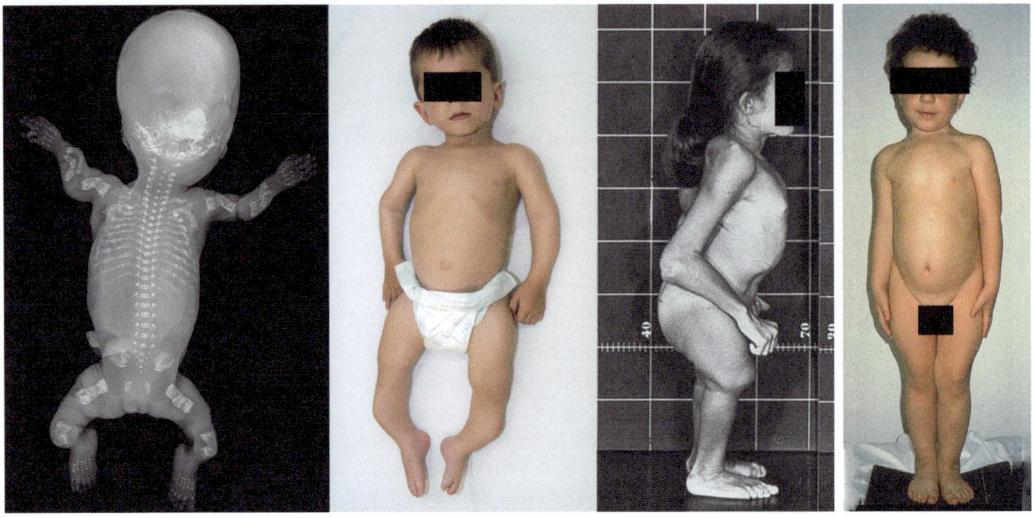

Fig. 23.2 Different forms of osteogenesis imperfecta. The left panel shows a 22-week-old fetus; the pregnancy was interrupted for multiple fractures and the diagnosis of lethal OI. Note the severe bowing of the limbs' bones, flat vertebrae, beaded ribs, and non-mineralized cranial vault. This severity corresponds to OI type 2. The second panel shows a 3-year-old boy with mild short stature and bowed legs; severity corresponds to OI type 4. The third panel shows an 11-year-old girl with severe short stature, deformity of the femurs, and short trunk; this corresponds to OI type III. The right panel shows a boy with mild deformity, normal stature, and a history of only a few fractures; this corresponds to OI type 1

with osteogenesis imperfecta and the Ehlers-Danlos syndrome. Fibroblasts from one baby with lethal OI were found to synthesize collagen chains that were shorter than normal; and cells from another lethal case were found to synthesize collagen 1 chains that contained a cysteine residue (normally not found in the helical portion of collagen). When gene cloning and sequencing techniques were applied to these cases, a multi-exon deletion was found in the first case (Chu et al. 1983), and a single base substitution leading to the replacement of a glycine with a cysteine was found in the second case[4]. While multi-exon deletions are rare, single nucleotide substitutions leading to the replacement of a glycine with another (larger, as glycine is the smallest) amino acid is the most common cause of OI – confirming the McKusick hypothesis.

What Are the Consequences of the Genetic Mutations in the Collagen 1 Genes on the Structure and Biosynthesis of the Collagen Protein? The fibrillary collagens are trimers, composed of three distinct procollagen chains. The procollagen chains are synthesized as individual polypeptide chains and inserted into the lumen of the endoplasmic reticulum (Fig. 23.3). When synthesis is completed, three procollagen chains are associated through their carboxy-polypeptides, and this initiates the "winding" of the chains into a triple-helical structure. The process of helix formation propagates from the carboxy-terminal end toward the amino-terminal end of the molecule. In the triple-helical structure that stretches for approx. 1000 residues, the three chains are tightly associated; the formation of the triple helix depends strictly on the presence of glycine (the smallest amino acid) in every third position (the "Gly-X-Y" amino acid sequence; Fig. 23.3). Only when the helix formation is complete, and the amino-propeptides interact to "lock" the structure, the mature procollagen molecule is secreted; then, specific peptides clip away the propeptides at both ends of the molecules, leaving the mature collagen molecule, that aggregates laterally with other molecules in a quarter-staggered array to give collagen fibers (Fig. 23.3). What happens when there is a struc-

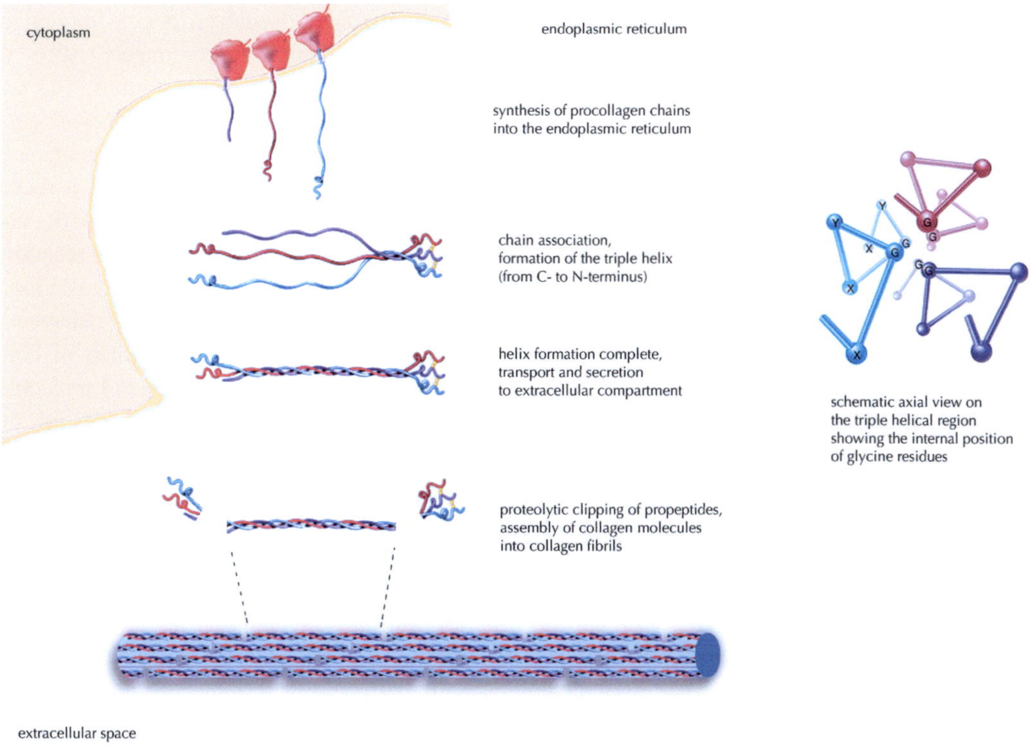

Fig. 23.3 Simplified scheme of the biosynthesis and secretion of collagen 1 in mesenchymal cells. The left part shows the sequential stages of synthesis, chain assembly, helix formation, secretion, cleavage of the C- and N-propeptides, and assembly of collagen molecules into collagen fibrils, and fibers are shown on the left. Secretion, propeptide cleavage, and fibril assembly are tightly coupled in finger-like, actin-rich collagen containing projections of the plasma membrane known as fibripositors (not shown here for simplicity). The scheme on the right shows an axial view on the triple-helical region of the collagen molecule. To allow for the tight positioning of the three chains, the small glycine residues are on the inside of the helix; the "X" and "Y" residues of the Gly-X-Y-repeating sequence are on the outside. The X and Y residues are often proline and hydroxyproline, respectively (see text)

tural abnormality, such as a substitution of a glycine with another amino acid, in the collagen chains? In this case, the formation of the helix is delayed; the procollagen molecule is retained in the ER for a longer time, leading to a "congestion" of the ER. A significant proportion of the affected molecules will never make it to secretion and be degraded intracellularly instead. When a helical structure is achieved and the molecule is secreted, it shows a distortion of its structure (a "kink") at the site of the substitution. Such a distorted molecule is unable to form the tightly packed collagen fibers, and its mere presence leads to the formation of very poor collagen fibers (Kadler et al. 1991). Thus, the overall consequences of the abnormal collagen chains are (1) congestion of the endoplasmic reticulum, (2) a decrease in the amount of collagen secreted by the cell (fibroblast, osteoblast, or other collagen 1-producing cell), and (3) the perturbation of the formation of collagen fibrils in the extracellular matrix because of the presence of molecules with abnormal conformation. This leads to a weakening of the fibrillary network that is the basis of normal, healthy bone formation. The study of the consequences of structural mutations on collagen biosynthesis has led to concepts of "protein suicide," "molecular havoc," and "dominant negative" that are fundamental in the appreciation of mutations in many other genes and proteins (Prockop 1984).

23.4 Collagen 1, Osteogenesis Imperfecta, and Medical Genetics

The collagen 1 mutations in osteogenesis imperfecta laid down some milestone concept in medical genetics. First, because mature collagen molecules are composed of three single polypeptide "chains" (for collagen 1, two alpha-1(I) chains and one alpha-2(I) chain) assembled as a triple helix, in the presence of one normal (wild-type) and one mutated allele, three quarters of the trimers would contain one or two mutant chains; this explained the "dominance" of the mutation. In the case of collagen 2, that is a pure homotrimer, the proportion of faulty trimers becomes seven-eighths. Second, because of the specific triple-helical structure that requires a glycine (the smallest amino acid) to allow for the snug fitting of the three chains, any substitution of glycine would perturb the helical structure and potentially cause disease. Thus, every third amino acid in the collagen chain is intolerant to change; this is much more than what is observed in most other proteins that, with less stringent globular conformations, are more tolerant to amino acid substitutions. The finding that a single nucleotide change at a heterozygous state was sufficient to produce the lethal OI phenotype was a powerful illustration of genetic dominance (Cohn et al. 1986). In 1990, a spectacular demonstration of gonadal mosaicism was given to account for recurrence of lethal OI in two babies who were born form the same father and two different mothers; the two babies were heterozygous for a pathogenic *COL1A1* mutation, and the same mutation was absent in the father's blood DNA but present in a fraction of DNA extracted from his sperm, indicating gonadal mosaicism (Cohn et al. 1990). Gonadal and somatic mosaicisms are two concepts that have become firmly established in medical genetics.

23.5 Osteogenesis Imperfecta: Collagen 1 and Beyond

While the discoveries of dominant collagen 1 mutations dominated the field in the 1980s and 1990s, the systematic diagnostic testing of individuals with OI did confirm that a fraction of affected individuals (approximately 10% of cases) did not have collagen mutations. This has led to the discovery of a large number of gene mutations in which can produce osteogenesis imperfecta or OI-like phenotypes (for a recent review, see Forlino and Marini (2016) and Table 23.2). Interestingly, some of the recessive OI genes are involved in the posttranslational modification and the processing of collagen 1 (see Table 23.2), and the pathogenesis of bone fragility revolves around the collagen 1 molecule. In contrast, other genes are involved in the homeostatic regulation of bone formation and resorption (Table 23.2). While individually much rarer, these genes have highlighted the many levels of complexity of bone biology; in synthesis, to make proper bones, good materials are needed to start with (the collagens).

Then, a specific cellular machinery (a set of specific enzymes and transport molecules) is needed to refine, secrete, and trim the collagens. Finally, a proper architectural information and tissue regulation are needed to coordinate the process of bone formation and remodeling and to assure the long-term homeostasis of the bone tissue by regulating deposition and resorption (which may be differentially regulated in cortical versus trabecular bone). In general, it is no longer useful to identify "types" of osteogenesis imperfecta based on single genes, with the possible exception of OI type 5 (progressive nature and hypertrophic callus; always associated with *IFITM5*) and OI type 6 (late-onset and progressive; always associated with *SERPINF*). While the "rarer" OI genes are biologically promising and perhaps therapeutically promising, the two genes for collagen 1, *COL1A1* and *COL1A2*, are responsible for approx. 90% of cases of OI, while all other genes together add up to the remaining 10% of cases. Because of the complex genetic basis of "brittle bone diseases," its elucidation over many years (still ongoing), and the fact that quite different phenotypes may arise from single genes, the older numbers in the OMIM catalog (such as MIM 166200 for OI type 1, and 166210 for OI type 2) are obsolete.

Table 23.2 Genes associated with genetic osteogenesis imperfecta and bone fragility phenotypes (non-exhaustive list)

Defects in collagen 1 structure and processing	*COL1A1*	Collagen type 1, alpha 1 chain
	COL1A2	Collagen type 1, alpha 2 chain
	BMP1	Bone morphogenic protein 1/procollagen C proteinase
Defects in posttranslational processing of collagens	*CRTAP*	Cartilage-associated protein
	LEPRE1/P3H1	Leucine proline-enriched proteoglycan 1/prolyl 3-hydroxylase 1
	PPIB	Peptidylprolyl isomerase B/cyclophilin B
	TMEM38B	Transmembrane protein 38 B
Defect in collagen transport, folding, and cross-linking	*SERPINH1*	Serpin peptidase inhibitor, clade H, member 1/heat shock protein 47
	FKBP10	FK506-binding protein 65
	PLOD2	Procollagen-lysine, 2-oxoglutarate 5-dioxygenase 2
Defects in bone tissue homeostasis (including signaling pathways and osteoblast differentiation)	*IFITM5*	Interferon-induced transmembrane protein 5
	SERPINF1	Pigment epithelium-derived factor
	SP7	Transcription factor 7/osterix
	WNT1	Wingless-type MMTV integration site family, member 1
	CREB3L1	cAMP responsive element-binding protein 3 like 1
	LRP5	Low density lipoprotein receptor-related protein 5
	PLS3	Plastin 3

23.6 Osteogenesis Imperfecta: What Are the Therapeutic Approaches?

As we have seen, the structural mutation of collagen 1 are strongly "dominant", in the sense that the presence of even a small proportion of abnormal molecules may interfere with the normal molecules in the formation of trimeric collagen molecules, result in the formation of "kinked" molecules, and prevent the formation of strong collagen fibers. On the other hand, a mere reduction in the quantity of collagen 1 produced (OI type 1) results in a comparatively mild disease. Therefore, the therapeutic strategy would be to abolish, or reduce, the production of mutant collagen 1 molecules. This is not an easy task, and no successful approach has been found to date. Inhibitory RNA and gene editing have not worked so far; maybe the CRISPR technology will be successful. Stem cell transplantation seems also not to be effective; in spite of some claims of success, it seems like in most instances, the osteoblasts in the bone are not replaced by donor-derived cells. The most widely used approach today is quite non-specific, namely, cyclic infusions with bisphosphonate type of drugs (alendronate, pamidronate) that promote bone mineralization in a non-specific way. The clinical effect is moderately positive, and in many children, bisphosphonate treatment may make the difference between being wheelchair-bound and being able to walk independently – a significant benefit. The next drug on the horizon are antibodies to sclerostin that have been shown to increase the mechanical strength of osteoporotic bones by promoting the formation of cortical bone. Clinical trials are underway. The surgical approach is controversial; the insertion of medullary rods, sometimes of the telescopic kind that can allow for some lengthening to accommodate the natural growth of bone, can stabilize bones but tends to result in a further loss of bone mass, making it difficult to remove them at a later stage; they are a last resort. Last but not least, physiotherapy and exercises are another mainstay of therapy to preserve or increase bone mass, to develop muscle strength, and to reduce pain. Periodic audiometry should be performed to assess for hearing loss caused by fragility of the ossicle chain in the middle ear as well as from progressive deformation of the cochlear structures.

23.7 Case Report: Patient 2

Like in the case of patient 1, routine sonographic examination at pregnancy week 26 showed that the long bones were too short. The thorax was smaller than normal. The skull was normal in size. Structure and mineralization of the skull and long bones were normal. After vaginal delivery at week 37, the newborn showed a moderate respiratory distress for which he received noninvasive ventilation support (continuously positive airway pressure, CPAP) for several hours, after which he recovered. The length at birth was 45 cm (normal, 48–52 cm), weight was 2.9 kg and head circumference was 35 cm (normal). Clinical findings included a short trunk and short limbs with a normal head, a "flat" face, and a midline cleft of the posterior palate. The diagnosis of a probable skeletal dysplasia was made. He was fed by a nasogastric tube until a palatal prosthesis was installed, allowing for autonomous sucking. After the parents were instructed on the handling of the prosthetic palatal plate, he was discharged in good conditions at 3 weeks of age. At age 6 months, the midline palatal cleft was closed surgically. His social and fine motor development was adequate, but there was muscle hypotonia and the gross motor development was delayed.

A radiographic survey at age 1 year showed a generalized delay in epiphyseal ossification; the proximal femoral epiphyses and the pubic bones were not ossified. The vertebral bodies were flattened. The diagnosis of congenital spondylo-epiphyseal dysplasia was made, and a gene panel for skeletal dysplasias confirmed the presence of a G > A transition in exon 49 of the *COL2A1* gene (c.3382G > A) predicting the substitution of glycine-1129 with serine (G1129S) in the triple-helical domain of the collagen 2 molecule. This finding confirmed the diagnosis of SEDC (spondylo-epiphyseal dysplasia congenital; MIM 183900) (Bonafe et al. 2015).

23.8 The Collagen 2 Disorder "Family"

We have seen that mutation that affect the structure of collagen 1 produce a range of phenotypes from mild to severe that all fall under the diagnosis of "osteogenesis imperfecta." Analogous mutations in the gene that codes for the alpha(I) chain of collagen 2 can also produce a range of clinical phenotypes that may range from mild to incompatible with life ("lethal"). However, unlike in osteogenesis imperfecta, these phenotypes have been given different names long before it was recognized that they all came from the same gene, *COL2A1*. Thus, the collagen 2 disorder "family" comprises achondrogenesis type 2, hypochondrogenesis, congenital spondylo-epiphyseal dysplasia, Kniest dysplasia, Torrance dysplasia, spondylo-peripheral dysplasia, Stickler syndrome, dominant SED with premature arthritis, and others (Bonafe et al. 2015). The "collagen 2 dysplasia family" is an example for a phenotypic spectrum resulting from mutations in one and a single gene (Spranger 1988). Figure 23.4 illustrates the phenotypic variability arising from mutations at the *COL2A1* locus. The molecular pathogenesis leading from mutations in the *COL2A1* gene to the phenotype of "chondrodysplasia" is similar to that seen for collagen 1 and fragile bones and will not be explored in detail here.

23.9 Two Principles Illustrated: Molecular Topography and Tissue-Specific Expression

Molecular Topography The phenotypic effects of mutations in the collagen 1 genes (*COL1A1* and *COL1A2*) and the collagen 2 gene (*COL2A1*) and the structural variations they determine are largely determined by the nature of the mutation and their position in the protein.

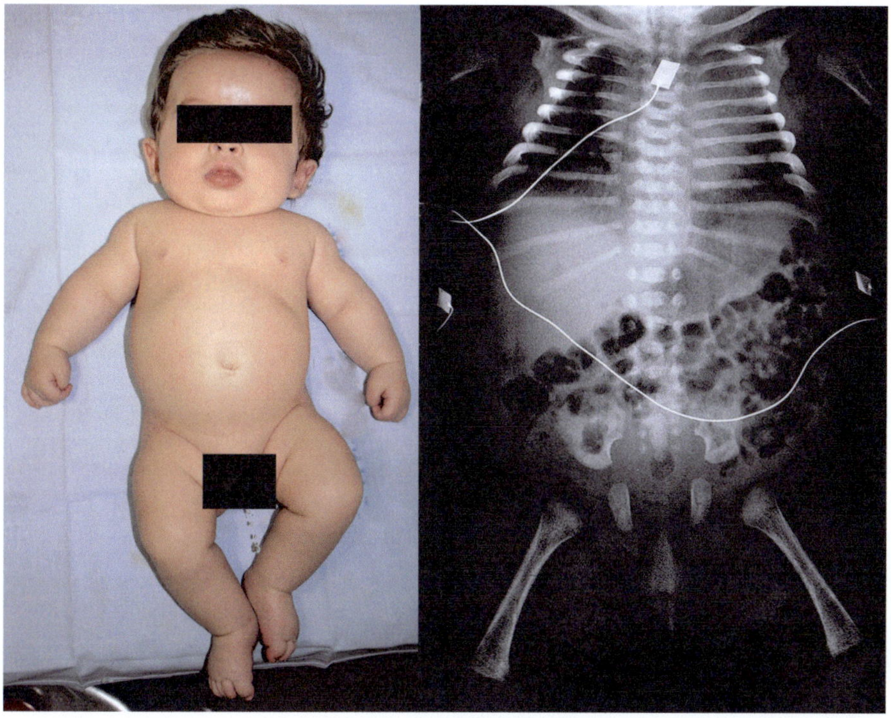

Fig. 23.4 Patient 2 is shown at age 4 months. Note the large forehead, the short trunk, and the short limbs with normal hands and feet. In the corresponding radiograph, the vertebrae are flattened resulting in shortening of the trunk; the proximal femurs are plump and the femoral heads are still unossified (they should be at this age); also, the pubic bones are not ossified. The findings are typical for congenital spondylo-epiphyseal dysplasia (SEDC)

1. Any mutation that determines a premature truncation of the polypeptide chain will result in "nonfunctional" chains because the carboxy-propeptide is necessary for chain association and trimer formation. Any polypeptide chain lacking that domain will be lost for collagen assembly. At the heterozygous state, such mutations will result in production of lower amounts of collagen but free of structural defects. The resulting phenotypes are mild: OI type 1 (for collagen 1) and the Stickler syndrome (for collagen 2).
2. Structural mutations within the triple-helical domain may produce, at the heterozygous state, a clinical phenotype of different severity according to the type of mutation and its position in the triple helix; for collagen 1, the phenotype may vary from mild (OI type 1) to OI of intermediate severity (OI type 4, OI type 3) to lethal (OI type 2); for collagen 2, the corresponding phenotypes range from lethal (achondrogenesis type 2) to severe (hypochondrogenesis, Kniest) to moderate or mild (SEDC and its variants). A further distinction can be made between glycine substitutions, non-glycine substitutions, and exon-skipping mutations (Terhal et al. 2015).
3. Mutations in the carboxy-propeptides may result in distinct phenotypes: a rare "dense bone type" of OI (for collagen 1) and Torrance dysplasia or spondylo-peripheral dysplasia (for collagen 2).
4. In collagen 1, mutations that produce the skipping of exon 6, which contains the cleaving site for the peptidase that cleaves the amino-propeptide to form the mature collagen molecule, do not result in a brittle bone phenotype but in a phenotype with joint dislocations and fragile skin (the type VII of the Ehlers-Danlos syndrome; see Table 23.3).

Table 23.3 A non-exhaustive overview on human collagens, their tissue-specific expression, and associated genetic phenotypes

Collagen type	Coding genes	Representative cell types and tissues	Associated genetic phenotypes
Collagen type 1	*COL1A1, COL1A2*	Bone (osteoblasts), skin, tendons and ligaments (fibroblasts), dentin (odontoblasts)	Osteogenesis imperfecta (many types), Ehlers-Danlos syndrome types VI and VII
Collagen type 2	*COL2A1*	Cartilage (chondrocytes), vitreous humor of the eye	Type 2 collagen dysplasia family (see text)
Collagen type 3	*COL3A1*	Skin, blood vessels (endothelial cells and fibroblasts), internal organs	Ehlers-Danlos syndrome type IV
Collagen type 4	*COL4A1, COL4A2, COL4A3, COL4A4, COL4A5, COL4A6* (different genes have different expression patterns)	Basement membranes of blood vessels; brain, kidney, inner ear	Vasculopathies: arterial tortuosity, susceptibility to hemorrhage, cerebral vasculopathy, cerebral bleedings with porencephaly; Alport syndrome; hematuria; isolated deafness
Collagen type 5	*COL5A1, COL5A2*	Skin, tendons and ligaments, interstitial tissues	Ehlers-Danlos syndrome types I and II
Collagen type 6	*COL6A1, COL6A2, COL6A3*	Most interstitial tissues (fibroblasts), striated muscle (myoblasts, myocytes)	Ullrich and Bethlem myopathies
Collagen type 7	*COL7A1*	Anchoring fibrils between the dermis (fibroblasts) and epidermis (keratinocytes)	Epidermolysis bullosa, dystrophic type
Collagen type 9	*COL9A1, COL9A2, COL9A3*	Cartilage, vitreous humor	Skeletal dysplasia (similar to collagen II)
Collagen type 10	*COL10A1*	Hypertrophic chondrocytes (growth plate of long bones)	Schmid-type metaphyseal chondrodysplasia
Collagen type 11	*COL11A1, COL11A2*	Cartilage, vitreous humor	Skeletal dysplasias (similar to collagen II)
Collagen type 12	*COL12A1*	Fibroblasts, myocytes (?)	Ullrich and Bethlem myopathies
Collagen type 17	*COL17A1*	Dermis and epidermis	Epidermolysis bullosa, junctional type
Collagen type 18	*COL18A1*	Eye, cranial vault (?)	Knobloch syndrome – myopia and occipital skull defects
Collagen type 27	*COL27A1*	Bone and cartilage (?)	Steel-type chondrodysplasia

Tissue-Specific Expression The two collagens, type 1 and type 2, are similar molecules, and we have seen that the pathogenic mechanisms associated with structural mutations are also similar. What explains the different phenotypes? The answer is their expression in different cell types and tissues. Collagen 1 is synthesized by fibroblasts, osteoblasts, odontoblasts, and tendon cells; collagen 2 is synthesized by chondrocytes and a few other cell types. Thus, collagen 1 mutations will affect bone tissue, teeth, skin, tendons, and the ossicle chain in the middle ear; collagen 2 mutations will affect the growing end of the bones, the joints, the intervertebral disks, the eye (the vitreous body contains collagen 2), and the cochlear structures in the inner ear. There are

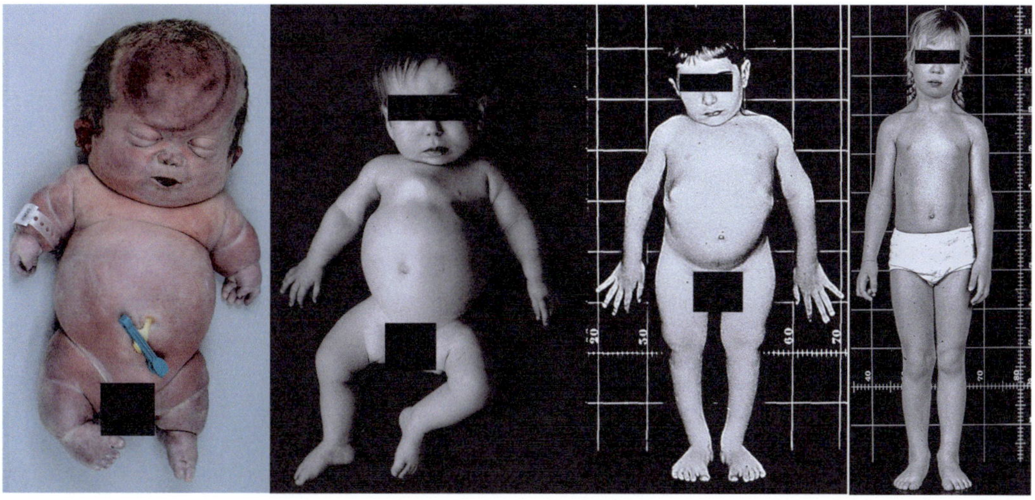

Fig. 23.5 Four individuals affected by different mutations in the *COL2A1 gene.* Left panel, a stillborn with lethal platyspondylic dysplasia, Torrance type; two center panels, a girl (aged 1 year) and a boy (aged 10 years) with congenital spondylo-epiphyseal dysplasia. Right panel: a 5-year-old girl with Stickler dysplasia

many other collagen types in the human body, and some of them have been associated with genetic disease (Table 23.2). In conclusion, both the molecular topography of a given protein and the tissue-specific expression of related genes in a "gene family" are widespread phenomena in human biology and must be taken into account to understand the phenotypic effect of mutation in similar genes (Fig. 23.5).

Questions

1. Collagen type 3 is another fibrillary collagen, structurally similar to collagens 1 and 2, that is mainly expressed in blood vessels and internal organ (intestine, uterus, and others). What might be the clinical consequences of structural mutations therein? Check your answer by conducting a search on OMIM (www.omim.org)!
2. Mutations in collagen type 11 and collagen type 9 may result in clinical phenotypes that are similar to distinguish from those produced by mutations in collagen type 2. What does this suggest regarding their expression? Is it possible that collagen fibers are composed by more than one type of collagen? Check your answers by a web search!
3. Can you explain why a substitution of a given glycine residue in the triple-helical domain of collagen type 1 may produce a phenotype of severe osteogenesis imperfecta, while substitution of the adjacent proline residue may be harmless? Remember the basis of the triple-helical structure of collagens!

References

Bonafe L, Cormier-Daire V, Hall C et al (2015) Nosology and classification of genetic skeletal disorders: 2015 revision. Am J Med Genet A 167a:2869–2892

Chu MI, Williams CJ, Pepe G, Hirsch JI, Prockop DJ, Ramirez F (1983) Internal deletion in a collagen gene in a perinatal lethal form of osteogenesis imperfecta. Nature 304:78–80

Cohn DH, Byers PH, Steinmann B, Gelinas RE (1986) Lethal osteogenesis imperfecta resulting from a single nucleotide change in one human pro alpha 1(I) collagen allele. Proc Natl Acad Sci U S A 83:6045–6047

Cohn DH, Starman BJ, Blumberg B, Byers PH (1990) Recurrence of lethal osteogenesis imperfecta due to parental mosaicism for a dominant mutation in a human type I collagen gene (Col1a1). Am J Hum Genet 46:591–601

Forlino A, Marini JC (2016) Osteogenesis imperfecta. Lancet 387:1657–1671

Kadler KE, Torre-Blanco A, Adachi E, Vogel BE, Hojima Y, Prockop DJ (1991) A type I collagen with substitution of a cysteine for glycine-748 in the alpha 1(I) chain copolymerizes with normal type I collagen and can generate fractallike structures. Biochemistry 30:5081–5088

McKusick VA (1972) Heritable disorders of connective tissue. C.V. Mosby, St. Louis

Prockop DJ (1984) Osteogenesis imperfecta: phenotypic heterogeneity, protein suicide, short and long collagen. Am J Hum Genet 36:499–505

Sillence DO, Senn A, Danks DM (1979) Genetic heterogeneity in osteogenesis imperfecta. J Med Genet 16:101–116

Spranger J (1988) Bone dysplasia 'families'. Pathol Immunopathol Res 7:76–80

Terhal PA, Nievelstein RJ, Verver EJ et al (2015) A study of the clinical and radiological features in a cohort of 93 patients with a COL2A1 mutation causing spondyloepiphyseal dysplasia congenita or a related phenotype. Am J Med Genet A 167a: 461–475

Part III

Others

Acute Kidney Injury: Transition to Chronic Kidney Disease

24

Daisuke Katagiri, Eisei Noiri, Ryo Matsuura, and Raymond C. Harris

Keywords
AKI · Acute kidney diseases (AKD) · cisplatin-induced AKI · L-FABP, liver fatty acid binding protein · VAP-1, vascular adhesion protein-1

24.1 Case Report

A 49-year-old man was admitted to the dermatology ward to receive chemoradiotherapy for squamous cell carcinoma (SCC) of the left inguinal lymph node. The patient had a "birthmark" at the same lesion, which had gradually formed to become a 2.0 × 4.0 cm tumor mass in the past 2 years. According to the biopsy, computed tomography (CT), and positron emission tomography (PET), this mass was diagnosed as stage 4 SCC. He had hypertension and received 20 mg olmesartan and 20 mg nifedipine. He had no specific family history and allergies. His social history indicated he had smoked 20 cigarettes a day and drunk 500 ml of Japanese vodka a day. He had no history of renal disease, and his estimated glomerular filtration rate (eGFR) was 115.8 ml/min/1.73 m^2 at admission. A 2.3 × 4.8 cm soft, reddish mass was found at the left inguinal lymph node.

The major laboratory findings at admission were as follows:

Blood test: white blood cell 8200/μL, blood hemoglobin 14.3 g/dL, platelet count 203,000/μL, total protein 7.1 g/dL, albumin 3.9 g/dL, lactate dehydrogenase 238 U/L, aspartate transaminase 21 U/L, alanine transaminase 20 U/L, γ-glutamyltransferase 278 U/L, alkaline phosphatase 386 U/L, total bilirubin 0.7 mg/dL, total cholesterol 179 mg/dL, calcium 9.0 mg/dL, blood urea nitrogen 11.6 mg/dL, serum creatinine 0.58 mg/dL, sodium 140 mEq/L, potassium 4.5 mEq/L, chloride 104 mEq/L, C-reactive protein 2.63 mg/dL, fasting blood glucose 166 mg/dL, and glycated hemoglobin (HbA_{1c}) 7.5%

D. Katagiri
Department of Nephrology and Endocrinology, University Hospital, The University of Tokyo, Tokyo, Japan

Department of Hemodialysis and Apheresis, 107 Laboratory, University Hospital, The University of Tokyo, Tokyo, Japan

Division of Nephrology, Vanderbilt University Medical Center, Nashville, TN, USA
e-mail: dkatagiri-tky@umin.ac.jp

E. Noiri (✉) · R. Matsuura
Division of Nephrology, Department of Hemodialysis and Apheresis, 107 Laboratory, University Hospital, The University of Tokyo, Tokyo, Japan
e-mail: noiri-tky@umin.ac.jp; rimatsuura-tky@umin.ac.jp

R. C. Harris
Division of Nephrology, Vanderbilt University Medical Center, Nashville, TN, USA
e-mail: raymond.harris@vanderbilt.edu

T. Oohashi et al. (eds.), *Human Pathobiochemistry*, https://doi.org/10.1007/978-981-13-2977-7_24

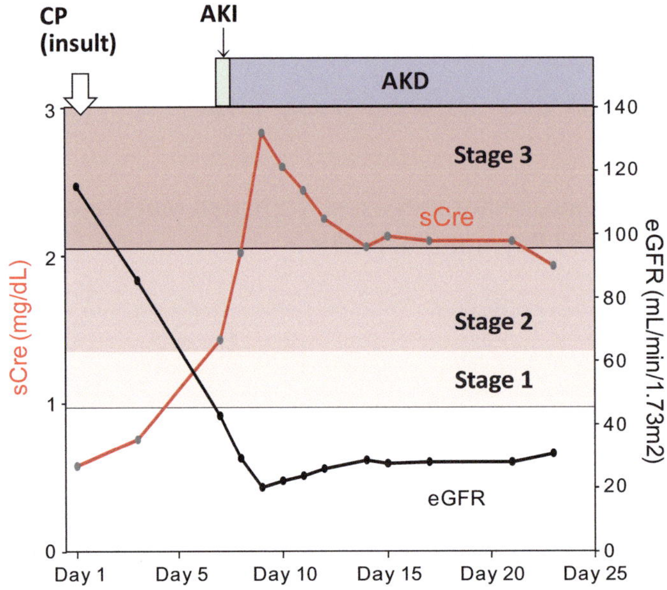

Fig. 24.1 Clinical course of the patient (during admission). CP was given this patient on day 1. On the 8th day, he was diagnosed as stage 2 AKI. Although CP was discontinued and IV fluids therapy was performed, AKI stage increases to stage 3 at day 10 (sCre 2.83 mg/dL). After 7 days duration of AKI, the persistent kidney damage had been recognized as stage 3 AKD but improved to stage 2 AKD, where it was sustained. The patient was discharged on the 26th day. *CP* cisplatin, *IV* intravenous, *AKI* acute kidney injury, *AKD* acute kidney disease. For stage calculation, see Tables 24.1 and 24.4

Urine test: protein (−), glucose (−), and occult blood (−)
Electrocardiography: within normal range and heart rate 83/min
Chest X-ray: cardiothoracic ratio 53% and cardiophrenic angles sharp

He began to receive cisplatin (CP) and 5-fluorourcil (5-FU) chemotherapy. 1000–3000 ml per day of intravenous (IV) fluids was given. Single 20 mg of furosemide was given on day 1 to avoid overhydrating, though his urine flow gradually decreased (3.10 ml/kg/h on day 1 and 0.67 ml/kg/h on day 7). He lost 4 kg in a week. On the 8th day, he was diagnosed with stage 2 acute kidney injury (AKI) due to increased serum creatinine (sCre). Following consultation with nephrology team and considering CP-induced AKI (CP-AKI), CP was discontinued, and he continued receiving an IV fluid replacement therapy. Olmesartan and nifedipine were discontinued as systolic blood pressures were 90 mmHg. Meanwhile he was newly diagnosed with type 2 diabetes. After referral to a diabetologist, he began to receive 50 mg daily vildagliptin. Although sCre began to decrease, it did not return to baseline. The patient was discharged on the 26th day (Fig. 24.1).

One month later after discharge, he received three doses of DTX + 5-FU chemotherapy and was considered in remission from his malignancy. Although he did not show any overt proteinuria during the follow-up period, his sCre did not return to baseline and remained 50–60 ml/min/1.73m^2 of eGFR for 5 years after discharge. This patient was considered to have developed CKD after an episode of CP-AKI (Fig. 24.2).

24.2 Diagnosis

24.2.1 Cisplatin-Induced AKI, Which Led to CKD

AKI is a syndrome characterized by rapid or sudden decrease in kidney function, often accompa-

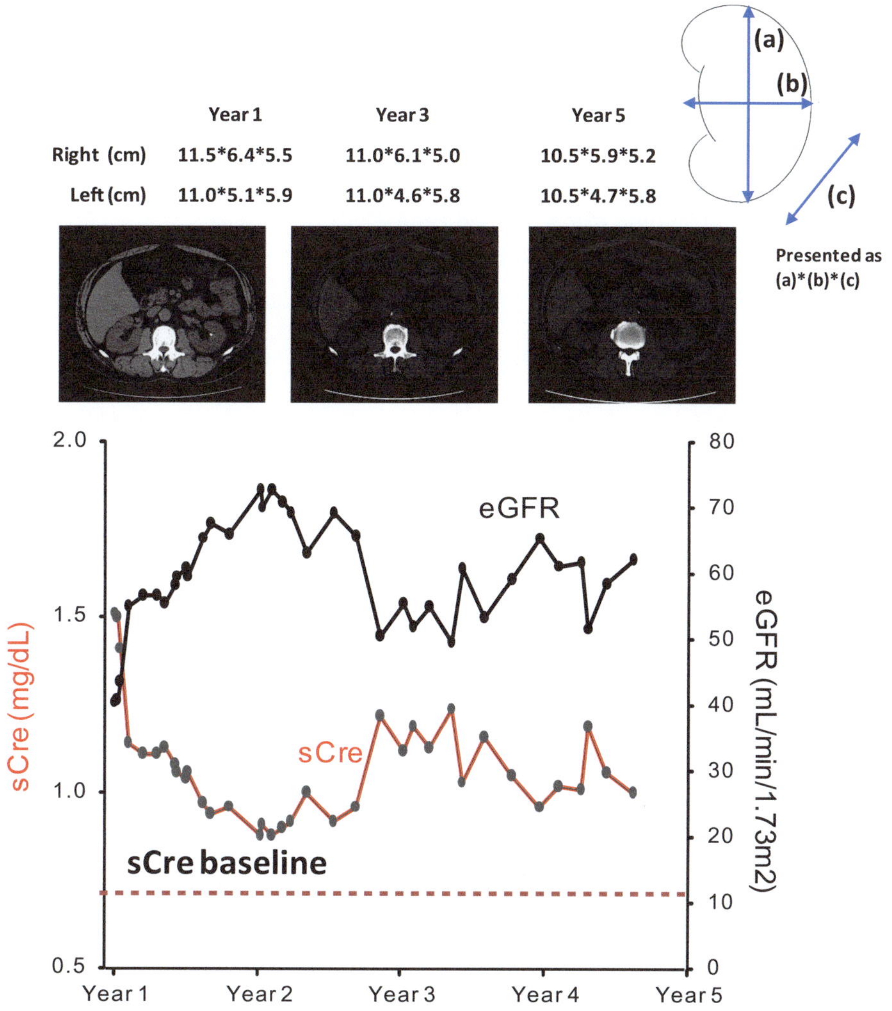

Fig. 24.2 Clinical course of the patient (5 years follow-up). His sCre did not return to baseline (0.58 mg/dL) and remains 50–60 ml/min/1.73 m^2 of eGFR for 5 years after discharge. This patient has been considered to develop CKD after CP-AKI. His follow-up CT presented slightly atrophied kidneys. Note kidneys in each CT presented as $(a)*(b)*(c)$. Year 5 means 4 years later after an admission of the presented case. *eGFR* estimated glomerular filtration rate, *CP-AKI* cisplatin-induced acute kidney injury, *CKD* chronic kidney disease, *CT* computed tomography

nied with oliguria, which happens within a few days. Ischemia-reperfusion injury (IRI), nephrotoxic agents, and sepsis are the major causes of AKI. There are few specific tests to establish the etiology beyond a rising sCre, indicating lack of kidney functioning. AKI has high morbidity/mortality in hospitalized patients, affects >13 million people per year around the world, and causes about 1.7 million deaths/year (Mehta et al. 2015). The risk of AKI among cancer patients has been reported to be 17.5% (1 year) and 27.0% (5 years) (Lam and Humphreys 2012).

It is widely accepted that the eGFR is the most useful kidney function index and that changes in sCre levels are surrogate functional biomarkers for changes in the GFR. Of note eGFR calculations are not valid in the setting of extremes of body mass or under 18-year-old and pregnancy. Two previous criteria for AKI, which were based on sCre levels and urine output, were proposed and validated: the Risk, Injury, Failure, Loss, End-Stage Renal Disease (RIFLE) criteria (Bellomo et al. 2004) and Acute Kidney Injury Network (AKIN) criteria (Mehta et al. 2007). In 2012, the Kidney Disease: Improving Global Outcomes (KDIGO) AKI Guideline Work Group proposed an integrated definition of AKI, based on sCre elevation and urine output decrease, as below (Kellum and Lameire 2013):

- "Increase in sCre by ≧0.3 mg/dL (26.5 μmol/L) within 48 h"
- "Increase in sCre to ≧1.5 times baseline, which have occurred within 7 days"
- "Urine volume <0.5 ml/kg/h for 6 h"

In this case, the patient sCre was increased from 0.58 (day 1) to 1.43 mg/dL (day 8; 2.4 times) in 7 days. According to the staging of AKI (Table 24.1), the patient was diagnosed as "stage 2 AKI." Nephrotoxicity has been attributed to anywhere from 8 to 60% of hospital-acquired AKI (Schetz et al. 2005). CP is a platinum compound and is known as an effective chemotherapeutic agent for many carcinomas, sarcomas, and lymphomas, though it has adverse effects of nephrotoxicity and ototoxicity (Pabla and Dong 2008). Exposure of tubular epithelial cells to CP activates complex signaling pathways that engender cell injury and death. In the CP-AKI model, the pro-apoptotic family Bax is activated and accumulates in the mitochondria in renal tubular cells (Wei et al. 2007; Katagiri et al. 2013). Generally this nephrotoxicity is reversible, though in some cases, such as in the patient described above, it can be persistent.

The majority of patients will recover their renal function following an episode of AKI, though some AKI survivors have a higher risk of developing chronic kidney disease (CKD) (Cerda et al. 2008). If AKI patients are diagnosed as acute tubular necrosis, or there are multiple causes of the AKI, they have a higher risk of subsequently developing CKD (Humphreys et al.

Table 24.1 Criteria of AKI (KDIGO 2012)

Stage	Serum creatinine	Urine output
1	1.5–1.9 times baseline or ≧0.3 mg/dL (≧26.5 μmol/L) increase	<0.5 ml/kg/h for 6–12 h
2	2.0–2.9 times baseline	<0.5 ml/kg/h for ≧12 h
3	3.0 times baseline, or increase in serum creatinine to ≧4.0 mg/dL (≧353.6 μmol/L), or initiation of renal replacement therapy	<0.3 ml/kg/h for ≧24 h or anuria for ≧12 h

Anuria is commonly defined in the adult population as a urine passage of less than 50–100 mL per day

Table 24.2 Criteria for CKD (either of the following present for >3 months) (KDIGO 2013)

Markers of kidney damage (one or more)	Albuminuria (AER ≧30 mg/24 h, ACR≧30 mg/g)
	Urine sediment abnormalities
	Electrolyte and other abnormalities due to tubular disorders
	Abnormalities detected by histology
	Structural abnormalities detected by imaging
	History of kidney transplantation
Decreased GFR	GFR <60 ml/min/1.73 m^2

AER albumin excretion rate, *ACR* albumin to creatinine ratio, *GFR* glomerular filtration rate

2016). Also, it is important to evaluate and follow AKI patient's pre-existing risk for CKD, such as advanced age, congestive heart failure, and uncontrolled diabetes. The present patient had newly diagnosed diabetes, though he has not developed overt proteinuria in response to the treatment. CKD is defined as structural/functional abnormalities that persist for >3 months. Patients with eGFR <60 ml/min/1.73 m^2 or with markers of kidney damage such as albuminuria, urine sediment, or electrolyte abnormalities, or a history of kidney transplantation, are defined as CKD (Table 24.2) (KDIGO 2013). To date, the

Table 24.3 Meaning and definition of AKI, AKD, and CKD (Chawla et al. 2017)

	AKI	AKD	CKD
Days post injury	0–7 days	7–90 days	90 days –
Meaning	Sudden decrease of kidney function or structure	Patients with AKD can progress to CKD	Abnormalities of kidney structure/function for >3 months
Definition	See Table 24.1	sCre >1.5 times baseline	See Table 24.2
Follow-up		Patient education, medication, reconciliation, nephrology referral	

AKI acute kidney injury, *AKD* acute kidney disease, *CKD* chronic kidney disease, *sCre* serum creatinine

Table 24.4 Suggested staging of AKD (Chawla et al. 2017)

Stage	Definition
3	sCre ≥3.0 times baseline or increase ≥4.0 mg/dL or renal replacement therapy
2	sCre ≥2.0 times baseline
1	sCre ≥1.5 times baseline
0	B/C: sCre <1.5 times baseline but not back to baseline and continued evidence of ongoing injury, repair, and/or regeneration
	C: sCre <1.5 times baseline but not back to baseline
	B: Continued evidence of ongoing injury, repair and/or regeneration, or indicators of loss of renal glomerular or tubular reserve
	A: Absence of criteria for B or C

AKD acute kidney disease, *sCre* serum creatinine

leading cause of CKD is diabetes in most developed countries, and both CKD and diabetes are well-known risk factors for cardiovascular disease. In a large cohort study, a 30–40% decline in eGFR after AKI can be a surrogate endpoint for end-stage kidney disease (ESKD) (Grams et al. 2016). The continuum of AKI to CKD may cause significant personal and economic strain.

Recently the term acute kidney disease (AKD) has been proposed by KDIGO to define the course after AKI in which the kidney injury is ongoing (Chawla et al. 2017). In this guideline, AKD is defined as a new concept to provide an integrated clinical approach to patients with kidney abnormalities of function and structure, whose kidney damage is more than 7 days but less than 3 months (Table 24.3). (KDIGO 2012).

Regardless of AKI severity, AKI episodes lasting over 72 h after the insult have been reported to be associated with poorer outcome than AKI that is rapidly ameliorated within 72 h (Chawla et al. 2017). AKI resolving within 48 h is often described as transient AKI. This patient was diagnosed as stage 2 AKI at 7 days after first CP treatment, which developed further into AKD stage 3 at day 10 (sCre 2.83 mg/dL). He had persistent kidney damage during AKD, which was sustained as stage 2 (Table 24.4) (Chawla et al. 2017). His sCre did not return to basal level for 30 days, even for 5 years after this AKI episode.

His follow-up CT presented slightly atrophied kidneys with partial renal capsular indentations (Fig. 24.2). Progression of CKD with persistent interstitial fibrosis (IF) is probable with this patient. His clinical course clearly indicates the continuum of AKI to CKD as a result of risk factors and disease modifiers and further predicts clinical outcome (Chawla et al. 2017).

24.3 Molecular Perspectives

Repeated or chronic kidney injury results in extracellular matrix (ECM) accumulation and tubular atrophy that eventually lead to hypoxia and IF in kidney. Proximal tubule injury beyond the potential for adaptive repair will arrest epithelial cells in the G2/M transition of the cell cycle and will enhance production of pro-fibrotic factors (Bonventre 2014; Yang et al. 2010). Injured epithelial cells produce a number of growth factors, such as epidermal growth factor, hepatocyte growth factor (HGF), and insulin-like growth factor 1. Increased levels of transforming growth factor (TGF)-β lead to upregulation of α-smooth muscle actin (SMA), F-actin, and collagen I by fibroblasts (Borges et al. 2013; Mack and Yanagita 2015). Glomerular injury followed by IF leads to a decrease of glomerular filtration rate. IF typically shows excessive fibrillar collagen accumulation.

CP-AKI is often seen after 8–10 days of CP administration in patients. In rodents, AKI induced by a single high dose of CP is a well-studied as dose dependently. A model of chronic kidney injury induced by a single CP injection presents with glomerular sclerosis, cyst formations, and interstitial fibrosis. We developed multiple, lower CP treatment method in mice as a potential clinically relevant AKD model (Katagiri et al. 2015). Since urinary liver fatty acid-binding protein (L-FABP), one of the novel AKI biomarkers currently reimbursed in Japan and CE marked in EU, shows promise as an AKI biomarker and has also been reported to assist in predicting CKD, we evaluated AKD in L-FABP transgenic (Tg) mice.

Male L-FABP-Tg mice heterozygous for human L-FABP (C57BL/6 background) were administered three CP doses (10 mg/kg; at 0, 1, 3 weeks) for 4 weeks. The mice were sacrificed at 4 weeks after the first CP administration. Increased BUN and elevated sCre were observed at 4 weeks. Urinary L-FABP significantly increased 1 week after every CP injection. L-FABP levels at 3 weeks decreased according to the skip of CP injection at 2 weeks, which did not return to baseline (Fig. 24.3c), suggesting a causal relation of each mild AKI and the development of CKD as a mild but a cumulative CP toxicity. Pathological analysis indicated tubule dilatation with brush border loss and moderate renal IF at 4 weeks (Fig. 24.3). This model is robust and relevant to clinical findings of AKI to CKD in our daily practice. This study also demonstrated urinary L-FABP as a promising early clinical biomarker predictable CP-AKI to CKD and AKD.

24.4 Therapy

Despite recent advances in renal replacement therapy, AKI is still associated with poor outcomes. As per KDIGO guidelines, personalized management and diagnostic steps according to each AKI stages have been suggested (Fig. 24.4) (KDIGO 2012). To predict AKI to CKD development (Fig. 24.5), it is important to distinguish rapid reversal (transient) or persistent AKI. As stated above, rapid reversal AKI, which shows quick recovery within 48 h, has a better renal outcome than persistent AKI that continues beyond 48 h.

Unfortunately, there have been no uniformly reliable biomarkers other than sCre that can evaluate kidney repair. Note that it is hard to distinguish functional recovery and structural reconstitution from decrease of sCre. When we treat patients with persistent AKI, careful monitoring including hemodynamic and volume status is required. Moreover, introduction of adequate renal replacement therapy (RRT) should be considered if complications of AKI such as electrolyte imbalance or fluid overload are severe. On the other hand, the limitations of sCre as a functional biomarker are well known. Delayed intervention with sCre-based diagnosis in intensive care unit (ICU) provides the impetus to searching for novel incisive damage AKI biomarkers. After more than a decade of intensive effort, several novel damage biomarkers have been developed and evaluated to facilitate early detection, differential diagnosis, and prognosis of AKI. These new biomarkers that have been identified and evaluated include neutrophil gelatinase-associated lidocaine (NGAL), kidney injury molecule-1

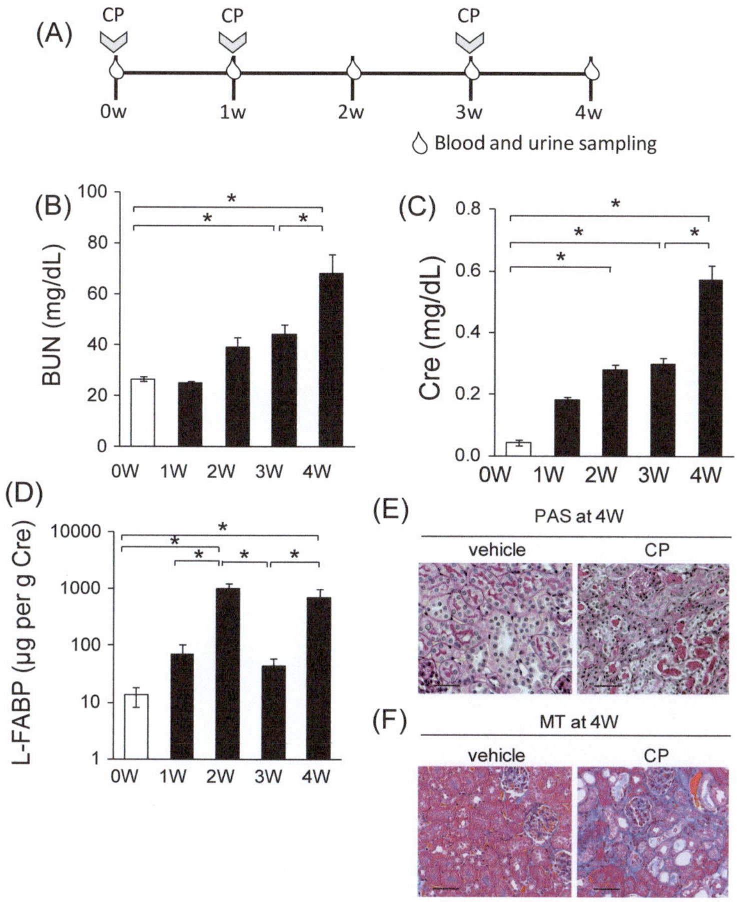

Fig. 24.3 AKD model and interstitial fibrosis by multiple CP treatment (Katagiri et al. 2015). (**a**) Male L-FABP Tg mice were administered three 10 mg/kg CP doses (at 0, 1, 3 weeks) for 4 weeks. $*P < 0.05$ vs. 0 week. CP, cisplatin; L-FABP, L-type fatty acid-binding protein; 4 W, 4 weeks. (**b–d**) After exposure to CP, levels of BUN, sCre, and urinary L-FABP were significantly higher at 4 weeks ($n = 7–8$). No further elevation of L-FABP was observed at 3 weeks, following the omission of CP at 2 weeks, but L-FABP also did not revert to baseline levels. (**e–f**) Exposure to CP caused brush border loss and interstitial fibrosis at 4 weeks after three CP injections (size bar = 50 μm). $*P < 0.05$. Data are expressed as mean values ± S.E.M. *CP* cisplatin, *sCre* serum creatinine, *BUN* blood urea nitrogen, *L-FABP* L-type fatty acid-binding protein, *W* weeks, *PAS* periodic acid-Schiff, *MT* Masson's trichrome. (http://dx.doi.org/10.1038/ki.2015.327)

(KIM-1), cystatin C, IL-18, urinary L-FABP, urine insulin-like growth factor-binding protein 7 (IGFBP7), and tissue inhibitor of metalloproteinase-2 (TIMP-2).

Duration of this short-term therapeutic window in AKI is highly important to prediction of prognosis for kidney recovery. Limitation of damage during AKI is crucial to tubular cell proliferation and angiogenesis of endothelial cells that lead to kidney repair after insult. As for nephrotoxic-related AKI, it is often difficult to clarify the exact nephrotoxic agents that have induced AKI. Administration of antimicrobials to septic patients, radiocontrast agents to acute coronary syndrome patients, and cisplatin to malignancy patients are examples of agents given to patients who are already have had higher risks for developing AKI. However, it is

Fig. 24.4 Stage-based AKI management (KDIGO 2012). Personalized management and diagnostic steps according to each AKI stages are suggested by KDIGO guidelines. Blue shading indicates actions that are equally appropriate at all stages, whereas red-graded shading indicates increasing priority as intensity increases. *AKI* acute kidney injury, *RRT* renal replacement therapy, *ICU* intensive care unit, *KDIGO* Kidney Disease: Improving Global Outcomes

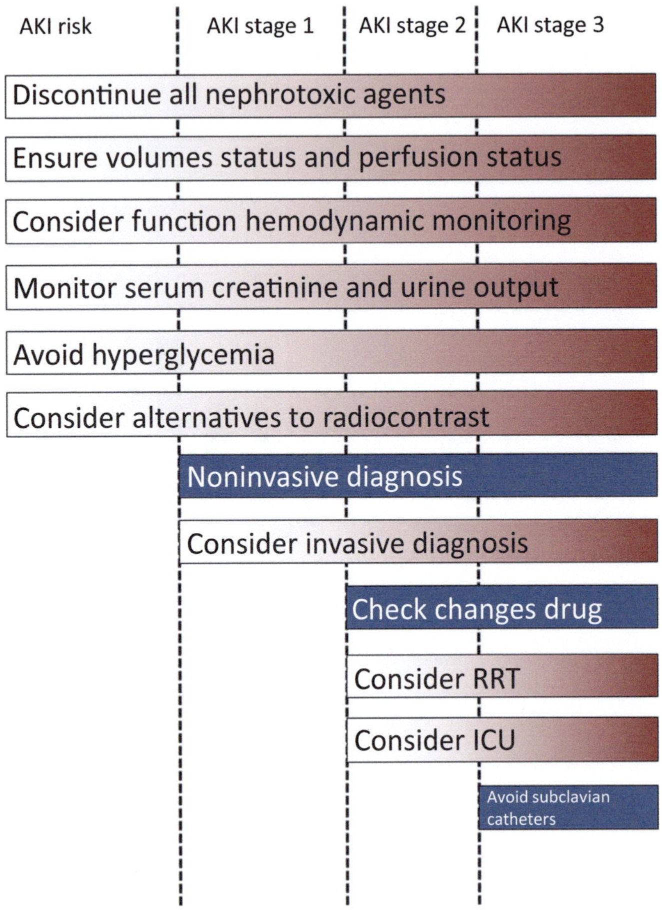

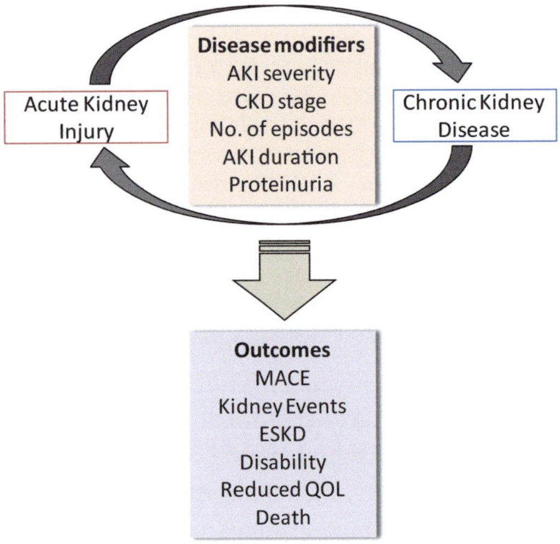

Fig. 24.5 Continuum of AKI and CKD (Chawla et al. 2017). AKI and CKD are not always separate disease states and tend to represent a continuum with patients in the setting of sustained AKI who have an increased risk of developing de novo/existing CKD. The various disease modifiers and risk factors might represent opportunities to intervene. *AKI* acute kidney injury, *CKD* chronic kidney disease, *MACE* major adverse cardiovascular events, *ESKD* end-stage kidney disease, *QOL* quality of life

safe to avoid an additional exposure to a suspected drug if patients develop AKI during treatment.

Vascular adhesion protein-1 (VAP-1) is a circulating and type-1 membrane-bound protein that is expressed predominantly in the endothelium and leukocytes and plays a dual role in mediating inflammation and reactive oxygen species (ROS) production. In our experiment, 1 week of VAP-1 inhibitor treatment during multiple CP injections attenuated biomarker elevation and pathological injury of AKD, with suppression of oxidative stress and IF development (Katagiri et al. 2015). VAP-1 inhibition is focused in nonalcoholic steatohepatitis (NASH) to ameliorate liver fibrosis recently. Above mentioned therapeutic intervention together with sensitive biomarker rather than sCre is necessary for the future pharmaceutical development in this disease entity similarly to pharmaceutical development in the oncology field.

End-of-Chapter Questions

1. Are novel AKI biomarkers useful to monitor the clinical course of AKD? Do AKI biomarkers have a role in indicating further deterioration of AKD as shown in our mice AKD model?
2. How do we evaluate a patient's basal kidney function if he/she already developed AKI without any recent medical records? Could any lab test/imaging/ etc. help in evaluating pre-existing CKD?
3. AKI combined with malignancy is expected to be increased in aging countries. What kind of risk factors should we care for AKI among these patients? (ex. anticancer drug, dehydration, infection…)

References

Bellomo R, Ronco C, Kellum JA et al (2004) Acute renal failure – definition, outcome measures, animal models, fluid therapy and information technology needs: the second international consensus conference of the Acute Dialysis Quality Initiative (ADQI) Group. Crit Care 8:R204–R212

Bonventre JV (2014) Primary proximal tubule injury leads to epithelial cell cycle arrest, fibrosis, vascular rarefaction, and glomerulosclerosis. Kidney Int Suppl 4:39–44

Borges FT, Melo SA, Ozdemir BC et al (2013) TGF-beta1-containing exosomes from injured epithelial cells activate fibroblasts to initiate tissue regenerative responses and fibrosis. J Am Soc Nephrol 24:385–392

Cerda J, Lameire N, Eggers P et al (2008) Epidemiology of acute kidney injury. Clin J Am Soc Nephrol 3:881–886

Chawla LS, Bellomo R, Bihorac A et al (2017) Acute kidney disease and renal recovery: consensus report of the Acute Disease Quality Initiative (ADQI) 16 Workgroup. Nat Rev Nephrol 13:241–257

Grams ME, Sang Y, Coresh J et al (2016) Candidate surrogate end points for ESRD after AKI. J Am Soc Nephrol 27:2851–2859

Humphreys BD, Cantaluppi V, Portilla D et al (2016) Targeting endogenous repair pathways after AKI. J Am Soc Nephrol 27:990–998

Katagiri D, Hamasaki Y, Doi K et al (2013) Protection of glucagon-like peptide-1 in cisplatin-induced renal injury elucidates gut-kidney connection. J Am Soc Nephrol 24:2034–2043

Katagiri D, Hamasaki Y, Doi K et al (2015) Interstitial renal fibrosis due to multiple cisplatin treatments is ameliorated by semicarbazide-sensitive amine oxidase inhibition. Kidney Int 89:374–385

KDIGO (2012) Clinical Practice Guideline for Acute Kidney Injury. Kidney Int Suppl 2:1–124

KDIGO (2013) Clinical practice guideline for the evaluation and management of chronic kidney disease. Kidney Int Suppl 3:1–116

Kellum JA, Lameire N (2013) Diagnosis, evaluation, and management of acute kidney injury: a KDIGO summary (Part 1). Crit Care 17:204

Lam AQ, Humphreys BD (2012) Onco-nephrology: AKI in the cancer patient. Clin J Am Soc Nephrol 7:1692–1700

Mack M, Yanagita M (2015) Origin of myofibroblasts and cellular events triggering fibrosis. Kidney Int 87:297–307

Mehta RL, Kellum JA, Shah SV et al (2007) Acute Kidney Injury Network: report of an initiative to improve outcomes in acute kidney injury. Crit Care 11:R31

Mehta RL, Cerda J, Burdmann EA et al (2015) International Society of Nephrology's 0by25 initiative for acute kidney injury (zero preventable deaths by 2025): a human rights case for nephrology. Lancet 385:2616–2643

Pabla N, Dong Z (2008) Cisplatin nephrotoxicity: mechanisms and renoprotective strategies. Kidney Int 73:994–1007

Schetz M, Dasta J, Goldstein S et al (2005) Drug-induced acute kidney injury. Curr Opin Crit Care 11:555–565

Wei Q, Dong G, Franklin J et al (2007) The pathological role of Bax in cisplatin nephrotoxicity. Kidney Int 72:53–62

Yang L, Besschetnova TY, Brooks CR et al (2010) Epithelial cell cycle arrest in G2/M mediates kidney fibrosis after injury. Nat Med 16:535–543, 531p following 143

Type I Interferonopathies: Common Pathological Features Between Congenital Infections and Genetic Disorders

25

Tatsuharu Sato, Masako Moriuchi, and Hiroyuki Moriuchi

Keywords
Aicardi-Goutieres syndrome · Cytomegalovirus · TORCH syndrome · TREX1 · Type I interferon

25.1 Case Reports

25.1.1 Case 1

A newborn Japanese boy was admitted to our neonatal intensive care unit (NICU) for respiratory failure soon after birth. He was born at 38 weeks of gestation. The pregnancy was complicated by pregnancy-induced hypertension and intrauterine growth retardation (IUGR). Meconium staining of the amniotic fluid was noted at birth. The Apgar scores were both 8 at 1 and 5 min. His birth weight was 2080 g (−2.3 standard deviations [SD]), height 43.5 cm (−2.6SD), and head circumstance 29.5 cm (−2.7SD). He developed generalized cyanosis and needed oxygen administration. The chest X-ray findings were compatible to meconium aspiration syndrome. He also had thrombocytopenia and was referred for findings in the right ear on newborn hearing screening. Cranial ultrasonography (US) showed enlarged ventricles, and computed tomography (CT) of the brain further demonstrated calcification of the basal ganglia and periventricular white matter (Fig. 25.1a). His respiratory condition improved over the next few days, and his platelets gradually increased; however, abnormal findings of the brain such as ventriculomegaly and calcification persisted. Sensorineural hearing loss (SNHL) of the right ear was confirmed by auditory brainstem response (ABR). He was discharged at 18 days of age.

25.1.2 Case 2

A 2-day-old Japanese boy was admitted to our NICU for hypotonia, loss of Moro reflex, and poor activity. He was born at 39 weeks and 5 days of gestation. His parents were second cousins. His mother developed premature rupture of the membranes (PROM) at 39 weeks and 5 days of gestation, and leakage of turbid amniotic fluid was noted. The Apgar scores were 8 and 9 at 1 and 5 min, respectively. His birth weight was 3388 g (+1.0SD), height 51.0 cm (+0.9SD), and head circumstance 33.5 cm (+0.1SD). Although his condition was stable at birth, he was hypotonic and gradually became weaker and less active. Blood tests at admission showed leukocytosis, thrombocytopenia, and a high

T. Sato · M. Moriuchi · H. Moriuchi (✉)
Department of Pediatrics, Nagasaki University Graduate School of Biomedical Sciences, Nagasaki, Japan
e-mail: a65-masa@nagasaki-u.ac.jp; hiromori@nagasaki-u.ac.jp

T. Oohashi et al. (eds.), *Human Pathobiochemistry*, https://doi.org/10.1007/978-981-13-2977-7_25

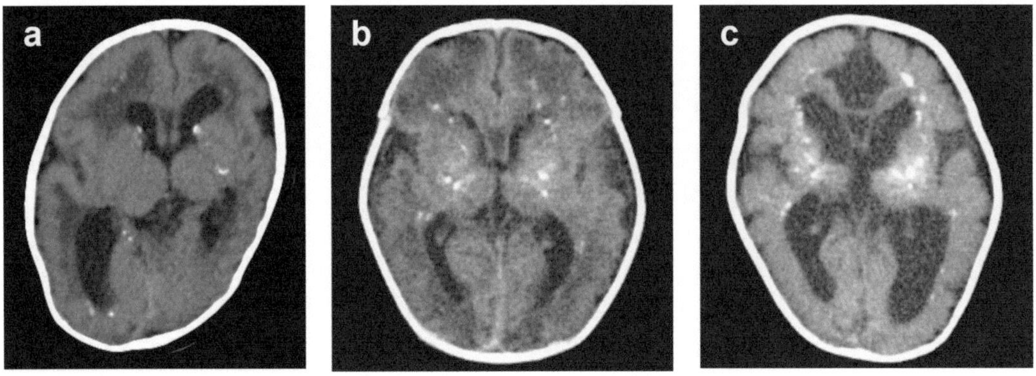

Fig. 25.1 Characteristic findings of brain images of patients with type I interferonopathies
(**a**) Brain CT scan of Case 1 with congenital cytomegalovirus infection, showing calcification of the basal ganglia and periventricular white matter, enlargement of the ventricles, diffuse low-density areas in the cerebral white matter, and widespread cortical thickness
(**b**) Brain CT scan of Case 2 with Aicardi-Goutieres syndrome on day 2, showing calcification of the basal ganglia and periventricular white matter and enlargement of the ventricles
(**c**) Brain CT scan of Case 2 at 3 months of age, showing the progression of calcification, ventriculomegaly, and brain atrophy

C-reactive protein (CRP) value. He was empirically given antibiotics, which had little effect on his clinical symptoms and abnormal laboratory findings. He repeatedly experienced aspiration due to severe hypotonia, necessitating tube feeding. Cranial US revealed enlarged ventricles, and CT of the brain on day 2 showed calcifications of the basal ganglia and periventricular white matter (Fig. 25.1b). He failed bilaterally on newborn hearing screening, and mild bilateral SNHL was confirmed by ABR later. He gradually became more active but remained hypotonic. He was discharged at 50 days of age with tube feeding.

25.2 Diagnosis

25.2.1 Case 1: Congenital Cytomegalovirus Infection

The problems of this patient are summarized as follows: symmetrical IUGR, enlarged ventricles, calcifications of the basal ganglia and white matter, thrombocytopenia at birth, and SNHL. All of those characteristics were indicative of TORCH complex (see below), especially congenital cytomegalovirus infection (CCMVI).

The diagnosis of CCMVI requires the detection of virus or viral DNA within the first 3 weeks of life, as neonatal infection during vaginal delivery or through breastfeeding is common. PCR is the preferred method of diagnosis because it is sensitive, rapid, and accurate. The presence of CMV-specific IgM antibody at birth also supports a diagnosis of CCMVI, but false-negative results are common, and false-positive results can occur as well.

Comprehensive serological tests were positive for CMV IgM antibody in his serum, and real-time PCR detected CMV-DNA in his urine collected at 6 days of age. Thus, a definite diagnosis of CCMVI was made.

25.2.1.1 Overview of TORCH Syndrome

Congenital infections account for 2–3% of all congenital anomalies. Some of the representative congenital infections are included in the TORCH complex or TORCH syndrome (Table 25.1), which stands for ***T****oxoplasma gondii*, **o**ther pathogens (listed below), ***R****ubella virus*, ***C****ytomegalovirus* (CMV), and ***H****erpes simplex virus* (HSV). Other pathogens include *Treponema pallidum* (syphilis), *varicella-zoster virus*, *human immunodeficiency virus*, *lymphocytic choriomeningitis virus*, and *Zika virus*.

One of the reasons why such taxonomically variable pathogens (from protozoa, spirochete,

Table 25.1 Members of TORCH complex

	Species	Family (subfamily)	Phylum/group	Domain
T	*Toxoplasma gondii*	*Toxoplasmatidae*	*Apicomplexa*	Eukaryota
O	*Treponema pallidum* (syphilis)	*Spirochaetaceae*	*Spirochaetes*	Bacteria
(other)	*Varicella-zoster virus*	*Herpesviridae (Alphaherpesvirinae)*	Group I: Ds DNA viruses	Viruses
	Human immunodeficiency virus	*Retroviridae*	Group VI: ssRNA-RT viruses	
	Lymphocytic choriomeningitis virus	*Arenaviridae*	Group V: (−)ssRNA viruses	
	Zika virus	*Flaviviridae*	Group IV: (+)ssRNA viruses	
R	*Rubella virus*	*Togaviridae*	Group IV: (+)ssRNA viruses	
C	*Human cytomegalovirus*	*Herpesviridae (Betaherpesvirinae)*	Group I: Ds DNA viruses	
H	*Herpes simplex virus*	Herpesviridae (Alphaherpesvirinae)	Group I: Ds DNA viruses	

and DNA viruses to RNA viruses) are combined into the single clinical entity of TORCH syndrome is the similarity of their clinical presentations (Table 25.2). Such similarities imply that many, if not all, of them result from host responses or chronic inflammation rather than direct damage by the pathogens themselves, as described later.

25.2.1.2 The C in TORCH Stands for CMV

CMV is thought to be the most common cause of TORCH syndrome and is the leading cause of neurodevelopmental problems in many developed countries. Although only 10–15% of children with CCMVI exhibit clinical signs at birth, even children who appear asymptomatic at birth are at risk for neurodevelopmental sequelae. Most children (60–90%) with symptomatic infection and 10–15% of asymptomatic infants develop one or more long-term neurological sequelae. CMV is currently the leading cause of nonhereditary SNHL in children. The frequency of hearing loss in children due to CCMVI is between 0.2 and 1.3/1000 live births.

Chronic inflammation due to fetal infection inhibits intrauterine growth of the fetus; therefore, infected newborns are often small for gestational age. Outside of the central nervous system (CNS), chronic inflammation occurs mainly in the lymphatic and mononuclear phagocyte systems, resulting in such common and transient clinical signs as hepatosplenomegaly, jaundice, anemia, thrombocytopenia, and blueberry muffin-like skin (subcutaneous bleeding and extramedullary hematopoiesis).

In contrast, manifestations involving the CNS and sensory organs typically persist for life and are sometimes delayed-onset, and progressive conditions that can cause serious neurodevelopmental sequelae include mental retardation, cerebral palsy, epilepsy, autism, SNHL, and ophthalmological problems.

Neuroimaging studies with magnetic resonance imaging and CT show a variety of abnormal findings, including intracranial calcifications, migrational abnormalities (lissencephaly, pachygyria, diffuse or focal polymicrogyria, schizencephaly, and cortical dysplasia), white matter abnormalities, periventricular cysts, cerebral atrophy, cerebellar hypoplasia, ventriculomegaly, ventricular adhesions, lenticulostriate vasculopathy, and destructive encephalopathy.

25.2.2 Case 2: Aicardi-Goutieres Syndrome

The problems of this patient are summarized as follows: septic appearance, enlarged ventricles, calcification of the basal ganglia and white matter, thrombocytopenia at birth, SNHL, persistently high levels of CRP, and recurrent aspiration due to severe hypotonia. Initially, we suspected sepsis from his perinatal history

Table 25.2 Clinical manifestations of type I interferonopathies

Clinical manifestation		Infectious				Genetic										
		CMV	Rubella virus	*T. gondii*	Zika virus	*TREX1*	*RNASEH2A/2B/2C*	*SAMHD1*	*ADAR1*	*IFIH1*	*DDX58*	*TMEM173*	*USP18*	*ISG15*	*ACP5*	*PSMB8*
CNS and sensory organs	Microcephaly	*	*	*	*	*	*	*	*	*			*		*	
	Intracranial calcification	*	*	*	*	*	*	*	*	*			*	*	*	*
	White matter abnormalities	*	*	*	*	*	*	*	*	*					*	
	Cerebral atrophy	*	*	*	*	*	*	*	*	*			*		*	
	Large vessel disease					*		*					*		*	
	Developmental delay/abnormalities	*	*	*	*	*	*	*	*	*			*		*	*
	Epilepsy	*	*	*	*	*	*	*	*	*			*	*	*	
	Spastic paraparesis				*		*		*	*					*	
	Chorioretinitis	*	*	*	*											
	Glaucoma		*			*	*	*		*	*					
	Cataract		*													
	Other ophthalmological problems		*	*	*											
	Sensorineural hearing loss	*	*	*	*	*										
Lymphatic and mononuclear phagocytic systems	Thrombocytopenia	*	*	*		*	*	*	*	*			*			
	Hepatosplenomegaly	*	*	*		*	*	*	*	*			*			*
	Leukemogenesis	(*)						*								
	Immunodeficiency													*		

Possibly auto-inflammatory or autoimmune disorders	Digital vasculitis and/or necrosis (chilblains)					*	*	*	*	*		*			*	*
	Livedo and/or skin mottling					*	*	*	*	*		*			*	*
	Recurrent fever					*	*	*	*	*		*			*	*
	Autoimmune disorders		*			*	*	*	*	*		*			*	
Joint, bone, teeth	Arthritis/joint contractures				*	*		*		*	*	*				*
	Skeletal anomalies										*					*
	Premature tooth loss/hypoplastic enamel	*	*							*						
Cardio-vascular system	Aortic/valvular/coronary lesions										*					
	Cardiomyopathy	(*)				*	*			*						*
Lungs	Interstitial pneumonitis	(*)										*				

(PROM and the leakage of turbid amniotic fluid), clinical symptom (septic appearance), and laboratory data (leukocytosis, thrombocytopenia, and high CRP levels). Subsequently, demonstration of brain lesions (ventriculomegaly and calcifications) and SNHL led us to another potential diagnosis of TORCH complex that might unitarily explain all of his problems listed above. However, the following examinations were all negative: *T. gondii* IgG/IgM, serological tests for syphilis, rubella IgM, CMV IgM, CMV DNA in urine, HSV IgG/IgM, and HSV DNA in the peripheral blood and cerebrospinal fluid (CSF).

Follow-up CT of the brain at 3 months of age showed the progression of calcification, ventriculomegaly, and brain atrophy (Fig. 25.1c). A CSF examination showed mild pleocytosis (8/μL) and a high protein concentration (175 mg/dL). Based on the aforementioned data, he was suspected of having an ongoing inflammatory process in the CNS. Therefore, we further investigated his CSF, noting a high level of interferon (IFN)-alpha at 67 IU/mL (normal range <5 IU/mL). In addition to the consanguineous marriage of his parents, we suspected genetic inflammatory disorders might be affecting the brain, including Aicardi-Goutieres syndrome (AGS), and detected a homozygous missense mutation of the *TREX1* gene encoding a 3-prime repair exonuclease with preferential activity on single-stranded (ss) DNA, as described below.

25.2.2.1 Overview of Pseudo-TORCH Syndrome

Not only intrauterine infections but also certain genetic conditions represent the main clinical features of TORCH syndrome, with the latter initially named "pseudo-TORCH syndrome." Jean Aicardi and Françoise Goutieres reported the first familial cases of early-onset progressive brain disease that had noteworthy clinical overlap with TORCH syndrome and chronic lymphocytosis in the CSF, suggestive of a chronic inflammatory process in the CNS (Crow et al. 2015).

Chronic inflammation has a strong and unique impact on the fetal brain, mainly because of the immunological characteristics of the CNS and vulnerability of the developing brain to the neurotoxic effects of type I IFN. The type I IFN family consists of multiple IFN-alpha, IFN-beta, IFN-delta, IFN-epsilon, IFN-kappa, IFN-tau, IFN-omega, and IFN-zeta/Limitin. They are produced following recognition of microbial products by cell surface and intracellular pattern recognition receptors. The synthesis of type I IFN in the CNS is barely seen in non-viral neurological diseases. However, Lebon et al. revealed the association of pseudo-TORCH syndrome, currently referred to as AGS, with an increased type I IFN activity in the CSF and serum (Lebon et al. 1988). In other words, AGS is both clinically and biochemically reminiscent of TORCH syndrome, and type I IFN plays a key role in the pathogenesis of both TORCH syndrome and AGS.

25.2.2.2 Aicardi-Goutieres Syndrome

AGS is a rare genetic neurodegenerative disorder, usually inherited in an autosomal recessive pattern, with a wide spectrum of clinical manifestations. The causative genes for AGS will be fully discussed later. Although patients with AGS usually exhibit severe intellectual and physical impairment and exhibit severe progressive microcephaly, variability in the severity of the neurologic outcome can be observed even among siblings who share the same genotype.

As seen in the present patient (Case 2), a subgroup (20%) of cases manifest themselves at birth or even in utero with a picture highly suggestive of congenital infection, including abnormal neurologic findings, hepatosplenomegaly, elevated liver enzymes, and thrombocytopenia. However, it is characteristic for AGS and different from TORCH syndrome that such neurological abnormalities as intracranial calcifications and microcephaly are progressive, while other symptoms, such as hepatosplenomegaly, elevated liver enzymes, and thrombocytopenia, are prolonged.

The majority (80%) of patients with AGS have normal perinatal and neonatal histories and present at variable durations after the first few weeks of life, frequently after a period of apparently normal development. Typically, they demonstrate progressive subacute encephalopathy of early onset, generally in the first year of life, which is characterized by pleocytosis and elevated levels of IFN-alpha in the CSF, alterations in the white matter of the brain, and intracranial calcifications. Over time, as many as 40% of patients with AGS develop chilblain skin lesions, ophthalmologic lesions, SNHL, and autoimmune disorders (Crow et al. 2015) (Table 25.2).

IFN signature genes are highly expressed in the peripheral blood of patients with AGS. CSF examinations typically show pleocytosis, increased IFN-alpha activity, and increased concentration of neopterin.

25.3 Biochemical and Molecular Perspectives

25.3.1 Multiple Neuropathological Mechanisms of CCMVI

The pathological mechanisms caused by CMV are variable and still remain obscure in many aspects, especially regarding the development of fetal brain injury. It has been shown that CMV preferentially replicates in developing glia and neurons, regardless of the adaptive immune system. CMV infection of the fetus can lead to direct and indirect effects. It may have direct cytopathic effects on developing neurons and glial cells or may induce vasculitis that results in the loss of the blood supply to the relevant regions in the developing brain. In addition to the aforementioned virally mediated effects, immunopathological mechanisms should be considered (Fig. 25.2).

The inflammatory response during the prenatal period will not only affect the brain development and make it vulnerable but also increase the risk of neurologic or neuropsychiatric diseases in childhood and adulthood (Hagberg et al. 2012) (Fig. 25.3), including but not limited to mental retardation, cerebral palsy, epilepsy, and SNHL. Congenital infections with rubella, CMV, *T. gondii*, or some other microbial agents have been associated with the development of autistic spectrum disorder (Sakamoto et al. 2015) or adult schizophrenia (Yolken and Torrey 2008). While the underlying mechanisms remain obscure, autoimmune or autoinflammatory responses have been hypothesized to play a role. A potential role for persistent infections or complications during the prenatal or perinatal period has been hypothesized, even in cases of Alzheimer's disease and other neurodegenerative disorders (Hagberg et al. 2012).

25.3.1.1 Disturbances in the Functions of Neural Stem Progenitor Cells (NSPCs)

Animal studies provided clues to improve our understanding of the neuropathogenesis of CCMVI. It was reported that mouse CMV (MCMV)-induced brain defects are caused by disturbances in the functions of NSPCs within the ventricular and subventricular zones, resulting in defects in their proliferation and differentiation (Tsutsui et al. 2008). Similarly, disturbances in the proliferation and differentiation of human NSPCs may underlie the pathophysiology of microcephaly due to human CMV (HCMV) infection.

It was also found that the injection of MCMV into the cerebral ventricles of mouse embryos disturbed the migration of neurons and caused a marked loss of neurons in neonatal brains. This animal study may explain the neuropathogenesis of neuronal migration disorders in patients with CCMVI, as described above. Adhesive interactions between the extracellular matrix, neurons, and radial glia are likely to play important roles in the process of neuronal migration. It is likely that the direct and/or indirect effects of CMV on the NSPC-dense ventricular and subventricular zones downregulate the expression of adhesion molecules, thereby inducing a defect in neuronal migration.

25.3.1.2 Hypoxic Ischemic Events

CCMVI can lead to hypoxic ischemic brain damage in several ways. Since endothelial cells are targets of CMV, the resultant cerebral vasculitis induces a loss of blood supply to the relevant regions in the developing brain.

CMV infection of the placenta can cause placental insufficiency, resulting in sequelae due to a defective oxygen and nutrient supply to the fetus,

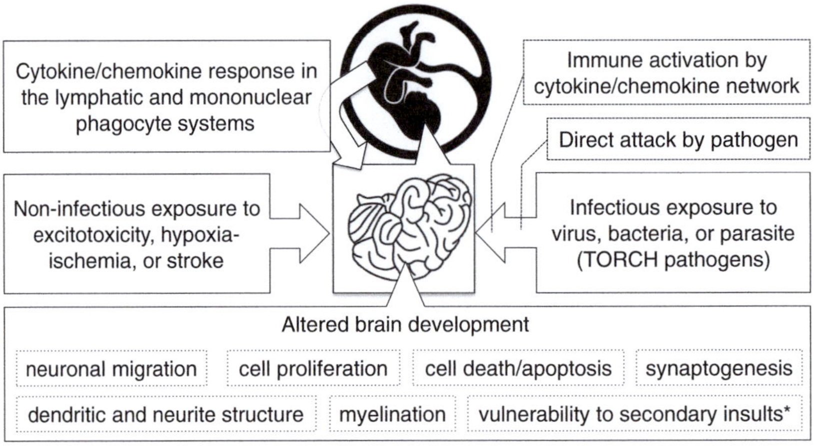

Fig. 25.2 Inflammatory, both infectious and noninfectious, exposure during the fetal period induces altered brain development
Intrauterine pathogens may directly injure the central nervous system but also damage the developing brain by immunopathological mechanisms. Immune activation through the cytokine/chemokine network can occur inside the central nervous system. In addition, cytokine/chemokine responses in the lymphatic and mononuclear phagocyte systems have a significant impact on the fetal brain
*Secondary insults: placental perfusion abnormalities, neonatal hypotension, or anything that results in hypoxia-ischemia

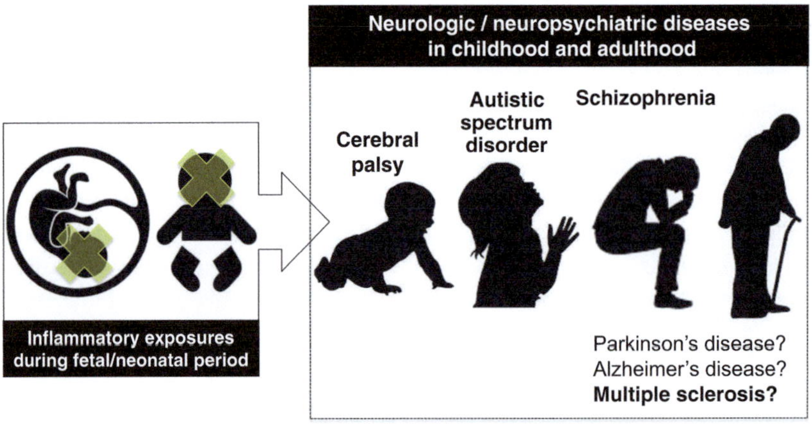

Fig. 25.3 Predisposition to neurologic/neuropsychiatric diseases in childhood and adulthood by inflammatory exposures to the developing brain during fetal/neonatal period
Epidemiological studies have suggested a potential link between infection/inflammation during the fetal period and an increased risk of developing autistic spectrum disorder or schizophrenia. In contrast, the involvement of prenatal/neonatal inflammatory exposure in the etiology of neurodegenerative disorders such as Parkinson's disease, Alzheimer's disease, and multiple sclerosis remains merely speculated (Hagberg et al. 2012)

such as fetal growth retardation and fetal asphyxia, which can also cause severe brain damage.

In addition, CCMVI-induced immunoactivation may increase the vulnerability of the fetal/neonatal brain to hypoxic ischemic events caused by other etiologies, as described later.

25.3.1.3 Immunopathological Events

CMV induces focal but widespread non-necrotizing encephalitis in the developing brain. In a murine model of CCMVI, lytic infection of neurons and glial cells with MCMV continued for only 3 weeks, but the pathohistological process lasted for several months afterward (Cekinovic et al. 2008), implying either viral persistence in the brain that constantly primes the immune response or the consequent development of immunopathology in the brain (Koontz et al. 2008).

The CNS is so called an "immunological sanctuary" into which immune cells do not migrate, with inflammatory mediators unable to cross through the intact blood-brain barrier; however, massive surges of cytokine levels, referred to as a cytokine storm, have been shown to break down the blood-brain barrier as well as to induce a variety of systemic manifestations including shock, disseminated intravascular coagulation, and multiorgan failure. In addition, immune responses do occur in the CNS, including the induction of a variety of cytokines and chemokines.

Since the fetal immune system is immature, prone to immune tolerance, and biased toward Th2, the intracellular TORCH pathogens, including CMV, persist under the weakened Th1 immunity. However, since innate immunity is competent, prolonged interactions between pathogens and the fetal immune system lead to chronic inflammation. Chronic neuroinflammation is caused by many kinds of cytokines, such as interleukin (IL)-10, IL-5, IL-4, and type I IFN; is mediated by astrocytes and microglia; and has active interactions with the somatic immune system during brain infection.

25.3.1.4 Synthesis of Type I IFN

Type I IFN is produced in congenital infections with CMV or rubella virus. In congenital rubella infection, the synthesis of type I IFN in the peripheral blood starts at the 20th week of pregnancy or sooner (Lebon et al. 1988) and continues beyond birth. Not only viruses but also the protozoa *T. gondii* can induce type I IFN (Han et al. 2014).

Although such pathogens are neurotropic, CNS disease is rare in immunocompetent hosts. In contrast, involvement of the CNS is common in congenital infections, and type I IFN is detected in the CSF derived from neonates with congenital infection (Dussaix et al. 1985). In animal studies, increased levels of type I IFN as well as a number of proinflammatory chemokines and cytokines are observed in MCMV-infected newborn mice (Koontz et al. 2008). Type I IFN, which may reduce CMV infections in the embryonic brain, is neurotoxic to the developing brain, and toll-like receptor 3 (TLR3) activation appears to increase the vulnerability of the fetal/neonatal brain to hypoxia-ischemia (Stridh et al. 2013).

25.3.2 Type I Interferonopathies (Too Much Type I IFN)

The persistent upregulation of type I IFN through nucleic acid sensing contributes greatly to immunopathogenesis of CCMVI, as described above, and AGS and other type I interferonopathies (see below).

25.3.2.1 Sensing and Signaling of Nucleic Acids

A huge amount of self nucleic acids is produced in the human body, including those derived from DNA replication and repair and those derived from endogenous retroelements. In addition, apoptosis occurs at a rate of approximately 10 billion cells per hour in the human body, releasing a huge amount of DNA. Furthermore, large scale of DNA degradation is observed in the differentiation processes of red blood cells, skin, and the optic lens.

It is critical for nucleic acid sensors to sense non-self (pathogen-derived) DNA and not to sense self (host-derived) nucleic acids. Otherwise, such inappropriate responses to endogenous DNA will lead to autoimmunity and neurotoxicity in the developing brain (Fig. 25.4). There are three major mechanisms for avoiding activation through nucleic acid sensors by host-derived DNA. First, the sensors/receptors appear to be able to preferentially recognize structures and modifications that are specific to pathogen-derived DNA (e.g., unmethylated CpG dinucleotides). Second, sensors/receptors are located only in certain subcellular compartments to which endogenous DNA has no access (e.g., endosomally located TLR9). Third, a variety of DNA-degrading enzymes (deoxyribonuleases or DNases) digest endogenous DNA restlessly and diligently so that the level of endogenous DNA

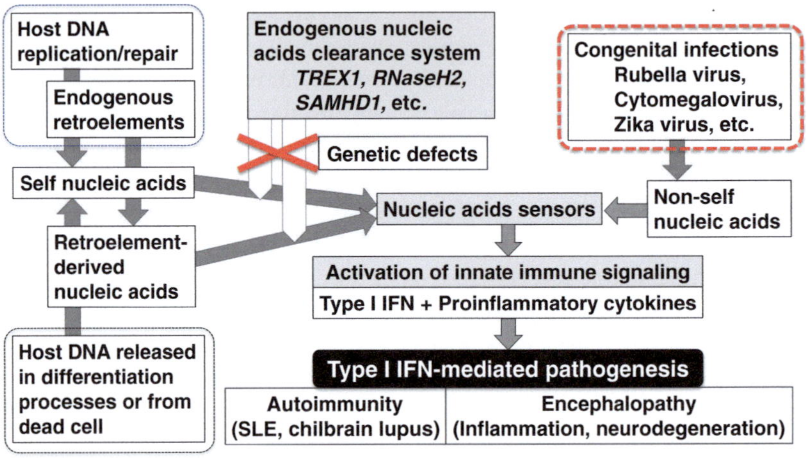

Fig. 25.4 Too much type I interferon (IFN) due to genetic or infectious etiologies
[Right] When a fetus is infected with any of the TORCH pathogens, he or she is exposed to pathogen-derived nucleic acids
[Left, upper] A huge amount of self nucleic acids is produced in the human body, including those derived from DNA replication and repair and those derived from endogenous retroelements
[Left, lower] Apoptosis occurs at a rate of approximately 10 billion cells per hour in the human body; therefore, a huge amount of DNA has to be promptly eliminated. In addition, large-scale DNA degradation is observed in the differentiation processes of red blood cells, skin, and the optic lens
[Center, upper] Endogenous nucleic acids are promptly cleared by the mechanisms described later to ensure that nucleic acid-sensing systems are not activated. However, genetic defects in any of the elements involved in the endogenous nucleic acid clearance system lead to the accumulation of nucleic acids, which activates sensors
[Center, lower] The activation of nucleic acid sensors is followed by the initiation of the innate immune signaling, producing type I IFN along with proinflammatory cytokines, both of which are toxic to the developing brain. Prolonged exposure to a substantial amount of self nucleic acids also results in the synthesis of autoantibodies against them, which in turn urges extracellular nucleic acids to be engulfed into immune cells that induce type I IFN through the activation of the nucleic acid-sensing systems

does not exceed the threshold of the sensors/receptors.

25.3.2.2 Regulation of Type I IFN Activity

The biological activity of type I IFN must be tightly regulated to avoid its detrimental effects. Type I IFN signaling can be hazardous if it occurs too much, for too long, or at the wrong time and place (such as the developing brain). The synthesis of type I IFN in the CNS is barely seen in non-viral neurological diseases. However, Lebon et al. revealed the association of pseudo-TORCH syndrome, currently referred to as AGS, with increased type I IFN activity in the CSF and serum (Lebon et al. 1988). In other words, AGS is both clinically and biochemically reminiscent of TORCH syndrome: type I IFN is a key player in the pathogenesis of both TORCH syndrome and AGS and is induced through the recognition of nucleic acids.

A clinically heterogenic group of Mendelian-inherited diseases linked to defective regulation of type I IFN have been identified and are collectively referred to as type I interferonopathies (Table 25.3). In this paper, we tentatively define them as genetic type I interferonopathies and TORCH syndrome as infectious type I interferonopathies.

In cases of infectious type I interferonopathies, non-self (exogenous or pathogen-derived) nucleic acids activate the innate immune signaling pathways, inducing type I IFN and proinflammatory cytokines. In contrast, self (endogenous or host-derived) nucleic acids induce type I IFN and proinflammatory cytokines in patients with genetic type I interferonopathies (Fig. 25.4).

25.3.2.3 Molecular Mechanisms in Genetic Type I Interferonopathies

To prevent the accumulation of endogenous nucleic acids, nucleic acid-sensing and metabolizing systems are established (Fig. 25.5b). For example, a huge amount of DNA and DNA:RNA hybrid molecules are derived from endogenous retroelements, accounting for almost 10% of the human genome and represent reminiscent retroviruses. They are recognized and processed by TREX1, RNase H2, and ADAR1. TREX1 also removes ssDNA derived from DNA replication/repair and is a component of an ER-associated complex called the SET complex, together with two other DNases (APE1 and NM23-H1). In response to oxidative stress induced by granzyme A, the SET complex translocates into the nucleus, where NM23-H1 promotes the ssDNA nicks, and TREX1 stops the DNA ends from re-annealing by removing bases from the free 3′ end, thus preventing DNA repair (Fig. 25.5a). If any of those factors is deficient, the accumulation of endogenous nucleic acids will constitutively exceed a threshold, activating the nucleic acid-sensing systems to induce innate immunity (Fig. 25.5b). Causative genes for type I interferonopathies and proposed molecular mechanisms are summarized in Table 25.3.

25.4 Therapy

25.4.1 Case 1: Congenital CMV Infection

Although we cannot expect to reverse any deficit that patients with CCMVI already have at birth, viral replication is robust even after birth, and some of the clinical manifestations are progressive or develop later in life. For example, up to 70% of CCMVI-related SNHL is progressive without any therapeutic intervention. Therefore, postnatal antiviral therapy is expected to be clinically beneficial. It was demonstrated that neonates with symptomatic congenital CMV infection had a better audiologic and developmental prognosis upon treatment with ganciclovir administered intravenously over a period of 6 weeks. Furthermore, longer-term antiviral therapy with valganciclovir, a prodrug of ganciclovir, orally administered over a period of 6 months was shown to provide better long-term benefit (Kimberlin et al. 2015).

This patient started oral valganciclovir at 12 days of life and completed a 6-month treatment without any adverse reaction. His right-side SNHL had been stable, and his psychomotor development had been slightly delayed but was catching up.

The pathogenesis of TORCH syndrome involves both direct invasion by the pathogen in question and immunopathological processes. Therefore, combination therapy of antimicrobial and immunosuppressive agents may be theoretically beneficial. For example, glucocorticoids (prednisone 1 mg/kg/day) are added to antiprotozoal agents for the treatment of congenital toxoplasmosis if CSF protein is >1 g/dL or active chorioretinitis threatens the vision. However, the efficacy and safety of glucocorticoids have not been investigated for the treatment of congenital CMV infection.

25.4.2 Case 2: Aicardi-Goutieres Syndrome

Although we cannot expect to reverse any deficit that patients with AGS already have at birth, some therapeutic options should be taken into consideration for several reasons. First, some of the clinical manifestations develop later in life. Second, some patients with AGS develop active regression over several months after onset. Third, the early diagnosis before the onset of symptoms is possible for familial cases of delayed-onset AGS. Fourth, there exists considerable phenotypic variation among familial cases. Based on the above, therapeutic intervention may be beneficial for many, if not all, patients with AGS.

Since AGS is considered an immunopathological disorder with autoinflammatory and autoimmune characteristics, a variety of immunomodulating therapies have been tested, including prednisone, azathioprine, methylpred-

Table 25.3 Causative genes for type I interferonopathies

Molecular mechanisms		Causative genes	Functions	Location	OMIM	Inheritance	Phenotype	
Overstimulation of the type I IFN response machinery by the abnormal accumulation of an endogenous nucleic acid ligand caused by loss-of-function mutations in		*TREX1*	Which encodes DNA 3′-repair exonuclease 1 that degrades DNA	3p21.31	225750	AR >> AD	Aicardi-Goutieres syndrome	AGS1
		RNase H2B	Which encode ribonuclease H2 (RNase H2) endonuclease complex that degrades DNA-RNA hybrid molecules	13q14.3	610181	AR		AGS2
		RNase H2C		11q13.1	610329	AR		AGS3
		RNase H2A		19p13.2	610333	AR		AGS4
		Deoxynucleoside triphosphate triphosphohydrolase and ribonuclease SAM domain and HD domain 1 (*SAMHD1*)	Which restricts the availability of cytosolic deoxynucleotides (dNTPs)	20q11.23	612952	AR		ASG5
Ultrasensitive or constitutive activation of nucleic acid receptor signaling to the type I IFN pathway caused by	Loss-of-function mutations in	Adenosine deaminase acting on RNA 1 (*ADAR1*)	Which edits endogenous dsRNA preventing its recognition by IFIH1	1q21.3	615010	AR > AD		AGS6
	Gain-of-function mutations in	IFN-induced helicase C domain-containing protein 1 (*IFIH1*)	Which encodes the cytosolic receptor for dsRNA	2q24	615846	AD		AGS7
		Dead box polypeptide 58 (*DDX58*)	Which encodes RIG-I, a pattern recognition receptor that is a sensor for viruses	9p21.1	609631	AD	Singleton-Merten syndrome 2	
Ultrasensitive or constitutive activation of non-nucleic acid receptor component (e.g., an adaptor molecule) of the IFN-induced signaling pathway caused by loss-of-function mutations in		Transmembrane protein 173 (*TMEM173*)	Which encodes stimulator of interferon genes protein [STING]	5q31.2	612374	AD	STING-associated vasculopathy, infantile-onset	
Defective negative regulation of a nucleic acid-dependent type I IFN response caused by loss-of-function mutations in		Ubiquitin-specific peptidase 18 (*USP18*)	Which encodes a negative regulator of type I IFN	22q111.21	607057	AR	Pseudo-TORCH syndrome 2	
		IFN-stimulated gene protein 15 (*ISG15*)	Which stabilizes USP18	1p36.33	147571	AR	Immunodeficiency 38 with basal ganglia calcification	

Unknown mechanisms of type I IFN deregulation, e.g., caused by loss-of-function mutations in	Phosphatase, acid, type 5, tartrate-resistant (*ACP5*)	Which encodes tartrate-resistant acid phosphatase [TRAP], possibly linked at least in part to an increased signaling through the TLRs	19p13.2	171640	AR	Spondyloenchondrodysplasia with immune dysregulation
	Proteosome subunit, beta-type, 8 (*PSMB8*)	Which encodes the inducible proteasome subunit-ß type 8 (ß5i) that plays an essential role in protein degradation and in antigen processing during infection	6p21.32	177046	AR	Autoinflammation, lipodystrophy, and dermatosis syndrome

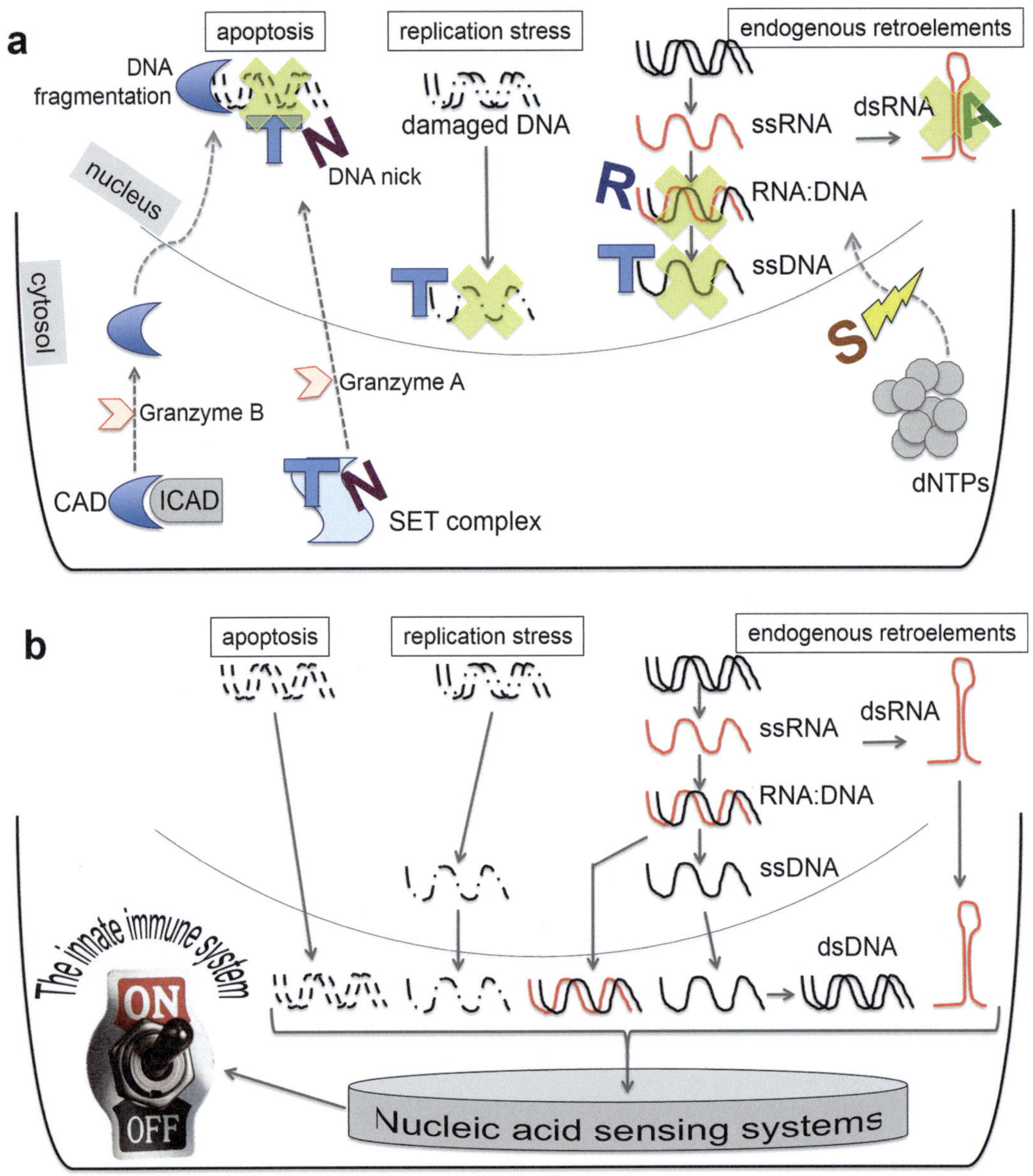

Fig. 25.5 Failure to prevent the accumulation of endogenous nucleic acids

(**a**) Preventing the accumulation of endogenous nucleic acids

[Right] DNA and RNA:DNA hybrid molecules derived from endogenous retroelements are digested by TREX1 and RNase H2, respectively. ADAR1 edits endogenous dsRNA preventing its recognition by IFIH1. SAMHD1 restricts the availability of cytosolic dNTPs

[Center] TREX1 also removes ssDNA derived from DNA replication/repair

[Left] Granzyme B promotes apoptosis by both caspase-dependent and caspase-independent pathways, and the former includes activation of caspase-activated DNase (CAD), which fragments chromosomal DNA into oligonucleosomes. In contrast, granzyme A promotes a caspase-independent apoptotic pathway by translocating SET complex into the nucleus where its component NM23-H1 promotes DNA nicks and TREX1 stops the DNA ends from re-annealing by removing bases from the free 3′ end, thus preventing DNA repair

T, TREX1. R, RNase H2. A, ADAR1. S, SAMHD1. N, NM23-H1

(**b**) The accumulation of endogenous nucleic acids switches on the innate immune system via nucleic acid-sensing systems.

Nucleic acids are immunogenic molecules that activate immune responses. If they are left unprocessed, immunopathological problems are inevitable

nisolone, intravenous immunoglobulin alone, and combinations thereof. Assessing the efficacy, however, is quite difficult, as those studies involved just individual cases, and the clinical courses of AGS are quite variable.

Although the present patient started daily administration of prednisone and tacrolimus as well as methylprednisolone pulse therapy every 6 weeks, it is also difficult to assess the efficacy of this regimen.

Additional Materials

1. Clicker/class-response type questions:

   Q1. Which of the following is included in TORCH syndrome pathogens?

   (a) *Toxocara canis*
   (b) *Orientia tsutsugamushi*
   (c) RS virus
   (d) *Campylobacter jejuni*
   (e) Herpes simplex virus

   Q2. Which is characteristic for Aicardi-Goutieres syndrome and different from TORCH syndrome?

   Intracranial calcification and microcephaly are:

   (a) Absent
   (b) Milder
   (c) Severer
   (d) Transient
   (e) Progressive

   Q3. What is TREX1?

   (a) A sensor of intracellular DNA
   (b) A nucleic acid-digesting enzyme
   (c) An adaptor of signals from DNA sensors
   (d) A transactivator of type I IFN expression
   (e) A transrepressor of type I IFN expression

2. Group discussion questions/activities:

Discussion-1. What could be the ideal therapeutic regimen for symptomatic congenital CMV infection? What agents would you like to have? When and how would you start the regimen?

Discussion-2. A mother whose first child was diagnosed with Aicardi-Goutieres syndrome with TREX1 deficiency turns out to be pregnant at 12 weeks after her last menstrual cycle. She is worried about the possibility of AGS for the fetus and asks you whether she should continue the pregnancy. What is your response?

End-of-Chapter Questions

Q1. Why do congenital infections with taxonomically variable pathogens, including protozoa (*T. gondii*), spirochete (*Treponema pallidum*), DNA viruses (e.g., CMV), and RNA viruses (e.g., rubella virus), share many clinical features?

Q2. Why is the developing brain so vulnerable to CMV infection?

Q3. Describe three major mechanisms for avoiding activation through nucleic acid sensors by host-derived DNA.

Q4. Summarize the physiological roles of TREX1 and the possible outcomes of TREX1 deficiency.

References

Cekinovic D, Golemac M, Pugele P, Tomac J, Cicinsain L, Slavuljica I, Bradford R, Misch S, Winkler TH, Mach M, Britt WJ, Jonjic S (2008) Passive immunization reduces murine cytomegalovirus-induced brain pathology in newborn mice. J Virol 82:12172–12180

Crow YJ, Chase DS, Schmidt JL et al (2015) Characterization of human disease phenotypes associated with mutations in *TREX1, RNASEHH2A, RNASEH2B, RNASEH2C, SAMHD1, ADAR,* and *IFIH1*. Am J Hum Genet A 167A:296–312

Dussaix E, Lebon P, Ponsot G, Huault G, Tardieu M (1985) Intrathecal synthesis of different alpha-interferons in children with various neurological disorders. Acta Neurol Scand 71:504–509

Hagberg H, Gressens P, Mallard C (2012) Inflammation during fetal and neonatal life: implications for neurologic and neuropsychiatric disease in children and adults. Ann Neurol 71:444–457

Han S-J, Melichar HJ, Coombes JL, Chan SW, Koshy AA, Boothroyd JC, Barton GM, Robey EA (2014) Internalization and TLR-dependent type I interferon production by monocytes in response to *Toxoplasma gondii*. Immunol Cell Biol 92:872–881

Kimberlin D, Jester PM, Sanchez PJ, Ahmed A, Arav-Boger R, Michaels MG et al (2015) Valganciclovir for symptomatic congenital cytomegalovirus disease. N Engl J Med 372:933–943

Koontz T, Bralic M, Tomac J, Pernjak-Pugel E, Bantug G, Jonjic S, Britt W (2008) Altered development of the brain after focal herpesvirus infection of the central nervous system. J Exp Med 205:423–435

Lebon P, Badoual J, Ponsot G, Goutieres F, Hemeury-Cukier F, Aicardi J (1988) Intrathecal synthesis of interferon-alpha in infants with progressive familial encephalopathy. J Neurol Sci 84:201–208

Sakamoto A, Moriuchi H, Matsuzaki J, Motoyama K, Moriuchi M (2015) Retrospective diagnosis of congenital cytomegalovirus infection in children with autism spectrum disorder but no other major neurologic deficit. Brain and Development 37:200–205

Stridh L, Mottahedin A, Johansson ME, Valdez RC, Northington F, Wang X, Mallard C (2013) Toll-like receptor-3 activation increases the vulnerability of the neonatal brain to hypoxia-ischemia. J Neurosci 33:12041–12051

Tsutsui Y, Kosugi I, Arai Y, Han GP, Li L, Kaneta M (2008) Roles of neural stem progenitor cells in cytomegalovirus infection of the brain in mouse models. Pathol Int 58:257–267

Yolken RH, Torrey EF (2008) Are some cases of psychosis caused by microbial agents? A review of the evidence. Mol Psychiatry 13:470–479

Epilepsy

26

Katsuhiro Kobayashi, Tomoyuki Akiyama, and Cristina Go

Keywords
Seizure · Electroencephalogram · Development · Epilepsy syndrome · Antiepileptic drug

26.1 Case Reports

26.1.1 Case 1

An 11-month-old girl was referred to a pediatric epilepsy clinic because of repetitive intractable seizures. Her initial seizure was a left-side-dominant convulsive status epilepticus (protracted seizure) that occurred at 3 months of age, lasted for 1.5 h, and was eventually stopped by intravenous infusion of diazepam. Thereafter the patient repeatedly had seizures that were often provoked by fever. Her seizures were mostly tonic (stiffening)-clonic (repetitively jerking) with loss of consciousness and were generalized or lateralized involving either side. The duration of seizures tended to be long and was over 1 h twice. Administration of various antiepileptic drugs (AEDs) was largely ineffective.

The patient was born to non-consanguineous Japanese parents after an uncomplicated pregnancy and delivery. The family history is unremarkable and her sibling was healthy. The growth and development were normal.

During the initial hospital admission, neurological examination revealed no abnormalities. She could stand without support and speak a few words, indicating her development was still normal at 11 months of age. The electroencephalogram (EEG) showed no epileptic discharges. Magnetic resonance imaging (MRI) of the brain was unremarkable. Also normal were blood sugar and ammonia; serum levels of electrolytes, calcium, phosphorus, magnesium, protein, aspartate aminotransferase (AST), alanine aminotransferase (ALT), creatine phosphokinase, etc.; cerebrospinal fluid studies; urinary organic acid profile; and chromosomal studies.

These initial clinical findings in infancy suggested the diagnosis of Dravet syndrome (also known as [aka] severe myoclonic epilepsy in infancy), a type of infantile-onset epileptic encephalopathy. The DNA extracted from peripheral blood cells from the patient was positive for a mutation of the *SCN1A* gene that encodes the α1 subunit of neuronal voltage-gated sodium channels (protein name: Na_v 1.1). Both parents were found negative for the mutation.

K. Kobayashi (✉) · T. Akiyama
Department of Child Neurology, Okayama University Graduate School of Medicine, Dentistry and Pharmaceutical Sciences, Okayama, Japan
e-mail: k_koba@md.okayama-u.ac.jp; takiyama@okayama-u.ac.jp

C. Go
Division of Neurology, Department of Paediatrics, University of Toronto, Toronto, ON, Canada
e-mail: cristina.go@sickkids.ca

T. Oohashi et al. (eds.), *Human Pathobiochemistry*, https://doi.org/10.1007/978-981-13-2977-7_26

After the initial admission, the patient continued to suffer from weekly seizures despite vigorous antiepileptic treatment. At the age of 1, she began to show erratic segmental myoclonia (irregular twitching movements) involving various parts of extremities. She also had additional types of seizures including massive myoclonic (jerking) seizures, atypical absence (brief consciousness-losing) seizures with myoclonic components, and focal impaired awareness seizures (aka complex partial seizures) that were characterized by consciousness disturbance lasting for a few minutes, head and eye deviation to one side, and cyanosis. In EEG, generalized and multifocal spike-and-wave complex appeared (Fig. 26.1). Her development gradually regressed, she developed autistic behaviors, and her gait is ataxic at 6 years of age. Antiepileptic treatment has been optimized including the combined administration of sodium valproate, clobazam, and stiripentol, which decreased the frequency of protracted seizures though she still has frequent seizures.

26.1.2 Case 2

A 5-year-old boy was referred to pediatric epilepsy clinic because of intractable seizures. He had no complications at birth, and his development during infancy and early childhood was normal. At 2 years and 9 months of age, he started having seizures with staring and stiffening of bilateral arms lasting 10 s. They occurred daily but were controlled by carbamazepine therapy. Two years later, his seizures recurred with stiffening of all limbs, eye deviation, and mouth pulling to the left, at times leading to falls. He continued to have several seizures per day despite aggressive medical treatment.

On examination, he was a right-handed and restless boy without abnormalities in general and neurological examinations.

Long-term video EEG recorded seizures originating from the right frontal and anterior temporal head regions, interictal epileptic discharges, and focal background slowing over the right

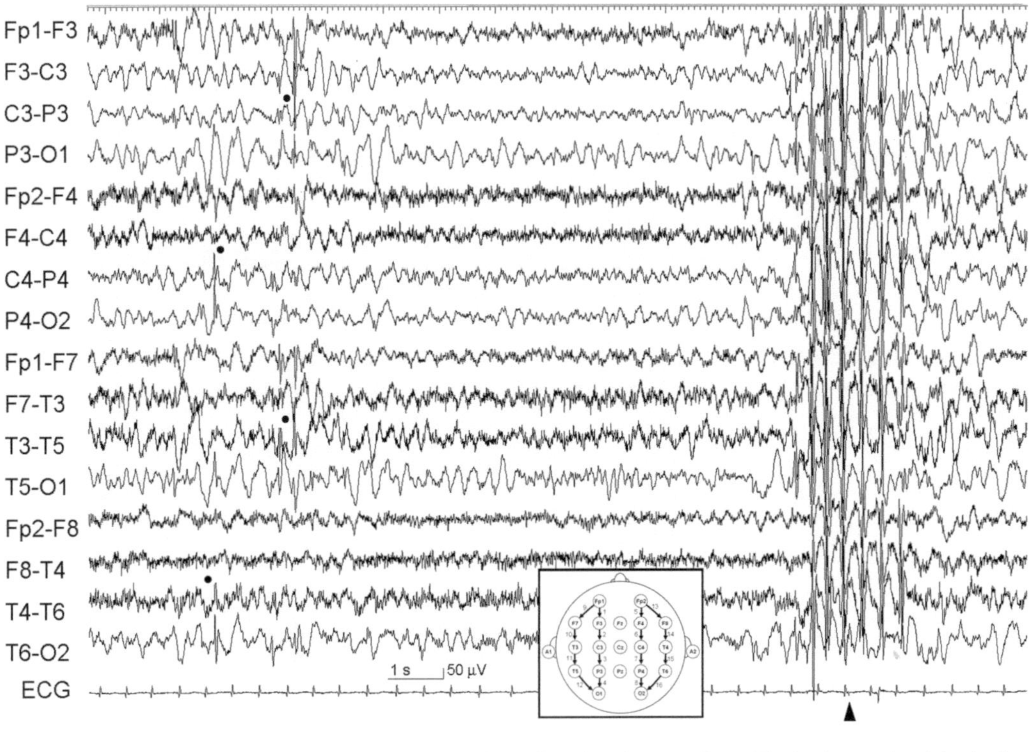

Fig. 26.1 Interictal EEG during wakefulness recorded from Case 1 at 4 years of age. The background activity is slow, and there are multifocal spike-wave discharges (dots) and a generalized spike-and-wave burst (arrowheads)

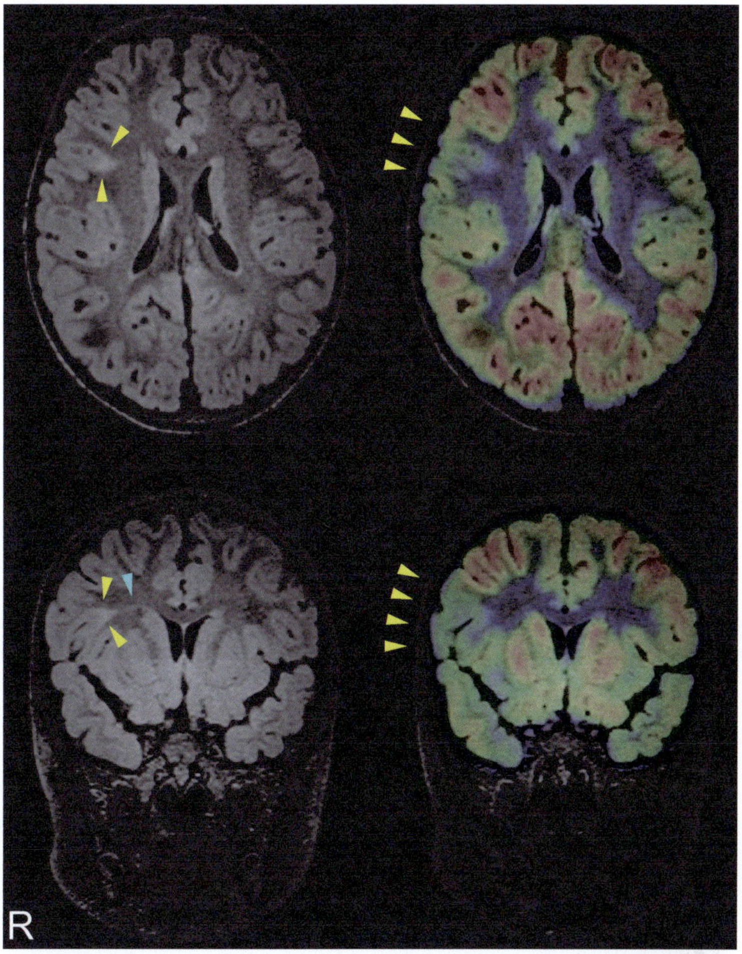

Fig. 26.2 Brain MRI and PET findings in Case 2. Fluid-attenuated inversion recovery images (left) demonstrate high-intensity signals at the gray and white matter junction at the right frontal operculum (yellow arrowheads). Coronal MRI (left, bottom) shows thin linear high-intensity signal extending toward the lateral ventricle from the gray and white matter junction (light-blue arrowhead). PET images (right) demonstrate hypometabolism at and around the MRI lesion

central and mid-temporal head regions. Brain MRI demonstrated high-intensity signals at the gray and white matter junction in the right frontal operculum extending to the lateral ventricle in fluid-attenuated inversion recovery images (Fig. 26.2). Fluorodeoxyglucose positron emission tomography demonstrated hypometabolism at and around the MRI abnormality (Fig. 26.2). Neuropsychological evaluation by the Wechsler Preschool and Primary Scale of Intelligence demonstrated that his full intelligence quotient (IQ) was 58.

He underwent intracranial electrode implantation covering the MRI lesion by subdural and intracerebral electrodes. Intracranial video EEG confirmed that his seizures originated from the MRI lesion and the patient underwent focal resection of the lesion. The pathology result demonstrated a disorganized lamination of the cerebral cortex with large dysmorphic neurons and balloon cells, consistent with focal cortical dysplasia (FCD) Type IIb. He has been seizure-free for 3 years since then and shows progress in language skills in the special needs class at the elementary school.

26.2 Diagnosis

26.2.1 Definition of Epilepsy and Seizure

Epilepsy is defined as a disorder of the brain characterized by an enduring predisposition to generate epileptic seizures (Fisher et al. 2005). An epileptic seizure is a transient occurrence of signs and/or symptoms due to abnormal excessive or synchronous neuronal activity in the brain that is expressed by epileptic discharges in EEG. The definition of epilepsy requires any of the following conditions: (1) at least two unprovoked seizures; (2) one unprovoked seizure and likely subsequent recurrence of similar seizures; and (3) diagnosis of an epilepsy syndrome, which is described below (Fisher et al. 2014).

There are various types of seizures with different symptomatology corresponding to the mode of ictal (seizure-related) brain involvement. Epileptic seizures are conceptualized as either generalized or focal (Berg et al. 2010; Fisher et al. 2017). Generalized seizures originate within, and rapidly engaging, bilaterally distributed neuronal networks in the brain. Focal seizures originate within networks limited to one cerebral hemisphere.

26.2.2 Seizure Classification

Representative seizure types based on clinical and EEG classification are as follows and listed in Table 26.1 (Commission 1981; Blume et al. 2001; Fisher et al. 2017).

A generalized tonic-clonic seizure (GTCS) is the most common type of generalized convulsive seizure, and it is clinically characterized by a combination of sudden loss of consciousness, tonic phase, and subsequent clonic movements and causes the patient a fall to the ground with occasional injuries. During the seizure, respiration is inhibited with resulting cyanosis, and urinary incontinence may occur. A GTCS usually lasts for a few minutes, and after the termination of seizure, the patient often goes into a deep sleep.

An absence seizure is another type of generalized seizure characterized by a sudden onset, interruption of ongoing activity, a blank stare, unresponsiveness, and an abrupt termination. The duration of absence seizure lasts from a few seconds to half a minute.

A myoclonic seizure is characterized by sudden, shock-like involuntary jerk(s), i.e., brief contraction(s) of muscle(s) or muscle group of variable topography (axial, proximal limb, distal).

A tonic seizure is a rigid, violent muscular contraction with loss of consciousness.

Table 26.1 Representative types of seizures

Seizure classification	Clinical symptomatology	Ictal EEG findings
Generalized seizure		
Tonic-clonic seizure	Sudden loss of consciousness, tonic contraction of muscles, and subsequent repetitive jerking movements	Abnormal generalized rhythm at ≥10 Hz during tonic phase; such rhythm interrupted by slow waves (run of polyspike-waves) during clonic phase
Absence seizure	Brief loss of consciousness with sudden onset and offset	Burst of generalized 3 Hz spike-and-wave complexes
Myoclonic seizure	Sudden brief jerk(s)	Generalized polyspike-wave burst
Clonic seizure	Repetitive jerks	Run of generalized polyspike-waves
Tonic seizure	Rigid, violent muscular contraction with loss of consciousness	Generalized low-amplitude fast activity or fast rhythm decreasing in frequency and increasing in amplitude
Atonic seizure	Sudden loss of muscle tone causing a fall	Polyspike-waves or flattening or low-amplitude fast activity
Focal seizure		
Focal aware seizure (motor onset or autonomic or behavior arrest or cognitive or emotional or sensory)		Variable types of continuous epileptic activity over the seizure onset zone with variable evolution and/or spread (unapparent ictal EEG findings in some focal seizures)
Focal impaired awareness seizure (may be accompanied by automatism)		
Focal to bilateral tonic-clonic seizure		
Epileptic spasms (infantile spasms)	Sudden flexion, extension, or mixed extension-flexion of proximal and truncal muscles usually occurring in clusters	High-amplitude slow waves, often with overriding fast components and/or succeeding amplitude reduction

An atonic seizure is characterized by a sudden loss or diminution in muscle tone that may cause a fall and resulting injury. Brief atonic seizures are also called "drop attacks" though drops may also occur by other types of seizures.

Focal seizures are described according to the degree of impairment during seizures (Berg et al. 2010). There are focal seizures involving subjective sensory or psychic phenomena only, and they are called "aura" when such a subjective ictal phenomenon precedes an observable seizure. Focal impaired awareness seizures may be accompanied by automatism (more or less coordinated, repetitive, purposeless, motor activity resembling voluntary movements). There are also focal seizures evolving to bilateral, convulsive seizures.

Epileptic spasms may be focal or generalized. This type of seizure is characterized by sudden flexion or extension of predominantly proximal and truncal muscles with duration of 1–2 s. Epileptic spasms usually occur in clusters. These are observed mostly during infancy and are therefore also called infantile spasms.

26.2.3 EEG

EEG is the most important clinical examination for diagnosis and management of epilepsy. The presence of epileptic discharges in the interictal (period during the interval between seizures) EEG indicates the diagnosis of epilepsy if the pattern and localization of given epileptic discharges are compatible with the assumed type of epilepsy or epilepsy syndrome. Epileptic discharges are observed over the background activity of EEG. The patterns of epileptic discharges and the background activity change considerably according to age, levels of arousal and sleep, medication, etc.

EEG represents amplified electric current that is generated from the neuronal system in the brain. The normal EEG background in adulthood and late childhood shows alpha rhythm with frequency at about 10 Hz during wakefulness and vertex sharp transients and 14 and 12 Hz bilateral sleep spindles during non-rapid eye movement (NREM) sleep. The background may be slow indicating brain pathology and/or functional disturbance caused by epileptic discharges.

The representative type of interictal epileptic discharge is a spike which is a sharp transient standing out from the background (Fig. 26.3). When a spike is immediately followed by a slow wave, it is called a spike-and-wave complex (aka spike-and-slow-wave complex, spike-wave). A similar sharp discharge with a duration >70 ms is named sharp wave. When a cluster of multiple spikes are followed by a slow wave, it is called a polyspike-and-wave complex (aka polyspike-wave). Epileptic discharges may be generalized

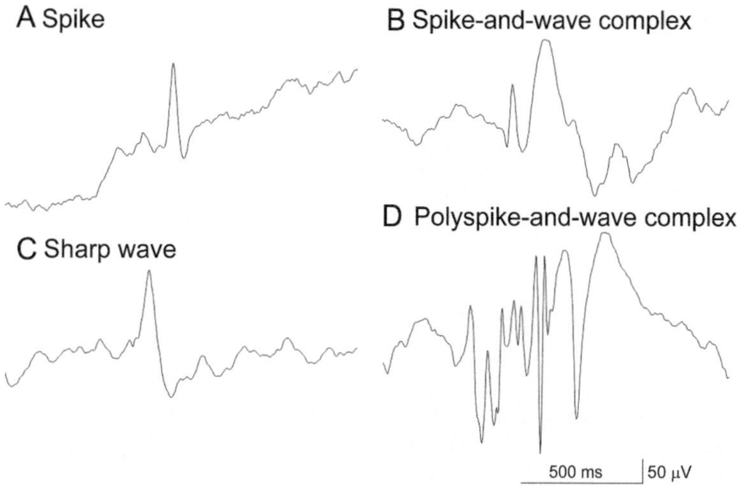

Fig. 26.3 Representative epileptic discharges in EEG

involving the bilateral cerebral hemispheres corresponding to generalized seizures or may be focal limited within a region corresponding to focal seizures.

The occurrence and patterns of epileptic discharges depend on states. Epileptic discharges may increase and, in case of focal discharges, tend to spread or generalize during sleep. Generalized spike-and-wave complexes, particularly 3 Hz spike-wave bursts in childhood absence epilepsy, may be dramatically provoked with induction of associated absence seizures by hyperventilation that causes hypocapnia. Intermittent photic stimulations may induce spike-and-wave complexes as photoparoxysmal responses in photosensitive epilepsies.

Ictal epileptic discharges are different from interictal discharges, and each type of seizure is associated with the own EEG pattern as listed in Table 26.1.

26.2.4 Etiological Classification

Etiology or underlying cause of epilepsy is classified as genetic, structural, metabolic, infectious, immune, or unknown cause (Berg et al. 2010; Scheffer et al. 2017). Genetic epilepsy is conceptually the direct result of a known or presumed genetic defect(s) in which seizures are the core symptom of the disorder. Regarding structural/metabolic etiology, there is conceptually a distinct structural or metabolic condition or disease that has been demonstrated to be associated with a substantially increased risk of developing epilepsy. Structural lesions include acquired disorders such as stroke, trauma, infection, and even lesions of genetic origin (e.g., tuberous sclerosis complex, malformations of cortical development).

26.2.5 Epilepsy Syndromes

Epilepsy syndromes are electroclinical syndromes defined based on combinations of clinical and EEG characteristics. Relevant clinical findings include seizure types, age at seizure onset, etiology, and prognosis. Representative syndromes are listed in Table 26.2. Patients may move from one syndrome to another during the evolution of their epileptic condition. Beside syndromes, there are clinically distinctive constellations that have characteristics based on specific lesions or other causes, and also epilepsies that do not exactly fit to any syndromes.

26.2.6 Diagnostic Procedure and Management

In the management of patients presenting with seizures (Fig. 26.4), seizures provoked by acute brain injuries, such as a head trauma and encephalitis, or systemic/metabolic abnormalities, including hypoglycemia and electrolyte disturbances, are symptomatic seizures and not regarded as epilepsy, and the original disorders should be treated along with the management of seizures. Febrile seizures, i.e., seizures provoked by fever in infancy and young childhood, are also distinct from epilepsy. There are events that mimic seizures, such as parasomnia (e.g., sleepwalking, sleep terror), tics, syncope, etc. Information regarding the symptomatology of events is most important for the differentiation between epileptic seizures and seizure mimics, and normal EEG findings may be also helpful in excluding non-epileptic disorders.

When the occurrence of epileptic seizures is confirmed, the seizure type should be diagnosed based on the ictal symptomatology and EEG findings. Etiology is investigated through family history, developmental history, physical and neurological examinations, neuroimaging studies, metabolic examinations, etc. The diagnosis of a certain epilepsy syndrome is established, if it is appropriately based on all these findings. A diagnosis of epilepsy requires the concordance between the observed clinical seizure type and the EEG findings, and mere observation of epileptic discharges does not prove this diagnosis. When the diagnosis of seizure types is difficult, long-term EEG-video monitoring may be performed to capture the seizures and aid in electroclinical correlation.

Table 26.2 Representative epilepsy syndromes

Epilepsy syndrome	Clinical characteristics	Interictal EEG characteristics
Benign familial neonatal epilepsy	Neonatal seizures that start mostly on the 2nd or 3rd day of life, frequently repeat over 7 days, and stop thereafter; occurrence in neurologically normal full-term babies with genetic background particularly *KCNQ2/3* mutations	Normal
West syndrome	Epileptic spasms (aka infantile spasms), sudden and brief contraction of axial and proximal limb muscles, usually repeat in cluster; variable etiologies including perinatal brain injuries, cortical dysplasia, etc.; developmental regression or stagnation with poor prognosis	Hypsarrhythmia (chaotic activity with random high-amplitude slow waves intermingled with multifocal spikes)
Dravet syndrome	Febrile and afebrile, generalized and unilateral, clonic or tonic-clonic seizures that occur in the first year of life in otherwise normal infants and are typically associated with myoclonic, atypical absence, and focal seizures later; resistant to treatment; developmental delay appearing with cognitive impairment; *SCN1A* mutations detected in most patients	Usually normal at onset; appearance of generalized and focal (usually multifocal) spike-and-wave complexes; photosensitivity often observed (provocation of photoparoxysmal responses with occasional induction of seizures)
Panayiotopoulos syndrome	Childhood focal epilepsy showing seizures with autonomic symptoms particularly vomiting; age-dependent appearance of infrequent, mainly nocturnal, seizures that may develop into status epilepticus, in children with no brain pathology between ages 3 and 6 years; spontaneous remission	Focal spikes at variable locations, particularly occipital regions; spike activation during sleep
Benign epilepsy with centrotemporal spikes	The most common childhood focal epilepsy; age-dependent appearance of infrequent, mostly nocturnal, brief hemifacial motor seizures and/or generalized convulsive seizures in children with no brain pathology between ages 3 and 13 years; seizure remission after adolescence; occasional development of linguistic, cognitive, and behavioral abnormalities	Unilateral or bilateral centrotemporal (Rolandic) spikes that are typically activated during sleep
Lennox-Gastaut syndrome	Representative intractable childhood epilepsy; multiple seizure types (tonic seizure as the main seizure type, others including atypical absence, myoclonic, atonic, tonic-clonic, and focal seizures); onset between ages 3 and 8 years, often following West syndrome; high seizure frequency and frequent status epilepticus; variable etiologies; developmental regression or stagnation with poor prognosis	Diffuse slow spike-and-wave complexes at 2–2.5 Hz over slow background; rapid rhythms or generalized paroxysmal fast activity at about 10 Hz during sleep
Juvenile myoclonic epilepsy	Myoclonic seizures appearing around puberty, with occasional association of tonic-clonic and absence seizures; genetic epilepsy with no structural brain pathology; seizure occurring shortly after awakening with precipitation by sleep deprivation	Generalized spike-waves and polyspike-waves; occasional photosensitivity

Treatment is selected according to the diagnosed epilepsy syndrome and/or seizure type(s). Each AED has its own way of action, and therefore its effect differs depending on epilepsy syndromes and seizure types. Medical treatment is usually started after the occurrence of at least two unprovoked seizures, but it may be started after a single seizure when its recurrence is likely (Fisher et al. 2014).

The effect of treatment for epilepsy should be evaluated in terms of seizure suppression, presence or absence of adverse effects, etc. If the present treatment is ineffective, then the treatment selection should be reconsidered.

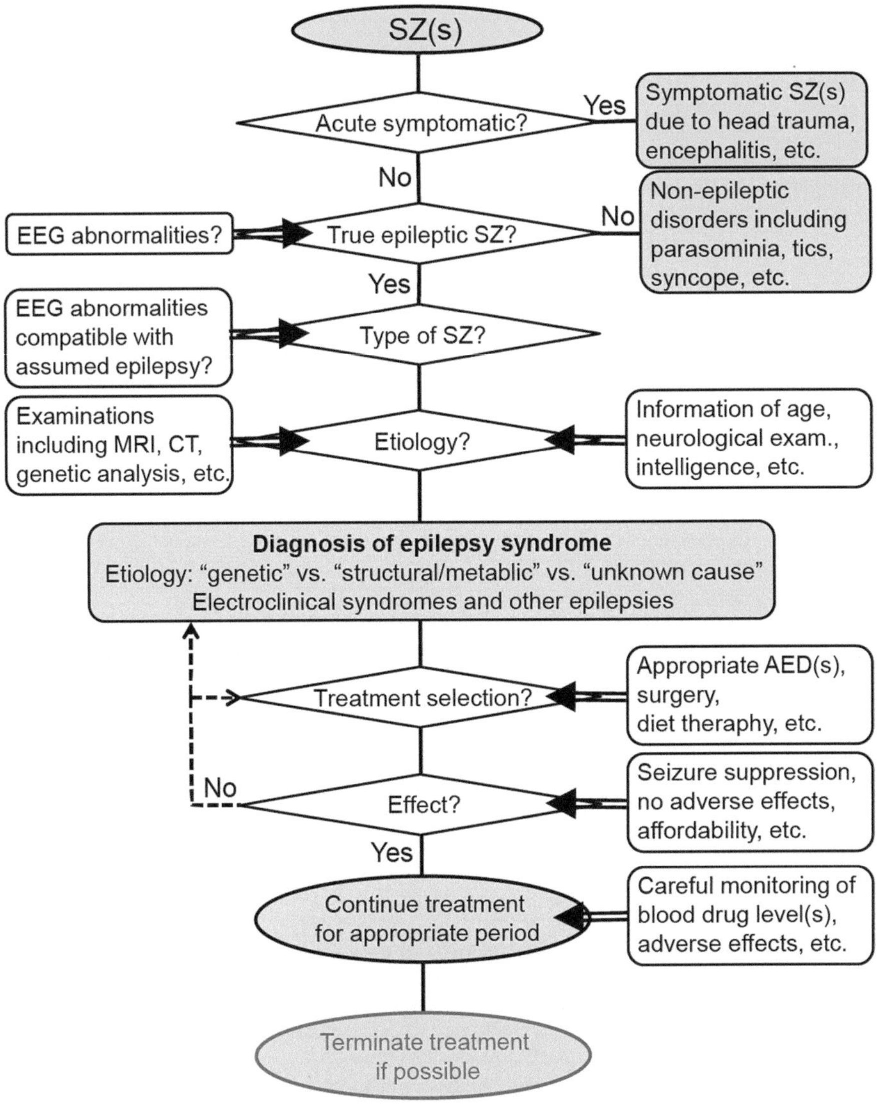

Fig. 26.4 A schematic flowchart illustrating the diagnosis and management of epilepsy. *SZ* seizure

Occasionally the very diagnosis of epilepsy may be changed by the observation that apparent seizures are actually psychogenic non-epileptic seizures (aka pseudo-seizures) through long-term EEG-video monitoring.

It is proposed that drug-resistant epilepsy is defined as failure of adequate trials of two tolerated, appropriately chosen, and used AED schedules (whether as monotherapies or in combination) to achieve sustained seizure freedom (Kwan et al. 2010). For patients with intractable epilepsy, treatment options other than AEDs include resective epilepsy surgery, neuromodulation with neurostimulation devices, and ketogenic diet therapy.

When the medical treatment is effective for epilepsy, it is continued for an appropriate period lasting for years with careful monitoring of blood level(s) of AED(s), the emergence of possible adverse effects, etc. Epilepsy is considered to be resolved for individuals who have remained seizure-free for the last 10 years and off antiseizure medicines for at least the last 5 years (Fisher et al. 2014).

26.3 Pathophysiological and Biochemical Perspectives

26.3.1 Epileptogenesis

Epileptogenesis refers to the process in which an initially normal brain develops the ability to generate spontaneous/unprovoked seizures. In many patients with an acute brain insult, such as traumatic brain injury, epilepsy develops after a latent period, i.e., a seizure-free interval that indicates the process of epileptogenicity and lasts for months to years between the brain insult and the onset of seizures, from the insult.

Epileptogenesis may be a dynamic process that progressively alters neuronal excitability, establishes critical interconnections, and perhaps requires intricate structural changes before the first spontaneous seizure occurs. It is suggested that epileptogenesis also occurs in genetic epilepsies, in which it is regulated, for example, by developmental programming of gene expression leading to abnormal circuitry during maturation (Zara and Bianchi 2009).

26.3.2 Channelopathies and Neurotransmitter-Related Dysfunction

There are certain proteins that are now known to be involved in epilepsy due to gene defects or immune dysfunction. Majority of these mutations linked to genetic epilepsy syndromes are related to genes that encode ion channels or associated molecules. Ion channels are critically important in determining membrane potential and therefore neuronal excitability. Generally, channels are transmembrane proteins that are composed of primary pore-forming subunits which pass ions through the membrane and associated proteins with regulatory functions (Bautista and Anderson 2015), as illustrated in Fig. 26.5. Many of these channels are also the targets of therapeutics in epilepsy.

Voltage-gated sodium channels (Na_v) are essential for the generation and propagation of action potentials in neurons. Excitatory postsynaptic membrane potential induced by the activation of glutamate receptors, particularly α-amino-3-hydroxy-5-methyl-4-isoxazole propionic acid (AMPA) receptors, causes subsequent opening of sodium channels to flux sodium ions inward to initiate an action potential. Of channelopathies involving sodium channels, the genetic mutations of *SCN1A* that encodes Na_v 1.1 are most representative and cause Dravet syndrome (indicated as Case 1). Mutations in *SCN2A* coding Na_v 1.2 are reported in relation with benign familial neonatal/infantile seizures and Dravet syndrome.

Potassium channels play an important role in the central nervous system (CNS) functions including development, learning, memory, and plasticity, as they reduce neuronal excitability with membrane repolarization by fluxing potassium ions outward. The channels are very diverse with many subunits. In the family of voltage-gated potassium channels (K_v), mutations in *KCNQ2* and *KCNQ3*, encoding $K_v7.2$ and $K_v7.3$, respectively, are found in association with benign familial neonatal epilepsy (BFNE), in which an autosomal dominant occurrence and spontaneous disappearance of clustering seizures are observed during the neonatal period. *KCNQ2* is also associated with neonatal epileptic encephalopathy.

Voltage-gated calcium channels (Ca_v) allow calcium ions to flux inward in association with depolarization and therefore mediate the CNS physiologic functions including the activation of intracellular signaling pathways, gene transcription, and neurotransmitter release. Calcium channels comprise high-voltage-activated (HVA) family (Ca_v1 [L-type] and Ca_v2 [P/Q-, N-, R-type]) and low-voltage-activated (LVA) family (Ca_v3 [T-type]). Ca_v1 channels are typically localized in the postsynaptic regions and contribute to calcium signaling and gene regulation. Ca_v2 channels play an important role in the regulation of presynaptic neurotransmitter release. Ca_v3 channels underlie a transient calcium current and are involved in the intrinsic thalamocortical oscillations related to the generation of spike-wave complex of absence seizures. The susceptibility to childhood absence epilepsy and

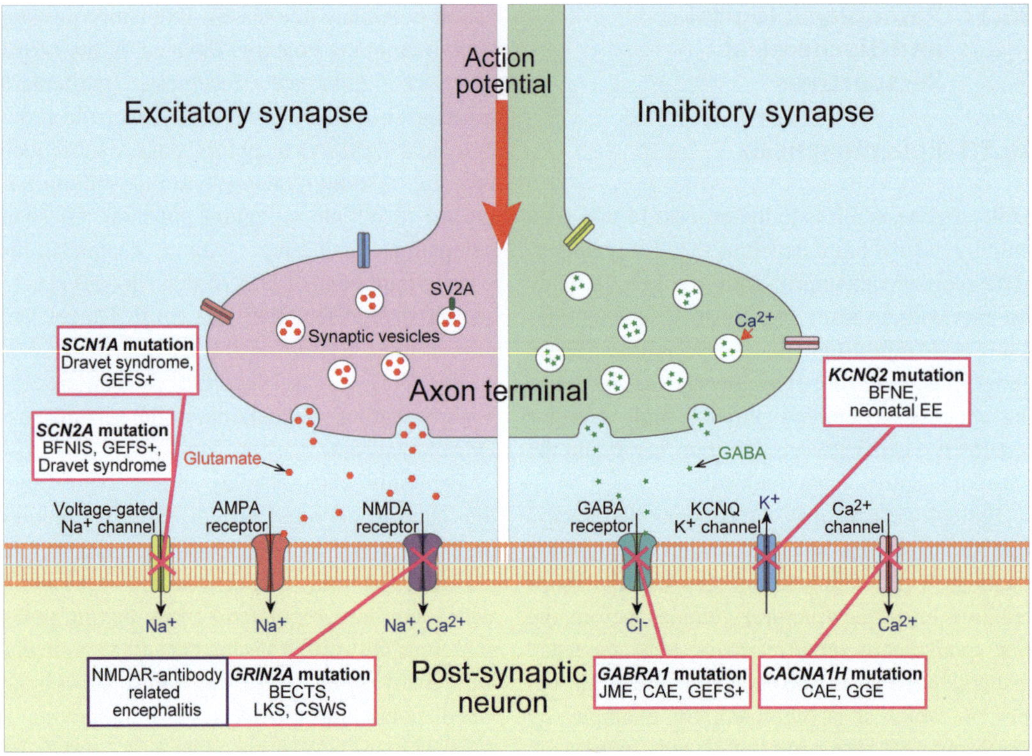

Fig. 26.5 A schema of a synapse indicating various channelopathies. The left- and right-hand side of the illustration shows the axon terminal of excitatory and inhibitory neuron, respectively. NMDAR, NMDA receptor; BFNE, benign familial neonatal epilepsy; BFNIS, benign familial neonatal/infantile seizures; EE, epileptic encephalopathy; GEFS+, genetic epilepsy with febrile seizures plus; GGE, genetic generalized epilepsy; CAE, childhood absence epilepsy; JME, juvenile myoclonic epilepsy; BECTS, benign epilepsy with centrotemporal spikes; LKS, Landau-Kleffner syndrome; CSWS, epileptic encephalopathy with continuous spike-and-wave during sleep

genetic epilepsies with various types of generalized seizures (collectively termed genetic generalized epilepsy) is associated with mutations of *CACNA1H* that encodes $Ca_v3.2$ channels. Mutations of *CACNA1A* gene is found to be linked to familial hemiplegic migraine.

Ligand-gated receptors (ion channels) flux ions in response to binding of the corresponding ligand to the extracellular domains of the pore-forming regions of channels. The first gene mutation identified with an epilepsy syndrome is *CHRNA4* which encodes neuronal nicotinic acetylcholine receptors that flux cations responding to the acetylcholine binding. This mutation is associated with autosomal dominant (sporadic in some cases) nocturnal frontal lobe epilepsy that is characterized by clustering seizures with a frontal lobe origin mimicking parasomnias during NREM sleep.

Of γ-aminobutyric acid (GABA) receptors, $GABA_A$ receptors are ionotropic and flux chloride ions responding to the GABA binding to induce fast inhibition of neurons. Various subunits are combined to form the receptor. Mutations in $GABA_A$ receptors involve *GABRA1* and *GABRA6* (encoding α1 and α6 subunits, respectively), *GABRG2* (γ2 subunit), etc. and are associated with a variety of genetic epilepsies including juvenile myoclonic epilepsy, childhood absence epilepsy, and Dravet syndrome.

Glutamate is the main excitatory neurotransmitter in the CNS, and representative types of ionotropic glutamate receptors include AMPA receptors and N-methyl-D-aspartate (NMDA) receptors. At resting membrane potential, NMDA receptors are inactive due to the voltage-dependent blockade of the channel pore by magnesium ions. Once the postsynaptic neuron is depolarized, NMDA receptors can be activated by the release of blockade. The activation of NMDA receptors allows influx of cations including calcium ions with resulting intracellular signaling. Mutations in *GRIN2A* coding NR2A subunit are associated with childhood epilepsies including benign epilepsy with centrotemporal spikes (BECTS) (Table 26.2) and related epilepsies.

It is of note that autoantibodies against NMDA receptor subunits are detected in various types of encephalitis and seizures, particularly anti-NMDA receptor encephalitis. This anti-NMDA receptor encephalitis is characterized by a progression of symptoms from memory deficits and seizures into a state of unresponsiveness. The disorder predominantly affects children and young adults and occurs in association with or without tumor (usually an ovarian teratoma) (Dalmau et al. 2008).

The leucine-rich glioma-inactivated protein 1 (LGI1) is linked to the voltage-gated potassium channel complex. Mutations in *LGI1* gene coding for LGI1 are related to autosomal dominant epilepsy with auditory features, while anti-LGI1 autoantibodies cause limbic encephalitis that is clinically characterized by psychosis, cognitive disturbances, and faciobrachial dystonic seizures and tends to occur in elderly patients.

26.3.3 Mesial Temporal Lobe (Hippocampal) Sclerosis

Mesial temporal lobe epilepsy (MTLE) associated with mesial temporal sclerosis (MTS) (aka hippocampal sclerosis) is a unique entity with a characteristic constellation of clinical and pathological findings. MTS is a histopathological term indicating neuronal loss and gliosis involving the hippocampus, often the amygdala, etc. This entity is important because it is the most common type of intractable temporal lobe epilepsy, and surgical treatment is very often effective to achieve seizure freedom (Cendes et al. 2012).

MTS is closely related to febrile seizures, particularly febrile status epilepticus (i.e., prolonged febrile seizure), during infancy and early childhood. Focal seizures originating from the sclerotic mesial temporal structures usually begin to occur from around late childhood or adolescence and are clinically characterized by initial sensory and/or psychic and/or autonomic symptoms (aura), typically arising epigastric sensation, and subsequent motionless staring, progressive clouding of consciousness, and automatisms, such as lip smacking and chewing. The duration of seizures is usually 1–2 min.

The typical interictal EEG findings in patients with MTLE are epileptic discharges over the unilateral anterior temporal region ipsilateral to the MTS. The ictal EEG recordings usually show regular rhythmic 5–9 Hz activity over the anterior temporal scalp region in the hemisphere ipsilateral to the MTS, but the ictal activity may be detected only by invasive intracranial electrodes in some patients.

26.3.4 Malformation of Cortical Development

There are a variety of malformations of the cerebral cortex that cause epilepsy as well as developmental and neurological disabilities. Such malformations of cortical development (MCD) can be detected through neuroimaging studies, particularly high-resolution MRI, by which MCD is found as abnormal distribution and depth of cortical sulci, cortical thickness, blurring of the boundaries between gray and white matter, variations in signal intensity, or a combination of these findings (Guerrini and Parrini 2012).

Malformations related to abnormal proliferation of neurons and glia include hemimegalencephaly and FCD. Hemimegalencephaly is characterized by an enlarged abnormal unilateral hemisphere that has wide convolutions with reduced sulci and thick cortex without laminar organization or clear demarcation between gray and white matter. In the histopathology of hemimegalencephaly, there are large dysmorphic neurons and, in a large proportion of patients, also "balloon cells" that present with a large cell body, lacking Nissl substance, and, often, multiple nuclei (Blümcke et al. 2011). When similar pathology is restricted to, say, a cerebral lobe or segment, the lesion is designated as FCD, as in Case 2. FCD is classified as follows (Blümcke et al. 2011). FCD Type I refers to isolated lesions, which present either as radial (FCD Type Ia) or tangential (FCD Type Ib) dyslamination of the neocortex, microscopically identified in one or multiple lobes. FCD Type II is an isolated lesion characterized by cortical dyslamination and dysmorphic neurons without (Type IIa) or with balloon cells (Type IIb). FCD Type III lesions are cortical dyslamination and/or dysmorphic neurons that occur in combination with hippocampal sclerosis (FCD Type IIIa), with epilepsy-associated tumors (FCD Type IIIb), adjacent to vascular malformations (FCD Type IIIc), whereas FCD Type IIId can be diagnosed in association with epileptogenic lesions acquired in early life (i.e., traumatic injury, ischemic injury, or encephalitis) (Blümcke et al. 2011). In the brain tissue of FCD Type II, de novo somatic mutations in mechanistic target of rapamycin (*mTOR*) have been reported (Lim et al. 2015). Although epilepsies associated with hemimegalencephaly and FCD tend to be medically intractable, surgical treatment is often effective.

Malformations caused by abnormal neuronal migration include lissencephaly, subcortical band heterotopia (SBH), gray matter heterotopia, and periventricular nodular heterotopia. Lissencephaly refers to a severe malformation of smooth brain with absent (agyria) or reduced (pachygyria) surface convolutions. In SBH, there is a heterotopic gray matter band under the cortical ribbon with an interspace of thin white matter band. Almost all patients with lissencephaly have epilepsy, particularly infantile spasms. Representative genes responsible for lissencephaly and SBH are *LIS1* and *DCX*.

Malformations related to abnormal cortical organization are represented by polymicrogyria, which refers to an excessive number of microconvolutions. Associated clinical manifestations vary from severe early-onset epileptic encephalopathies to minimal impairment depending on the pathology and distribution.

26.3.5 Metabolic and Neurodegenerative Diseases Underlying Epilepsy

Metabolic diseases (inborn errors of metabolism) may cause epileptic seizures due to various mechanisms such as defects in energy production and accumulation of neurotoxic metabolites. The age of onset is usually the neonatal period or infancy, but they can occur at older ages. Although these diseases are rare, some are treatable, and correct diagnosis in epileptic patients may help improve seizures by specific treatment.

Various metabolic pathways can be affected. Metabolic defects of amino acids, organic acids, fatty acids, and urea cycle are well known and can present in early life. Defects in organelles such as mitochondrial disorders, lysosomal disorders, and peroxisomal disorders consist of numerous subtypes of diseases and can occur at any age. Metabolic defects that involve vitamins and cofactors (pyridoxine-dependent epilepsy, etc.) are very rare; however, many of them can be treated with supplementation of deficient vitamins or cofactors. Menkes disease is caused by a defect of intestinal copper absorption and manifests symptoms in early infancy. Transport defects (glucose transporter 1 deficiency, etc.) cause depletion of essential nutrients limited to the CNS leading to epileptic seizures and other neurological symptoms.

Neurodegenerative diseases (without known specific metabolic defects) mainly involving gray matter can present with epileptic seizures and developmental regression. Rett syndrome, caused by mutations of *MECP2* gene on the X chromosome, is mostly seen in girls and presents with regression of motor development, loss of social skills, and acquired microcephaly after 5 months of age. It is also characterized by loss of purposeful use of hands, stereotypical hand movements, autistic behavior, and epilepsy. Lafora disease, Unverricht-Lundborg disease, dentatorubral-pallidoluysian atrophy, and juvenile Huntington disease can manifest progressive myoclonus epilepsy (PME), a clinical group characterized by myoclonus, epilepsy, progressive cerebellar ataxia, and dementia. PME can be caused by metabolic disorders including neuronal ceroid lipofuscinosis, sialidosis, and mitochondria disorders.

26.3.6 Animal Models of Epilepsy

There are a variety of animal models of epilepsy and seizures (Chauvette et al. 2016). Development of animal model is important to understand the mechanisms of epileptogenesis, to elucidate the long-lasting effects of seizures on cognitive functions, and to find the best treatment.

With respect to the kindling model, kindling refers to increased tendency of the occurrence of seizures that follows repeated electrical or chemical stimulations of certain regions of the animal brain. After repeated stimulations, the brain becomes sensitized and can generate a seizure even only by a weak stimulation.

Among the pharmacological models, pentylenetetrazole (PTZ) is a $GABA_A$ receptor antagonist and acutely induces absence-like seizures at low doses and clonic and tonic seizures at increased doses. As a model of temporal lobe epilepsy, intracerebral injection of kainic acid or pilocarpine at or around the unilateral hippocampus induces status epilepticus and subsequent spontaneous recurrent seizures after a latent period.

Mutations of various genes associated with human genetic epilepsies may be induced to animals to explore genotype-phenotype correlation, pathomechanisms in neural networks, and treatment strategies in these epilepsies.

26.4 Therapy

26.4.1 Antiepileptic Medications

AEDs suppress seizures by interactions with various cellular targets. The actions on these targets are categorized as follows (Rogawski and Cavazos 2015): (1) modulation of voltage-gated ion channels, including sodium, calcium, and potassium channels; (2) enhancement of GABA inhibition through the effects on $GABA_A$ receptors, the GABA transporter 1 (GAT-1), or GABA transaminase; (3) direct modulation of synaptic release through the effects on components of release machinery; and (4) inhibition of synaptic excitation mediated by ionotropic glutamate receptors, including AMPA receptors. The effects of interactions are to modify the bursting properties of neurons and to reduce synchronization in the neuronal system.

AEDs that interact with voltage-gated sodium channels to protect against seizures (sodium channel blockers) are often used to control focal seizures and generalized tonic-clonic seizures and include phenytoin, carbamazepine, lamotrigine, oxcarbazepine, and lacosamide. These drugs can exacerbate absence epilepsy perhaps due to their potentiation of $GABA_A$ activation although the actual mechanism of action is not clear (Perucca et al. 1998; Zheng et al. 2009).

Voltage-gated potassium channels in the nervous system are positively modulated by ezogabine (retigabine) to repolarize the depolarized neuronal membrane potential and to inhibit focal seizures.

T-type calcium channels (Ca_v3) are important in the intrinsic thalamocortical oscillations to generate spike-and-wave complex. These channels are inhibited by ethosuximide, which preferentially suppresses absence seizures.

The effects of $GABA_A$ receptors are augmented by allosteric modulators including benzodiazepines (e.g., diazepam, lorazepam, clonazepam, midazolam) and barbiturates (e.g., phenobarbital). Benzodiazepines are specific for synaptic $GABA_A$ receptors containing the $\gamma 2$ subunit. They have a broad-spectrum antiepileptic action and are often used to suppress prolonged seizures or status epilepticus.

GABA transaminase is responsible for the metabolic inactivation of GABA and is irreversibly inhibited by vigabatrin (γ-vinyl GABA) with resulting elevation of brain GABA levels. Vigabatrin is effective for infantile spasms and focal seizures, particularly spasms associated with tuberous sclerosis complex, although it is associated with increased risk of retinal toxicity and visual field constriction; therefore patients have to be monitored for these risks.

SV2A is a membrane glycoprotein observed in the secretory vesicles of neurons and is the molecular target of levetiracetam, which is effective for focal-onset seizures, generalized tonic-clonic seizures, and myoclonic seizures.

Perampanel is a selective antagonist for AMPA receptors that are cation channels responsible for fast synaptic excitation, and it is effective for focal seizures and tonic-clonic seizures.

Topiramate has multiple actions to protect against seizures, such as actions on voltage-activated sodium channels, $GABA_A$ receptors, AMPA receptors, and carbonic anhydrase. It is a broad-spectrum AED that exerts effects for focal-onset seizures, generalized tonic-clonic seizures, and myoclonic seizures as well as intractable epilepsies including Lennox-Gastaut syndrome, West syndrome, and Dravet syndrome. Zonisamide has some similarities with topiramate and has in addition an inhibitory effect on T-type voltage-gated calcium channels.

Sodium valproate (valproic acid) is a short-chain, branched fatty acid and is very potent for the treatment of generalized seizures including absence, myoclonic, and tonic-clonic seizures. The precise mechanisms of action of valproate, however, are poorly understood, though they may be partly related to GABA systems.

Adrenocorticotropin (ACTH) is highly effective for infantile spasms or West syndrome, but its mechanisms of action are not yet understood. Glucocorticoids are also effective, and their synthesis is stimulated by ACTH. The effects may be related to anti-inflammatory action or a stimulation of neurosteroid synthesis but should be elucidated in the future.

26.4.2 Surgical Treatment

A subset of intractable epilepsy can be controlled or ameliorated by epilepsy surgery, as in Case 2.

Once epilepsy in a patient is considered to be intractable based on treatment failure with two appropriate antiepileptic drugs, the next step is a thorough workup to evaluate whether the patient is a surgical candidate. This process requires multimodal investigations including detailed history taking, long-term video EEG, brain MRI with epilepsy-dedicated protocol, positron emission tomography, single-photon emission tomography (interictal and ictal), magnetoencephalography, and neuropsychological evaluation. If the results of these investigations are concordant, indicating a single epileptic focus, the patient may be a surgical candidate for curative surgery. Even when these results are discordant, the patient can be considered for palliative surgery.

Curative surgery usually requires intracranial video EEG to narrow down the localization of the epileptogenic focus and mapping of the eloquent cortices (e.g., primary motor cortex) that may be located nearby. One of the exceptions is MTLE caused by MTS, which does not always need invasive EEG monitoring as long as the results of noninvasive investigations are concordant. Epilepsy associated with brain tumor, in particular with well-demarcated lesion, may not require invasive tests if the tumor is away from the eloquent areas. The surgical outcomes for MTLE with MTS and epilepsies with well-defined focal brain lesion (e.g., tumor) are good (60–90%) in general. Extratemporal epilepsies, especially neuroimaging-negative cases, are associated with poorer surgical outcome (seizure freedom in 30–50%).

Palliative surgery consists of corpus callosotomy and vagus nerve stimulator implantation. Corpus callosotomy is effective for generalized seizures, in particular for atonic seizures. Complete corpus callosotomy can be done to young children without significant neurological sequelae. Older children (>10 years old) may develop disconnection syndrome after complete corpus callosotomy (Cross et al. 2006); therefore, anterior two-thirds corpus callosotomy is preferred in older children. Vagus nerve stimulation (VNS) therapy utilizes a stimulator device implanted under the skin of the chest wall and connected to the vagus nerve in the left side of the neck via an electric wire. The device sends regular electric stimulation to the left vagus nerve (Normal mode). The mechanism by which VNS ameliorate epileptic seizures is not clearly understood. VNS is effective for various types of epileptic seizures, especially for focal seizures. Patients with VNS can use magnet for on-demand stimulation (magnet mode), and this is especially helpful for patients who can feel auras at the beginning of epileptic seizures.

26.4.3 Diet Therapy

Historically, it was known that starvation could suppress epileptic seizures. In starvation, the ketone body is produced by degradation of fat in the adipose tissue, and it is presumed that ketosis reduces seizure frequency, although the mechanism is not well understood. Diet therapy aims to simulate starvation to increase the production of the ketone body while keeping adequate caloric intake to maintain appropriate body growth. It is the first-line treatment for glucose transporter 1 deficiency. Epileptic encephalopathies such as West syndrome and Dravet syndrome are also good indications. It is contraindicated in some metabolic disorders, especially fatty acid metabolism disorders.

Classic ketogenic diet, medium chain triglyceride (MCT) ketogenic diet, and modified Atkins diet are commonly used. Classic and MCT ketogenic diet is low-carbohydrate and high-fat diet. Daily intake of carbohydrate, protein, and fat is specified and must be strictly followed with advice from experienced dietitians. Because MCT is more ketogenic than regular fat, restriction of carbohydrate is less strict in the MCT diet than in classic ketogenic diet. Deficient nutrients such as vitamins and carnitine should be supplemented. Modified Atkins diet restricts only carbohydrate intake (15–20 g/day) and thus provides flexibility for menu construction, which is advantageous for older patients.

Patients on diet therapy need regular evaluation, and continuous support from dietitians is essential to make the diet therapy successful. When the diet therapy is effective, it is usually continued for 2–3 years and later on gradually tapered off, except for patients with glucose transporter 1 deficiency.

Questions

1. What symptoms do patients with epilepsy experience during seizures?
2. How should a patient with suspected seizures be managed?
3. How is EEG examination used in the diagnosis of epilepsy?
4. Discuss the pathophysiology involved in the following epilepsy syndromes:
   (1) Dravet syndrome
   (2) Childhood absence epilepsy
5. What handicaps do patients with epilepsy experience as direct or indirect consequences of seizures, particularly when epilepsy is hard to control?

References

Bautista JF, Anderson A (2015) Genetics of the epilepsies. In: Wyllie E, Gidal BE, Goodkin HP, Loddenkemper T, Sirven JI (eds) Wyllie's treatment of epilepsy, 6th edn. Walters Kluwer, Philadelphia, pp 31–43

Berg AT, Berkovic SF, Brodie MJ, Buchhalter J, Cross JH, van Emde Boas W, Engel J, French J, Glauser TA, Mathern GW, Moshé SL, Nordli D, Plouin P, Scheffer IE (2010) Revised terminology and concepts for organization of seizures and epilepsies: report of the ILAE commission on classification and terminology, 2005–2009. Epilepsia 51(4):676–685

Blümcke I, Thom M, Aronica E, Armstrong DD, Vinters HV, Palmini A, Jacques TS, Avanzini G, Barkovich AJ, Battaglia G, Becker A, Cepeda C, Cendes F, Colombo N, Crino P, Cross JH, Delalande O, Dubeau F, Duncan J, Guerrini R, Kahane P, Mathern G, Najm I, Ozkara C, Raybaud C, Represa A, Roper SN, Salamon N, Schulze-Bonhage A, Tassi L, Vezzani A, Spreafico R (2011) The clinicopathologic spectrum of focal cortical dysplasias: a consensus classification proposed by an ad hoc Task Force of the ILAE Diagnostic Methods Commission. Epilepsia 52:158–174

Blume WT, Lüders HO, Mizrahi E, Tassinari C, van Emde Boas W, Engel J Jr (2001) Glossary of descriptive terminology for ictal semiology: report of the ILAE task force on classification and terminology. Epilepsia 42(9):1212–1218

Cendes F, Kahane P, Brodie M, Andermann F (2012) The mesio-temporal lobe epilepsy syndrome. In: Bureau M, Genton P, Dravet C, Delgado-Escueta A, Tassinari CA, Thomas P, Wolf P (eds) Epileptic syndromes in infancy, childhood and adolescence, 5th edn. John Libbey Eurotext, Montrouge, pp 383–399

Chauvette S, Soltani S, Seigneur J, Timofeev I (2016) In vivo models of cortical acquired epilepsy. J Neurosci Methods 260:185–201

Commission on Classification and Terminology of the International League Against Epilepsy (1981) Proposal for revised clinical and electroencephalographic classification of epileptic seizures. Epilepsia 22(4):489–501

Cross JH, Jayakar P, Nordli D, Delalande O, Duchowny M, Wieser HG, Guerrini R, Mathern GW (2006) Proposed criteria for referral and evaluation of children for epilepsy surgery: recommendations of the Subcommission for Pediatric Epilepsy Surgery. Epilepsia 47:952–959

Dalmau J, Gleichman AJ, Hughes EG, Rossi JE, Peng X, Lai M, Dessain SK, Rosenfeld MR, Balice-Gordon R, Lynch DR (2008) Anti-NMDA-receptor encephalitis: case series and analysis of the effects of antibodies. Lancet Neurol 7(12):1091–1098

Fisher RS, van Emde Boas W, Blume W, Elger C, Genton P, Lee P, Engel J Jr (2005) Epileptic seizures and epilepsy: definitions proposed by the International League Against Epilepsy (ILAE) and the International Bureau for Epilepsy (IBE). Epilepsia 46(4):470–472

Fisher RS, Acevedo C, Arzimanoglou A, Bogacz A, Cross JH, Elger CE, Engel J Jr, Forsgren L, French JA, Glynn M, Hesdorffer DC, Lee BI, Mathern GW, Moshé SL, Perucca E, Scheffer IE, Tomson T, Watanabe M, Wiebe S (2014) A practical clinical definition of epilepsy. Epilepsia 55(4):475–482

Fisher RS, Helen Cross J, French JA, Higurashi N, Hirsch E, Jansen FE, Lagae L, Moshé SL, Peltola J, Perez ER, Scheffer IE, Zuberi SM (2017) Operational classification of seizure types by the international league against epilepsy: position paper of the ILAE commission for classification and terminology. Epilepsia 58(4):522–530

Guerrini R, Parrini E (2012) Epilepsy and malformations of the cerebral cortex. In: Bureau M, Genton P, Dravet C, Delgado-Escueta A, Tassinari CA, Thomas P, Wolf P (eds) Epileptic syndromes in infancy, childhood and adolescence, 5th edn. John Libbey Eurotext, Montrouge, pp 607–629

Kwan P, Arzimanoglou A, Berg AT, Brodie MJ, Allen Hauser W, Mathern G, Moshé SL, Perucca E, Wiebe S, French J (2010) Definition of drug resistant epilepsy: consensus proposal by the ad hoc task force of the ILAE commission on therapeutic strategies. Epilepsia 51(6):1069–1077

Lim JS, Kim WI, Kang HC, Kim SH, Park AH, Park EK, Cho YW, Kim S, Kim HM, Kim JA, Kim J, Rhee H, Kang SG, Kim HD, Kim D, Kim DS, Lee JH (2015) Brain somatic mutations in MTOR cause focal cortical dysplasia type II leading to intractable epilepsy. Nat Med 21(4):395–400

Perucca E, Gram L, Avanzini G, Dulac O (1998) Antiepileptic drugs as a cause of worsening seizures. Epilepsia 39:5–17

Rogawski MA, Cavazos JE (2015) Mechanisms of action of antiepileptic drugs. In: Wyllie E, Gidal BE, Goodkin HP, Loddenkemper T, Sirven JI (eds) Wyllie's treatment of epilepsy, 6th edn. Walters Kluwer, Philadelphia, pp 522–529

Scheffer IE, Berkovic S, Capovilla G, Connolly MB, French J, Guilhoto L, Hirsch E, Jain S, Mathern GW, Moshé SL, Nordli DR, Perucca E, Tomson T, Wiebe S, Zhang Y-H, Zuberi SM (2017) ILAE classification of the epilepsies: position paper of the ILAE commission for classification and terminology. Epilepsia 58(4):512–521

Zara F, Bianchi A (2009) The impact of genetics on the classification of epilepsy syndromes. Epilepsia 50(Suppl 5):11–14

Zheng T, Clarke AL, Morris MJ, Reid CA, Petrou S, O'Brien TJ (2009) Oxcarbazepine, not its active metabolite, potentiates GABAA activation and aggravates absence seizures. Epilepsia 50:83–87

27 Hemophagocytic Lymphohistiocytosis

Kyung-Nam Koh

Keywords
Hemophagocytic lymphohistiocytosis (HLH) · Perforin/granzyme pathway · Familial hemophagocytic lymphohistiocytosis (FHL) · Secondary hemophagocytic lymphohistiocytosis (secondary HLH) · Hyperinflammation

27.1 Case Report

A previously healthy 11-month-old girl was transferred to the emergency room (ER) with a 7-day history of high fever and progressive lethargy. Her parents reported that she had developed mild cough and rhinorrhea 1 week before presentation. These symptoms were followed by severe illness with high fever, distended abdomen, skin rash, and poor oral intake. She was empirically started on broad-spectrum antibiotics at the referring hospital before transfer.

Upon arrival at the ER, the patient appeared acutely ill. She was febrile (39.2 °C), tachycardic, and tachypneic. Physical examination was notable for generalized petechiae and substantial hepatosplenomegaly. A complete blood count test demonstrated pancytopenia (white blood cells, 3290/μL; hemoglobin, 7.9 g/dL; platelets, 34,000/μL; and neutrophils, 691/μL). A peripheral blood smear did not reveal any malignant cells. The patient also had elevated serum levels of aspartate aminotransferase (985 U/L; reference range, 20–60 U/L), alanine aminotransferase (440 U/L; reference range, 5–45 U/L), bilirubin (2.5 mg/dL; reference range, <1.0 mg/dL), triglycerides (372 mg/dL; reference range, <150 mg/dL), ferritin (12,530 ng/mL; reference range, 10–60 ng/mL), and lactate dehydrogenase (620 U/L; reference range, 150–500 U/L). A coagulation profile indicated normal prothrombin time and activated partial thromboplastin time with hypofibrinogenemia (110 mg/dL; reference range, 150–400 mg/dL). Chest radiography revealed bilateral pleural effusions, and computerized tomography (CT) scans of the abdomen and pelvis showed multiple, prominent mesenteric lymph nodes, ascites, and hepatosplenomegaly. A CT scan of the head was normal with no evidence of an intracranial mass or acute bleeding. Bone marrow examination was performed on hospital day 1 and demonstrated hypercellular marrow with 90% cellularity and remarkable hemophagocytosis without any obvious evidence of malignant cells (Fig. 27.1).

The combination of the patient's clinical features (i.e., fever and splenomegaly) and laboratory findings (i.e., pancytopenia, hypertriglyc-

K.-N. Koh (✉)
Division of Pediatric Hematology/Oncology, Department of Pediatrics, Asan Medical Center Children's Hospital, University of Ulsan College of Medicine, Seoul, South Korea
e-mail: pedkkn@amc.seoul.kr

T. Oohashi et al. (eds.), *Human Pathobiochemistry*, https://doi.org/10.1007/978-981-13-2977-7_27

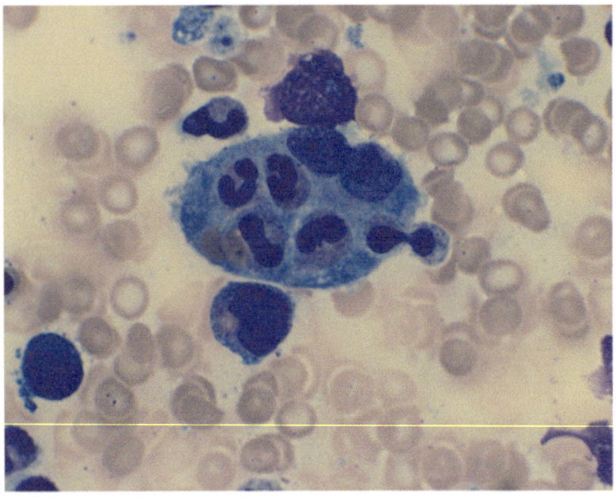

Fig. 27.1 Hemophagocytosis seen in bone marrow aspirates

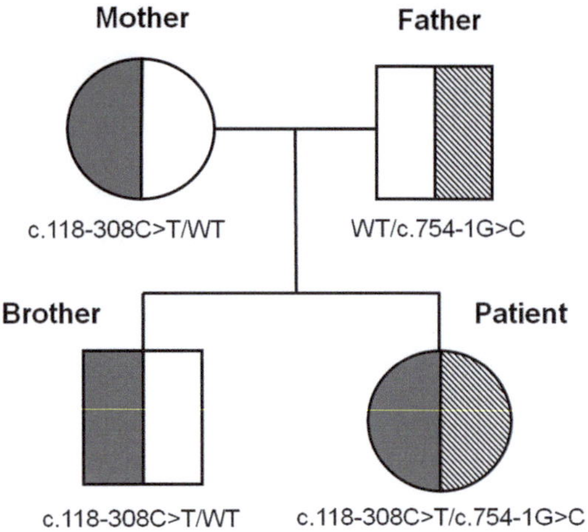

Fig. 27.2 Genetic pedigree of *UNC13D* mutations in the patient and her family

eridemia, hypofibrinogenemia, elevated ferritin level, and hemophagocytosis in the bone marrow) met the revised diagnostic criteria for hemophagocytic lymphohistiocytosis (HLH) proposed by the Histiocyte Society (Henter et al. 2007). The patient was administered dexamethasone (10 mg/m^2), etoposide (150 mg/m^2), and cyclosporine on hospital day 2, according to the HLH-2004 treatment protocol imparted by the Histiocyte Society (Henter et al. 2007).

The results of additional tests were as follows: Serum level of soluble interleukin-2 receptor alpha (IL-2RA) was 22,400 U/mL (reference range, 334–3026 U/mL). Natural killer (NK) cell activity was reportedly decreased. A blood culture was negative for microorganisms. Polymerase chain reaction (PCR) studies of respiratory viruses were positive for parainfluenza virus and rhinovirus. Serologic and/or PCR studies for mycoplasma, Epstein–Barr virus (EBV), and cytomegalovirus were negative. Cerebrospinal fluid (CSF) examination was unremarkable. Brain magnetic resonance imaging revealed no significant findings. Genetic testing for HLH confirmed heterozygous mutations in the *UNC13D* gene (c.118_308C > T and c.754_1G > C). Genetic testing of the parents and a sibling revealed them as heterozygous mutation carriers (Fig. 27.2). After 1 week of treatment, the patient improved in general condition and laboratory findings. After 8 weeks of initial therapy, the patient showed a complete response.

Because familial HLH was diagnosed, the patient and her family were allowed to decide to proceed with allogeneic hematopoietic stem cell transplantation (HSCT). She was kept on continuation treatment with dexamethasone, etoposide, and cyclosporine before receiving HSCT.

She received an allogeneic HSCT from an unrelated donor. The conditioning regimen was reduced-intensity and consisted of fludarabine (150 mg/m^2), melphalan (140 mg/m^2), and alemtuzumab. Cyclosporine and short-term methotrexate were used for graft-versus-host-disease prophylaxis. Allogeneic HSCT was uneventful with successful engraftment. She remains asymptomatic 3 years after the initial presentation with complete donor chimerism.

27.2 Diagnosis

HLH is a disorder of immune cell regulation. It is characterized by distinctive clinical features that are indicative of excessive inflammation and tissue damage and is often associated with genetic defects affecting the cytotoxic function of cytotoxic T cells and NK cells. HLH can be classified into two major forms: primary/genetic and secondary/acquired. According to the classification of human primary immunodeficiencies by the International Union of Immunological Societies Expert Committee on Primary Immunodeficiency in 2014, primary/genetic HLH is further classified as familial HLH (FHL) syndromes and lymphoproliferative syndromes (Table 27.1) (Al-Herz et al. 2014). FHL syndromes are then further classified into subtypes with and without hypopigmentation. FHL without hypopigmentation, in which HLH is the defining clinical manifestation, has five subtypes (FHL1, FHL2, FHL3, FHL4, and FHL5), which are classified by their causative mutated genes. In FHL with hypopigmentation, HLH is a component of syndromic disease and is usually accompanied by other clinical manifestations, such as defects in pigmentation. This form of HLH includes Chédiak–Higashi syndrome, Griscelli syndrome type 2, and Hermansky–Pudlak syndrome type 2. Lymphoproliferative syndromes can be X-linked (e.g., XLP and XMEN syndrome) or autosomal recessive (e.g., *ITK* deficiency and *CD27* deficiency).

Secondary HLH is not associated with any known genetic abnormality or immunodeficiency syndrome. It is thought to be a kind of reactive phenomenon and can be associated with infection (e.g., infection-associated hemophagocytic syndrome), malignant neoplasms (e.g., malignancy-associated hemophagocytic syndrome), and rheumatic/autoimmune diseases (e.g., macrophage activation syndrome) (Table 27.2).

It is challenging to make a timely diagnosis of HLH in clinical settings because a single diagnostic indicator for HLH other than a genetic diagnosis does not exist. Patients with HLH can present with several clinical features that overlap with those of infection, inflammatory conditions, and malignant neoplasms and are accompanied by various abnormal laboratory tests. Each of these individual findings may not be specific for HLH but collectively can contribute to a diagnosis of HLH. Therefore, diagnostic criteria consisting of clinical and laboratory findings are required to diagnose HLH. The HLH-2004 diagnostic criteria suggested by the Histiocyte Society are most commonly used (Table 27.3) (Henter et al. 2007). Diagnosis of HLH can be established by molecular/genetic findings that are consistent with HLH or when five of the eight criteria (two clinical symptoms, one histopathologic finding, and five laboratory findings) are fulfilled.

The clinical symptoms of HLH can be attributed to severe inflammation, hypercytokinemia, and infiltration of histiocytes into organs. Most patients present with a prolonged and unexplainable fever that is frequently proceeded by suspicious viral infections. Splenomegaly is a direct result of infiltration by activated lymphocytes and macrophages. It is present in 80–100% of patients at diagnosis and is usually progressive (George 2014; Koh et al. 2015).

Hemophagocytosis is a self-destructive phenomenon, consisting of phagocytosis of erythrocytes, leukocytes, platelets, and their precursors in the bone marrow and other tissues by activated histiocytes. Hemophagocytosis is a striking finding in patients with HLH, but it is neither very sensitive nor specific for diagnosis of HLH. Hemophagocytosis is not a defining feature of HLH but is rather a component of the diagnostic criteria. Hemophagocytosis in the bone marrow may not be apparent during the early phase of disease, and therefore the absence of hemophagocytosis does not preclude the diagnosis of HLH. However, bone marrow examination is mandatory not only to search for hemophagocytosis but also to differentiate HLH from hematologic malignancies.

Cytopenias can be attributed to high concentrations of tumor necrosis factor alpha (TNF-α) and interferon gamma (IFN-γ), as well as direct hemophagocytosis (George 2014). Bicytopenia or pancytopenia is observed in most cases of HLH. In some cases, cytopenia may not be present during the early period but eventually appears as the disease progresses.

Hypertriglyceridemia occurs secondary to decreased lipoprotein lipase activity, which is initiated by increased TNF-α and IFN-γ levels (George 2014). Activated macrophages secrete plasminogen activators that lead to high plasmin levels, hyperfibrinolysis, and hypofibrinogen-

Table 27.1 The classification of primary HLH according to the updated classification of human primary immunodeficiencies by the International Union of Immunological Societies Expert Committee on Primary Immunodeficiency, 2014

HLH type	Gene	Protein	Function	Inheritance	Clinical features
1. Familial HLH (FHL) syndromes					
(1) FHL syndromes without hypopigmentation					
FHL1	Unknown	Unknown		AR	HLH is the defining feature
FHL2	*PRF1*	Perforin	Cytolytic pore formation	AR	HLH is the defining feature
FHL3	*UNC13D*	Munc13-4	Vesicle priming for fusion	AR	HLH is the defining feature
FHL4	*STX11*	Syntaxin-11	Vesicle fusion with the cell membrane	AR	HLH is the defining feature
FHL5	*STXBP2*	Munc18-2	Vesicle fusion with the cell membrane	AR	HLH is the defining feature
(2) FHL syndromes with hypopigmentation					
Griscelli syndrome, type 2	*RAB27A*	Rab27a	Vesicle docking to the cell membrane	AR	Partial albinism, hepatosplenomegaly, HLH, cytopenias
Chédiak–Higashi syndrome	*LYST*	LYST	Intracellular protein trafficking	AR	Partial albinism, recurrent infections, hepatosplenomegaly, HLH, giant lysosomes, neutropenia, cytopenias, bleeding tendency, neurologic dysfunction
Hermansky–Pudlak syndrome, type 2	*AP3B1*	AP3B1	Intracellular protein trafficking	AR	Partial albinism, recurrent infections, pulmonary fibrosis, increased bleeding, neutropenia, HLH
2. Lymphoproliferative syndrome					
(1) X-linked					
XLP1	*SH2D1A*	SAP	Intracellular signaling	XR	Triggered by EBV infection, HLH, lymphoproliferation, aplastic anemia, lymphoma, hypogammaglobulinemia, absent iNKT cells
XLP2	*BIRC4*	XIAP	Inhibition of apoptosis	XR	EBV infection, splenomegaly, lymphoproliferation, HLH, colitis, IBD, hepatitis, low iNKT cells
XMEN syndrome	*MAGT1*	Magnesium transporter 1	T-cell activation via TCR	XR	Combined immunodeficiency, lymphoma
(2) Autosomal recessive					
ITK deficiency	*ITK*	ITK	TCR-mediated activation	AR	EBV-associated B-cell lymphoproliferation, lymphoma, normal or decreased IgG
CD27 deficiency	*CD27*	CD27	Lymphocyte costimulatory molecule	AR	Triggered by EBV infection, HLH, aplastic anemia, lymphoma, hypogammaglobulinemia, low iNKT cells

AP3B1 adaptor-related protein complex 3 beta 1 subunit; *AR* autosomal recessive; *EBV* Epstein–Barr virus; *FHL* familial hemophagocytic lymphohistiocytosis; *HLH* hemophagocytic lymphohistiocytosis; *IBD* inflammatory bowel disease; *ITK* interleukin-2-inducible T-cell kinase; *LYST* lysosomal trafficking regulator; *iNKT cells* invariant natural killer T cells; *SAP* signaling lymphocytic activation molecule-associated protein; *TCR* T-cell receptor; *XIAP* X-linked inhibitor of apoptosis protein; *XLP* X-linked lymphoproliferative disease; *XMEN* X-linked immunodeficiency with magnesium defect, Epstein-Barr virus infection, and neoplasia; *XR* X-linked recessive

Table 27.2 Subtypes of secondary hemophagocytic lymphohistiocytosis and their associated conditions

1. Infection-associated hemophagocytic syndrome
Virus-associated hemophagocytic syndromes: EBV, CMV, HSV, HIV, influenza virus, parainfluenza virus, etc.
Other infections associated with hemophagocytic syndromes: bacterial, parasitic, fungal
2. Malignancy-associated hemophagocytic syndrome
Hematologic malignancies: lymphoma, leukemia
Solid tumors: prostate cancer, lung cancer, hepatocellular carcinoma, etc.
3. Macrophage activation syndromes (associated with rheumatic/autoimmune disease)
Systemic idiopathic juvenile arthritis
Still disease
Systemic lupus erythematosus
Kawasaki disease
Rheumatoid arthritis

CMV cytomegalovirus, *EBV* Epstein–Barr virus, *HIV* human immunodeficiency virus, *HSV* herpes simplex virus

Table 27.3 Diagnostic criteria of hemophagocytic lymphohistiocytosis proposed in the HLH-2004 treatment protocol by the Histiocyte Society (Henter et al. 2007)

A diagnosis of HLH can be established if one of either parameter (1 or 2) below is fulfilled:
1. A genetic diagnosis consistent with HLH
2. Diagnostic criteria for HLH fulfilled (five out of the eight criteria below):
Clinical criteria
Fever
Splenomegaly
Histopathologic criteria
Hemophagocytosis in the bone marrow, spleen, or lymph nodes with no evidence of malignant neoplasm
Laboratory criteria
Cytopenias (affecting ≥ 2 of 3 lineages in the peripheral blood): hemoglobin (<9.0 g/dL), platelets (<100,000/μL), neutrophils (<1000/μL) (in infants <4 weeks: hemoglobin <10.0 g/dL)
Hypertriglyceridemia and/or hypofibrinogenemia: fasting triglycerides ≥ 3.0 mmol/L (i.e., ≥ 265 mg/dL), fibrinogen ≤ 150 mg/dL
Low or absent NK-cell activity (according to local laboratory reference)
Ferritin ≥ 500 ng/mL
Soluble CD25 (i.e., soluble IL-2 receptor) ≥ 2400 U/mL

HLH hemophagocytic lymphohistiocytosis, *NK* natural killer

emia (Rosado and Kim 2013). Such biochemical dysfunctions are present in 40–70% of cases at diagnosis (Koh et al. 2015; George 2014).

Ferritin is a clinically useful screen for the early diagnosis of HLH. Ferritin production is upregulated in macrophages in response to inflammatory cytokines (Rosado and Kim 2013). Ferritin is an acute-phase reactant and is nonspecifically elevated in other inflammatory conditions, including infection, cancer, and autoimmune diseases. According to the HLH-2004 criteria, a ferritin value greater than 500 ng/mL is a positive diagnostic parameter. Although this value is highly sensitive, it may be unspecific for HLH. Marked hyperferritinemia (ferritin >2000 ng/mL up to 10,000 ng/mL) was demonstrated to be specific for HLH, especially in pediatric patients (Lehmberg et al. 2014).

Serum levels of soluble IL-2RA (also called soluble CD25) reflect T-cell activation and may be elevated in many clinical conditions, such as autoimmune diseases, malignant neoplasms, and HLH. Soluble IL-2RA can serve as a useful marker for diagnosis and follow-up of HLH because its levels are closely associated with that of disease activity.

NK-cell activity is a marker of cytotoxic function, which is markedly reduced or absent in patients with primary HLH. NK-cell activity normalizes only after allogeneic HSCT. In contrast, patients with secondary HLH who have low NK-cell activity at presentation experience normalized NK-cell activity later. NK-cell activity can be measured by chromium-51 (^{51}Cr) release assays or flow cytometric assays. As ^{51}Cr release assays require the use of radioactive chromium, are cost prohibitive, and often exhibit interlaboratory variability, these assays are not preferred in clinical settings. Flow cytometric assays are a more preferred method to measure NK-cell activity in clinical practice. However, the results from these assays should be interpreted with caution because they may vary, depending on sample quality and the expertise of the examiners.

In addition to the abnormalities listed as components of the HLH diagnostic criteria, many other clinical features and laboratory abnormalities can be observed at presentation. HLH may affect many other organ systems in addition to the hematopoietic system, including the liver, kidneys, heart, and central nervous system (CNS). CNS involvement is especially frequent at presentation. CNS involvement is defined by either the presence of neurologic symptoms such as seizure, inconsolable irritability, and mental status changes or by abnormalities found in CSF examination or neuroimaging. CNS involvement is reported to be one of the most important prognostic indicators of HLH (Koh et al. 2015; Horne et al. 2008). Other laboratory abnormalities include coagulopathy and increased levels of lactate dehydrogenase, bilirubin, and transaminases. Although not a component of the diagnostic criteria of HLH, abnormalities in liver function tests are frequently observed in HLH and are usually progressive if proper management is not initiated (Koh et al. 2015; Jordan et al. 2011). Therefore, HLH should be considered in the differential diagnosis of patients who present with prolonged fever and hepatitis of unknown causes (Ryu et al. 2013).

Other ancillary tests such as flow cytometric assays for intracellular perforin expression, granule release assays for evaluation of granule-mediated cytotoxicity, and assays for expression of SLAM-associated protein (SAP) and X-linked inhibitor of apoptosis (XIAP) can be helpful in the diagnosis of HLH and rapid categorization of HLH subtypes. These assays can also be used to guide the priority of genetic testing (Morimoto et al. 2016; Sieni et al. 2014).

Some tests required for HLH diagnosis are not readily available in individual hospitals, and genetic tests or flow cytometric assays are available only in some tertiary referral hospitals. HLH is often rapidly progressive and fulminant. It is, therefore, very important to make a timely diagnosis in the early stages of the disease. Unexplained cytopenias and hepatitis accompanied by fever and marked elevation of ferritin without evidence of malignant disease can be clinical clues of HLH (Jordan et al. 2011). An algorithmic approach is useful for the prompt suspicion of HLH and proper initiation of early treatment (Fig. 27.3).

27.3 Molecular Pathophysiology

HLH is a hyperinflammatory syndrome due to a highly stimulated but ineffective immune process. Primary/genetic HLH is caused by genetic defects of the effector cell, leading to impaired secretion of cytotoxic granules. A cytotoxic granule is a type of specialized lysosome that acts as a secretory vesicle containing perforin and granzyme, which efficiently kill targeted cells. Several intracellular steps are required for granule-mediated cytotoxic pathways in cytotoxic T cells and NK cells, including the production of perforin and granzyme, the packaging of these molecules into cytotoxic granules, and the intracellular transportation, docking, priming, and fusion of the granules with cell membranes (Ishii 2016; Jordan et al. 2011; Seo 2015). Primary HLH is associated with mutations in the genes that encode the proteins involved in each step of the pathway (Fig. 27.4). During conventional immune regulation, antigen-presenting cells (APCs) activate cytotoxic T cells in response to a viral infection. The activated cytotoxic T cells then secrete inflammatory cytokines, which activate macrophages. In turn, the activated macrophages produce cytokines, which activate additional cytotoxic T cells and themselves. As the cytotoxic T cells become further activated and antigen presentation is no longer required, they selectively eliminate the APCs that continue to present virus-related antigens through granule-mediated cell lysis. However, in HLH, these APCs are not eliminated by cytotoxic T cells because of impaired cytotoxic granule secretion. This produces APCs that continue to present antigens to T cells, resulting in the uncontrolled over-activation of T cells. This leads to the release of large amounts of IFN-γ by the T cells, which in turn continues to activate APCs that release large amounts of cytokines. This vicious cycle of immune dysregulation causes an ineffective and uncontrolled cytokinemia, thereby leading to a systemic hyperinflammatory response in primary HLH (Morimoto et al. 2016). The causative genes and their associated proteins and functions in primary HLH are summarized in Table 27.1.

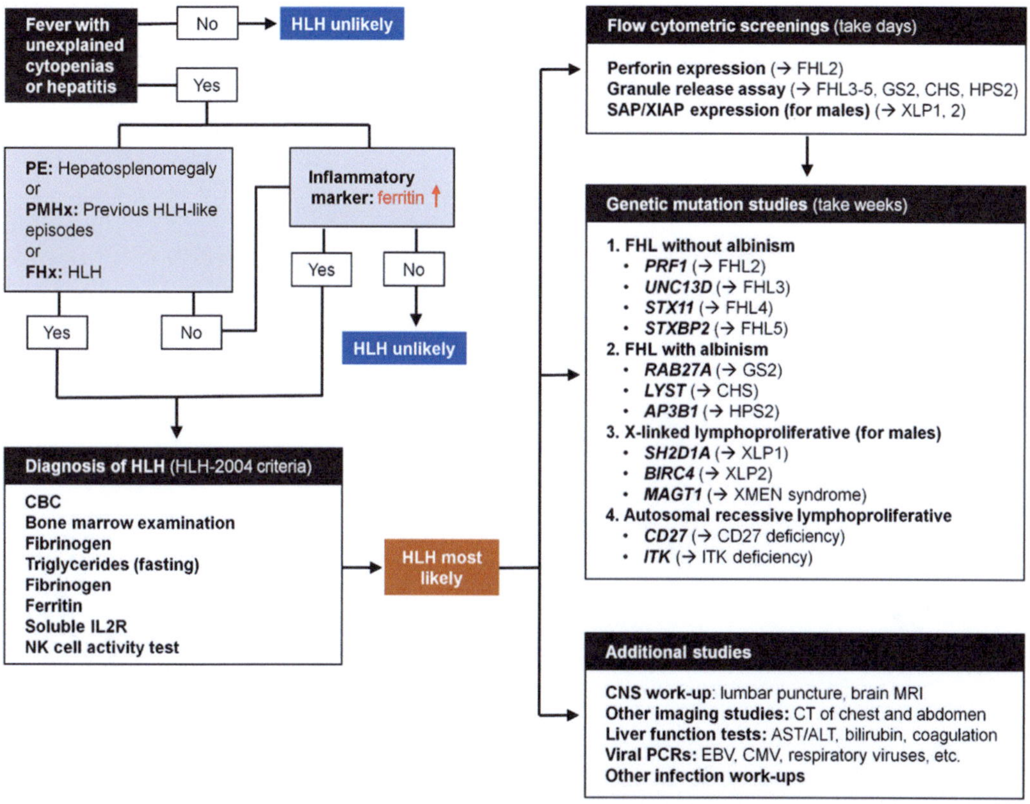

Fig. 27.3 The HLH diagnostic algorithm. It is important to recognize clinical features suggesting HLH (fever with unexplained cytopenias or hepatitis accompanied by hepatosplenomegaly) to make a timely diagnosis of HLH. Ferritin is a very useful screen for early diagnosis of HLH. Flow cytometric assays are helpful in the rapid diagnosis of HLH and can guide the priority of genetic testing. Genetic testing should be recommended for all patients meeting HLH diagnostic criteria, irrespective of age and clinical presentation

Although both primary and secondary HLH are characterized by uncontrolled hypercytokinemia, secondary HLH is not associated with any obvious genetic abnormality. The pathophysiology of secondary HLH is not fully defined. Historically, primary and secondary HLH were considered two distinct syndromes. However, recent understanding of HLH pathogenesis has blurred this distinction, suggesting that the clinical spectrum of HLH occurs across a continuum of genetic risk. Severe (null) mutations, mild hypomorphic (usually missense) mutations/polymorphisms, and even complex polygenic traits appear to determine an individual's risk for developing HLH in response to immune stimuli (Risma and Jordan 2012). Specifically, severe mutations underlie early-onset, primary HLH, and polygenic traits are more likely to be associated with secondary HLH.

27.3.1 Familial HLH Syndromes Without Hypopigmentation

The FHL syndromes without hypopigmentation are often referred to simply as familial HLH. They are inherited in an autosomal recessive manner and comprise five subtypes (FHL1, FHL2, FHL3, FHL4, and FHL5). Approximately 70% of FHL cases result from mutations in two genes, *PRF1* and *UNC13D*, which define the FHL2 and FHL3 subgroups, respectively.

FHL1: The mutation driving FHL1 occurs in chromosome 9 (9p21.3-22), although the exact protein and gene remain to be identified.

FHL2: The mutation driving FHL2 resides in the *PRF1* gene on chromosome 10 (10q21-22). It

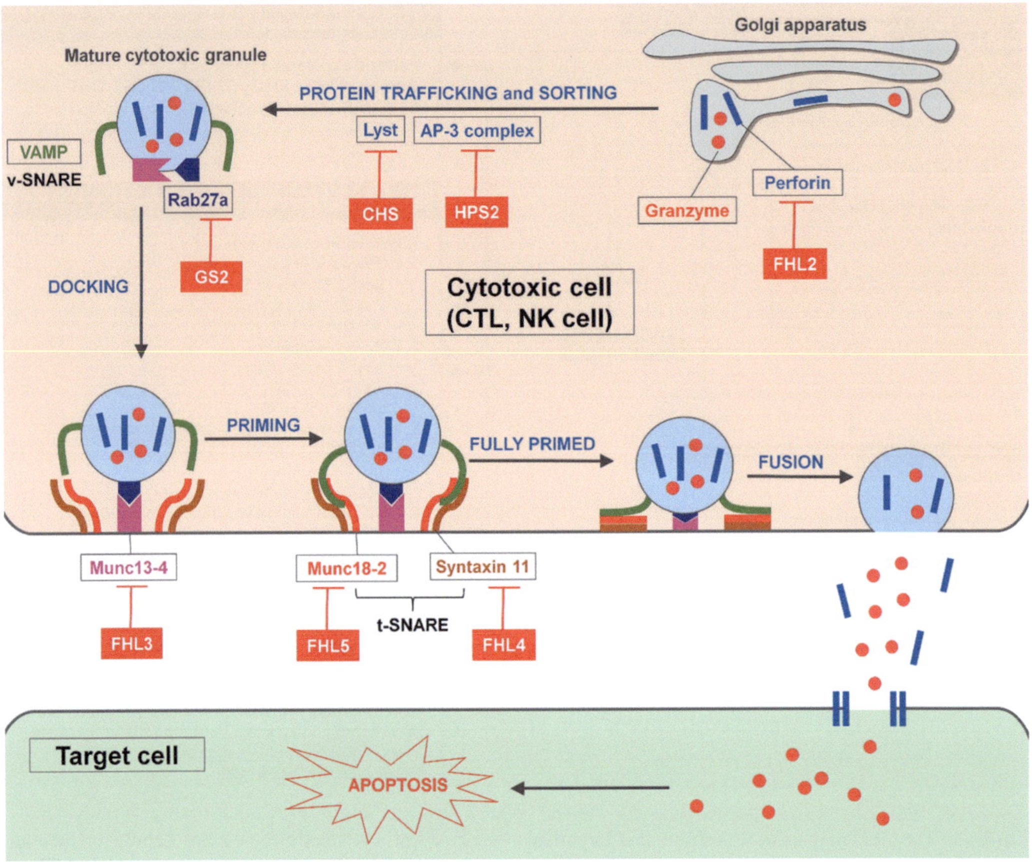

Fig. 27.4 Schematic overview of the genetic defects and pathway of granule-mediated cytotoxic events that cause primary HLH. Several intracellular steps are required for granule-mediated cytotoxicity in cytotoxic T cells and NK cells, including production of cytotoxic granules, intracellular transportation of granules, docking, priming, and fusion with the plasma membrane. Primary HLH is associated with mutations in the genes that encode the proteins involved in each step of the pathway. Abbreviations: *AP-3* adaptor protein 3, *CHS* Chédiak–Higashi syndrome, *CTL* cytotoxic T lymphocyte, *FHL* familial hemophagocytic lymphohistiocytosis, *GS2* Griscelli syndrome type 2, *HPS* Hermansky–Pudlak syndrome, *LYST* lysosomal trafficking regulator, *NK* natural killer, *SNARE* soluble N-ethylmaleimide-sensitive factor attachment protein receptor, *v-SNARE* vesicle membrane SNARE, *t-SNARE* target membrane SNARE, *VAMP* vesicle-associated membrane protein

accounts for 20–50% of FHL cases, depending on the study cohort (Sieni et al. 2014; Ishii 2016). The *PRF1* gene encodes the cytolytic, pore-forming protein perforin, which is synthesized and stored in cytotoxic lymphocytes and NK cells along with granzyme. Perforin monomers form a polymeric pore structure in target cell plasma membranes, through which granzymes enter and induce cell death. Mutations in *PRF2* result in reduced, absent, or functional impairment of perforin (Sieni et al. 2014; Zhang et al. 2014).

FHL3: FHL3 is caused by mutations in the *UNC13D* gene. It accounts for 30–40% of FHL cases but varies widely depending on the geographic area and ethnic group (Sieni et al. 2014; Koh et al. 2015). *UNC13D* encodes Munc13-4, which is involved in the priming of cytotoxic granules and their fusion with cell membranes. Priming is an additional biochemical event that enables granules to fuse with the plasma membrane. Therefore, Munc13-4 loss of function results in the impaired release of perforin and granzyme to the target cell (Sieni et al. 2014; Zhang et al. 2014).

FHL4: Approximately 10–20% of FHL cases include mutations in the *STX11* gene, which encodes syntaxin-11. This protein belongs to the soluble N-ethylmaleimide-sensitive factor attachment protein receptor (SNARE) family of vesicular proteins. The two types of SNARE proteins include vesicle membrane SNAREs (v-SNAREs) found on the surface of transport vesicles and target membrane SNAREs (t-SNAREs). SNARE proteins act like ropes that facilitate the transport of cytotoxic granules to their destination by binding tightly to the plasma membrane, thus assisting vesicle fusion with cell membranes. Syntaxin-11 is a t-SNARE, and mutations in *STX11* result in decreased or absent expression, leading to defects in vesicle endocytosis and exocytosis (Ishii 2016; Sieni et al. 2014; Zhang et al. 2014).

FHL5: FHL5 is caused by mutations in the *STXBP2* gene, which encodes syntaxin-binding protein 2 (also called Munc18-2). This protein binds to syntaxin-11 to form a t-SNARE complex, thus regulating vesicle membrane fusion. Mutations in *STXBP2* result in impaired cytotoxic function, similar to that observed in FHL4 (Ishii 2016; Sieni et al. 2014).

27.3.2 Familial HLH Syndromes with Hypopigmentation

Griscelli syndrome type 2 (GS2): GS2 is characterized by immune system abnormalities, in addition to hypopigmented skin and hair. Patients with GS2 carry mutations in the *RAB27A* gene. The Rab-27A protein product belongs to the Rab family, which is part of the Ras superfamily of small GTPases. Rab-27A interacts with Munc13-4 during the docking of cytotoxic granules to the plasma membrane. Mutations in *RAB27A* disable this function, resulting in severe degranulation defects in NK and cytotoxic T cells (Sieni et al. 2014; Zhang et al. 2014).

Chédiak–Higashi syndrome (CHS): CHS is characterized by oculocutaneous albinism, neuropathy, coagulopathy, and immune dysregulation. CHS is caused by mutations in the *LYST* gene, which encodes the lysosomal trafficking regulator LYST. *LYST* mutations cause defects in the production of cytotoxic granules, leading to the appearance of dysfunctional giant lysosomes (Sieni et al. 2014; Zhang et al. 2014). Therefore, exocytosis of cytotoxic granules is impaired in CHS, resulting in defective cytotoxic function.

Hermansky–Pudlak syndrome type 2 (HPS2): Hermansky–Pudlak syndrome includes a clinically homogeneous but genetically heterogeneous group of nine autosomal recessive genetic disorders. They share the clinical features of partial oculocutaneous albinism and bleeding disorders. Among these disorders, HPS type 2 is the only subtype that includes both the congenital neutropenia and impaired cytotoxic activity associated with HLH. HPS2 is caused by mutations in the *AP3B1* gene, which encodes the beta-3A subunit of the adaptor protein 3 (AP-3) complex. The AP-3 complex controls trafficking of proteins from the Golgi apparatus to lysosomes. Defects in the AP-3 complex result in the missorting of proteins, thereby impairing the function of secretory lysosomes (Sieni et al. 2014; Zhang et al. 2014).

27.3.3 X-Linked Lymphoproliferative Disease (XLP)

XLP is an X-linked inherited immune disorder that conveys a high risk of developing HLH, which is triggered by EBV infection in most cases. The two types of XLP, XLP1 and XLP2, are caused by different mutations.

XLP1: XLP1 results from loss-of-function mutations in the *SH2D1A* gene. The gene encodes the SAP protein, named for its association with SLAM proteins, which refer to signaling lymphocytic activation molecules. SAP is an adaptor protein that interacts predominantly with the NK-cell receptor 2B4, which is a member of SLAM family. SAP operates as a molecular switch that converts 2B4 from an inhibitory receptor into an activating one. Thus, in the absence of SAP, 2B4 delivers inhibitory instead of activating signals, resulting in impaired cytotoxicity (Zhang et al. 2014; Sieni et al. 2014).

XLP2: Patients with XLP2 carry mutations in the *BIRC4* gene, which encodes the XIAP protein. XIAP belongs to a family of apoptotic suppressor proteins. *BIRC4* mutations are purported to increase the susceptibility of lymphocytes to activation-induced cell death, thereby reducing the size of the active cell population. This appears inconsistent with the paradigm of HLH, which emphasizes the role of overactivated T cells. Nevertheless, patients with XIAP deficiency can exhibit defects in cytotoxicity similar to that of patients with XLP1 resulting from SAP deficiency (Sieni et al. 2014). The pathophysiologic mechanism that accounts for the occurrence of HLH in XLP2 is not fully understood.

27.4 Treatment

The initial goals of HLH treatment are to suppress the life-threatening inflammatory process that underlies HLH and to kill pathogen-infected APCs, thereby eliminating the stimulus for the ongoing but ineffective activation of cytotoxic cells. In addition, a curative HSCT should be provided to patients with primary or relapsed/refractory HLH to exchange the defective immune system with appropriately functioning cells. To achieve these goals, prompt initiation of immunosuppressive and proapoptotic chemotherapy is mandatory. Because initial treatment is similar in patients with both primary and secondary HLH and the disease is rapidly progressive, treatment should be promptly started for patients who are suspected to have HLH, rather than waiting for the results of genetic testing.

The HLH-94 and subsequent HLH-2004 protocols proposed by the Histiocyte Society are the most commonly used treatment protocols (Fig. 27.5) (Henter et al. 2007). The HLH-94 protocol recommended an 8-week induction therapy with dexamethasone, etoposide, and intrathecal methotrexate. After induction therapy, patients with resolved secondary HLH were weaned off

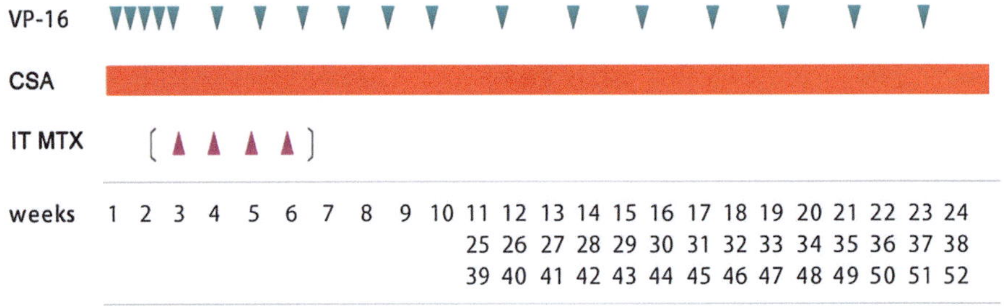

Fig. 27.5 Overview of therapeutic guidelines from the HLH-2004 treatment protocol recommended by the Histiocyte Society

therapy, and patients with primary HLH or persistent secondary HLH were directed to continue therapy as a bridge to allogeneic HSCT. In the HLH-2004 trial, the protocol was modified to achieve a better remission rate by using cyclosporine during the induction therapy phase and by adding hydrocortisone to intrathecal therapy. The HLH-94 trial resulted in an estimated 5-year survival of 54% ± 6%, which was a remarkable improvement for an otherwise fatal disease (Trottestam et al. 2011). The results of the HLH-2004 trial have not yet been published.

An alternative protocol combining anti-thymocyte globulin (ATG) as a frontline therapy with corticosteroids, cyclosporine, and intrathecal therapy showed promising results (Mahlaoui et al. 2007). On the basis of these results, a phase II clinical trial testing the efficacy of ATG in combination with dexamethasone and etoposide was recently opened in the United States (Filipovich and Chandrakasan 2015).

Allogeneic HSCT is generally recommended as a curative option in patients with genetically verified disease, a family history of HLH, or persistent or reactivated HLH (Seo 2015). In the HLH-94 and HLH-2004 protocols, busulfan-based myeloablative chemotherapy was suggested as a conditioning regimen. However, this conditioning regimen resulted in extensive transplant-related mortality (TRM). Recently, a reduced-intensity conditioning regimen using alemtuzumab (a monoclonal antibody targeting CD52 on mature lymphocytes), fludarabine, and melphalan has improved outcomes and reduced TRM (Marsh et al. 2010).

Some patients with refractory, progressive disease may require salvage or alternative therapeutic approaches. However, the evidence supporting effective salvage therapy is minimal (Marsh et al. 2016). ATG can be used as either a first-line or second-line therapy. Alemtuzumab is suggested to play a therapeutic role in the treatment of refractory HLH. Rituximab, which targets $CD20^+$ B cells, is reportedly efficacious in EBV-associated HLH and XLP. The use of anakinra (a recombinant human interleukin-1 receptor antagonist) is also reported in patients with rheumatologic/autoimmune disorder-associated HLH or macrophage activation syndrome. The IFN-γ blocking monoclonal antibody emapalumab is a promising novel agent that is expected to have a putative therapeutic role as frontline therapy or salvage therapy (Filipovich and Chandrakasan 2015).

In summary, HLH is a hyperinflammatory syndrome that can be caused either by known genetic abnormalities (primary HLH) or may be triggered by infection, cancer, or autoimmune disease in patients with no known genetic cause (secondary HLH). Because initial presentations may be nonspecific, high-suspicion and systematic clinical, immunologic, and genetic work-ups are required. Prompt initiation of proper treatment is mandatory for survival. Considerable progress has been made exploring the molecular/immunologic drivers of HLH pathophysiology and managing HLH treatment. Future directions include identifying improved biomarkers or diagnostic/prognostic markers, establishing better methods of genetic screening, and introducing novel targeted agents.

End-of-Chapter Questions

1. Describe the diagnostic criteria of HLH, and explain why each component occurs in patients with HLH.
2. Describe the perforin/granzyme pathway of immune effector cells and the genetic mutations that cause primary HLH.
3. What kind of triggers or conditions can be associated with secondary HLH?
4. Explain the goals of HLH treatment according to the pathophysiology of HLH.
5. What are the differences in the treatment strategies between primary and secondary HLH?

References

Al-Herz W, Bousfiha A, Casanova JL, Chatila T, Conley ME, Cunningham-Rundles C, Etzioni A, Franco JL, Gaspar HB, Holland SM, Klein C, Nonoyama S, Ochs HD, Oksenhendler E, Picard C, Puck JM, Sullivan K, Tang MLK (2014) Primary immunodeficiency diseases: an update on the classification from the international union of immunological societies expert committee for primary immunodeficiency. Front Immunol 5:162

Filipovich AH, Chandrakasan S (2015) Pathogenesis of hemophagocytic lymphohistiocytosis. Hematol Oncol Clin North Am 29:895–902

George MR (2014) Hemophagocytic lymphohistiocytosis: review of etiologies and management. J Blood Med 5:69–86

Henter JI, Horne A, Aricó M, Egeler RM, Filipovich AH, Imashuku S, Ladisch S, McClain K, Webb D, Winiarski J, Janka G (2007) HLH-2004: diagnostic and therapeutic guidelines for hemophagocytic lymphohistiocytosis. Pediatr Blood Cancer 48:124–131

Horne A, Trottestam H, Aricò M, Egeler RM, Filipovich AH, Gadner H, Imashuku S, Ladisch S, Webb D, Janka G, Henter JI (2008) Frequency and spectrum of central nervous system involvement in 193 children with haemophagocytic lymphohistiocytosis. Br J Haematol 140:327–335

Ishii E (2016) Hemophagocytic lymphohistiocytosis in children: pathogenesis and treatment. Front Pediatr 4:1–9

Jordan MB, Allen CE, Weitzman S, Filipovich AH, Mcclain KL (2011) How I treat hemophagocytic lymphohistiocytosis. Blood 11815(118):4041–4052

Koh KN, Im HJ, Chung NG, Cho B, Kang HJ, Shin HY, Lyu CJ, Yoo KH, Koo HH, Kim HJ, Baek HJ, Kook H, Yoon HS, Lim YT, Kim HS, Ryu KH, Seo JJ (2015) Clinical features, genetics, and outcome of pediatric patients with hemophagocytic lymphohistiocytosis in Korea: report of a nationwide survey from Korea Histiocytosis Working Party. Eur J Haematol 94:51–59

Lehmberg K, Mcclain KL, Janka GE, Allen CE (2014) Determination of an appropriate cut-off value for ferritin in the diagnosis of hemophagocytic lymphohistiocytosis. Pediatr Blood Cancer 61:2101–2103

Mahlaoui N, Ouachee-Chardin M, de Saint Basile G, Neven B, Picard C, Blanche S, Fischer A (2007) Immunotherapy of familial hemophagocytic lymphohistiocytosis with antithymocyte globulins: a single-center retrospective report of 38 patients. Pediatrics 120:e622–e628

Marsh RA, Vaughn G, Kim M, Li D, Jodele S, Joshi S, Mehta PA, Davies SM, Jordan MB, Bleesing JJ, Filipovich AH (2010) Reduced-intensity conditioning significantly improves survival of patients with hemophagocytic lymphohistiocytosis undergoing allogeneic hematopoietic cell transplantation. Blood 116:5824–5831

Marsh RA, Jordan MB, Talano J-A, Nichols KE, Kumar A, Naqvi A, Vaiselbuh SR (2016) Salvage therapy for refractory hemophagocytic lymphohistiocytosis: a review of the published experience. Pediatr Blood Cancer 64:1–7

Morimoto A, Nakazawa Y, Ishii E (2016) Hemophagocytic lymphohistiocytosis: pathogenesis, diagnosis, and management. Pediatr Int 58:817–825

Risma K, Jordan M (2012) Hemophagocytic lymphohistiocytosis: updates and evolving concepts. Curr Opin Pediatr 24:9–15

Rosado FGN, Kim AS (2013) Hemophagocytic lymphohistiocytosis. Am J Clin Pathol 139:713–727

Ryu J-M, Kim KM, Oh SH, Koh KN, Im HJ, Park C-J, Chi H-S, Seo JJ (2013) Differential clinical characteristics of acute liver failure caused by hemophagocytic lymphohistiocytosis in children. Pediatr Int Off J Jpn Pediatr Soc 55:748–752

Seo JJ (2015) Hematopoietic cell transplantation for hemophagocytic lymphohistiocytosis: recent advances and controversies. Blood Res 50:131–139

Sieni E, Cetica V, Hackmann Y, Coniglio ML, Da Ros M, Ciambotti B, Pende D, Griffiths G, Arico M (2014) Familial hemophagocytic lymphohistiocytosis: when rare diseases shed light on immune system functioning. Front Immunol 5:1–19

Trottestam H, Horne AC, Arico M, Egeler RM, Filipovich AH, Gadner H, Imashuku S, Ladisch S, Webb D, Janka G, Henter JI (2011) Chemoimmunotherapy for hemophagocytic lymphohistiocytosis: long-term results of the HLH-94 treatment protocol. Blood 118:4577–4584

Zhang L, Zhou J, Soko L (2014) Hereditary and acquired hemophagocytic lymphohistiocytosis. Cancer Control 21:301–312

Hepatitis C Virus Infection

28

Etsuko Iio and Yasuhito Tanaka

Keywords
HCV · DAA · IFN · HCC · SVR

28.1 Case Report

A 76-year-old man who had been diagnosed with HCV infection at the age of 55 was suspected of suffering from hepatocellular carcinoma on ultrasound (US) examination. He had received four interferon (IFN)-based treatments, but HCV had not been eradicated. He was then treated with direct-acting antiviral agents (DAAs) at the age of 74. A combination of daclatasvir and asunaprevir for 24 weeks achieved a sustained virological response (SVR). However, US examination 9 months after DAA treatment (3 months after SVR) revealed a hepatic tumor.

Physical findings were as follows: height 154 cm, weight 57 kg, blood pressure 124/71 mm Hg, and pulse 63/min. There was no anemia or jaundice in the conjunctiva. The abdomen was soft with no tenderness. There were no other symptoms. Subsequently, biochemical values were determined (Table 28.1). Hepatic function and tumor markers including alpha-fetoprotein (4.3 ng/ml, normal range up to 10.0 ng/ml) and prothrombin induced by vitamin K absence II (34 mAU/ml, normal range up to 40 mAU/ml) were normal. The indocyanine green retention rate at 15 min (ICG-R15) was 11.2% (normal, up to 15.0%) and the K value was 0.148 (normal, 0.168–0.206). Measurement of ICG kinetics is the most common and easy test for the perioperative assessment of liver function in case of surgery or liver transplantation. A space-occupying lesion (SOL) showing a mosaic-echo pattern with a peripheral low echo area (capsule) was found in segment (S) 5 by US in this patient. Gadolinium ethoxybenzyl diethylenetriaminepentaacetic acid (Gd-EOB-DTPA)-enhanced magnetic resonance imaging (MRI) showed a 31 × 23 mm tumor with poor EOB intake, contrast enhancement at early phase, and washout at delayed phase in liver S5 close to the gallbladder (Fig. 28.1a, b). Computed tomography (CT) during arteriography (CTA) showed a hypervascular nodule (Fig. 28.2a), and CT during arterial portography (CTAP) showed a hypovascular nodule (Fig. 28.2b).

28.2 Diagnosis

Based on these findings, hepatocellular carcinoma (HCC) was diagnosed after HCV eradication (defined as having achieved SVR) and demonstrated by partial resection of the right hepatic

E. Iio · Y. Tanaka (✉)
Department of Virology, Liver Unit, Nagoya City University Graduate School of Medical Sciences, Nagoya, Japan
e-mail: eiio@med.nagoya-cu.ac.jp; ytanaka@med.nagoya-cu.ac.jp

T. Oohashi et al. (eds.), *Human Pathobiochemistry*, https://doi.org/10.1007/978-981-13-2977-7_28

Table 28.1 Clinical characteristics of the patient prior to DAA treatment and at the time of HCC occurrence

Number	Reference value	Before DAA treatment	HCC occurrence
WBC (/μL)	3600–9600	3600	5800
Hemoglobin (g/dL)	13.2–17.2	14.1	15.6
Platelet count ($\times 10^4$/μL)	14.8–33.9	8.8	10.9
Albumin (g/dL)	4.0–5.0	3.7	4.2
AST (IU/L)	13–33	29	21
ALT (IU/L)	6–30	25	20
γ-GTP (IU/L)	10–47	20	13
Total bilirubin (mg/dL)	0.3–1.2	0.6	1.3
Creatinine (mg/dL)	0.6–1.0	0.89	1.03
Prothrombin time (%)	70.0–130.0	80.9	81.7
AFP (ng/ml)	<10.0	13.5	4.3
PIVKA-II (mAU/ml)	<40	29	34
HCV RNA (logIU/mL)	Negative	6.0	ND
HAV Ab-IgM	Negative	NT	ND
HBs Ag	Negative	NT	ND

DAA direct-acting antiviral, *WBC* white blood cell, *HCC* hepatocellular carcinoma, *AST* aspartate aminotransferase, *ALT* alanine aminotransaminase, *γ-GTP* γ-glutamyl transpeptidase, *AFP* α-fetoprotein, *PIVKA-II* protein induced by vitamin K absence II, *HAV Ab* hepatitis A virus antibody, *HBs Ag* hepatitis B surface antigen, *NT* not tested, *ND* not detected

lobe. The resected gross specimen of the tumor was well circumscribed with a soft whitish necrotic area. The surrounding areas were non-cirrhotic (Fig. 28.3). Histologically, moderately differentiated HCC with foci of necrotic tissue was found. The non-tumorous area showed chronic hepatitis with moderate bridging fibrosis, interface hepatitis, and portal tract inflammatory cell infiltration. The patient progressed satisfactorily after surgery and was discharged from hospital.

28.2.1 Differential Diagnosis

HCC can be diagnosed using medical imaging such as US, CT, and MRI. Serum markers (AFP, PIVKA-II) are also helpful if these levels are elevated. However, normal AFP levels are present in about 30% of patients at the time of diagnosis, especially with small HCC. The tumor biopsy can be performed if the diagnosis of HCC remains unclear by noninvasive procedure. Differential diagnostic criteria for HCC rather than other hepatocytic lesions include focal nodular hyperplasia, dysplastic nodules and hepatocellular adenoma, and other malignant tumors, such as intrahepatic cholangiocarcinoma, combined hepatocellular and cholangiocarcinoma, and malignant hepatic lymphoma. Thus, medical imaging and histopathology with small samples of biopsied tissue are valuable for diagnosis of HCC in clinical practice.

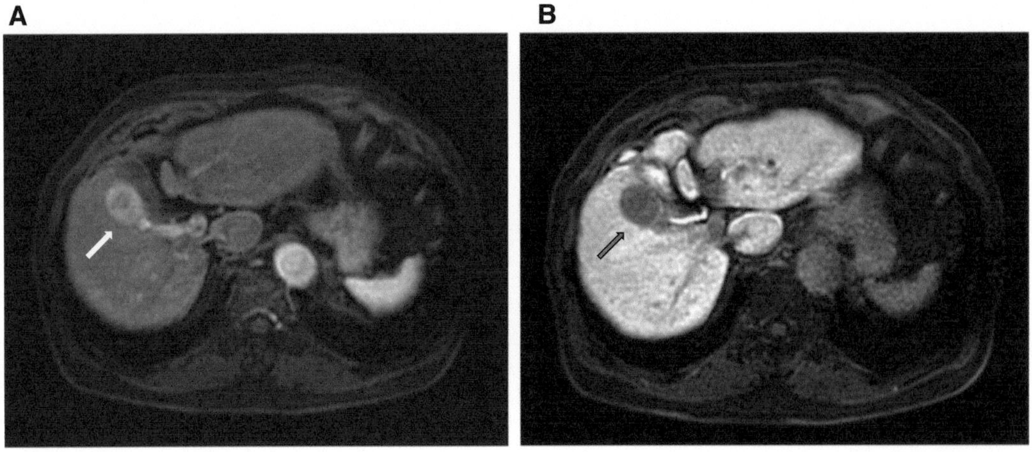

Fig. 28.1 Gadolinium ethoxybenzyl diethylenetriamine-pentaacetic acid (Gd-EOB-DTPA)-enhanced magnetic resonance imaging (MRI). A 25 mm high-intensity nodule (arrow) was detected at segment 5 in the arterial phase (**a**) and hypointensity in the hepatobiliary phase (**b**)

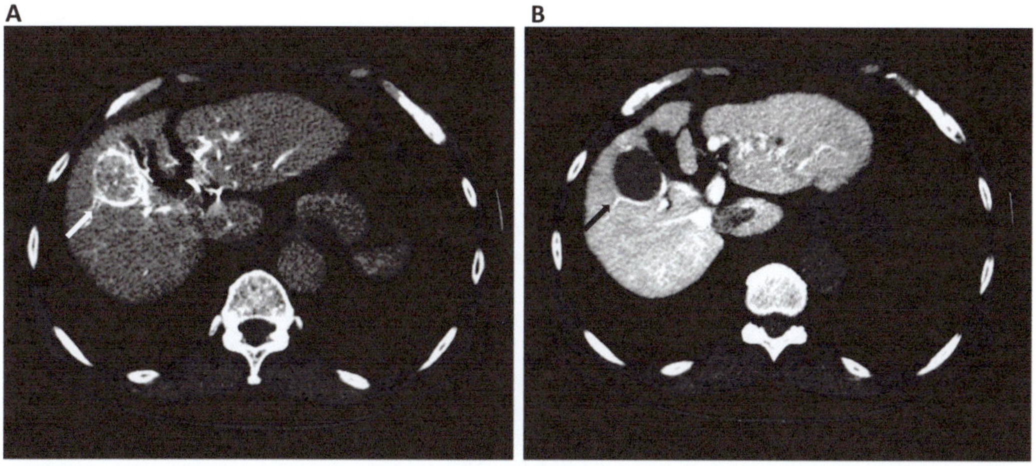

Fig. 28.2 CT during arteriography/CT during arterial portography (CTA/CTAP). CTA showing hypervascular nodule (white arrow) (**a**) and CTAP showing hypovascular nodule (black arrow) in contact with the gallbladder (**b**)

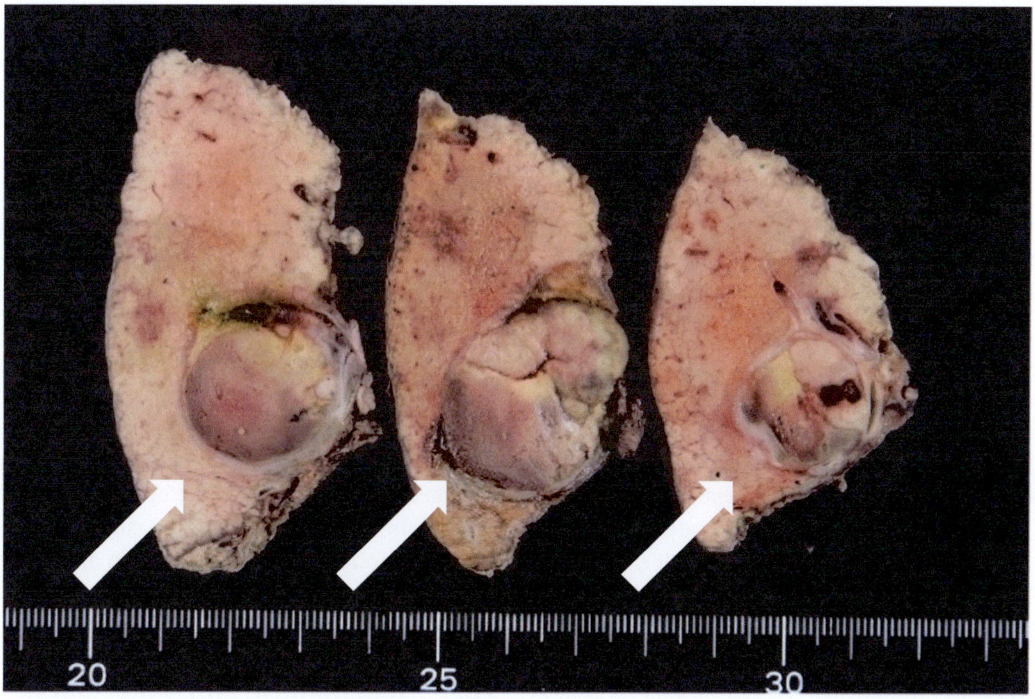

Fig. 28.3 Pathological specimen. The resected specimen appeared to be an encapsulated nodular tumor (arrow) with a maximum diameter of 25 mm. The tumor consisted of viable and necrotic areas

28.2.2 Molecular Genetics

In the present case, IFN-based treatment did not eradicate HCV, but subsequently SVR was achieved on treatment with DAAs approved in many countries. However, this patient developed HCC early after SVR. The AFP level at the time of HCC diagnosis was lower than at the start of treatment with DAAs (4.3 ng/ml versus 13.5 ng/ml). HCC was detected by abdominal US screening. The AFP level was normal at that time. The recent clinical development of DAAs

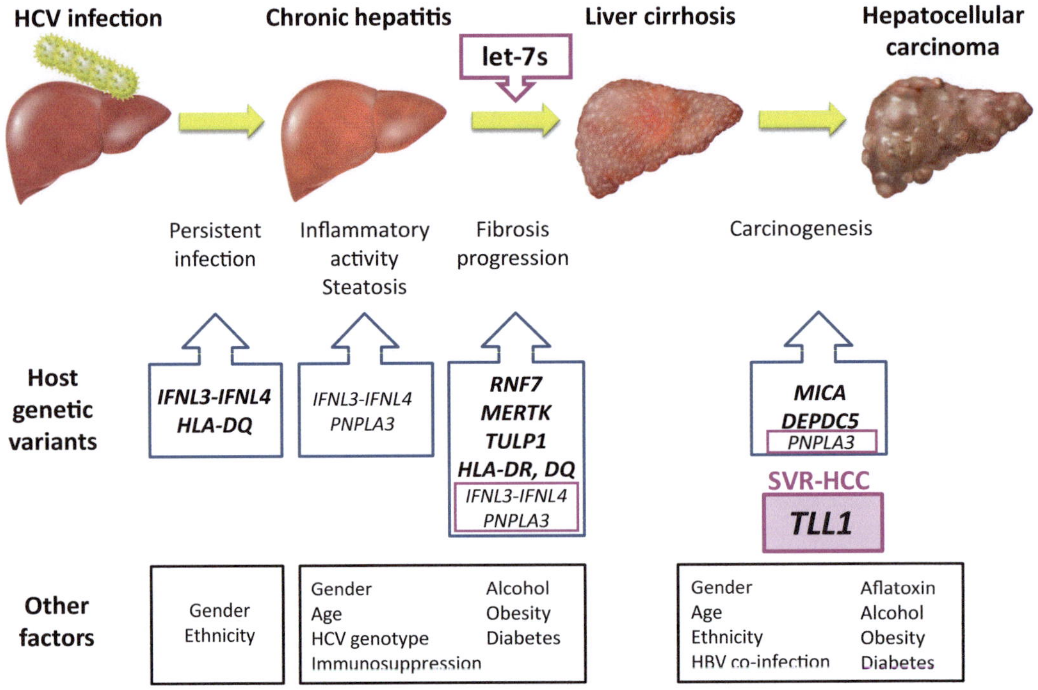

Fig. 28.4 Host genetic factors associated with clinical course of HCV infection

for treating HCV infection represents a major therapeutic advance. IFN-free regimens combining oral DAAs are available for HCV patients, and high rates of SVR are achieved. However, although the risk of HCC decreases after HCV eradication, the annual rate of HCC development remains at about 1.5–8.0% after SVR. Therefore, HCC occurrence after achieving SVR is still a critical issue even in the DAA era. Previous studies have identified several risk factors for developing HCC after achieving SVR, such as older age, male gender, advanced hepatic fibrosis, lower platelet count and albumin level, higher α-fetoprotein (AFP) level, alcohol intake, and the complication of diabetes. Recently, tolloid-like 1 (*TLL1*) gene polymorphisms were reported to be related to the development of HCC after SVR. The *TLL1* gene on chromosome 4 shows a strong association with the development of HCC in patients who achieved SVR on treatment for chronic HCV infection, at a genome-wide level of significance (odds ratio = 2.37, $P = 2.66 \times 10^{-8}$) (Fig. 28.4) (Matsuura et al. 2017). It would be important to identify which patients are at high risk of developing HCC within the large cohort of patients who achieve SVR on DAA therapy, so that special attention could be given to screening for cancer in these individuals. The goal of treatment for these patients thus includes not only eradication of the virus but also prevention of cancer development and improvement of prognosis by inhibiting the progression of hepatic fibrosis. Together, this suggests that long-term follow-up is required even after virus eradication.

28.3 Biochemical and Molecular Perspectives

28.3.1 Acute Hepatitis C

HCV infection can occur via contact with blood or other bodily fluids and can lead to acute hepatitis after a latency period of 2–14 weeks. HCV causes acute hepatitis with a milder impact on liver function than hepatitis A or B virus (HAV, HBV) and rarely results in fulminant hepatitis. HCV infection is associated with few subjective symptoms and is often asymptomatic but becomes chronic in around 70% of cases. Eradication of the virus after acute hepatitis is related to *IL28B* gene polymorphisms in that the incidence of chronic infection is reported to be higher in patients with an *IL28B* (rs8099917) TG/GG genotype than in those with a TT genotype (Thomas et al. 2009).

Acute hepatitis C can be associated with various extrahepatic comorbidities, including autoimmune thyroiditis, cryoglobulinemia, chronic glomerulonephritis, skin and mouth lichen planus, Sjogren's syndrome, malignant lymphoma, and others. These comorbidities influence the prognosis or QOL in patients with hepatitis C.

28.3.2 Chronic Hepatitis C

The estimated 170 million carriers worldwide persistently infected with HCV make hepatitis C a global health problem affecting a significant proportion of the world's population. Only 20–30% of HCV-infected individuals recover spontaneously, while 70–80% develop chronic infection. In approximately 30% of patients with chronic hepatitis, liver damage may continue to progress and result in liver cirrhosis 20–30 years after infection. Moreover, the incidence of HCC is known to be associated with the degree of liver fibrosis. The actual incidence of HCC in patients with chronic hepatitis C stratified according to different degrees of liver fibrosis is 0.5%/year in F0–1, 1.5%/year in F2, 5%/year in F3, and as much as 7–8%/year in F4. The frequency with which HCC develops from chronic hepatitis is approximately 10% over 10 years but becomes close to 50% in patients with liver cirrhosis. Additionally, as liver cirrhosis progresses, it leads to an irreversible decompensated stage of cirrhosis associated with the development of esophagogastric varices due to portal hypertension, as well as jaundice and hepatic encephalopathy. Factors that promote the progress of hepatic fibrosis include primary infection at older age, being male, alcohol use, diabetes, HBV coinfection, and HIV coinfection.

Virological Perspective HCV is an enveloped, positive-sense, single-stranded RNA virus. The HCV genome consists of >9500 nucleotides, flanked by 5′ and 3′ untranslated regions (UTRs) encoding a polyprotein precursor of about 3000 amino acids. The precursor is cleaved into at least ten different proteins comprising the structural proteins E1, E2, and p7, as well as the nonstructural proteins NS2, NS3, NS4A, NS4B, NS5A, and NS5B. Infection with HCV is closely associated with chronic hepatitis, liver cirrhosis, and hepatocellular carcinoma (Saito et al. 1990). HCV proteins recruit different host proteins for efficient viral propagation (Okamoto et al. 2006) and chronic infection. The viral serine protease NS3 participates in the processing of HCV nonstructural proteins through a non-covalent interaction with cofactor NS4A and plays a crucial role in HCV replication. In addition, the HCV NS3/4A protease plays an important role in inhibiting both the host innate immune response and apoptosis induced by HCV infection through cleavage of IPS-1 (also called MAVS, Cardif, or VISA), a mitochondria-resident adaptor molecule of retinoic acid-inducible gene I.

Based on phylogenetic analysis of genomic regions and the complete genome of HCV, six genotypes (HCV genotypes 1–6) have been described and further classified into numerous subtypes (e.g., HCV subtypes 1a, 1b) (Simmonds et al. 2005). There are many subtypes varying in geographical distribution and transmission patterns. Subtypes 1a, 1b, 2a, 2b, and 3a are distributed globally and account for the majority of HCV infections worldwide. Genotypes 1, 2, and 4 appear to be endemic in West and Central Africa and in the Middle East, while divergent endemic strains of genotypes 3 and 6 are found in Southeast Asia.

28.4 Therapy

Beginning in 2004, the standard of care (SOC) for chronic hepatitis C was 24 or 48 weeks of therapy with pegylated interferon-alfa (PEG-IFN) and ribavirin (RBV). Response to therapy is variable, and viral and host characteristics can influence whether patients achieve an SVR, defined by undetectable HCV RNA 6 months after completing the therapy. A proportion of patients who fail to achieve SVR has a transient virological response (TVR), defined by undetectable HCV RNA during treatment but with relapse and renewed presence of viral RNA after discontinuation of treatment. Finally, some patients fail to achieve an SVR at all (non-virological responders, NVR, defined by detectable viremia during and after treatment).

In 2009, four independent groups simultaneously published the results of genome-wide association studies (GWAS) to assess the role of genetic variation in response to PEG-IFN/RBV for CHC patients (Ge et al. 2009; Tanaka et al. 2009). All four studies revealed a strong association between genetic polymorphisms in or near the *IL28B* (*IFNL3-IFNL4)* locus and treatment-induced HCV clearance. Patients homozygous for the favorable allele had a two- to threefold higher rate of achieving SVR. As a result of this discovery, prediction of treatment outcomes, especially non-responsiveness to PEG-IFN/RBV, has been greatly improved by genotyping for *IFNL3-IFNL4* SNPs, enabling personalized medicine to be developed for CHC.

DAA therapies directly inhibit specific steps in the HCV viral life cycle, with targets including the NS3/4A protease, NS5B polymerase, and NS5A phosphoprotein that are essential for viral replication. To date, the first-generation protease inhibitors, telaprevir and boceprevir, have been approved, and several clinical trials of new DAAs have been started since 2013. It appears that although the *IFNL3-IFNL4* genotype is associated with treatment efficacy on PEG-IFN/RBV plus telaprevir therapy (Akuta et al. 2010) or IFN-free combination DAA regimens, its impact may be less important in other settings. Thus, the recently-approved IFN-free regimen of sofosbuvir (SOF) (NS5B polymerase inhibitor) with daclatasvir (NS5A inhibitor) achieved extremely high SVR rates, regardless of previous treatment exposures, prior responses to treatment, or *IFNL3-IFNL4* genotype. Hence, the effect of *IFNL3-IFNL4* genotype on treatment efficacy is low in more highly active regimens combining two or three DAAs; however, pretreatment *IFNL3-IFNL4* genotyping might still be useful for regimen selection or when deciding on shorter treatment duration. Studying resistance-associated variants (RAVs) in the NS3/NS5A region for their influence on treatment outcome with daclatasvir (DCV) and asunaprevir (ASV) therapy (first approved in Japan for chronic hepatitis C including liver cirrhosis in 2014) indicated that a history of simeprevir therapy and preceding daclatasvir/asunaprevir therapy may have resulted in the emergence of multiple RAVs. At present, new IFN-free combination treatment regimens such as SOF-based regimens or glecaprevir/pibrentasvir are available as valuable options for each HCV genotype (Recommendations by European Association for the Study of the Liver on Treatment of Hepatitis C 2018 as shown in Table 28.2). Japanese guidelines for HCV therapy were also updated in May 2018 (http://www.jsh.or.jp/files/uploads/HCV_GL_ver6.1_May30.pdf).

Table 28.2 Recommendations by the European Association for the Study of the Liver for Treatment of Hepatitis C 2018

Patients	Prior treatment experience	SOF/VEL	GLE/PIB	SOF/VEL/VOX	SOF/LDV	GZR/EBR	OBV/PTV/r + DSV
GT 1a	Treatment-naïve	12 weeks	8 weeks (CH) 12 weeks (LC)	No	8–12 weeks (CH) 12 weeks (LC)	12 weeks (HCV RNA ≤800,000 IU/ml)	No
	Treatment-experienced	12 weeks	8 weeks (CH) 12 weeks (LC)	No	No	12 weeks (HCV RNA ≤800,000 IU/ml)	No
GT 1b	Treatment-naïve	12 weeks	8 weeks (CH) 12 weeks (LC)	No	8–12 weeks (CH) 12 weeks (LC)	8 weeks (F0–F2) 12 weeks (F3–F4)	8 weeks (F0–F2) 12 weeks (F3–F4)
	Treatment-experienced	12 weeks	8 weeks (CH) 12 weeks (LC)	No	12 weeks	12 weeks	12 weeks
GT 2	Treatment-naïve	12 weeks	8 weeks (CH) 12 weeks (LC)	No	No	No	No
	Treatment-experienced	12 weeks	8 weeks (CH) 12 weeks (LC)	No	No	No	No
GT 3	Treatment-naïve	12 weeks (CH) No (LC)	8 weeks (CH) 12 weeks (LC)	12 weeks (LC)	No	No	No
	Treatment-experienced	12 weeks (CH) No (LC)	12 weeks (CH) 16 weeks (LC)	12 weeks (LC)	No	No	No
GT 4	Treatment-naïve	12 weeks	8 weeks (CH) 12 weeks (LC)	No	12 weeks	12 weeks (HCV RNA ≤800,000 IU/ml)	No
	Treatment-experienced	12 weeks	8 weeks (CH) 12 weeks (LC)	No	No	No	No
GT 5	Treatment-naïve	12 weeks	8 weeks (CH) 12 weeks (LC)	No	12 weeks	No	No
	Treatment-experienced	12 weeks	8 weeks (CH) 12 weeks (LC)	No	No	No	No
GT 6	Treatment-naïve	12 weeks	8 weeks (CH) 12 weeks (LC)	No	12 weeks	No	No
	Treatment-experienced	12 weeks	8 weeks (CH) 12 weeks (LC)	No	No	No	No

Treatment recommendations for HCV-infected patients with hepatitis C, including treatment-naïve patients (defined as patients who have never been treated for their HCV infection) and treatment-experienced patients (defined as patients who were previously treated with pegylated IFN-alpha and ribavirin; pegylated IFN-alpha, ribavirin, and sofosbuvir; or sofosbuvir and ribavirin)

DAA, direct-acting antiviral, *DSV* dasabuvir, *EBR* elbasvir, *GLE* glecaprevir, *GT* genotype, *GZR* grazoprevir, *HCV* hepatitis C virus, *HIV* human immunodeficiency virus, *LDV* ledipasvir, *OBV* ombitasvir, *PIB* pibrentasvir, *PTV* paritaprevir, *r* ritonavir, *SOF* sofosbuvir, *VEL* velpatasvir, *VOX* voxilaprevir, *CH* chronic hepatitis, *LC* compensated (Child-Pugh A) cirrhosis

References

Akuta N et al (2010) Amino acid substitution in hepatitis C virus core region and genetic variation near the interleukin 28B gene predict viral response to telaprevir with peginterferon and ribavirin. Hepatology 52:421–429

Ge D et al (2009) Genetic variation in IL28B predicts hepatitis C treatment-induced viral clearance. Nature 461:399–401

Matsuura K et al (2017) Genome-wide association study identifies TLL1 variant associated with development of hepatocellular carcinoma after eradication of hepatitis C virus infection. Gastroenterology 152(6):1383–1394

Okamoto T et al (2006) Hepatitis C virus RNA replication is regulated by FKBP8 and Hsp90. EMBO J 25:5015–5025

Saito I et al (1990) Hepatitis C virus infection is associated with the development of hepatocellular carcinoma. Proc Natl Acad Sci U S A 87:6547–6549

Simmonds P et al (2005) Consensus proposals for a unified system of nomenclature of hepatitis C virus genotypes. Hepatology 42:962–973

Tanaka Y et al (2009) Genome-wide association of IL28B with response to pegylated interferon-alpha and ribavirin therapy for chronic hepatitis C. Nat Genet 41:1105–1109

Thomas DL et al (2009) Genetic variation in IL28B and spontaneous clearance of hepatitis C virus. Nature 461:798–801

Substance Abuse Emergencies 29

Hiromichi Naito, Tetsuya Yumoto, and Clifton W Callaway

Keywords
Illegal · Addiction · Drug · Intoxication · Overdose

29.1 Case Report

A 35-year-old female was found unresponsive by her family on her bed surrounded by empty packets of pills. Her family called an ambulance; upon arrival of the emergency medical service personnel, she remained unresponsive but was breathing regularly. Her eyes were miotic and she had mild oral secretions. Gradually, her breathing became stertorous, and a nasal tube was introduced; her respiration was assisted by bag-valve-mask (BVM) ventilation.

H. Naito (✉)
Department of Emergency Healthcare and Disaster Medicine, Okayama University Graduate School of Medicine Dentistry and Pharmaceutical Sciences, Okayama, Japan
e-mail: naito.hiromichi@okayama-u.ac.jp

T. Yumoto
Department of Emergency and Critical Care Medicine, Okayama University Graduate School of Medicine Dentistry and Pharmaceutical Sciences, Okayama, Japan

C. W. Callaway
Department of Emergency Medicine, University of Pittsburgh School of Medicine, Pittsburgh, PA, USA

On arrival at the emergency department, an intravenous fluid line was established. Her blood pressure was 80/40 mmHg, heart rate 45 beat/min, and respiratory rate 6 breath/min. The oxygen saturation of her peripheral artery (SpO_2) remained 90%, although she was assisted by BVM with 100% oxygen. Crackling rales were recognized on auscultation. Her Glasgow Coma Scale (GCS) was 3: E1, V1, M1 with consistent stertorous breathing. Based on her clinical history, some identifiable labels on the pill packets, and a urine drug screening test, she was diagnosed with a mixed drug overdose including opioid and other unidentified drugs. Intravenous naloxone 0.2 mg was administered repeatedly, but the patient's condition did not improve. This lack of response was probably due both to a high opioid dosage and other co-ingested substances.

The emergency physician decided to intubate the patient. She was intubated with a 7.0 mm endotracheal tube. Mechanical ventilation was initiated. Blood gas analysis still showed respiratory acidosis (pH 7.30, $PaCO_2$ 50 mmHg), although oxygenation improved (PaO_2 80 mmHg on fraction of inspired oxygen 50%). A chest X-ray revealed right upper lobe infiltrate due to aspiration. There was no ST elevation , QRS, or QT prolongation on 12-lead electrocardiogram (ECG). Complete blood count and biochemistry results were almost normal. A computerized tomography (CT) scan of the brain was performed to exclude intracranial lesions: the results

T. Oohashi et al. (eds.), *Human Pathobiochemistry*, https://doi.org/10.1007/978-981-13-2977-7_29

were normal. Her urine drug screen test indicated that opioids were present.

The patient was monitored in intensive care unit. Supportive care, including ventilation and intravenous fluid, was continued. The patient awoke after 2 days without sequela and was extubated. Her vital signs were now stable, and she was asymptomatic on observation. She was given the opportunity to initiate a behavior modification program and rehabilitation.

29.2 Diagnosis

29.2.1 Identification/Recognition of Overdose from Abused Substances

Diagnosis is usually made based on the history of abuse, clinical course, and exclusion of other causes. Signs and symptoms of poisoning typically develop within a brief time after exposure, progress within minutes to an hour, and then gradually resolve over hours to days. The scene in which the patient was found (discarded packages, spread pills, or injection scarring) can be helpful for diagnosis. The time and route of intake are useful for estimating the absorption speed and amount of the substance taken, as well as for recognizing related complications. Valuable information (time when last seen, statements about the patient's plans) may be obtained from family members, acquaintances, and emergency medical service providers. Identification or estimation of the toxicity-inducing agent is helpful but often difficult to obtain. Regional or national poison centers, systematic databases and textbooks, or internet search engines can be used to identify the contents of unknown pills and other products.

Physical examination, including mental status (confusion, hallucinations, hyperactivity, central nervous system (CNS) depression, seizure, miosis, mydriasis) and respiratory and circulatory signs (bradypnea, tachypnea, dyspnea, hyperthermia, hypothermia, hypertension, hypotension, tachycardia, bradycardia, arrhythmia), provides the most actionable information for treating toxic patients. These specific symptoms associated with exposure give a hint to the agent; that is called "toxidrome." It may sometimes be difficult to recognize the symptoms of "substance abuse" at the first clinical examination. Substance abuse should always be considered in patients with an unclear cause of changes in mental status or metabolic abnormalities. In contrast, other causes of altered mental status, such as hypoglycemia, CNS pathology (infection, bleeding, tumor), or psychiatric disease, should always be considered for differential diagnosis.

29.2.2 Clinical Examination

Toxic severity depends on both the dose of exposure and the type of abused substance. Time from the exposure should also be considered. Medical staff should always consider a worst-case scenario with maximal toxicity. The severity of poisoning is primarily determined by physical examination (Table 29.1). Patients with moderate-to-severe poisoning should undergo complete blood tests, blood gas analysis, ECG, and chest X-ray.

29.2.3 Clinical Tests

For patients presenting with moderate-to-severe symptoms, the essential laboratory evaluation should include blood glucose level and electrolyte content. Because alcohol use is often coincident with other substance abuse and because alcohol causes similar CNS depression, a blood ethanol level is useful. Two substances with severe delayed toxicity but no clinical signs are acetaminophen and aspirin: testing for levels of these drugs in a mixed overdose can allow early treatment before toxicity develops. In addition, ongoing clinical care may be guided by a complete blood count, routine serum screening (aspartate aminotransferase, AST; alanine aminotransferase, ALT; blood urea nitrogen, BUN; amylase; creatine phosphokinase, CPK), coagulation studies, and blood gas sampling, anion gap, osmolar gaps, serum ketone level, and serum

Table 29.1 Physiologic grading of the severity of poisoning

Severity	Signs and symptoms	
	Stimulant poisoning	Depressant poisoning
Grade 1	Agitation, anxiety, diaphoresis, hyperreflexia, mydriasis, tremors	Ataxia, confusion, lethargy, weakness, verbal, able to follow commands
Grade 2	Confusion, fever, hyperactivity, hypertension, tachycardia, tachypnea	Mild coma (nonverbal but responsive to pain); brainstem and deep tendon reflexes intact
Grade 3	Delirium, hallucinations, hyperpyrexia, tachyarrhythmias	Moderate coma (respiratory depression, unresponsive to pain); some but not all reflexes absent
Grade 4	Coma, cardiovascular collapse, seizures	Deep coma (apnea, cardiovascular depression); all reflexes absent

Source: Irwin and Rippe's Intensive Care Medicine 7th ed

lactate level. Consider cardiac enzymes (troponin) if chest pain or cardiovascular signs are part of the presentation.

ECG (12-lead) should always be obtained, because lethal ventricular tachycardia (VT) or ventricular fibrillation (VF) may result from stimulant abuse or from drugs that cause sodium blockade. Atrioventricular block (AV block), QRS prolongation, and bradycardia should be noted as potential precursors to lethal arrhythmia. Urine immunoassay screening may be helpful for the diagnosis but may sometimes result in false-positive or false-negative results; moreover, they can detect only a small fraction of chemicals. In general, clinical examination usually creates suspicion for all of the drugs that are identified on urine immunoassay, making urine screen only a confirmatory test. Quantitative serum levels of agents are not always available.

29.2.4 Imaging

A chest radiograph is indicated in patients with hypoxemia, hypotension, abnormal cardiac rhythm, or CNS depression. Aspiration pneumonitis is a common complication due to disturbance in the CNS (Fig. 29.1). The chest X-rays may show infiltrates following inhalation of gases, fumes, or vapors. As central nervous system hemorrhage can be a complication of substance abuse, can mimic an overdose, or may result from falling while intoxicated, a CT scan of the brain should be obtained if CNS depression, seizures, or focal signs and symptoms are present.

Radiopaque substances, such as packets of drugs and enteric coated tablets, are sometimes identified by X-rays or CTs (Fig. 29.1), and this information may be helpful when making treatment decisions.

29.3 Clinical and Chemical Manifestations

In this article, we discuss four main categories of abused substances—**(1)** opioids, **(2)** sedatives, **(3)** stimulants, and **(4)** hallucinogens—based on similarities in their effects. There are, however, many differences, even within the same category. The characteristics of the major substances are shown in Table 29.2.

29.3.1 Opioids

Opioids are a group of substances that act on opioid receptors to produce morphine-like analgesic effects. They bind to and activate opioid receptors in the central nervous system (Fig. 29.2). Opioids may be administered through intravenous, intramuscular, subcutaneous, inhalational, oral, and mucosal routes. Plasma concentrations increase within seconds after intravenous use and over a few hours after oral abuse, although decreased gastrointestinal motility may prolong the absorption, especially with orally administered sustained-release formulations. Hepatic enzymes metabolize opioids, and the metabolites are eliminated through renal excretion. The most

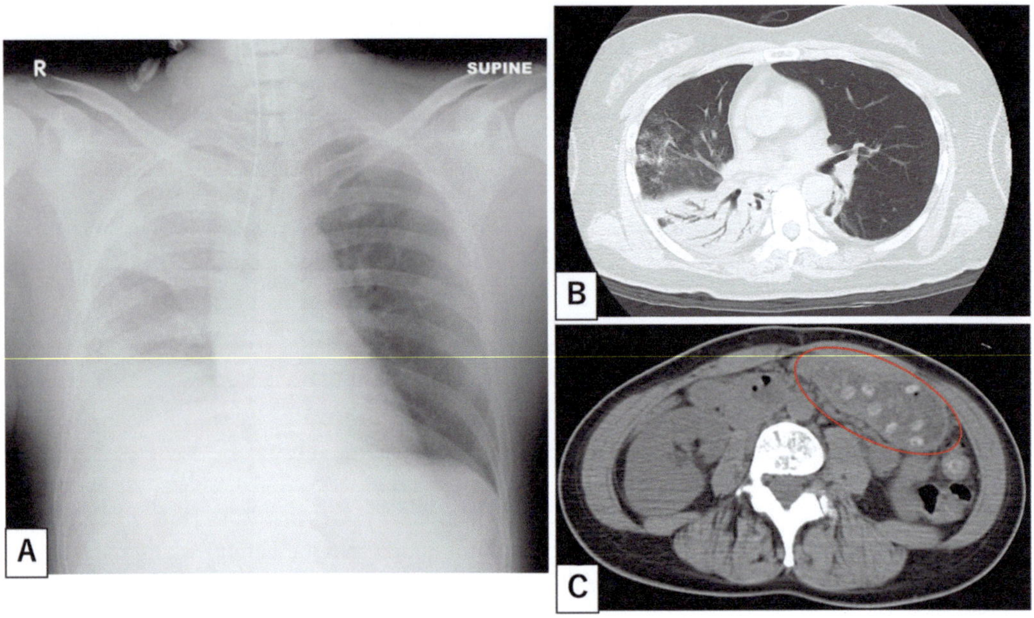

Fig. 29.1 Thirty-five-year-old female with aspiration pneumonitis of the right lung. (**a**, **b**) Right upper lobe infiltrates due to aspiration (**a**, chest X-ray; **b**, CT). (**c**) Radiopaque tablets identified in the stomach (CT)

critical adverse effect of opioids is respiratory depression, due to reduced carbon dioxide sensitivity of medullary chemoreceptors. Opioids remain the most common cause of drug-related emergency department visits in the United States (Watson et al. 2003). Combination with other agents (benzodiazepines or other types of opioids, etc.) or consumption with alcohol may exacerbate respiratory depression.

29.3.1.1 Morphine

Morphine is a major pain-relief agent used for acute and chronic pain; it has a high potential for addiction and abuse. Morphine mainly acts on μ-opioid and δ-opioid receptors, which are distributed in the brain and spinal cord. Morphine is administered through oral, intramuscular, subcutaneous, intravenously, intramedullary, or rectal routes. Respiratory depression or death may occur with a large overdose in the absence of medical treatment. Other major symptoms include miosis, euphoria, dysphoria, nausea, vomiting, constipation, myoclonus, hypotension, pruritus, and pulmonary edema. Withdrawal symptoms may occur if the dose is reduced in patients with chronic abuse.

29.3.1.2 Oxycodone

Oxycodone is a semisynthetic opioid prescribed for moderate-to-severe pain. Oxycodone has euphoric effects similar to other opioids, which makes it a very common agent of abuse, especially in the United States. Oxycodone is commonly consumed via the oral route, since sustained-release oxycodone is widely available. Crushing the tablet, however, makes it possible to be absorbed through insufflation or intravenously. Respiratory depression or death may occur with a large oxycodone overdose.

29.3.1.3 Diamorphine (Heroin)

Diamorphine, known as heroin, is an opiate commonly abused for its euphoric effects. Heroin has higher lipophilicity than morphine, which makes it easier to cross the blood–brain barrier. The potency of heroin is three times that of morphine. Heroin is usually available as a white or brown powder. Heroin can be absorbed through the intravenous route, insufflation, smoking (vaporizing it to inhale the fumes), the oral route, and anal/vaginal absorption. The purity of illegal agents ranges from 5% to 90%. Adulterants may cause additional toxicity.

Table 29.2 Characteristics of major substances

Abuse agent	Street names	Major symptoms associated with abuse
Opioids		
Morphine	M, Emma, white stuff	Miosis, euphoria, nausea, vomiting, constipation, myoclonus, hypotension, pruritis, respiratory depression
Oxycodone	Perks, Roxy	Miosis, euphoria, nausea, vomiting, constipation, myoclonus, hypotension, pruritis, respiratory depression
Diamorphine (heroin)	H, junk, smack	Miosis, euphoria, nausea, vomiting, constipation, myoclonus, hypotension, pruritis, respiratory depression
Fentanyl	China white, murder 8	Miosis, euphoria, nausea, vomiting, constipation, myoclonus, hypotension, pruritis, respiratory depression
Pentazocine	Ts, poor Man's heroin	Euphoria, dysphoria, nausea, vomiting, myoclonus, respiratory depression
Buprenorphine	Bupe, oranges	Euphoria, dysphoria, nausea, vomiting, myoclonus, respiratory depression
Meperidine	Demmies, pain killer	Euphoria, dysphoria, nausea, vomiting, myoclonus, respiratory depression, seizure, hyperthermia, serotonin syndrome
Tramadol	Ultras, chill pills	Euphoria, dysphoria, nausea, vomiting, myoclonus, respiratory depression, seizure, serotonin syndrome
Benzodiazepines	Downers, sleeping pills	Coma, ataxia, respiratory depression
Barbiturates	Tooies, barbs	Coma, ataxia, respiratory depression
Amphetamines	Adam, ecstasy, speed, ice	Mydriasis, anxiety, agitation, diaphoresis, tachycardia, arrhythmia, hypertension, hyperthermia, seizure
Cocaine	Base, C, candy	Mydriasis, anxiety, agitation, diaphoresis, tachycardia, arrhythmia, hypertension, hyperthermia, seizure, rhabdomyolysis
Lysergic acid diethylamide	A, paper acid, cubes	Mydriasis, distorted perception, anxiety, agitation, insomnia, tremor, tachycardia, hypertension, hyperthermia
Phencyclidine	Angel dust, magic sheets	Nystagmus, distorted perception, anxiety, agitation, ataxia, tachycardia, hyperthermia
Ketamine	K, cat valium	Nystagmus, coma, distorted perception, agitation, ataxia, nausea, vomiting, tachycardia, hypertension
Mescaline	Beans, Cactus	Mydriasis, distorted perception, euphoria, anxiety, tachycardia, diarrhea, vomiting
N, N-dimethyltryptamine	Businessman's LSD	Mydriasis, nystagmus, distorted perception, ataxia, nausea, vomiting, tachycardia, hypertension
Pyilocybin	Boomers, magic mushroom	Mydriasis, distorted perception, tremor, tachycardia, hypertension, hyperthermia
Marijuana	Mary Jane, hash, weed	Distorted perception, euphoria, tachycardia

Respiratory depression may cause death; co-ingestion with alcohol, benzodiazepines, or other types of opioids will increase the risk of death. Pulmonary edema is also common with heroin overdose.

29.3.1.4 Fentanyl

Fentanyl is a synthesized opioid commonly used for general anesthesia during surgery and pain management. Fentanyl has a high affinity to μ-receptors and has 200 times the potency of morphine. Fentanyl is administered through an intravenous route during anesthesia. Percutaneous absorption patches are used for pain management. Fentanyl and its derivatives are used as substitutes for heroin and other agents; this makes potent of the agents stronger and more dangerous. Fentanyl patches are sometimes intentionally ingested or smoked. In recent years, many chemical derivatives of fentanyl have been sold illegally. The major adverse event is respiratory depression.

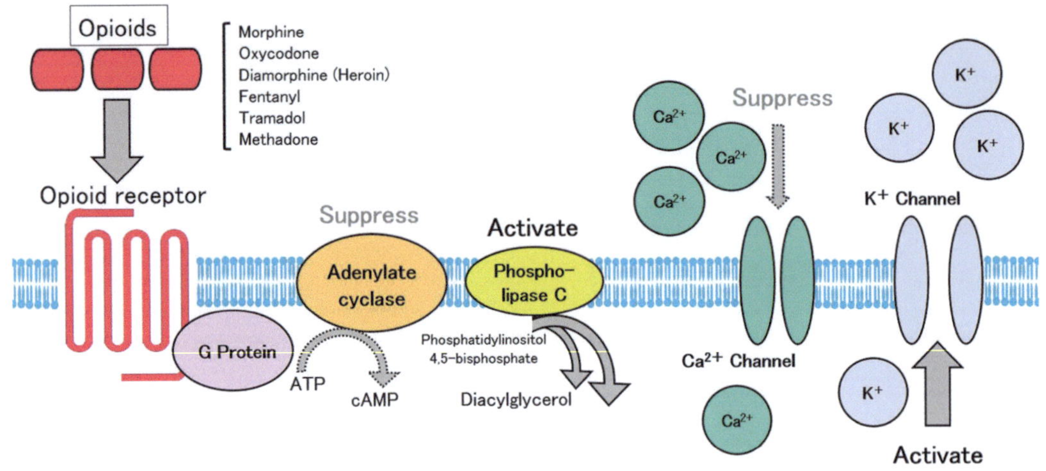

Fig. 29.2 Intracellular signal transduction pathway of the opioid receptor in the brain is shown in the figure. All mu(μ), delta(δ), and kappa(κ) receptors are inhibitory G protein-coupled receptor. After being activated by exogenous agonists, such as morphine, similar cell signaling is expected in all receptors. Ca^{2+} channel and adenylate cyclase are suppressed; K^+ channel and phospholipase C are activated by the opioid receptor

29.3.1.5 Tramadol

Tramadol is medically prescribed for pain relief. It is usually combined with acetaminophen to improve its effect. Tramadol binds to the μ-receptor. It also inhibits the reuptake of serotonin/norepinephrine and blocks the descending neural pathways. Respiratory depression and other side effects are milder than those observed with morphine. Unique side effects of tramadol include seizures and serotonin syndrome.

29.3.1.6 Methadone

Methadone is a long-acting opioid that is often prescribed for medically assisted treatment of opioid dependence and rarely for long-term pain relief. Methadone binds to the μ-receptor. It therefore causes respiratory depression and other signs similar to other opioids. Unique features of methadone include its long half-life (30–50 h) and its ability to prolong the QT interval on ECG. Methadone is usually taken orally. Because of the long half-life, overdose with methadone may require prolonged support of respiration or continuous intravenous infusions of naloxone.

29.3.2 Sedatives

Sedatives are a diverse group of drugs prescribed for the treatment of conditions such as insomnia, anxiety, agitation, muscle rigidity, seizures, and withdrawal syndromes and during anesthesia. They have the common feature of causing general central CNS depression. Sedatives are used illicitly and remain commonly abused (Byatt and Volans 1984). The most ubiquitous sedative is ethanol. Most of the toxic effects of other sedatives can be conceptualized as analogous to ethanol intoxication.

29.3.2.1 Benzodiazepines (BZDs)

BZDs work by binding to the BZD receptors located at the α-subunit of the gamma-aminobutyric acid (GABA) receptor complex in the CNS and peripheral tissues. The effect of BZDs on the $GABA_A$ receptor complex is to increase the frequency of Cl^- ion influx through the receptor channel, finally causing CNS depression. BZDs can be absorbed orally or taken intravenously. Most BZDs are metabolized in the liver with metabolites excreted by the kidney. Lorazepam is primarily excreted by the kidney.

BZDs vary widely in their distribution into fat and their rates of elimination. Acute intoxication causes CNS depression; severity commonly depends on the range of coma. Respiratory depression is generally mild. However, BZDs impair protective airway reflexes, which can lead to aspiration, stertorous breathing, and airway occlusion. Other effects include hypotension and circulatory depression. Aspiration pneumonitis is the most common complication in patients with coma; other major complications in coma patients include hypothermia and pressure-induced blisters/sores. Milder features include ataxia, slurred speech, and lethargy.

29.3.2.2 Barbiturates

Barbiturates are agents that cause CNS depression. They produce a wide spectrum of effects and are used as anxiolytics, hypnotics, anesthetics, and anticonvulsants. Barbiturates bind to the $GABA_A$ receptor. Barbiturates open the ion channels and facilitate Cl^- ion flux. Barbiturates are commonly absorbed orally or intravenously. Absorption of the agents is fast. Short-acting barbiturates (lipophilic) rapidly distribute to the CNS and then redistribute to other tissues, resulting in a rapid onset and short duration of action. Barbiturates are metabolized by the hepatic cytochrome P450 enzyme. Symptoms of abuse commonly include ataxia, nystagmus, and drowsiness. In most severe cases, respiratory and cardiovascular depression and coma occur.

29.3.3 Stimulants

Stimulants are drugs that stimulate the central nervous system. They usually increase alertness and physical activity. They include amphetamines, cocaine, and related derivatives.

29.3.3.1 Amphetamines

Amphetamines include large class of phenylethylamine derivatives. Amphetamine increases monoamine and excitatory neurotransmitter activity in the CNS, increasing the catecholamine neurotransmitters, norepinephrine, and dopamine and working as a strong CNS stimulant. They have a strong arousal effect, form addiction, and sometimes cause schizophreniform-like mental excitement symptoms. Amphetamines are medically used as weight loss agents or to treat narcolepsy and attention deficit disorder. Amphetamine derivatives are illegally manufactured, especially hallucinogenic synthesized forms called "designer amphetamines." Amphetamines are commonly abused by ingestion, intravenous injection, inhalation, or nasal insufflation. Serious life-threatening symptoms include arrhythmia, severe hypertension, severe hyperthermia, and seizures. Intracranial hemorrhage should be ruled out in cases of CNS depression (Chaudhuri and Salahudeen 1999). Other symptoms include sympathomimetic excess, presenting as tachycardia, mydriasis, diaphoresis, and parkinsonism. Milder abuse patients present to the hospital as restless, anxious, agitated, and sometimes combative.

A number of amphetamine derivatives also have hallucinogenic properties. This is often related to an increased affinity for serotonin receptors or effects on serotonergic neurotransmitter activity in the CNS. These drugs can create a mixed stimulant-hallucinogen toxidrome or a serotonin syndrome (hyperthermia, hyperreflexia, agitated delirium). Examples include methylenedioxyamphetamine, methylenedioxymethamphetamine (MDMA, known as ecstasy), and 4-methyl-2-5dimethoxyamphetamine. MDMA has been used recreationally as a "party drug" for decades and has many reports of causing hyperthermic agitated delirium in overdose.

29.3.3.2 Cocaine

Cocaine has strong sympathomimetic properties and is a CNS stimulant. Cocaine acts by inhibiting the reuptake of neurotransmitters (serotonin, dopamine, norepinephrine) in the central and sympathetic nervous system. Cocaine can be absorbed through the nasal mucosa, ingested (gastrointestinal mucosa absorption), smoked (alveolar absorption), insufflated (nasal mucosal absorption), and intravenously injected. Severe overdose can occur in body packers (McCarron & Wood 1983) and body stuffers due to packet rupture or leakage. Cocaine is rapidly metabolized by liver esterase and plasma cholinesterase and by nonenzymatic hydrolysis. Signs and symptoms of acute poisoning include elevated pulse, blood pressure, respirations, and temperature. The severity of toxicity can be determined by these abnormal vital signs. The most critical symptoms are myocardial ischemia (Hollander JE 1995) and cardiovascular collapse, including bradycardia, tachyarrhythmia, or severe hypotension. Cocaine-induced chest pain is common, related to the combination of drug-induced coronary artery vasospasm and systemic hypertension. Theoretically, clinicians should be cautious in treating cocaine-related cardiac symptoms with beta-blockers or other antihypertensive drugs without also addressing the strong alpha-adrenergic and vasospastic effects of cocaine. Usually cocaine-induced derangements of hemodynamics are brief and self-limited. Hyperthermia and subsequent rhabdomyolysis or vital organ failure can also occur. Milder manifestations include anxiety, agitation, euphoria, headache, tremors, twitching, nausea, vomiting, diaphoresis, mydriasis, pallor, confusion, hallucinations, hyperactivity, clonus, increased muscle tone, abdominal cramps, and seizures.

29.3.4 Hallucinogens

A hallucinogen is an abuse agent that can cause hallucinations, perceptual anomalies, and interfere with CNS activity. Auditory and visual hallucinations occur following abuse. The pharmacology of each hallucinogen differs, but the behavioral effects produced by hallucinogens are very similar. Hallucinogen effects include distorted sensory perception, altered mood, distorted reality, and depersonalization. The most dangerous adverse events of hallucinogens are accidents or suicidal impulses, due to impaired judgments (Paul and Smith 1999).

29.3.4.1 Lysergic Acid Diethylamide (LSD)

Lysergic acid diethylamide (LSD) is one of the most potent hallucinogenic agents. LSD is a clear or white material commonly synthesized by reacting diethylamine with lysergic acid. LSD can be obtained from natural products, such as morning glory seeds (ipomoea tricolor, argyreia nervosa, etc.) or ergot fungus. It is available in liquid, powders, tablet "microdots," capsules, gelatin square "window panes," or square-dipped blotter paper forms.

LSD acts as an agonist of serotonin receptors in the CNS. The effects of LSD are visual hallucinations and distortions (known as "trips") and vary greatly depending on the amount taken, individual personality, or the surrounding environment. Abusers may experience extreme change in mood, disassociation from usual activities, terrifying thoughts and feelings, etc. Affected abusers may have impaired judgment, endangering themselves in accidents, including death. Other symptoms include mydriasis, anxiety, loss of appetite, insomnia, tremor, tachycardia, hypertension, and hyperthermia. Some abusers report occasional, brief recurrent perceptual distortions months or years after drug use that are called "flashbacks."

29.3.4.2 Phencyclidine (PCP)

Phencyclidine (PCP) is an abuse agent that causes dissociative hallucinogenic effects. PCP acts as an N-methyl-D-aspartate (NMDA) receptor antagonist. PCP is available in powder or liquid form and can be smoked, orally ingested, insufflated, or injected. PCP is commonly sprayed onto a leaf and smoked with other abuse substances. The major adverse events of PCP are aggressive behavior and suicidal impulses. Individuals under the influence of PCP may be unaware of injuries and dissociated from pain, resulting in great difficulty for emergency medical service or public safety to control their behavior. Impaired perception combined with stimulation can result in dangerous and violent agitation. Because of these effects, the medical use of PCP was switched to ketamine, which is structurally similar with milder CNS stimulation. Symptoms of use include distorted perception, delusions, agitation, dysarthria, ataxia, euphoria, nystagmus, hyperthermia, and tachycardia. High doses of PCP may cause convulsions.

29.3.4.3 Ketamine

Ketamine is an NMDA receptor antagonist and commonly used as a general anesthetic, analgesic, and antidepressant. When used in anesthetic doses, airway reflex is usually preserved, and respiratory depression is rare. Ketamine is sometimes abused and may cause accidents, suicides, or other unpredictable events. Ketamine induces a trance-like state, along with providing sedation and analgesia. Ketamine can be orally ingested, insufflated, or injected. Severe symptoms of abuse include hypertension, tachycardia, CNS depression, hypertension, and rigidity/rhabdomyolysis. Milder symptoms include distorted perception, agitation, ataxia, euphoria, nystagmus, nausea, and vomiting. Urinary tract effects, such as irritation and urge incontinence, are well known in chronic abuse patients.

29.3.4.4 N,N-Dimethyltryptamine (DMT)

N,N-Dimethyltryptamine (DMT) is a hallucinogenic agent that has been historically used in rituals and healing purposes. Even today, it is abused to obtain psychedelic experiences. DMT can be biosynthesized from naturally grown plants (*Psychotria*, *Phalaris*, *Acacia*, and other species). It can also be synthesized from tryptamine and is available as a white crystal. DMT can be inhaled, insufflated, or injected; however, DMT is only orally active when it is combined with a monoamine oxidase inhibitor, because it is degraded by intestinal monoamine oxidase in the absence of a monoamine oxidase inhibitor. DMT has a rapid onset with a short duration. One of the well-known hallucinogenic effects of DMT is similar to a "near-death experience," such as presence of a light, a feeling of detachment from the body, or feelings of levitation. Other symptoms include nystagmus, mydriasis, ataxia, tachycardia, hypertension, nausea, and vomiting. Coma and respiratory arrest can occur after high-dose abuse.

29.3.4.5 Marijuana

Marijuana is an abuse agent that has been historically used because of its euphoric effects. Marijuana is obtained from the leaves of the hemp plant (*Cannabis* species); dried resin from the flower of this plant is called hashish. The active ingredient of the marijuana is δ-9-tetrahydrocannabinol; marijuana contains 1 to 5% δ-9-tetrahydrocannabinol. Since absorption is more potent and stable when smoked, marijuana is often smoked, but it can also be orally ingested. Lethal adverse effects are rare. Less serious effects that often result in seeking medical attention are anxiety and vomiting. An uncontrolled panic or paranoia is self-limited. Abusers are often afraid that the drug has caused them to have palpitations or other harm. Intractable vomiting is a paradoxical effect, because cannabinoids are usually associated with increased appetite. When cannabinoid hyperemesis does occur, it often responds to a hot bath or hot shower (mechanism not known) or to antidopaminergic antiemetic drugs. Other adverse effects include pneumothorax from holding smoke against a closed glottis. Other symptoms include euphoria, paranoia, tachycardia, impaired memory, and impaired motor coordination.

29.3.5 Other Drugs

Many other drugs and substances are abused. Anticholinergic drugs include many antihistamines. These drugs can cause tachycardia and delirium, with risk of seizures at high doses. A variety of antitussive medications produce alcohol-like intoxication when taken at high doses and can cause CNS depression in overdose. Formulations of medicines that contain acetaminophen are particularly perilous because of the potential for delayed hepatotoxicity. Many plants and herbal remedies produce CNS effects when ingested or smoked, and in many cases the pharmacology of these substances is unknown. Poison centers and medical toxicological evaluation can help sort out these toxidromes.

Prescribed drugs may be abused for purpose of self-harm or for recreational purposes. Antidepressant medicines in particular may be available to persons with high risk of abuse. Tricyclic antidepressants produce anticholinergic effects (tachycardia) and CNS depression and are potentially lethal if they cause cardiac sodium channel blockade (QRS prolongation on ECG) and dysrhythmia. Selective serotonin uptake inhibitors can cause serotonin syndromes (hyperreflexia, hyperthermia, delirium) that require prolonged supportive care.

29.4 Treatment

Although there are differences in treatment depending on the substances abused, in this chapter, we present basic treatment concepts. The major four components of overdose (abuse) treatments are **(1)** supportive care, **(2)** prevention of chemical absorption, **(3)** enhancement of chemical elimination, and **(4)** reversal of poisoning by use of antidotes.

29.4.1 Supportive Care

Supportive care includes monitoring (including vital signs, behavior, cardiac rhythm, pulse oximetry, urine output, etc.) and the maintenance of breathing, circulation, CNS function, and temperature management. It is the primary therapy for most patients. Generally, need for chemical absorption, enhanced elimination, or use of antidotes is rare.

29.4.1.1 Respiratory Care

Oxygenation should be monitored by pulse oximetry in patients with moderate-to-severe symptoms. Supplemental oxygen should be administered to maintain the SpO_2 over 90% (PaO_2 over 60 mmHg). Blood gas sampling should be conducted in patients with dyspnea. BVM ventilation with oxygen is a critical initial intervention in patients with respiratory depression (bradypnea/apnea). Prophylactic intubation may be required to prevent aspiration of gastric contents in patients with coma. Endotracheal intubation with sedation may be required for patients with extreme physiologic stimulation.

29.4.1.2 Cardiovascular Care

Patients with moderate-to-severe symptoms should be monitored with continuous ECG. Adequate blood pressure, pulse, and rhythm should be maintained. Hypertension should be treated by sedation with a benzodiazepine. Antihypertensive therapy such as an α-blocker or a peripheral vasodilator should be considered, especially in cases of stimulant abuse. Caution is warranted for use of beta-blockade in stimulant abuse, because these drugs might decrease cardiac inotropy and chronotropy in the setting of excessive afterload (alpha-stimulation), leading to hemodynamic collapse. In cases of QRS widening and ventricular tachycardia, correction of electrolyte and metabolic abnormalities should be conducted. When QRS prolongation is thought to be related to sodium channel blockade (cocaine, tricyclic antidepressants), consider sodium bicarbonate or hypertonic sodium chloride therapy. Hypotension should be treated with intravenous fluid and vasopressors (dopamine or norepinephrine). In patients who are unresponsive to conventional therapy, mechanical cardiac support or cardiopulmonary bypass should be considered. If hypertension/hypotension is accompanied by chest pain, acute coronary syndrome should be suspected. Cocaine-related chest pain often occurs in absence of coronary artery disease, but it also can provoke symptoms when disease is present. Patients with evidence of myocardial ischemia provoked by substance abuse should receive all of the standard treatments for this situation.

29.4.1.3 Neurological Care

Care of CNS depression is supportive. If CNS depression persists, CT scans should be obtained to rule out intracranial lesion including intracranial hemorrhage. Careful attention should be paid to respiratory depression.

Benzodiazepines may be considered for CNS overstimulation. Seizures may occur due to the abuse agent; they should be immediately treated with sedatives and anticonvulsants and be monitored by electroencephalograph to prevent subsequent neurological damage. Attention should also be paid to neurological damage due to hyperthermia (Callaway CW and Clark RF 1994). In some cases of hyperactivity or seizures, cooling measures are needed with deeper sedation and intubation. Rhabdomyolysis may occur, especially in these cases; intravenous fluids are recommended to treat rhabdomyolysis, and caution should be paid to acute renal failure.

29.4.2 Prevention of Chemical Absorption

Routine use of ipecac syrup (to induce vomiting) or gastric lavage is not recommended (Höjer J, et al. 2013). While used in the past, these interventions have been found to have minimal effects on drug levels and to create high risk of aspiration. Decontamination can be considered for patients who present within an hour of exposure to life-threatening amounts of substances. Endoscopy may be useful for removal of foreign bodies or breaking up drug masses. Oral-activated charcoal is another method of decontamination. The optimal dosage of activated charcoal is 1 to 2 g/kg of body weight. Activated charcoal is not recommended for patients in a coma (who are not intubated but in whom the cough reflex is lost). Whole bowel irrigation should be considered when ingestion of enteric-coated or sustained-release formulations is strongly suspected (Thanacoody R, et al. 2015). It should also be considered for body packers or body stuffers.

29.4.3 Enhancement of Chemical Elimination

The elimination of some of substances can be promoted by diuresis or alkalization of the urine (urine pH above 7.5). Extracorporeal techniques (hemodialysis, hemoperfusion, and hemofiltration) can accelerate the elimination of some substances (phenobarbital and phenytoin carbamazepine) and should be considered in patients with respiratory or cardiovascular instability.

29.4.4 Reversal of Poisoning by Antidotes

Antidotes counteract the effects of substances by competing for the target sites. Not all substances have effective antidotes. Naloxone and flumazenil are two major antidotes used in abuse patients. While naloxone is widely used, flumazenil has potential for adverse effects and is used much more selectively.

29.4.4.1 Naloxone

Naloxone is a rapid-acting μ-receptor antagonist. Naloxone can be safely administered to neutralize respiratory depression. Attention should be paid to chronic opioid dependents, however, because rapid infusion of naloxone may induce withdrawal symptoms for these patients. Use an initial intravenous dose of 0.2 mg; if agent effects persist, repeated doses should be administered. Overdoses with high-potency synthetic drugs such as fentanyl derivatives may require very high doses of naloxone.

29.4.4.2 Flumazenil

Flumazenil is a selective benzodiazepine receptor antagonist. Flumazenil can neutralize CNS and respiratory depression. Attention, however, should be paid to withdrawal symptoms in patients with chronic use of benzodiazepines or even chronic alcoholism (Spivey 1992). In those patients, flumazenil can precipitate intractable seizures. Because benzodiazepine overdose is likely to occur in patients with chronic use and because benzodiazepines cause minimal respiratory depression, supportive care is usually safer than flumazenil administration. Use an initial intravenous dose of 0.2 mg; if agent effects persist, repeated doses should be administered to a maximum dose of 3 mg.

29.5 Complications

Complications can be acute or subacute after abuse. They may result from effects of adulterant and other co-ingested substances (other than the

Table 29.3 Abstinence, maintenance, or reversal agents used for substance use disorder

	Abstinence therapy	Maintenance therapy	Reversal for acute symptom
Opioid	–	Methadone, buprenorphine/naloxone, oxycodone/naloxone	Naloxone
Cocaine	Ondansetron, buspirone	–	–
Methamphetamine	Ondansetron, buspirone	–	–
Benzodiazepine	–	Lower-dose diazepam for withdrawal	Flumazenil

main abused agents). Common complications due to CNS depression are **1.** aspiration pneumonitis (Isbister et al. 2004), **2.** accidental hypothermia, and **3.** pressure-induced blisters/sores/sometimes compartment syndrome. Some drug categories are likely to have specific complications. For example, stimulants can result in cerebrovascular (subarachnoid or intracerebral hemorrhage and cerebral vasculitis) and cardiovascular involvement (myocardial infarctions, aortic dissection, and kidney/bowel ischemia). Opioids cause a higher incidence of respiratory depression. Other complications include pulmonary edema, anoxia/anoxic brain injury, hallucination-associated trauma, rhabdomyolysis, and multiple organ failure from shock/hyperthermia.

29.6 Prevention of Re-Exposure

Patients with substance use disorder are at an increased risk of numerous health-related problems or complications including infections (Degenhardt, et al. 2013, Devlin and Henry 2008). Substance use disorder patients should be monitored during the time they are at risk for withdrawal. They should be given the opportunity for rehabilitation and behavior modification. Pharmacotherapies are used to promote or maintain abstinence. For example, opioid agonist treatment with buprenorphine or methadone is used for opioid abuse and is associated with cessation and decreased complications. Representative abstinence or maintenance agents are shown in Table 29.3.

End-of-Chapter Questions

1. When you encounter case of suspected drug abuse toxicity, what information should be obtained from family members or paramedics?
2. What will be the treatment strategy for an opioid abuse patient?
3. What will be the treatment strategy for a cocaine (stimulant) abuse patient?
4. How could the use of an antidote affect the treatment after abuse? What is the method (dosage) of application?
5. What are the common complications associated with abuse?

References

Ball IM, Linden CH (2011) Pharmacology, overdoses, and poisonings. In: Irwin RS, Rippe JM (eds) Irwin & Rippe's intensive care medicine, 7th edn. Lippincott Williams & Wilkins, Philadelphia, p 1314

Byatt C, Volans G (1984) ABC of poisoning. Sedative and hypnotic drugs. Br Med J 289:1214–1217

Callaway CW, Clark RF (1994) Hypothermia in psychostimulant overdose. Ann Emerg Med 24:68–76

Chaudhuri C, Salahudeen AK (1999) Massive intracerebral hemorrhage in an amphetamine addict. Am J Med Sci 317:350–352

Degenhardt L, Whiteford HA, Ferrari AJ, Baxter AJ, Charlson FJ, Hall WD, Freedman G, Burstein R, Johns N, Engell RE, Flaxman A, Murray CJ, Vos T (2013) Global burden of disease attributable to illicit drug use and dependence: findings from the global burden of disease study 2010. Lancet 382:1564–1574

Devlin RJ, Henry JA (2008) Clinical review: major consequences of illicit drug consumption. Crit Care 12:202

Höjer J, Troutman WG, Hoppu K, Erdman A, Benson BE, Mégarbane B, Thanacoody R, Bedry R, Caravati EM (2013) Position paper update: ipecac syrup for gastrointestinal decontamination. Clin Toxicol (Phila) 51:134–139

Hollander JE (1995) The management of cocaine-associated myocardial ischemia. N Engl J Med 333:1267–1272

Isbister GK, Downes F, Sibbritt D, Dawson AH, Whyte IM (2004) Aspiration pneumonitis in an overdose population: frequency, predictors, and outcomes. Crit Care Med 32:88–93

McCarron MM, Wood JD (1983) The cocaine body packer syndrome. JAMA 250:1417

Paul BD, Smith ML (1999) LSD–an overview on drug action and detection. Forensic Sci Rev 11:157–174

Spivey WH (1992) Flumazenil and seizures: analysis of 43 cases. Clin Ther 14:292–305

Thanacoody R, Caravati EM, Troutman B, Höjer J, Benson B, Hoppu K, Erdman A, Bedry R, Mégarbane B (2015) Position paper update: whole bowel irrigation for gastrointestinal decontamination of overdose patients. Clin Toxicol (Phila) 53:5–12

Watson WA, Litovitz TL, Klein-Schwartz W, Rodgers GC Jr, Youniss J, Reid N, Rouse WG, Rembert RS, Borys D (2003) 2003 annual report of the American Association of Poison Control Centers Toxic Exposure Surveillance System. Am J Emerg Med 22:335–404

Index

T. Oohashi et al. (eds.), *Human Pathobiochemistry*, https://doi.org/10.1007/978-981-13-2977-7

MIX
Papier aus verantwortungsvollen Quellen
Paper from responsible sources
FSC® C105338

If you have any concerns about our products,
you can contact us on
ProductSafety@springernature.com

In case Publisher is established outside the EU,
the EU authorized representative is:
Springer Nature Customer Service Center GmbH
Europaplatz 3, 69115 Heidelberg, Germany

Printed by Libri Plureos GmbH
in Hamburg, Germany